iOS 18 Programming for Beginners

Ninth Edition

Learn iOS development with Swift 6, Xcode 16, and iOS 18 – your path to App Store success

Ahmad Sahar

iOS 18 Programming for Beginners

Ninth Edition

Senior Publishing Product Manager: Larissa Pinto
Acquisition Editor – Peer Reviews: Swaroop Singh
Project Editor: K. Loganathan
Content Development Editor: Matthew Davies
Copy Editor: Safis Editing
Technical Editor: Simanta Rajbangshi
Proofreader: Safis Editing
Indexer: Pratik Shirodkar
Presentation Designer: Rajesh Shirsath
Developer Relations Marketing Executive: Sohini Ghosh

First published: December 2016
Second edition: January 2018
Third edition: December 2018
Fourth edition: January 2020
Fifth edition: November 2020
Sixth edition: December 2021
Seventh edition: November 2022
Eighth edition: October 2023
Ninth edition: November 2024

Production reference: 1041224

Published by Packt Publishing Ltd.
Grosvenor House
11 St Paul's Square
Birmingham
B3 1RB, UK.

ISBN 978-1-83620-489-3

www.packt.com

Contributors

About the author

Ahmad Sahar is a trainer, presenter, and consultant at Tomafuwi Productions. He is an Apple Certified Trainer for app development in Swift, and he has conducted training courses on for over ten years iOS mobile application development, Flutter mobile application development, and how to use Apple software and hardware. He also currently holds App Development with Swift Associate and Certified User certifications.

He is a member of the DevCon iOS online community in Malaysia and has conducted presentations and talks for this group. In his spare time, he likes building and programming LEGO MINDSTORMS® robots.

To my mother, Sharifah; my father, Sahar; my two sisters, Ainol Shareha and Ainol Shaharina; and my beloved wife, Oni – thank you for your love and support.

About the reviewer

Ian Lockett is a mobile application developer with over 18 years of experience in the software industry. He has spent the last decade developing iOS and Android apps for a wide range of clients and industries. Most of Ian's experience is in iOS development using Swift and Objective-C, but he has also created and maintained Android apps over the last few years using Kotlin.

Ian prides himself on attention to detail and delivering high-quality software. Ian lives in the UK with his wife and two children and now works for a large finance company having worked as a freelance/contract developer for the past seven years.

Join us on Discord!

Read this book alongside other users, experts, and the author himself. Ask questions, provide solutions to other readers, chat with the author via Ask Me Anything sessions, and much more. Scan the QR code or visit the link to join the community.

`https://packt.link/ios-Swift`

Table of Contents

Chapter 11: Building Your User Interface 155

Chapter 12: Finishing Up Your User Interface 187

Chapter 13: Modifying App Screens 199

Chapter 16: Passing Data between View Controllers 285

Chapter 17: Getting Started with Core Location and MapKit 309

Chapter 18: Getting Started with JSON Files 343

Chapter 19: Getting Started with Custom Views 361

Chapter 20: Getting Started with the Camera and Photo Library 377

Chapter 21: Getting Started with Search 391

Chapter 22: Getting Started with Collection Views 403

Part 4: Features 437

Chapter 23: Getting Started with SwiftData 439

Chapter 24: Getting Started with SwiftUI 455

Chapter 25: Getting Started with Swift Testing 481

Chapter 26: Getting Started with Apple Intelligence 489

Preface

Welcome to *iOS 18 Programming for Beginners*. This book is the ninth edition of the *iOS Programming for Beginners* series, and has been fully updated for iOS 18, macOS 15.0 Sequioa, and Xcode 16.

In this book, you will build a journal app called *JRNL*. You will start off by exploring Xcode, Apple's programming environment, also known as its **Integrated Development Environment** (IDE). Next, you will start learning the foundations of Swift, the programming language used in iOS apps, and see how it is used to accomplish common programming tasks.

Once you have a solid foundation of using Swift, you will start creating the user interface of the *JRNL* app. During this process, you will work with storyboards and connect your app's scenes together using segues.

With your user interface complete, you will then add code to implement your app's functionality. To start, you'll learn how to display data using a table view. Next, you'll learn how to add data to your app, and how to pass data between view controllers. After that, you'll learn how to determine your device location and display annotations on a map. You'll then learn how to persist app data using JSON files create custom views, and add photos from the camera or photo library. Finally, you'll make your app work on devices with larger screens, such as an iPad or Mac, by implementing a collection view in place of a table view.

You now have a complete app, but how about adding the latest iOS 18 features? You'll start by learning about SwiftData, which allows you to describe data models and manipulate model instances using regular Swift code. Next, you will learn how to develop apps using SwiftUI, a great new way of developing apps for all Apple platforms. After that, you'll learn how to test your code using Swift Testing, and how to bring Apple Intelligence features into your apps.

Finally, you'll learn how to test your app with internal and external testers and get it into the App Store.

Who this book is for

This book is tailored for individuals with minimal coding experience who are new to the world of Swift and iOS app development. A basic understanding of programming concepts is recommended.

What this book covers

Chapter 1, Exploring Xcode, takes you through a tour of Xcode and talks about all the different parts that you will use throughout the book.

Chapter 2, Simple Values and Types, deals with how values and types are implemented by the Swift language.

Chapter 3, Conditionals and Optionals, shows how `if` and `switch` statements are implemented, and how to implement variables that may or may not have a value.

Chapter 4, Range Operators and Loops, shows how to work with ranges and the different ways loops are implemented in Swift.

Chapter 5, Collection Types, covers the common collection types, which are arrays, dictionaries, and sets.

Chapter 6, Functions and Closures, covers how you can group instructions together using functions and closures.

Chapter 7, Classes, Structures, and Enumerations, talks about how complex objects containing state and behavior are represented in Swift.

Chapter 8, Protocols, Extensions, and Error Handling, talks about creating protocols that complex data types can adopt, extending the capabilities of existing types, and how to handle errors in your code.

Chapter 9, Swift Concurrency, introduces you to the concepts of parallel and asynchronous programming, and shows you how you can implement them in your app.

Chapter 10, Setting Up the User Interface, deals with creating the *JRNL* app and setting up the initial screen the users will see.

Chapter 11, Building Your User Interface, covers setting up the main screen for the *JRNL* app.

Chapter 12, Finishing Up Your User Interface, covers setting up the remaining screens for the *JRNL* app.

Chapter 13, Modifying App Screens, is about configuring each screen of the app in a storyboard.

Chapter 14, Getting Started with MVC and Table Views, covers working with a table view and how you can use it to display a list of items.

Chapter 15, Getting Data into Table Views, concerns the incorporation of data into table views using an array as a data source.

Chapter 16, Passing Data between View Controllers, teaches you how to add data entered using a view controller to an array, and how to pass data from the array to another view controller.

Chapter 17, Getting Started with Core Location and MapKit, deals with working with Core Location and MapKit to determine your device location and add annotations to a map.

Chapter 18, Getting Started with JSON Files, involves learning how to store and retrieve user data using a JSON file.

Chapter 19, Getting Started with Custom Views, teaches you how to create and use a custom view that displays a star rating.

Chapter 20, Getting Started with the Camera and Photo Library, talks about how to get photos from your camera or photo library into your app.

Chapter 21, Getting Started with Search, teaches you how to implement a search bar for your main screen.

Chapter 22, Getting Started with Collection Views, shows you how to implement collection views in place of table views to suit devices with larger screens, such as a Mac or iPad.

Chapter 23, Getting Started with SwiftData, deals with implementing Apple's new SwiftData framework to persist data on your app.

Chapter 24, Getting Started with SwiftUI, introduces building an app using Apple's new SwiftUI technology.

Chapter 25, Getting Started with Swift Testing, teaches you how to test your code using Swift Testing.

Chapter 26, Getting Started with Apple Intelligence, shows you how to add Apple Intelligence features to your app.

Chapter 27, Testing and Submitting Your App to the App Store, concerns how to test and submit your apps to the App Store.

To get the most out of this book

This book has been completely revised for iOS 18, macOS 15.0 Sequioa, Xcode 16, and Swift 6. *Part 4* of this book also covers the latest technologies introduced by Apple during WWDC 2024, which are SwiftData, SwiftUI, Swift Testing, and Apple Intelligence.

To complete all the exercises in this book, you will need:

- A Mac computer running macOS 14.0 Sonoma, macOS 15.0 Sequioa, or later
- Xcode 16.0 or later

To check if your Mac supports macOS 15.0 Sequioa, see this link: `https://www.apple.com/my/macos/macos-sequoia-preview/`. If your Mac is supported, you can update macOS using **Software Update** in **System Preferences**.

To get the latest version of Xcode, you can download it from the Apple App Store. Most of the exercises can be completed without an Apple Developer account and use the iOS Simulator. If you wish to test the app you are developing on an actual iOS device, you will need a free or paid Apple Developer account.

The following chapter requires a paid Apple Developer account: *Chapter 27, Testing and Submitting Your App to the App Store.* Instructions on how to get a paid Apple Developer account are included.

Download the example code files

You can download the example code files for this book from GitHub at `https://github.com/PacktPublishing/iOS-18-Programming-for-Beginners-Ninth-Edition`. If there's an update to the code, it will be updated in the GitHub repository.

We also have other code bundles from our rich catalog of books and videos available at `https://github.com/PacktPublishing/`. Check them out!

Code in Action

Visit the following link to check out videos of the code being run:

`https://www.youtube.com/playlist?list=PLeLcvrwLe185EJSoURfHhSHfbPFkiZl6m`

Download the color images

We also provide a PDF file that has color images of the screenshots/diagrams used in this book. You can download it here: `https://packt.link/gbp/9781836204893`.

Conventions used

There are a number of text conventions used throughout this book.

`CodeInText`: Indicates code words in text, database table names, folder names, filenames, file extensions, pathnames, dummy URLs, user input, and Twitter handles. Here is an example: "So, this is a very simple function, named `serviceCharge()`."

A block of code is set as follows:

```
class ClassName {
   property1
   property2
   property3
   method1() {
      code
   }
   method2() {
      code
   }
}
```

When we wish to draw your attention to a particular part of a code block, the relevant lines or items are set in bold:

```
let cat = Animal()
cat.name = "Cat"
cat.sound = "Mew"
cat.numberOfLegs = 4
cat.breathesOxygen = true
print(cat.name)
```

Bold: Indicates a new term, an important word, or words that you see onscreen. For example, words in menus or dialog boxes appear in the text like this. Here is an example: "Launch **Xcode** and click **Create a new Xcode project**:"

Important notes

appear like this.

Tips

appear like this.

Get in touch

Feedback from our readers is always welcome.

General feedback: If you have questions about any aspect of this book, mention the book title in the subject of your message and email us at `customercare@packtpub.com`.

Errata: Although we have taken every care to ensure the accuracy of our content, mistakes do happen. If you have found a mistake in this book, we would be grateful if you would report this to us. Please visit `www.packtpub.com/support/errata`, selecting your book, clicking on the Errata Submission Form link, and entering the details.

Piracy: If you come across any illegal copies of our works in any form on the Internet, we would be grateful if you would provide us with the location address or website name. Please contact us at `copyright@packt.com` with a link to the material.

If you are interested in becoming an author: If there is a topic that you have expertise in and you are interested in either writing or contributing to a book, please visit `authors.packtpub.com`.

Leave a review!

Thank you for purchasing this book from Packt Publishing—we hope you enjoy it! Your feedback is invaluable and helps us improve and grow. Once you've completed reading it, please take a moment to leave an Amazon review; it will only take a minute, but it makes a big difference for readers like you.

`https://packt.link/r/1836204892`

Scan the QR code below or visit the link to receive a free ebook of your choice.

`https://packt.link/NzOWQ`

Download a free PDF copy of this book

Thanks for purchasing this book!

Do you like to read on the go but are unable to carry your print books everywhere?

Is your eBook purchase not compatible with the device of your choice?

Don't worry, now with every Packt book you get a DRM-free PDF version of that book at no cost.

Read anywhere, any place, on any device. Search, copy, and paste code from your favorite technical books directly into your application.

The perks don't stop there, you can get exclusive access to discounts, newsletters, and great free content in your inbox daily.

Follow these simple steps to get the benefits:

1. Scan the QR code or visit the link below:

`https://packt.link/free-ebook/9781836204893`

2. Submit your proof of purchase.
3. That's it! We'll send your free PDF and other benefits to your email directly.

Part 1

Swift

Welcome to *Part 1* of this book. In this part, you will begin by exploring Xcode, Apple's programming environment, which is also known as the **Integrated Development Environment** (IDE). After that, you will start learning the foundations of Swift 6, the programming language used in iOS apps, and see how it is used to accomplish common programming tasks.

This part comprises the following chapters:

- *Chapter 1, Exploring Xcode*
- *Chapter 2, Simple Values and Types*
- *Chapter 3, Conditionals and Optionals*
- *Chapter 4, Range Operators and Loops*
- *Chapter 5, Collection Types*
- *Chapter 6, Functions and Closures*
- *Chapter 7, Classes, Structures, and Enumerations*
- *Chapter 8, Protocols, Extensions, and Error Handling*
- *Chapter 9, Swift Concurrency*

By the end of this part, you'll understand the process of creating an app and running it on Simulator or a device, and you'll have a working knowledge of how to use the Swift programming language in order to accomplish common programming tasks. This will prepare you for the next chapter and will also enable you to create your own Swift programs. Let's get started!

1

Exploring Xcode

Welcome to *iOS 18 Programming for Beginners*. I hope you will find this a useful introduction to creating and publishing iOS 18 apps on the App Store.

In this chapter, you'll download and install **Xcode** on your Mac. Then, you'll explore the Xcode user interface. After that, you'll create your first **iOS app** and run it in **Simulator**. Finally, you'll run your app on an **iOS device**.

By the end of this chapter, you will know how to create an iOS app, how to run it in Simulator, and how to run it on an iOS device.

The following topics will be covered in this chapter:

- Downloading and installing Xcode from the App Store
- Exploring the Xcode user interface
- Running your app in Simulator
- Running your app on an iOS device

Technical requirements

To do the exercises for this chapter, you will need the following:

- An Apple Mac computer (Apple Silicon or Intel) running macOS 14 Sonoma or macOS 15 Sequoia
- An Apple Account (if you don't have one, you will create one in this chapter)
- Optionally, an iOS device running iOS 18

The Xcode project for this chapter is in the `Chapter01` folder of the code bundle for this book, which can be downloaded here:

`https://github.com/PacktPublishing/iOS-18-Programming-for-Beginners-Ninth-Edition`

Check out the following video to see the code in action:

`https://youtu.be/g3mNosIoR8E`

You'll start by downloading Xcode, Apple's **integrated development environment** (IDE) for developing iOS apps from the App Store, in the next section.

The total size of the download is very large (2.98 GB for Xcode and 8.36 GB for iOS 18 Simulator), so it may take a while to download. Ensure that you have enough disk space prior to downloading.

Downloading and installing Xcode from the App Store

Xcode is Apple's IDE for developing macOS, iOS, iPadOS, watchOS, tvOS, and visionOS apps. You'll need to download and install Xcode on your Mac prior to writing your first app. Follow these steps:

1. On your Mac, choose **App Store** from the **Apple** menu.
2. In the search field in the top-right corner, type `Xcode` and press the *Return* key.
3. You'll see **Xcode** in the search results. Click **Get** and then click **Install.**
4. If you have an Apple Account, type it in the text field and enter your password when prompted. If you don't have one, click the **Create Apple Account** button and follow the step-by-step instructions to create one:

Sign in to download from the App Store.
If you have an Apple Account, sign in with it here. If you have used the iTunes Store or iCloud, for example, you have an Apple Account. If you don't have an Apple Account, click Create Apple Account.
Email or Phone Number
Forgot password?
Create Apple Account
Cancel
Sign In

Figure 1.1: Apple account creation dialog box

You can see more information on how to create an Apple Account using this link: `https://support.apple.com/en-us/108647#appstore`.

5. Once Xcode has been installed, launch it. You'll see a license agreement screen. Click **Agree**:

Xcode and Apple SDKs Agreement
You must agree to the license agreements below in order to use Xcode.
Xcode and Apple SDKs Agreement
PLEASE SCROLL DOWN AND READ ALL OF THE FOLLOWING TERMS AND CONDITIONS CAREFULLY BEFORE USING THE APPLE SOFTWARE OR APPLE SERVICES. THIS IS A LEGAL AGREEMENT BETWEEN YOU AND APPLE. BY CLICKING "AGREE" OR BY DOWNLOADING, USING OR COPYING ANY PART OF THIS APPLE SOFTWARE OR USING ANY PART OF THE APPLE SERVICES, YOU ARE AGREEING ON YOUR OWN BEHALF AND/OR ON BEHALF OF YOUR COMPANY OR ORGANIZATION TO THE TERMS AND CONDITIONS STATED BELOW. IF YOU DO NOT OR CANNOT AGREE TO THE TERMS OF THIS AGREEMENT, YOU CANNOT USE THIS APPLE SOFTWARE OR THE APPLE SERVICES. DO NOT DOWNLOAD OR USE THIS APPLE SOFTWARE OR APPLE SERVICES IN THAT CASE.
IMPORTANT NOTE: USE OF APPLE SOFTWARE IS GOVERNED BY THIS AGREEMENT AND IS AUTHORIZED ONLY FOR EXECUTION ON AN APPLE-BRANDED PRODUCT RUNNING MACOS. ANY OTHER DOWNLOAD OR USE OF APPLE SOFTWARE IS NOT AUTHORIZED AND IS IN BREACH OF THIS AGREEMENT.
Save...
Disagree
Agree

Figure 1.2: License agreement screen

6. You'll be prompted to enter your Mac's administrator **username** and **password**. Once you have done so, click **OK**:

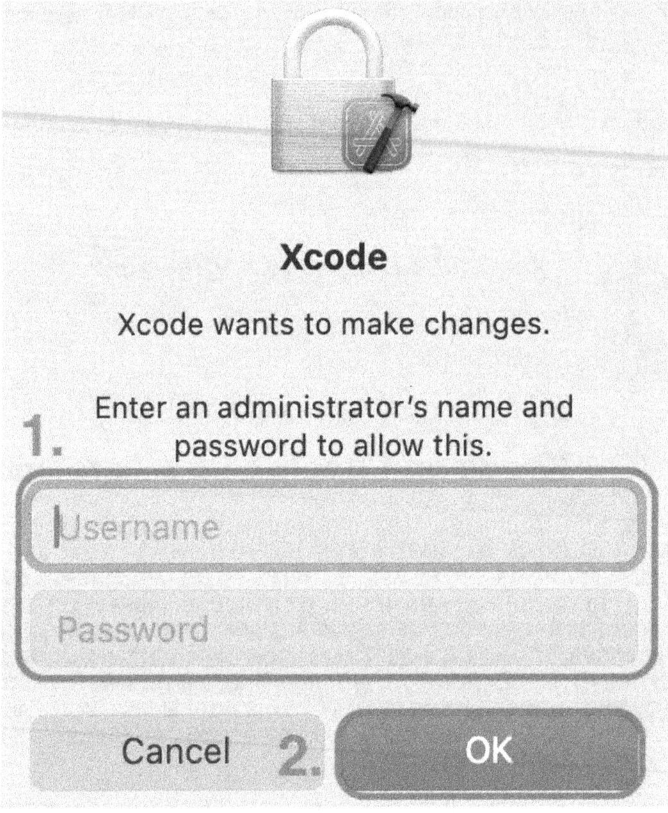

Figure 1.3: Prompt for administrator username and password

7. You'll see a screen showing you the available development platforms. You just need macOS and iOS for now. Tick **iOS 18.0**, leave all other options unticked, and click **Download & Install**:

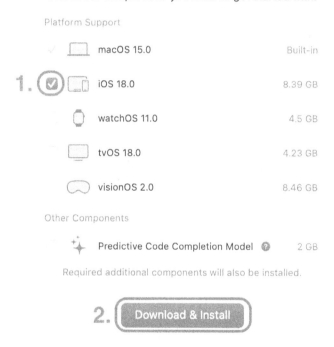

Figure 1.4: Development platforms screen

8. Xcode will prompt you to relaunch to use updated frameworks. Click **Relaunch Xcode**:

Xcode must relaunch to continue.
Xcode needs to relaunch to use updated frameworks.
Relaunch Xcode

Figure 1.5: Relaunch Xcode prompt

9. You'll see a **What's New in Xcode** screen. Click **Continue**:

What's New in Xcode

Code Assistance
Transform your ideas into code faster than ever with predictive code completion.

Swift 6
Write safer code with compile-time data-race safety. Write better tests with Swift Testing and keep them organized in the Test Navigator.

Previews
See design changes more rapidly, share test data across your previews, and add dynamic properties for more interactive previews.

Build & Diagnostics
Experience improved debugging and faster builds with explicit modules. Get deeper insights into your app with the thread performance checker and the flame graph view in Instruments.

Complete feature list >

Some features may not be available for all regions or on all Apple devices.

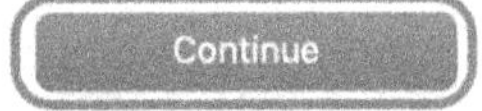
Continue

Figure 1.6: What's New in Xcode screen

10. You'll see the **Welcome to Xcode** screen. Click **Create New Project...** in the left-hand pane:

Figure 1.7: Welcome to Xcode screen

11. Xcode will start to download **iOS 18.0 Simulator** automatically. Note that you will not be able to run any apps on Simulator until this process has been completed:

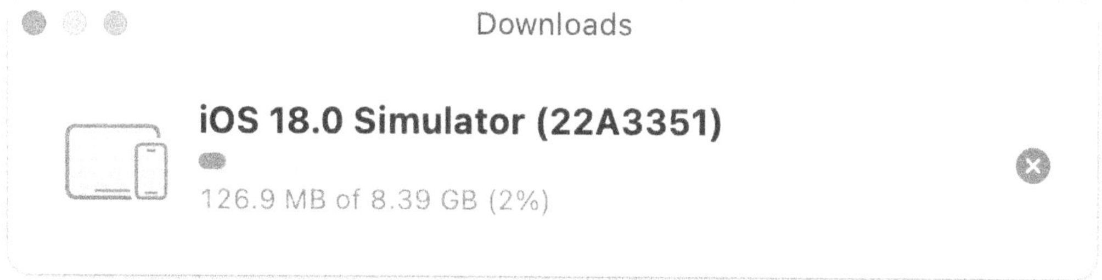

Figure 1.8: Simulator download progress bar

12. You'll see the new project screen as follows. In the **Choose a template for your new project** section, select **iOS**. Then choose **App** and click **Next**:

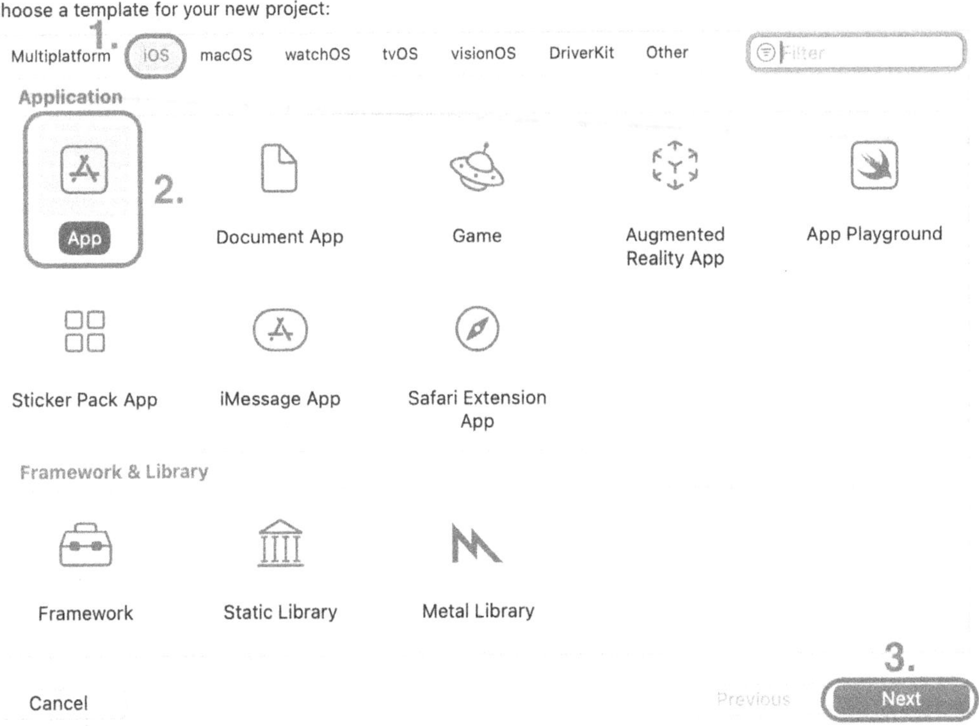

Figure 1.9: New project screen

13. You'll see the **Choose options for your new project** screen:

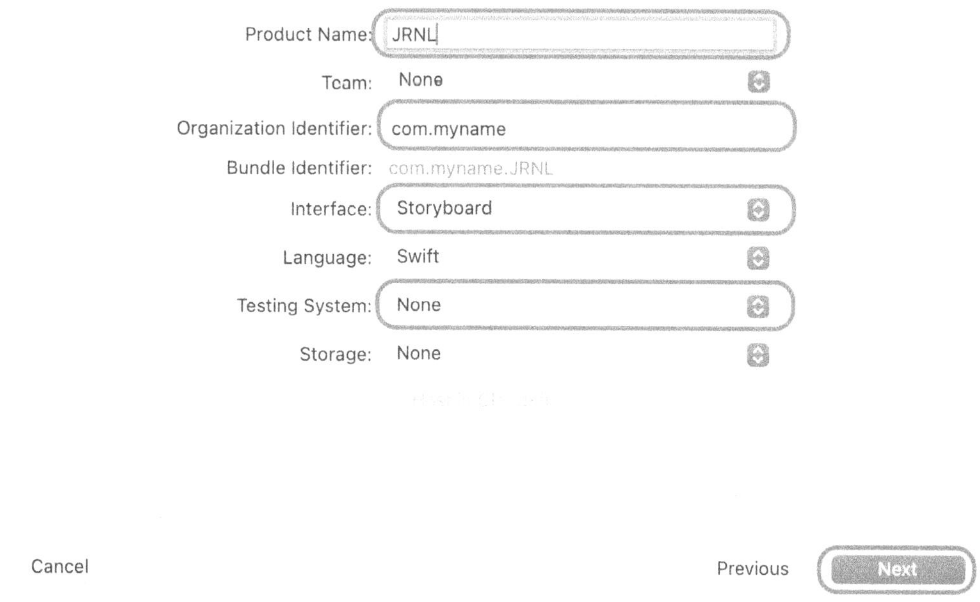

Figure 1.10: Choose options for your new project screen

Configure the options as follows:

- **Product Name:** The name of your app. Enter `JRNL` in the text field.
- **Organization Identifier:** Used to create a unique identifier for your app on the App Store. Enter `com.myname` for now. This is known as the reverse domain name notation format and is commonly used by iOS developers.
- **Interface:** The method used to create the user interface for your app. Set this to **Storyboard.**
- **Testing System:** The testing system you will use. You will learn about this in *Chapter 25, Getting Started with Swift Testing*. Set it to **None** for now.

Leave the other settings at their default values. Click **Next** when done.

14. You'll see a **Save** dialog box. Choose a location to save your project, such as the **Desktop** or **Documents** folder, and click **Create**:

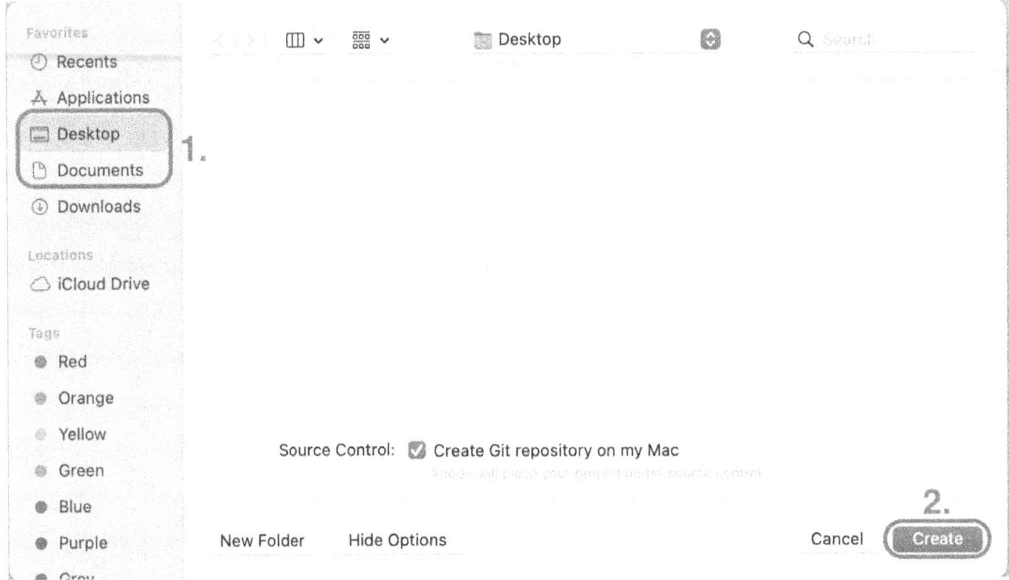

Figure 1.11: Save dialog box

15. You'll see a dialog box saying **Git Repository Creation Failed**. Click **Fix**.

The reason why you see this dialog box is because the **Source Control** checkbox in the **Save** dialog box was ticked. Apple recommends that **Source Control** be turned on. **Source Control** is outside the scope of this book but if you wish to learn more about version control and Git, see this link: https://git-scm.com/video/what-is-version-control.

16. You will see the **Source Control** screen as follows:

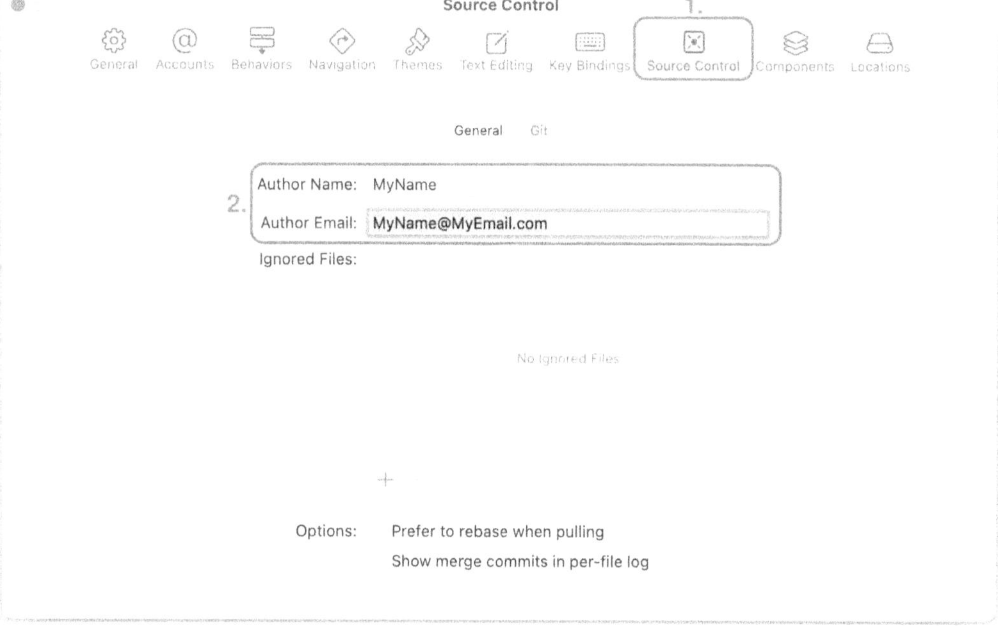

Figure 1.12: Source Control preference screen

Enter the following information:

- **Author Name**: Your own name
- **Author Email**: Your email address

Close the **Source Control** screen by clicking the close button in the top-left corner when done. The Xcode main window will appear.

Fantastic! You have now successfully downloaded and installed Xcode and created your first project. In the next section, you will learn about the Xcode user interface.

Exploring the Xcode user interface

You've just created your first Xcode project! As you can see, the Xcode user interface is divided into several distinct parts, as shown here:

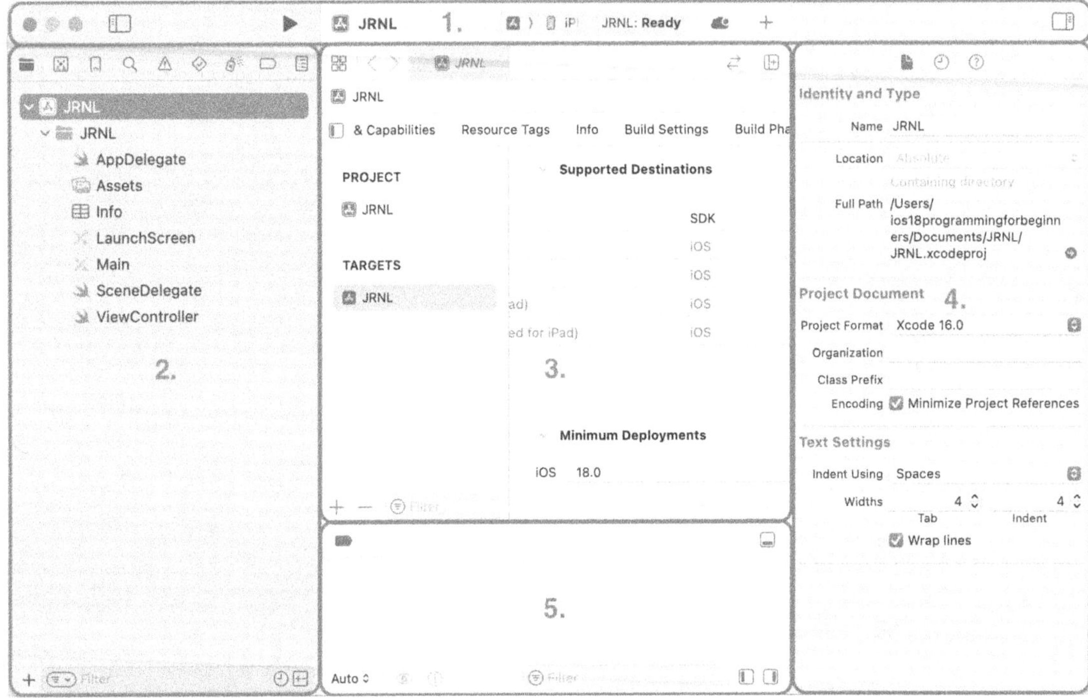

Figure 1.13: Xcode user interface

Let's look at each part in more detail. The following description corresponds to the numbers shown in the preceding screenshot:

- **Toolbar (1)**: Used to build and run your apps, and view the progress of running tasks.
- **Navigator area (2)**: Provides quick access to the various parts of your project. The **Project navigator** is displayed by default.
- **Editor area (3)**: Allows you to edit source code, user interfaces, and other resources.
- **Inspector area (4)**: Allows you to view and edit information about items selected in the **Navigator area** or **Editor area.**
- **Debug area (5)** – Contains the **debug bar**, the **variables view**, and the **Console.** The **Debug area** is toggled by pressing *Shift* + *Command* + *Y*.

Next, let's examine the toolbar more closely. The left side of the toolbar is shown here:

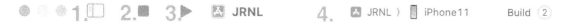

Figure 1.14: Xcode toolbar (left side)

Let's look at each part in more detail. The following descriptions correspond to the numbers shown in the preceding screenshot:

- **Navigator button** (1) – Used to display and hide the Navigator area.
- **Stop button** (2) – Only appears next to the Run button when the app is running. Stops the currently running app.
- **Run button** (3) – Used to build and run your app.
- **Scheme menu** (4) – Shows the specific scheme to build your project (**JRNL**) and the destination to run your app on (**iPhone SE (3rd generation)**). Schemes and destinations are distinct. Schemes specify the settings for building and running your project. Destinations specify installation locations for your app and exist for Simulator and physical devices.
- **Activity view** (5) – Displays the progress of running tasks.

The right side of the toolbar is shown here:

Figure 1.15: Xcode toolbar (right side)

Let's look at each part in more detail. The following descriptions correspond to the numbers shown in the preceding screenshot:

- **Xcode Cloud button** (1) – Allows you to sign in to Xcode Cloud, a continuous integration and delivery service built into Xcode.
- **Library button** (2) – Displays user interface elements, code snippets, and other resources.
- **Inspector button** (3) – Used to display and hide the Inspector area.

Don't be overwhelmed by all the different parts, as you'll learn about them in more detail in the upcoming chapters. Now that you are familiar with the Xcode interface, you will run the app you just created in Simulator, which displays a representation of an iOS device.

Running your app in Simulator

Simulator is downloaded and installed after you install Xcode. It provides a simulated iOS device so that you can see what your app looks like and how it behaves, without needing a physical iOS device. It can model all the screen sizes and resolutions for both iPad and iPhone so you can test your app on multiple devices easily.

To run your app in Simulator, follow these steps:

1. Click the Destination pop-up menu to view a list of simulated devices. Choose **iPhone SE (3rd generation)** from this menu:

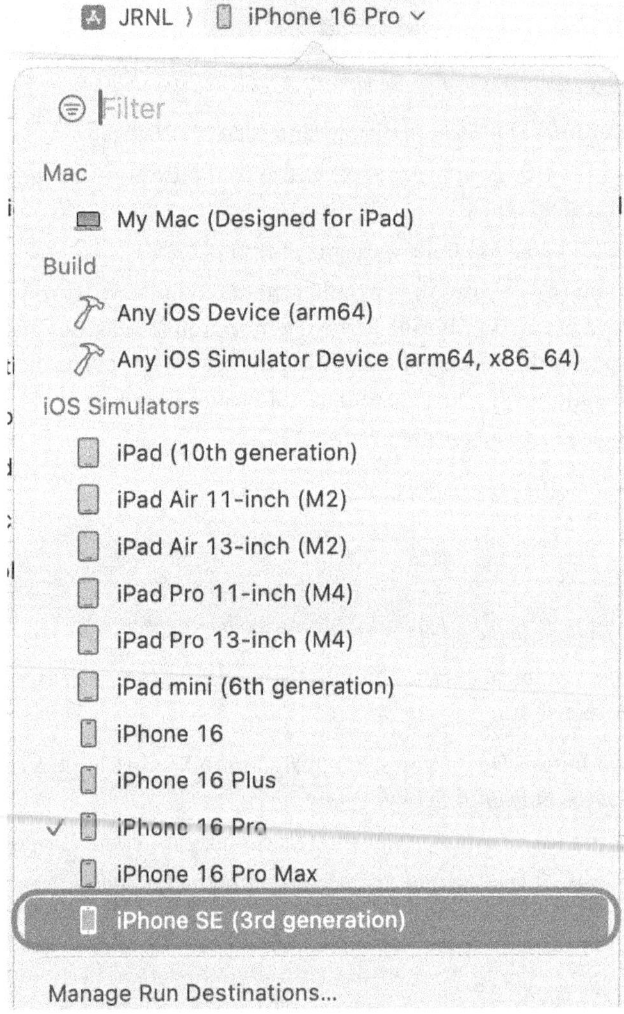

Figure 1.16: Xcode Destination pop-up menu with iPhone SE (3rd generation) selected

In your own projects, you should pick whichever simulator you require. That said, if you want to match the screenshots in this book exactly, use the **iPhone SE (3rd generation)** simulator. This simulator also has a home button, so it is easier to get to the home screen.

2. Click the Run button to install and run your app on the currently selected simulator. You can also use the *Command* + *R* keyboard shortcut.

3. Simulator will launch and show a representation of an iPhone SE (3rd generation). Your app displays a white screen, as you have not yet added anything to your project:

Figure 1.17: Simulator displaying your app

4. Switch back to Xcode and click on the Stop button (or press *Command* + .) to stop the currently running project.

You have just created and run your first iOS app in Simulator! Great job!

The Destination menu has a section showing physical devices connected to your Mac and a **Build** section. You may be wondering what they are used for. Let's look at them in the next section.

Understanding the Build section

You learned how to choose a simulated device in the Destination menu to run your app in the previous section. In addition to the list of simulated devices, this menu also has a section showing physical devices connected to your Mac, and a **Build** section.

These allow you to run apps on actual Mac or iOS devices and prepare apps for submission to the App Store.

Click the Destination menu in the toolbar to see the physical device and **Build** sections at the top of the menu:

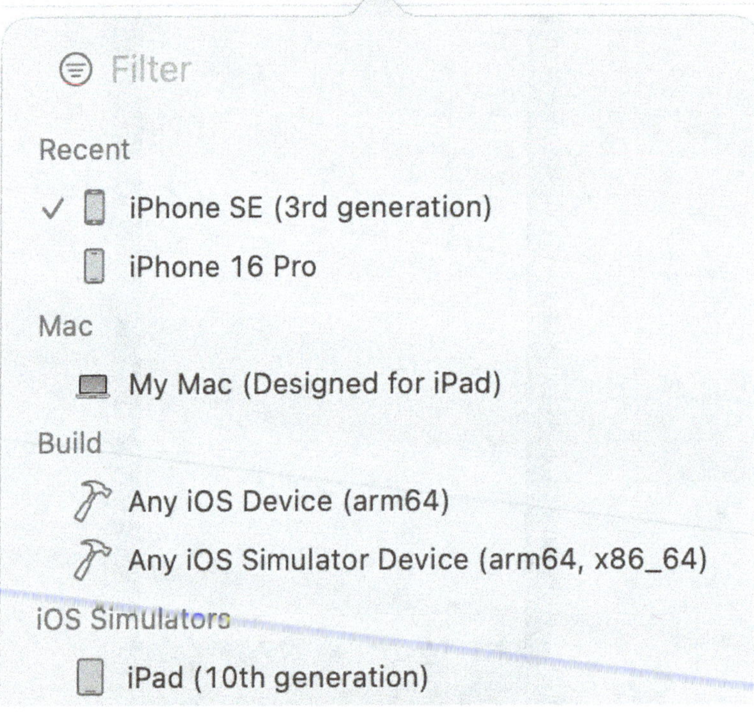

Figure 1.18: Xcode Destination menu showing device and Build sections

If you have an Apple Silicon Mac, the physical device section will display text stating **My Mac (Designed for iPad)**, because Apple Silicon Macs can run iOS apps. Otherwise, **No Devices** will be displayed. If you were to plug in an iOS device, it would appear in this section, and you would be able to run the apps you develop on it for testing. Running your apps on an actual device is recommended as Simulator will not accurately reflect the performance characteristics of an actual iOS device and does not have hardware features that actual devices have.

The **Build** section has two menu items, **Any iOS Device(arm64)** and **Any iOS Simulator Device (arm64, x86_64)**. These are used when you need to archive your app prior to submitting it to the App Store. You'll learn how to do this in *Chapter 27, Testing and Submitting Your App to the App Store*.

Now let's see how to build and run your app on an actual iOS device. Most of the instructions in this book do not require you to have an iOS device though, so if you don't have one, you can skip the next section and go straight to *Chapter 2, Simple Values and Types*.

Running your app on an iOS device

Although you'll be able to go through most of the exercises in this book using Simulator, it is recommended to build and test your apps on an actual iOS device, as Simulator will not be able to simulate some hardware components and software APIs.

For a comprehensive look at all the differences between Simulator and an actual device, see this link: `https://help.apple.com/simulator/mac/current/#/devb0244142d`.

In addition to your device, you'll need an Apple Account (used to automatically create a free Apple developer account) or a paid Apple developer account to build and run your app on your device. You can use the same Apple Account that you used to download Xcode from the App Store. To run your app on an iOS device, follow these steps:

1. Use the cable that came with your iOS device to connect your device to your Mac, and make sure the iOS device is unlocked.
2. Your Mac will display an **Allow Accessory to Connect** alert. Click **Allow**.
3. Your iOS device will display a **Trust This Computer** alert. Tap **Trust** and key in your device passcode when prompted. Your iOS device is now connected to your Mac and will appear in Xcode's Destination menu.
4. Choose **Window | Devices and Simulators** in the Xcode menu bar. You will see a window displaying a message saying **Developer Mode disabled**:

Figure 1.19: Xcode Devices and Simulators window showing Developer Mode disabled

Developer Mode was introduced by Apple during their Worldwide Developers Conference in 2022 (WWDC 2022) and is required to install, run, and debug your apps on devices running iOS 16 or greater.

To watch a WWDC 2022 video on Developer Mode, click this link: https://developer.apple.com/videos/play/wwdc2022/110344/.

5. To enable Developer Mode on your iOS device, go to **Settings** | **Privacy & Security**, scroll down to the **Developer Mode** item, and tap it:

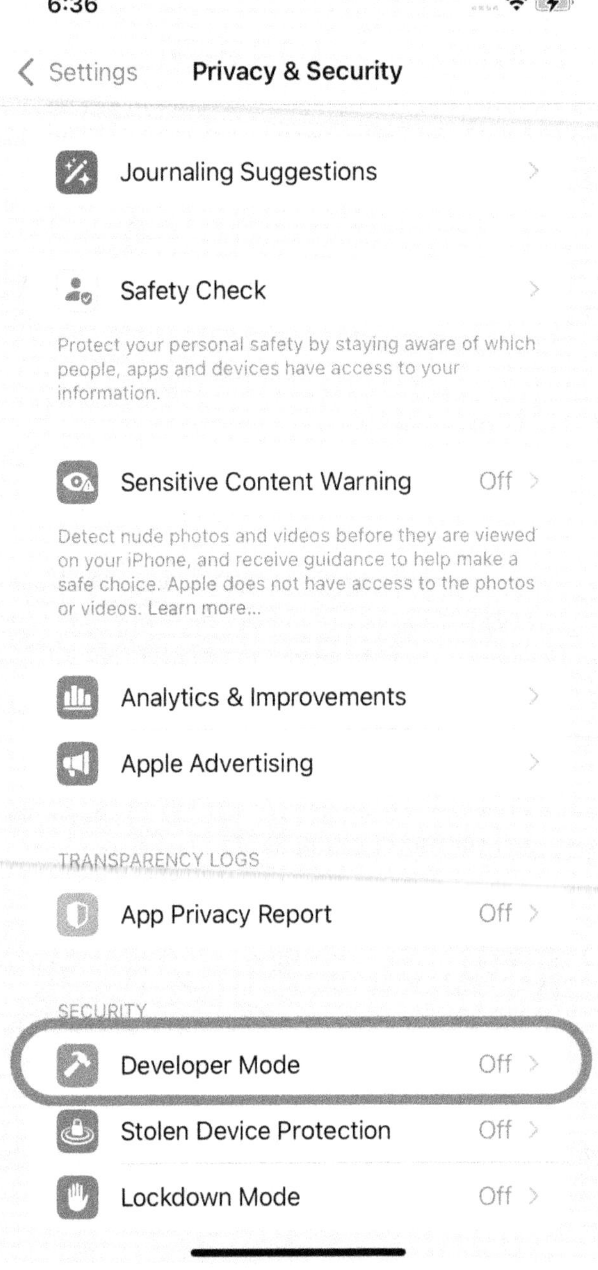

Figure 1.20: Privacy & Security screen showing Developer Mode

6. Turn the **Developer Mode** switch on:

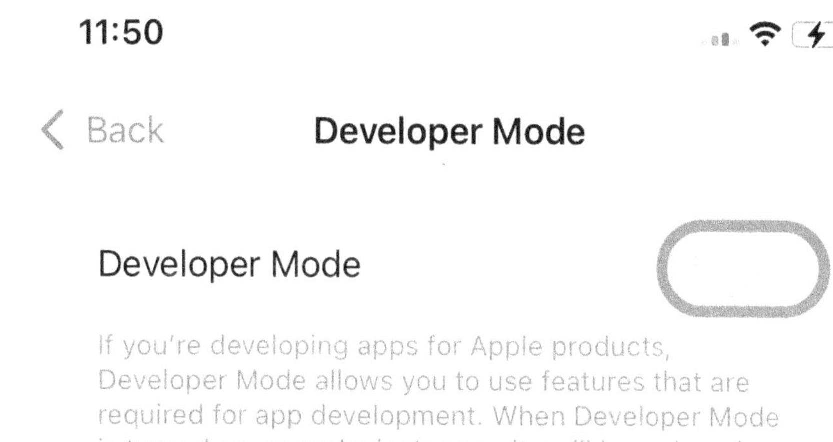

Figure 1.21: Developer Mode switch

7. An alert will appear to warn you that Developer Mode reduces the security of your iOS device. Tap the alert's **Restart** button.
8. After your iOS device restarts and you unlock it, confirm that you want to enable **Developer Mode** by tapping **Enable** and entering your iOS device's passcode.
9. The Devices and Simulators window will display a **Preparing iPhone** message. Wait a few minutes, then verify that the Devices and Simulators window no longer displays the **Developer Mode disabled** text:

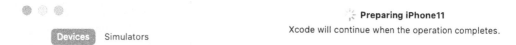

Figure 1.22: Xcode Devices and Simulators window showing Preparing iPhone message

Your iOS device is now ready to install and run apps from Xcode.

10. In Xcode, choose your iOS device from the Destination menu.

11. Run the project by clicking the Run button (or use *Command* + *R*). You will get the following error in Xcode's **Signing & Capabilities** panel: **Signing for "JRNL" requires a development team:**

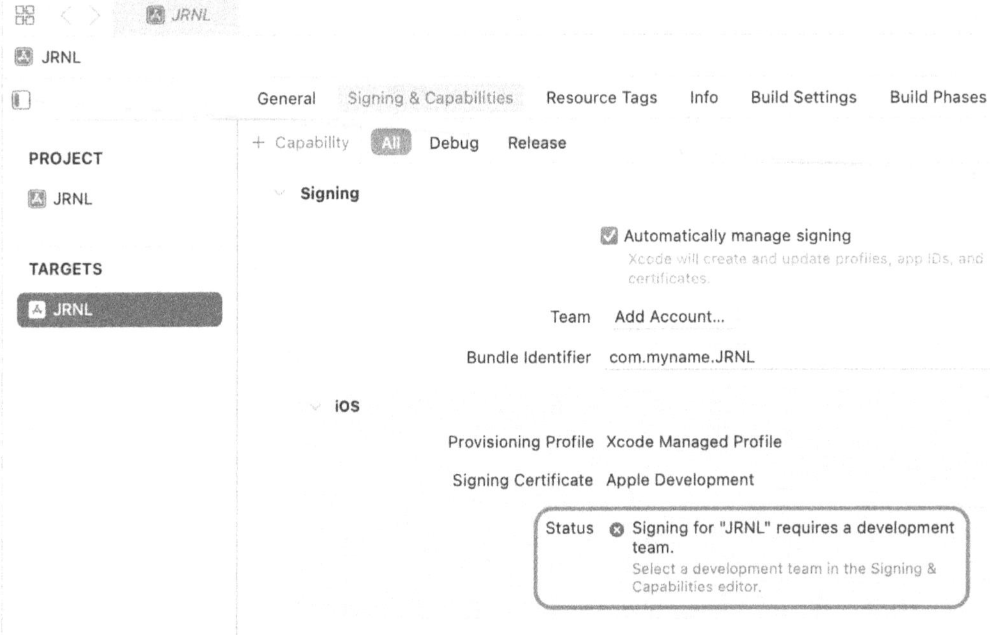

Figure 1.23: Xcode Signing & Capabilities panel

This is because a digital certificate is required to run the app on an iOS device, and you need to add a free or paid Apple developer account to Xcode so the digital certificate can be generated.

Using an Apple Account to create a free developer account will allow you to test your app on an iOS device, but it will only be valid for 7 days. Also, you will need a paid Apple developer account to distribute apps on the App Store. You'll learn more about this in *Chapter 27, Testing and Submitting Your App to the App Store*.

Certificates ensure that the only apps that run on your device are the ones you authorize. This helps to protect against malware. You can also learn more about them at this link: `https://help.apple.com/xcode/mac/current/#/dev60b6fbbc7`.

12. Click the **Add Account...** button:

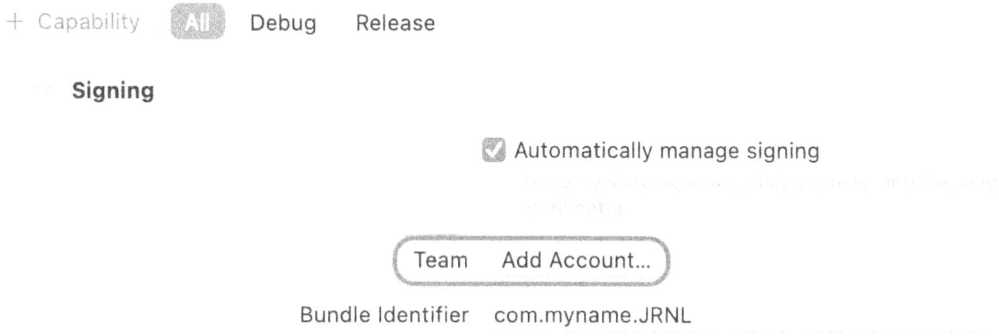

Figure 1.24: Xcode Signing & Capabilities pane with the Add Account... button selected

13. The Xcode **Settings** window appears with the **Accounts** pane selected. Enter your Apple Account and click **Next**:

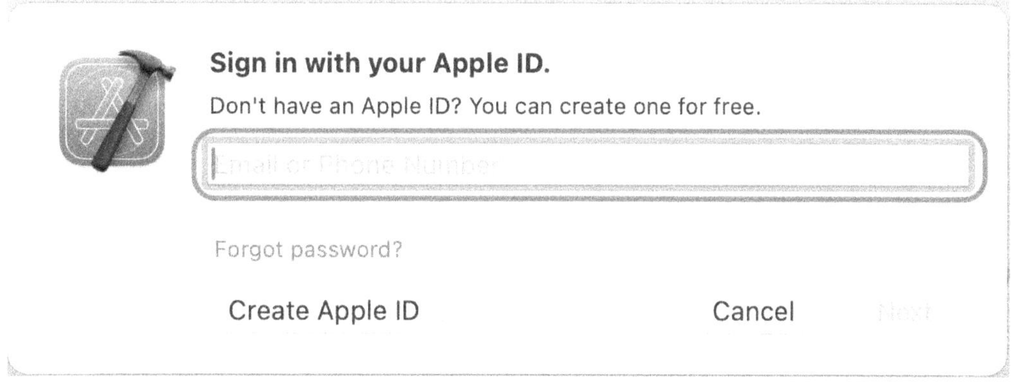

Figure 1.25: Apple Account creation dialog box

Note that you can create a different Apple Account if you wish, using the **Create Apple ID** button.

You can also access the Xcode settings by choosing **Settings** in the Xcode menu.

14. Enter your password when prompted. After a few minutes, the **Accounts** pane will display your account settings:

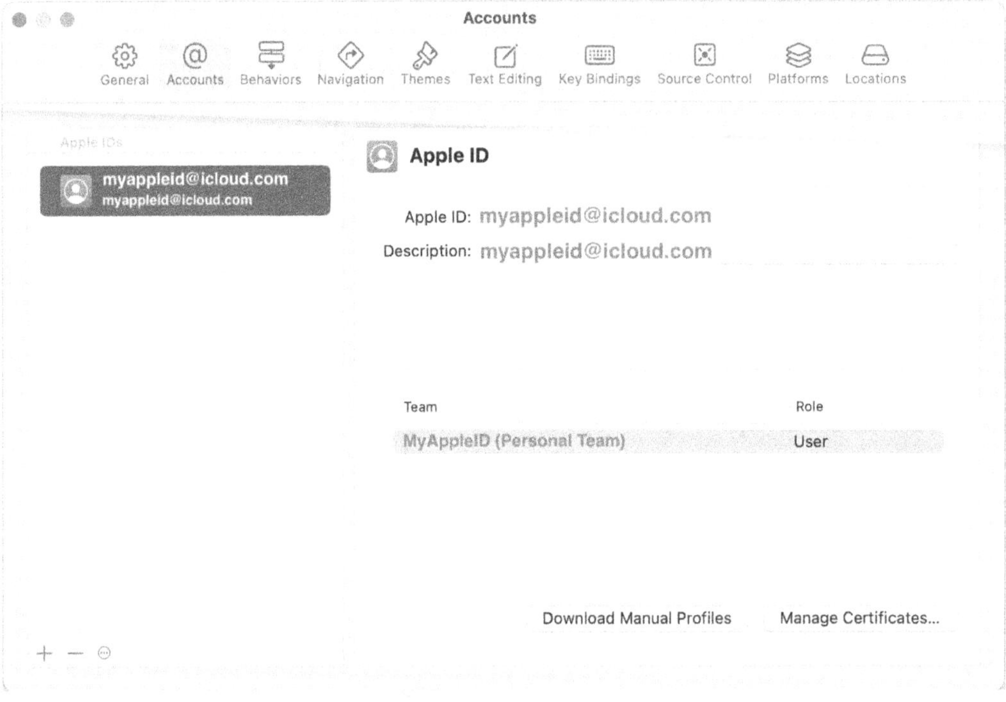

Figure 1.26: Accounts pane in Xcode preferences

15. Close the **Settings** window when you're done by clicking the red close button in the top-left corner.
16. In Xcode's editor area, click **Signing & Capabilities**. Make sure **Automatically manage signing** is ticked and **Personal Team** is selected from the **Team** pop-up menu:

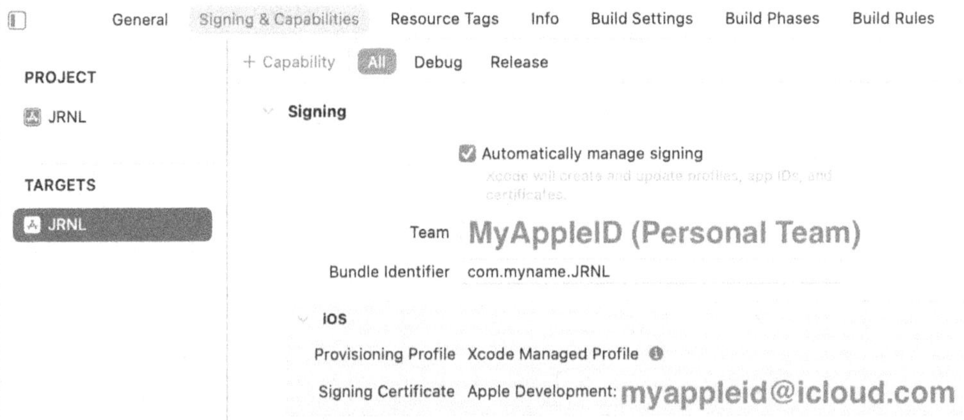

Figure 1.27: Xcode Signing & Capabilities pane with account set

17. If you still see errors on this screen, try changing your **Bundle Identifier** by typing some random characters into it, for example, `com.myname6712.JRNL`.
18. Build and run your app. If you are prompted for a password, enter your Mac's login password and click **Always Allow**.
19. Your app will be installed on your iOS device. However, it will not launch, and you will see the following message:

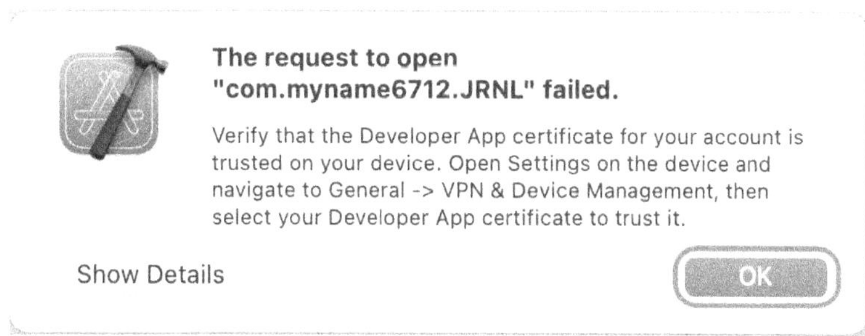

Figure 1.28: Could not launch "JRNL" dialog box

20. This means you need to trust the certificate that has been installed on your device. You'll learn how to do this in the next section.

Trusting the Developer App certificate on your iOS device

A **Developer App certificate** is a special file that gets installed on your iOS device along with your app. Before your app can run, you need to trust it. Follow these steps:

1. On your iOS device, tap **Settings** | **General** | **VPN & Device Management**:

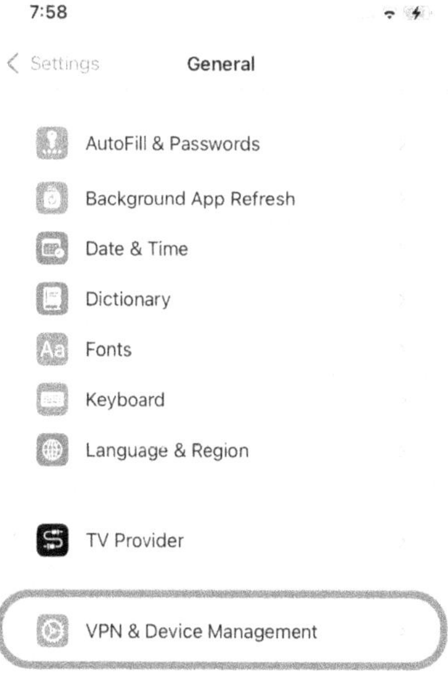

Figure 1.29: VPN & Device Management setting in Settings

2. Tap your Apple Account:

DEVELOPER APP

myappleid@icloud.com

Not Trusted

Figure 1.30: Your Apple Account in Device Management settings

3. Tap **Trust**:

Trust "myappleid@icloud.com"

Figure 1.31: Trust button

4. Tap **Allow**:

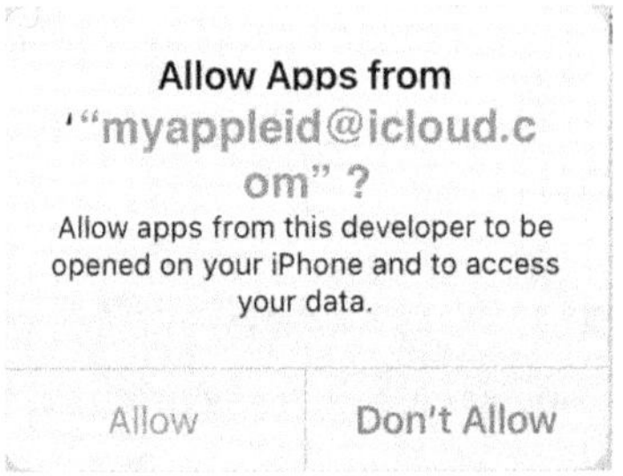

Allow Apps from
"myappleid@icloud.c
om" ?
Allow apps from this developer to be
opened on your iPhone and to access
your data.
Allow
Don't Allow

Figure 1.32: Allow dialog box

You should see the following text, which shows the app is now trusted:

Apps from developer "Apple Development: myappleid@icloud.com (XXXXXX)" ɾ ɑro trusted on this iPhone and will be trusted until all apps from the developer are deleted.

Delete App

Figure 1.33: Device management section with trusted certificate

5. Click the Run button in Xcode to build and run again. You'll see your app launch and run on your iOS device.

Congratulations! You have successfully run your app on an actual iOS device!

Summary

In this chapter, you learned how to download and install Xcode on your Mac. Then, you familiarized yourself with the different parts of the Xcode user interface. After that, you created your first iOS app, selected a simulated iOS device, and built and ran the app in Simulator. Finally, you learned how to connect an iOS device to Xcode via USB so that you can run your app on it.

In the next chapter, we'll start exploring the Swift language using Swift Playgrounds, and learn how simple values and types are implemented in Swift.

Join us on Discord!

Read this book alongside other users, experts, and the author himself. Ask questions, provide solutions to other readers, chat with the author via Ask Me Anything sessions, and much more. Scan the QR code or visit the link to join the community.

`https://packt.link/ios-Swift`

2

Simple Values and Types

Now that you have had a short tour of Xcode in the previous chapter, let's look at the Swift programming language, which you will use to write your app.

First, you'll explore **Swift playgrounds**, interactive environments where you can type in Swift code and have the results displayed immediately. Then, you'll study how Swift represents and stores various types of data. After that, you'll look at some cool Swift features, such as **type inference** and **type safety**, which help you to write code more concisely and avoid common errors. Finally, you'll learn how to perform common operations on data and how to print messages to the Debug area to help you troubleshoot issues.

By the end of this chapter, you should be able to write simple programs that can store and process letters and numbers.

The following topics will be covered:

- Introducing Swift playgrounds
- Exploring data types
- Exploring constants and variables
- Understanding type inference and type safety
- Exploring operators
- Using the `print()` statement

For more information about the latest version of the Swift language, visit `https://docs.swift.org/swift-book/documentation/the-swift-programming-language/`.

Technical requirements

To do the exercises in this chapter, you will need the following:

- An Apple Mac computer running macOS 14 Sonoma or macOS 15 Sequoia
- Xcode 16 installed (refer to *Chapter 1, Exploring Xcode,* for instructions on how to install Xcode)

The Xcode playground for this chapter is in the `Chapter02` folder of the code bundle for this book, which can be downloaded here:

`https://github.com/PacktPublishing/iOS-18-Programming-for-Beginners-Ninth-Edition`

Check out the following video to see the code in action:

`https://youtu.be/1SBD_Wacpdc`

In the next section, you'll create a new playground, where you can type in the code presented in this chapter.

Introducing Swift playgrounds

Playgrounds are interactive coding environments. You type code in the left-hand pane, and the results are displayed immediately in the right-hand pane. It's a great way to experiment with code and explore the iOS SDK.

SDK is an acronym for software development kit. To learn more about the iOS SDK, visit `https://developer.apple.com/ios/`.

Let's start by creating a new playground and examining its user interface. Follow these steps:

1. To create a playground, launch Xcode and choose **File | New | Playground...** from the Xcode menu bar:

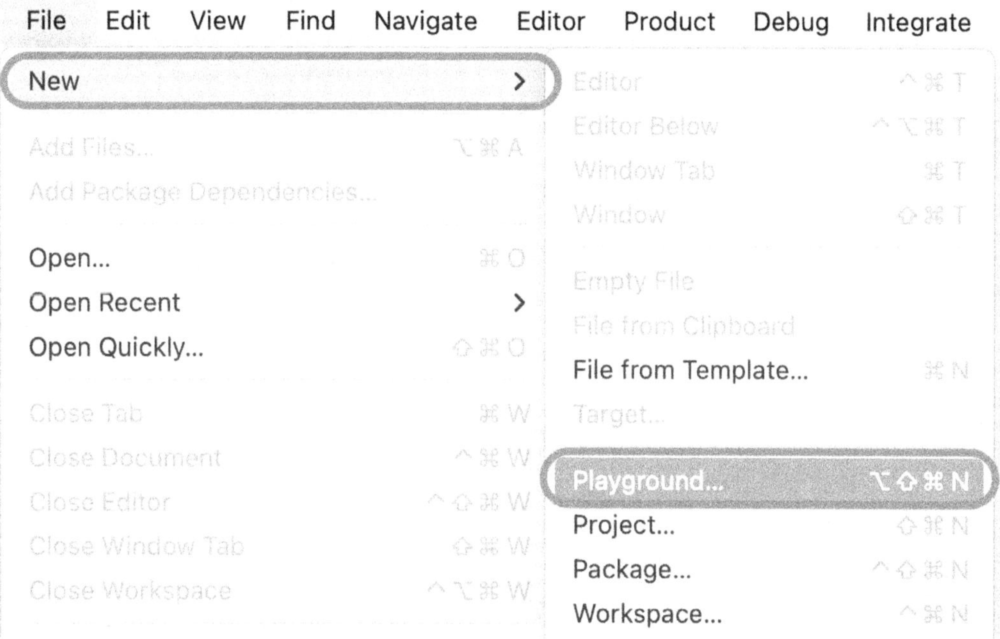

Figure 2.1: Xcode menu bar with File | New | Playground... selected

2. The **Choose a template for your new playground:** screen appears. **iOS** should already be selected. Choose **Blank** and click **Next:**

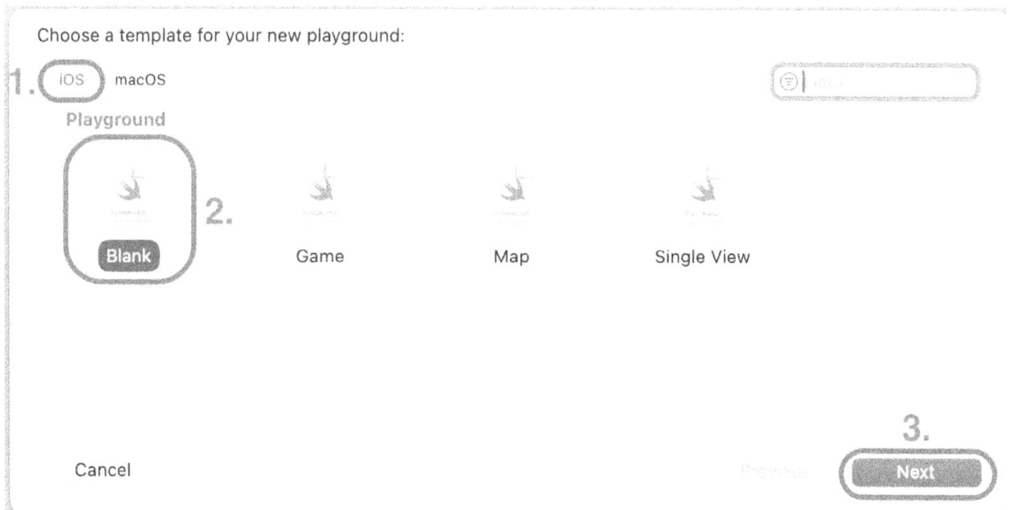

Figure 2.2: The Choose a template for your new playground: screen

3. Name your playground `SimpleValues` and save it anywhere you like. Click **Create** when done:

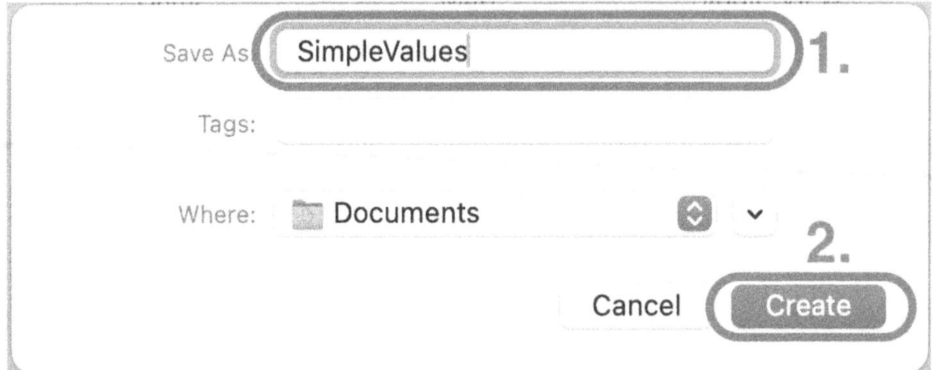

Figure 2.3: Save dialog box

4. You'll see the playground on your screen:

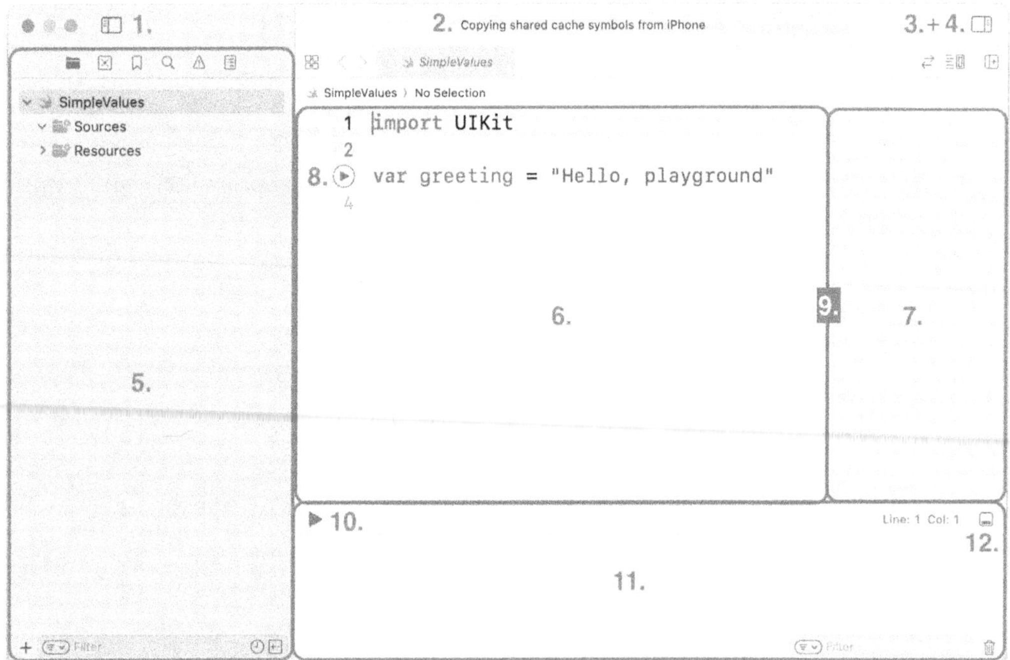

Figure 2.4: Xcode playground user interface

As you can see, it's much simpler than an Xcode project. Let's look at the interface in more detail:

- **Navigator button (1)**: This shows or hides the Navigator area.
- **Activity View (2)**: This shows the current operation or status.
- **Library button (3)**: This displays code snippets and other resources.
- **Inspector button (4)**: This shows or hides the Inspector area.

- **Navigator area** (5): This provides quick access to various parts of your project. The Project navigator is displayed by default.
- **Editor area** (6): You write code here.
- **Results area** (7): This provides immediate feedback on the code you write.
- **Run button** (8): This executes code from a selected line.
- **Border** (9): This border separates the Editor and Results areas. If you find that the results displayed in the Results area are truncated, drag the border to the left to increase its size.
- **Run/Stop button** (10): This executes or stops the execution of all code in the playground.
- **Debug area** (11): This displays the results of the `print()` command.
- **Debug button** (12): This shows and hides the Debug area.

You may find the text in the playground too small and hard to read. Let's see how to make it larger in the next section.

Customizing fonts and colors

Xcode has extensive customization options available. You can access them in the **Settings...** menu. If you find that the text is small and hard to see, follow these steps:

1. Choose **Settings...** from the Xcode menu to display the Settings window.
2. In the Settings window, click **Themes** and choose **Presentation (Light)** to make the text larger and easier to read:

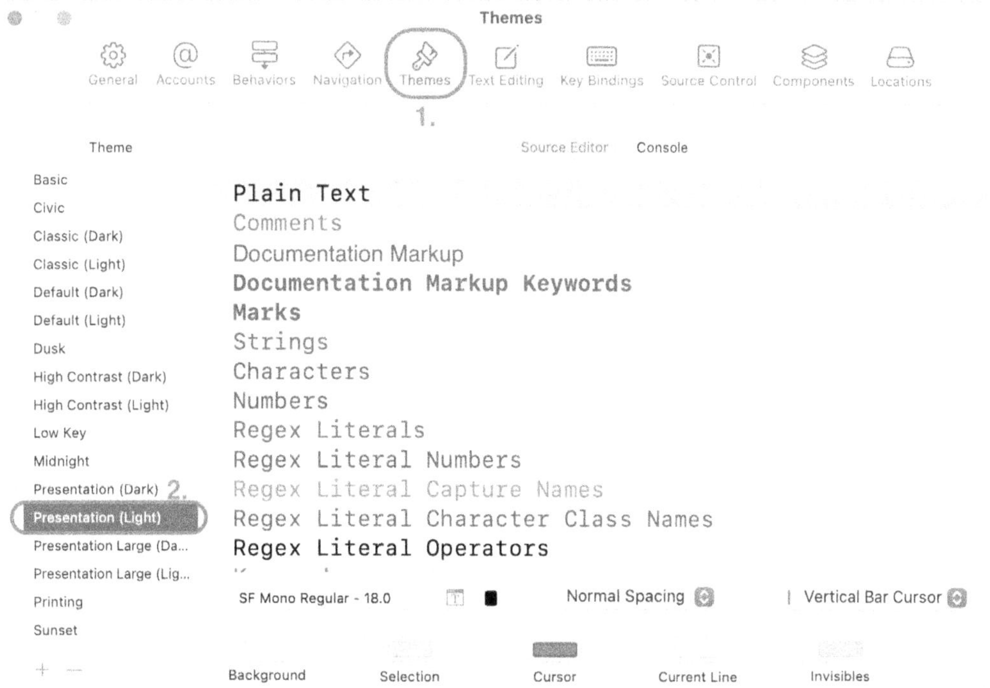

Figure 2.5: Xcode settings window with the Themes pane selected

3. Close the Settings window to return to the playground. Note that the text in the playground is larger than before. You can also try the other themes if you wish.

Now that you've customized the fonts and colors to your liking, let's see how to run playground code in the next section.

Running playground code

Your playground already has an instruction in it. To execute the instruction, follow these steps:

1. Click the Run button to the left of the instruction. After a few seconds, you will see `"Hello, playground"` displayed in the Results area:

Figure 2.6: Playground showing "Hello, playground" in the Results area

You can also use the Run/Stop button in the bottom-left corner or use the keyboard shortcut *Command* + *Shift* + *Return* to run all the code in your playground.

2. To prepare the playground for use in the remainder of this chapter, delete the `var greeting = "Hello, playground"` instruction from the playground. As you go along, type the code shown in this chapter into the playground, and click the Run button to the left of the last line to run it.

Let's dive into the simple data types used in Swift in the next section.

Exploring data types

All programming languages can store numbers, words, and logic states, and Swift is no different. Even if you're an experienced programmer, you may find that Swift represents these values differently from other languages that you may be familiar with.

For more information on data types, visit `https://docs.swift.org/swift-book/documentation/the-swift-programming-language/thebasics`.

Let's walk through the Swift versions of **integers**, **floating-point numbers**, **strings**, and **Booleans** in the next sections.

Representing integers

Let's say you want to store the following:

- The number of restaurants in a city
- Passengers in an airplane
- Rooms in a hotel

You would use integers, which are numbers without a fractional component (including negative numbers).

Integers in Swift are represented by the `Int` type.

Representing floating-point numbers

Let's say you want to store the following:

- Pi (3.14159...)
- Absolute zero (-273.15°C)

You would use floating-point numbers, which are numbers with a fractional component.

The default type for floating-point numbers in Swift is `Double`, which uses 64 bits, including negative numbers. You can also use `Float`, which uses 32 bits, but `Double` is the default representation.

Representing strings

Let's say you want to store the following:

- The name of a restaurant, such as "Bombay Palace"
- A job description, such as "Accountant" or "Programmer"
- A kind of fruit, such as "banana"

You would use Swift's `String` type, which represents a sequence of characters and is fully Unicode-compliant. This makes it easy to represent different fonts and languages.

To learn more about Unicode, visit this link: `https://home.unicode.org/basic-info/faq/`.

Representing Booleans

Let's say you want to store answers to simple yes/no questions, such as the following:

- Is it raining?
- Are there any available seats at the restaurant?

For this, you use Boolean values. Swift provides a `Bool` type that can be assigned `true` or `false`.

Now that you know how Swift represents these common data types, let's try them out in the playground you created earlier in the next section.

Using common data types in the playground

Anything that you type into a playground will be executed, and the results will appear in the Results area. Let's see what happens when you type numbers, strings, and Boolean values into your playground and execute it. Follow these steps:

1. Type the following code into the Editor area of your playground:

   ```
   // SimpleValues
   42
   -23

   3.14159
   0.1
   -273.15

   "hello, world"
   "albatross"

   true
   false
   ```

   Note that any line with // in front of it is a comment. Comments are a great way to create notes or reminders for yourself and will be ignored by Xcode.

2. Click the Run button to the left of the last line to run your code.
3. Wait a few seconds. Xcode will evaluate your input and display results in the Results area, as follows:

   ```
   42
   -23

   3.14159
   0.1
   -273.15

   "hello, world"
   "albatross"
   true
   false
   ```

Note that comments do not appear in the Results area.

Cool! You have just created and run your first playground. Let's look at how to store different data types in the next section.

Exploring constants and variables

Now that you know about the simple data types that Swift supports, let's look at how to store them so that you can perform operations on them later.

You can use **constants** or **variables** to store values. Both are containers that have a name, but a constant's value can only be set once and cannot be changed after it is set, whereas a variable's value can be changed at any time.

You must declare constants and variables before you use them. Constants are declared with the `let` keyword while variables are declared with the `var` keyword.

Let's explore how constants and variables work by implementing them in our playground. Follow these steps:

1. Add the following code to your playground to declare three constants:

```
let theAnswerToTheUltimateQuestion = 42
let pi = 3.14159
let myName = "Ahmad Sahar"
```

2. Click the Run button to the left of the last line to run it. In each case, a container is created and named, and the assigned value is stored in it.

You may have noticed that the names of constants and variables shown here start with a lowercase letter, and if there is more than one word in the name, every subsequent word starts with a capital letter. This is known as **camel case**. Doing this is strongly encouraged, as most experienced Swift programmers adhere to this convention.

Note that a sequence of characters enclosed by double quotation marks, `"Ahmad Sahar"`, is used to assign the value for `myName`. These are known as **string literals**.

3. Add the following code after the constant declarations to declare three variables and run it:

```
var currentTemperatureInCelsius = 27
var myAge = 50
var myLocation = "home"
```

Like constants, a container is created and named in each case, and the assigned value is stored in it.

The stored values are displayed in the Results area.

4. The value of a constant can't be changed once it is set. To test this, add the following code after the variable declarations:

```
let isRaining = true
isRaining = false
```

As you type the second line of code, a pop-up menu will appear with suggestions:

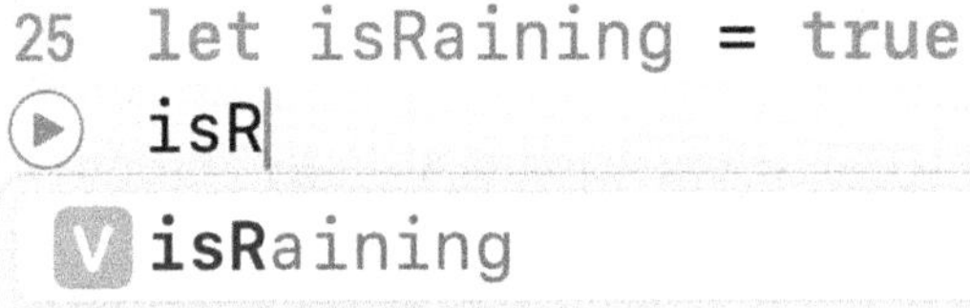

Figure 2.7: Autocomplete pop-up menu

Use the up and down arrow keys to choose the `isRaining` constant and press the *Tab* key to select it. This feature is called **autocomplete** and helps to prevent typing mistakes when you enter code.

5. When you have finished typing, wait a few seconds. On the second line, you'll see an error notification (a red circle with a white dot in the middle) appear:

```
24
25 let isRaining = true
   isRaining = false   Cannot assign to value: 'isRaining' is a 'let' const...
27
```

Figure 2.8: Error notification

This means there is an error in your program, and Xcode thinks it can be fixed. The error appears because you are trying to assign a new value to a constant after its initial value has been set.

6. Click the red circle to expand the error message. You'll see the following box with a **Fix** button:

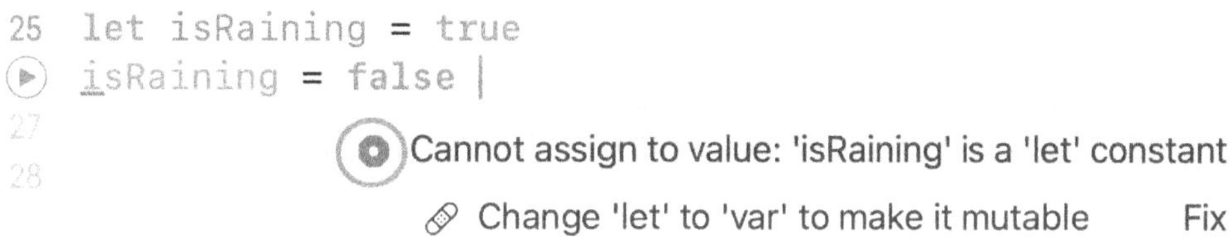

```
let isRaining = true
isRaining = false
```

Cannot assign to value: 'isRaining' is a 'let' constant

Change 'let' to 'var' to make it mutable Fix

Figure 2.9: Expanded error notification

   Xcode tells you what the problem is (**Cannot assign to value: 'isRaining' is a 'let' constant**) and suggests a correction (**Change 'let' to 'var' to make it mutable**). "Mutable" just means that the value can be changed after it has been set initially.

7. Click the **Fix** button. You'll see that the `isRaining` constant declaration has been changed to a variable declaration:

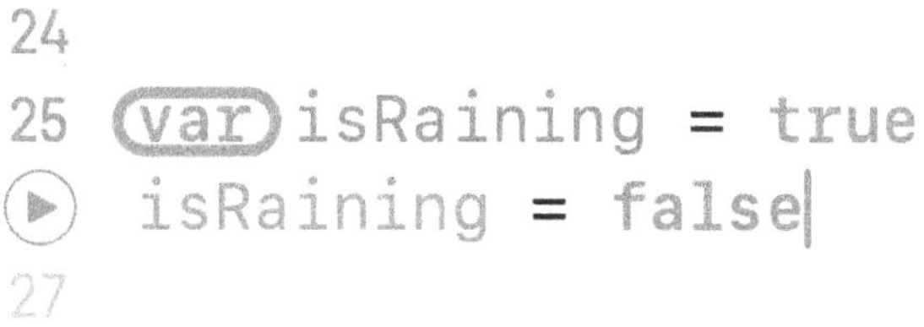

```
var isRaining = true
isRaining = false
```

Figure 2.10: Code with a fix applied

Since a new value can be assigned to a variable after it has been created, the error is resolved. Do note, however, that the suggested correction might not be the best solution. As you gain more experience with iOS development, you'll be able to determine the best course of action.

If you look at the code you typed in, you might wonder how Xcode knows the type of data stored in a variable or constant. You'll learn how that is done in the next section.

Understanding type inference and type safety

In the previous section, you declared constants and variables and assigned values to them. Swift automatically determines the constant or variable type based on the value provided. This is called **type inference**. You can see the type of a constant or variable by holding down the *Option* key and clicking its name. To see this in action, follow these steps:

1. Add the following code to your playground to declare a string and run it:

```
let cuisine = "American"
```

2. Hold down the *Option* key and click `cuisine` to reveal the constant type. You should see the following:

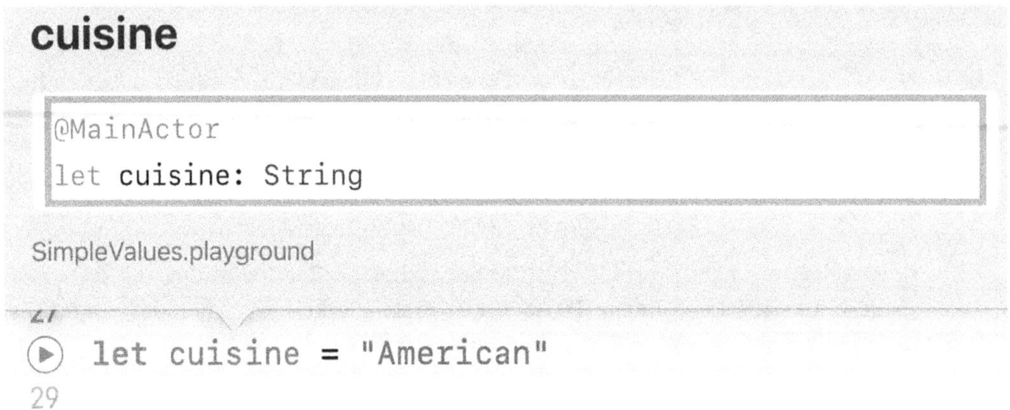

Figure 2.11: Type declaration displayed

As you can see, `cuisine`'s type is `String`.

What if you want to set a specific type for a variable or constant? You'll see how to do that in the next section.

Using type annotation to specify a type

You've seen that Xcode tries to automatically determine the data type of a variable or constant based on the value provided. However, at times, you may wish to specify a type instead of letting Xcode do it for you. To do this, type a colon (`:`) after a constant or variable name, followed by the desired type. This is known as **type annotation.**

Add the following code to your playground to declare a variable, `restaurantRating`, of type `Double`, and click the Run button to run it:

```
var restaurantRating: Double = 3
```

Here, you specified that `restaurantRating` has a specific type, `Double`. Even though you assigned an integer to `restaurantRating`, it will be stored as a floating-point number.

In the next section, you'll learn how Xcode helps you reduce the number of errors in your program by enforcing type safety.

Using type safety to check values

Swift is a type-safe language. It checks to see whether you're assigning values of the correct type to variables and flags mismatched types as errors. Let's see how this works by following these steps:

1. Add the following statement to your playground to assign a string to `restaurantRating` and run it:

```
restaurantRating = "Good"
```

2. You'll see an error notification (a red circle with an x inside it). The x means Xcode can't suggest a fix for this. Click on the red circle.
3. Since you are trying to assign a string to a variable of type `Double`, the following error message is displayed:

```
var restaurantRating: Double = 3
restaurantRating = "Good"
                          Cannot assign value of type 'String' to type 'Double'
```

Figure 2.12: Expanded error notification with no fix

4. Comment out the line by typing // before it, as shown here:

```
// restaurantRating = "Good"
```

The red circle disappears as there are no more errors in your program.

Selecting lines of code and typing *Command* + / will comment them out.

Now that you know how to store data in constants and variables, let's look at how to perform operations on them in the next section.

Exploring operators

You can perform arithmetic, comparison, and logical operations in Swift. **Arithmetic operators** are for common mathematical operations. **Comparison** and **logical operators** check an expression's value and return `true` or `false`.

For more information on operators, visit `https://docs.swift.org/swift-book/documentation/the-swift-programming-language/basicoperators`.

Let's look at each operator type in more detail. You'll start with arithmetic operators (addition, subtraction, multiplication, and division) in the next section.

Using arithmetic operators

You can perform mathematical operations on integer and floating-point numbers by using the standard arithmetic operators shown here:

+	Addition
-	Subtraction
*	Multiplication
/	Division

Figure 2.13: Arithmetic operators

Let's see how these operators are used. Follow these steps:

1. Add the following code to add arithmetic operations to your playground:

```
let sum = 23 + 20
let result = 32 - sum
let total = result * 5
let divide = total / 10
```

2. Run the code. The results displayed in the Results area will be `43`, `-11`, `-55`, and `-5`, respectively. Note that 55 divided by 10 returns 5 instead of 5.5, as both numbers are integers.
3. Operators can only work with operands of the same type. Enter the following code and run it to see what happens if the operands are of different types:

```
let a = 12
let b = 12.0
let c = a + b
```

You'll get an error message (**Binary operator '+' cannot be applied to operands of type 'Int' and 'Double'**). This is because `a` and `b` are different types. Note that Xcode can't fix this automatically, so it does not display any fix-it suggestions.

4. To fix the error, modify the program as follows:

```
let c = Double(a) + b
```

`Double(a)` gets the value stored in `a` and creates a floating-point number from it. Both operands are now of the same type, and now you can add the value in `b` to it. The value stored in `c` is `24.0`, and `24` will be displayed in the Results area.

Now that you know how to use arithmetic operators, you'll look at compound assignment operators (+=, -=, *=, and /=) in the next section.

Using compound assignment operators

You can perform an operation on a value and assign the result to a variable using the compound assignment operators shown here:

+=	Adds a value and assigns the result to the variable
-=	Subtracts a value and assigns the result to the variable
*=	Multiplies by the value and assigns the result to the variable
/=	Divides by the value and assigns the result to the variable

Figure 2.14: Compound assignment operators

Let's see how these operators are used. Add the following code to your playground and run it:

```
var d = 1
d += 2
d -= 1
```

The `d += 2` expression is shorthand for `d = d + 2`, so the value in `d` is now `1 + 2`, and `3` will be assigned to `d`. In the same way, `d -= 1` is shorthand for `d = d - 1`, so the value in `d` is now `3 - 1`, and `2` will be assigned to `d`.

Now that you are familiar with compound assignment operators, let's look at comparison operators (==, /=, >, <, >=, and <=) in the next section.

Using comparison operators

You can compare one value to another using comparison operators, and the result will be true or false. You can use the following comparison operators:

==	Equal to
!=	Not equal to
>	Greater than
<	Less than
>=	Greater than or equal to
<=	Less than or equal to

Figure 2.15: Comparison operators

Let's see how these operators are used. Add the following code to your playground and run it:

```
1 == 1
2 != 1
2 > 1
1 < 2
1 >= 1
2 <= 1
```

Let's see how this works:

- `1 == 1` returns `true` because 1 is equal to 1.
- `2 != 1` returns `true` because 2 is not equal to 1.
- `2 > 1` returns `true` because 2 is greater than 1.
- `1 < 2` returns `true` because 1 is less than 2.
- `1 >= 1` returns `true` because 1 is greater than or equal to 1.
- `2 <= 1` returns `false` because 2 is not less than or equal to 1.

The returned Boolean values will be displayed in the Results area.

What happens if you want to check more than one condition? That's where logical operators (**AND**, **OR**, and **NOT**) come in. You'll learn about those in the next section.

Using logical operators

Logical operators are handy when you deal with two or more conditions. For example, if you are at a convenience store, you can pay for items if you have cash or a credit card. OR is the logical operator in this case.

You can use the following logical operators:

&&	Logical AND - returns true only if all conditions are true
||	Logical OR - returns true if any condition is true
!	Logical NOT - returns the opposite Boolean value

Figure 2.16: Logical operators

To see how these operators are used, add the following code to your playground and run it:

```
(1 == 1) && (2 == 2)
(1 == 1) && (2 != 2)
(1 == 1) || (2 == 2)
(1 == 1) || (2 != 2)
```

```
(1 != 1) || (2 != 2)
!(1 == 1)
```

Let's see how this works:

- `(1 == 1) && (2 == 2)` returns `true` as both operands are `true`, so `true AND true` returns `true`.
- `(1 == 1) && (2 != 2)` returns `false` as one operand is `false`, so `true AND false` returns `false`.
- `(1 == 1) || (2 == 2)` returns `true` as both operands are `true`, so `true OR true` returns `true`.
- `(1 == 1) || (2 != 2)` returns `true` as one operand is `true`, so `true OR false` returns `true`.
- `(1 != 1) || (2 != 2)` returns `false` as both operands are `false`, so `false OR false` returns `false`.
- `!(1 == 1)` returns `false` as `1==1` is `true`, so `NOT true` returns `false`.

The returned Boolean values will be displayed in the Results area.

So far, you've only worked with numbers. In the next section, you'll see how you can perform operations on words and sentences, which are stored as strings using Swift's `String` type.

Performing string operations

As you saw earlier, a string is a series of characters. They are represented by the `String` type, and they are fully Unicode-compliant.

For more information on strings, visit `https://docs.swift.org/swift-book/documentation/the-swift-programming-language/stringsandcharacters`.

Let's learn about some common string operations. Follow these steps:

1. You can join two strings together using the + operator. Add the following code to your playground and run it:

   ```
   let greeting = "Good" + " Morning"
   ```

   The values of the string literals `"Good"` and `" Morning"` are joined together, and `"Good Morning"` is displayed in the Results area.

2. You can combine strings with constants and variables of other types by making them strings as well. To change a constant, `rating`, into a string, enter the following code and run it:

   ```
   let rating = 3.5
   var ratingResult = "The restaurant rating is " + String(rating)
   ```

   The `rating` constant contains `3.5`, a value of type `Double`. Putting `rating` in between the brackets of `String()` gets the value stored in `rating` and creates a new string based on it, `"3.5"`, which is combined with the string in the `ratingResult` variable, returning the string `"The restaurant rating is 3.5"`.

3. There is a simpler way of combining strings called **string interpolation**. String interpolation is done by typing the name of a constant or variable between "\(" and ")" in a string. Enter the following code and run it:

```
ratingResult = "The restaurant rating is \(rating)"
```

As in the previous example, the value in `rating` is used to create a new string, `"3.5"`, returning the string `"The restaurant rating is 3.5"`.

So far, you can see the results of your instructions in the Results area. However, when you write your app using Xcode, you won't have access to the Results area that you see in your playground. To display the contents of variables and constants while your program runs, you'll learn how to print them to the Debug area in the next section.

Using the print() statement

As you saw in *Chapter 1, Exploring Xcode*, an Xcode project does not have a Results area like a playground does, but both the project and playground have a Debug area. Using the `print()` statement will print anything between the brackets to the Debug area.

The `print()` statement is a function. You'll learn more about functions in *Chapter 6, Functions and Closures.*

Add the following code to your playground and click the **Run** button to run it:

```
print(ratingResult)
```

You'll see the value of `ratingResult` appear in the Debug area:

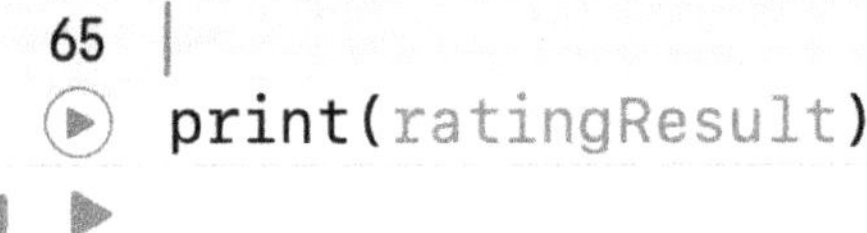

Figure 2.17: The Debug area showing the result of the print() statement

When you're just starting out, feel free to use as many `print()` statements as you like. It's a really good way to understand what is happening in your program.

Summary

In this chapter, you learned how to create and use playground files, which allow you to explore and experiment with Swift.

You saw how Swift represents different types of data, and how to use constants and variables. This enables you to store numbers, Boolean values, and strings in your program.

You also learned about type inference, type annotation, and type safety, which help you to write code concisely and with fewer errors.

You looked at how to perform operations on numbers and strings, which lets you perform simple data processing tasks.

You learned how to fix errors, and how to print to the Debug area, which is useful when you're trying to find and fix errors in the programs that you write.

In the next chapter, you'll look at **conditionals** and **optionals**. Conditionals deal with making logical choices in your program, and optionals deal with cases where a variable may or may not have a value.

Join us on Discord!

Read this book alongside other users, experts, and the author himself. Ask questions, provide solutions to other readers, chat with the author via Ask Me Anything sessions, and much more. Scan the QR code or visit the link to join the community.

`https://packt.link/ios-Swift`

3

Conditionals and Optionals

In the previous chapter, you looked at data types, constants, variables, and operations. At this point, you can write simple programs that process letters and numbers. However, programs don't always proceed in sequence. Oftentimes, you will need to execute different instructions based on a condition. Swift allows you to do this by using **conditionals**, and you will learn how to use them in this chapter.

Another thing you may have noticed is that, in the last chapter, each variable or constant was immediately assigned a value. What if you require a variable where the value may not be present initially? You will need a way to create a variable that may or may not have a value. Swift allows you to do this by using **optionals,** and you will also learn about them in this chapter.

By the end of this chapter, you should be able to write programs that do different things based on different conditions and handle variables that may or may not have a value.

The following topics will be covered:

- Introducing conditionals
- Introducing optionals and optional binding

Please spend some time understanding optionals. They can be daunting for the novice programmer, but as you will see, they are an important part of iOS development.

Technical requirements

The Xcode playground for this chapter is in the `Chapter03` folder of the code bundle for this book, which can be downloaded here:

`https://github.com/PacktPublishing/iOS-18-Programming-for-Beginners-Ninth-Edition`

Check out the following video to see the code in action:

`https://youtu.be/f90pTabsOgc`

Create a new playground and name it `ConditionalsAndOptionals`. You can type in and run all the code in this chapter as you go along. You'll start by learning about conditionals.

Introducing conditionals

At times, you'll want to execute different code blocks based on a specific condition, such as in the following scenarios:

- Choosing between different room types at a hotel. The price for bigger rooms would be higher.
- Switching between different payment methods at an online store. Different payment methods would have different procedures.
- Deciding what to order at a fast-food restaurant. Preparation procedures for each food item would be different.

To do this, you would use conditionals. In Swift, this is implemented using the `if` statement (for a single condition) and the `switch` statement (for multiple conditions).

For more information on conditionals, visit `https://docs.swift.org/swift-book/documentation/the-swift-programming-language/controlflow`.

Let's see how `if` statements are used to execute different tasks depending on a condition's value in the next section.

Using if statements

An `if` statement executes a block of code if a condition is `true`, and optionally, another block of code if the condition is `false`. It looks like this:

```
if condition {
  code1
} else {
  code2
}
```

Let's implement an `if` statement now to see this in action. Imagine that you're programming an app for a restaurant. The app would allow you to check if a restaurant is open, search for a restaurant, and check to see if a customer is over the drinking age limit. Follow these steps:

1. To check if a restaurant is open, add the following code to your playground. Run it to create a constant and execute a statement if the constant's value is `true`:

   ```
   let isRestaurantOpen = true
   if isRestaurantOpen {
     print("Restaurant is open.")
   }
   ```

First, you created a constant, `isRestaurantOpen`, and assigned `true` to it. Next, you have an `if` statement that checks the value stored in `isRestaurantOpen`. Since the value is `true`, the `print()` statement is executed and **Restaurant is open** is printed in the Debug area.

2. Try changing the value of `isRestaurantOpen` to `false` and running your code again. As the condition is now `false`, nothing will be printed to the Debug area.
3. You can also execute statements if a value is `false`. Let's say the customer has searched for a particular restaurant that is not in the app's database, so the app should display a message to say that the restaurant is not found. Type in the following code to create a constant and execute a statement if the constant's value is `false`:

```
let isRestaurantFound = false
if isRestaurantFound == false {
  print("Restaurant was not found")
}
```

The constant `isRestaurantFound` is set to `false`. Next, the `if` statement is checked. The `isRestaurantFound == false` condition returns `true`, and **Restaurant was not found** is printed in the Debug area.

You can also use `!isRestaurantFound` in place of `isRestaurantFound == false` to check the condition.

4. Try changing the value of `isRestaurantFound` to `true`. As the condition is now `false`, nothing will be printed to the Debug area.
5. To execute one set of statements if a condition is `true`, and another set of statements if a condition is `false`, use the `else` keyword. Type in the following code, which checks if a customer at a bar is over the drinking age limit:

```
let drinkingAgeLimit = 21
let customerAge = 23
if customerAge < drinkingAgeLimit {
  print("Under age limit")
} else {
  print("Over age limit")
}
```

Here, `drinkingAgeLimit` is assigned the value `21` and `customerAge` is assigned the value `23`. In the `if` statement, `customerAge < drinkingAgeLimit` is checked. Since 23 < 21 returns `false`, the `else` statement is executed and **Over age limit** is printed in the Debug area. If you change the value of `customerAge` to `19`, `customerAge < drinkingAgeLimit` will return `true`, so **Under age limit** will be printed in the Debug area.

Up to now, you have only been dealing with single conditions. What if there are multiple conditions? That's where `switch` statements come in, and you will learn about them in the next section.

Using switch statements

To understand `switch` statements, let's start by implementing an `if` statement with multiple conditions first. Imagine that you're programming a traffic light. There are three possible colors for the traffic light—red, yellow, or green—and you want something different to happen based on the color of the light. To do this, you can nest multiple `if` statements together. Follow these steps:

1. Add the following code to your playground to implement a traffic light using multiple `if` statements and run it:

```
var trafficLightColor = "Yellow"
if trafficLightColor == "Red" {
  print("Stop")
} else if trafficLightColor == "Yellow" {
  print("Caution")
} else if trafficLightColor == "Green" {
  print("Go")
} else {
  print("Invalid color")
}
```

   The first `if` condition, `trafficLightColor == "Red"`, returns `false`, so the `else` statement is executed. The second `if` condition, `trafficLightColor == "Yellow"`, returns `true`, so **Caution** is printed in the Debug area and no more `if` conditions are evaluated. Try changing the value of `trafficLightColor` to see different results.

   The code used here works, but it's a little hard to read. In this case, a `switch` statement would be more concise and easier to comprehend. A `switch` statement looks like this:

```
switch value {
case firstValue:
  code1
case secondValue:
  code2
default:
  code3
}
```

   The value is checked and matched to a case, and the code for that case is executed. If none of the cases match, the code in the `default` case is executed.

2. Here's how to write the `if` statement shown earlier as a `switch` statement. Type in the following code and run it:

```
trafficLightColor = "Yellow"
switch trafficLightColor {
case "Red":
  print("Stop")
case "Yellow":
  print("Caution")
case "Green":
  print("Go")
default:
  print("Invalid color")
}
```

The code here is much easier to read and understand when compared to the previous version. The value in `trafficLightColor` is `"Yellow"`, so `case "Yellow":` is matched and **Caution** is printed in the Debug area. Try changing the value of `trafficLightColor` to see different results.

There are two things to remember about `switch` statements:

- `switch` statements in Swift do not fall through the bottom of each case and into the next one by default. In the example shown previously, once `case "Yellow":` is matched, `case "Red":`, `case "Green":`, and `default:` will not execute.
- `switch` statements must cover all possible cases. In the example shown previously, any `trafficLightColor` value other than `"Red"`, `"Yellow"`, or `"Green"` will be matched to `default:` and **Invalid color** will be printed in the Debug area.

This concludes the section on `if` and `switch` statements. In the next section, you'll learn about optionals, which allow you to create variables without initial values, and **optional binding**, which allows instructions to be executed if an optional has a value.

Introducing optionals and optional binding

Up until now, every time you have declared a variable or constant, you have assigned a value to it immediately. But what if you want to declare a variable first and assign a value later? In this case, you would use optionals.

For more information on optionals, visit `https://docs.swift.org/swift-book/documentation/the-swift-programming-language/thebasics`.

Let's learn how to create and use optionals and see how they are used in a program. Imagine you're writing a program where the user needs to enter the name of their spouse. Of course, if the user is not married, there would be no value for this. In this case, you can use an optional to represent the spouse's name.

An optional may have one of two possible states. It can either contain a value or not contain a value. If an optional contains a value, you can access the value inside it. The process of accessing an optional's value is known as **unwrapping** the optional. Let's see how this works by following these steps:

1. Add the following code to your playground to create a variable and print its contents:

```
var spouseName: String
print(spouseName)
```

2. Since Swift is type-safe, an error will appear, **(Variable 'spouseName' used before being initialized)**.
3. To resolve this issue, you could assign an empty string to `spouseName`. Modify your code as shown:

```
var spouseName: String = ""
```

This makes the error go away, but an empty string is still a value, and `spouseName` should not have a value.

4. Since `spouseName` should not have a value initially, let's make it an optional. To do so, type a question mark after the type annotation and remove the empty string assignment:

```
var spouseName: String?
```

You'll see a warning because `spouseName` is now an optional string variable instead of a regular string variable, and the `print()` statement is expecting a regular string variable.

```
var spouseName: String?
print(spouseName)
```

Expression implicitly coerced from 'String?' to 'Any'

Figure 3.1: Warning notification

Even though there is a warning, ignore it for now and run your code. The value of `spouseName` is shown as **nil** in the Results area, and **nil** is printed in the Debug area. `nil` is a special keyword that means the optional variable `spouseName` has no value.

5. The warning appears because the `print` statement is treating `spouseName` as being of type `Any` instead of `String?`. Click the yellow triangle to display possible fixes, and choose the first fix:

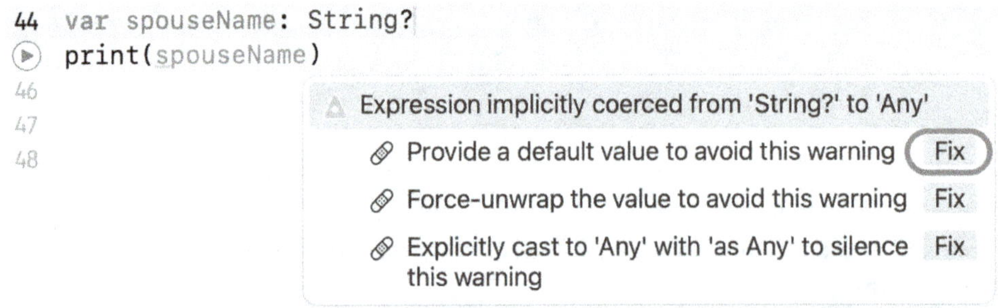

```
var spouseName: String?
print(spouseName)
```

Expression implicitly coerced from 'String?' to 'Any'

Provide a default value to avoid this warning Fix

Force-unwrap the value to avoid this warning Fix

Explicitly cast to 'Any' with 'as Any' to silence this warning Fix

Figure 3.2: Expanded warning notification with the first fix highlighted

The statement will change to `print(spouseName ?? default value)`. Note the use of the `??` operator. This means that if `spouseName` does not contain a value, a default value that you provide will be used instead in the `print` statement.

6. Replace the default value placeholder with `"No value in spouseName"` as shown. The warning will disappear. Run your program again and **No value in spouseName** will appear in the Debug area:

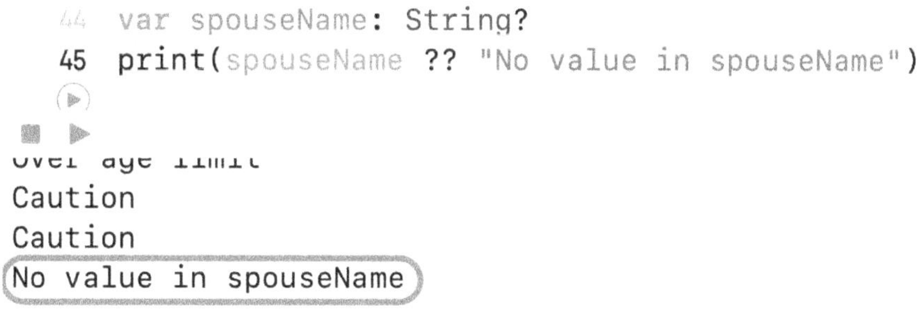

```
var spouseName: String?
print(spouseName ?? "No value in spouseName")
```

```
Caution
Caution
No value in spouseName
```

Figure 3.3: Debug area showing the default value

7. Let's assign a value to `spouseName`. Modify the code as shown:

```
var spouseName: String?
spouseName = "Nia"
print(spouseName ?? "No value in spouseName")
```

When your program runs, **Nia** appears in the Debug area.

8. Add one more line of code to join `spouseName` to another string as shown:

```
print(spouseName ?? "No value in spouseName")
let greeting = "Hello, " +  spouseName
```

You'll get an error, and the Debug area displays the error information and where the error occurred. This happens because you can't join a regular string variable to an optional using the + operator. You will need to unwrap the optional first.

9. Click on the red circle to display possible fixes, and you'll see the following:

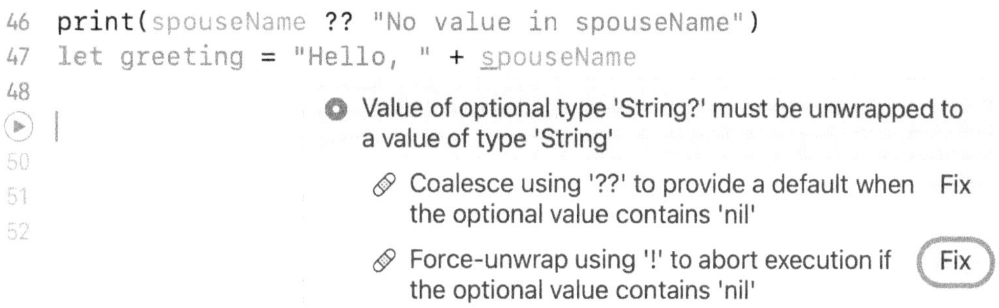

```
print(spouseName ?? "No value in spouseName")
let greeting = "Hello, " + spouseName
```

Value of optional type 'String?' must be unwrapped to a value of type 'String'

Coalesce using '??' to provide a default when the optional value contains 'nil' Fix

Force-unwrap using '!' to abort execution if the optional value contains 'nil' Fix

Figure 3.4: Expanded error notification with the second fix highlighted

The second fix recommends **force-unwrapping** to resolve this issue. Force-unwrapping unwraps an optional whether it contains a value or not. It works fine if `spouseName` has a value, but if `spouseName` is nil, your code will crash.

10. Click the second fix, and you'll see an exclamation mark appear after `spouseName` in the last line of code, which indicates the optional is force-unwrapped:

```
let greeting = "Hello, " + spouseName!
```

11. Run your program, and you'll see `Hello, Nia` assigned to `greeting`, as shown in the Results area. This means that `spouseName` has been successfully force-unwrapped.
12. To see the effect of force-unwrapping a variable containing `nil`, set `spouseName` to `nil`:

```
spouseName = nil
```

Your code crashes and you can see what caused the crash in the Debug area:

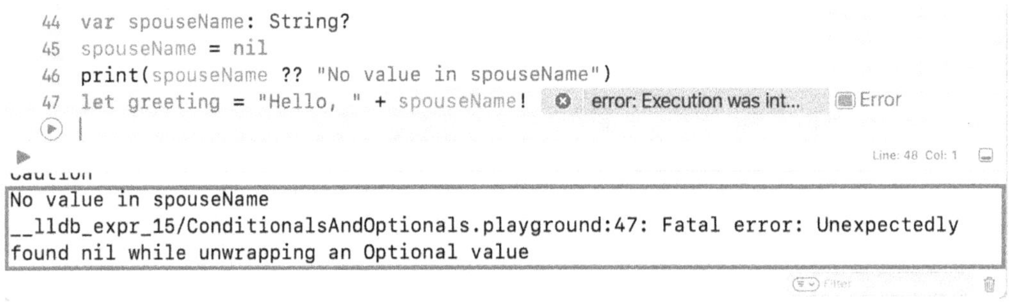

```
var spouseName: String?
spouseName = nil
print(spouseName ?? "No value in spouseName")
let greeting = "Hello, " + spouseName!    error: Execution was int...    Error
```

```
No value in spouseName
__lldb_expr_15/ConditionalsAndOptionals.playground:47: Fatal error: Unexpectedly
found nil while unwrapping an Optional value
```

Figure 3.5: Crashed program with details in the Debug area

Since `spouseName` is now `nil`, the program crashed while attempting to force-unwrap `spouseName`.

A better way of handling this is to use optional binding. In optional binding, you attempt to assign the value in an optional to a temporary variable (you can name it whatever you like). If the assignment is successful, a block of code is executed.

13. To see the effect of optional binding, modify your code as follows:

```
spouseName = "Nia"
print(spouseName ?? "No value in spouseName")
if let spouseTempVar = spouseName {
  let greeting = "Hello, " + spouseTempVar
  print(greeting)
}
```

Hello, Nia will appear in the Debug area. Here's how it works. If `spouseName` has a value, it will be unwrapped and assigned to a temporary constant, `spouseTempVar`, and the `if` statement will return `true`. The statements between the curly braces will be executed and the constant greeting will then be assigned the value `Hello, Nia`. Then, **Hello, Nia** will be printed in the Debug area. Note that the temporary variable `spouseTempVar` is not an optional. If `spouseName` does not have a value, no value can be assigned to `spouseTempVar` and the `if` statement will return `false`. In this case, the statements in the curly braces will not be executed at all.

14. You can also write the code in the previous step in a simpler way as follows:

    ```
    spouseName = "Nia"
    print(spouseName ?? "No value in spouseName")
    if let spouseName {
       let greeting = "Hello, " + spouseName
       print(greeting)
    }
    ```

    Here, the temporary constant is created with the same name as the optional value and will be used in the statements between the curly braces.

15. To see the effect of optional binding when an optional contains `nil`, assign `nil` to `spouseName` once more:

    ```
    spouseName = nil
    ```

You'll notice that nothing appears in the Debug area, and your program no longer crashes, even though `spouseName` is `nil`.

This concludes the section on optionals and optional binding, and you can now create and use optional variables. Awesome!

Summary

You're doing great! You learned how to use `if` and `switch` statements, which means you are now able to write your own programs that do different things based on different conditions.

You also learned about optionals and optional binding. This means you can now represent variables that may or may not have a value and execute instructions only if a variable's value is present.

In the next chapter, you will study how to use a range of values instead of single values, and how to repeat program statements using loops.

Join us on Discord!

Read this book alongside other users, experts, and the author himself. Ask questions, provide solutions to other readers, chat with the author via Ask Me Anything sessions, and much more. Scan the QR code or visit the link to join the community.

https://packt.link/ios-Swift

4

Range Operators and Loops

In the previous chapter, you looked at conditionals, which allow you to do different things based on different conditions, and optionals, which enable you to create variables that may or may not have a value.

In this chapter, you will learn about **range operators** and **loops**. Range operators allow you to represent a range of values by specifying the start and end values for a range. You'll learn about the different types of range operators. Loops allow you to repeat an instruction or a sequence of instructions over and over. You can repeat a sequence a fixed number of times or until a condition is met. You'll learn about the different types of loops used to accomplish this.

By the end of this chapter, you'll have learned how to use ranges and create and use the different types of loops (`for-in`, `while`, and `repeat-while`).

The following topics will be covered in this chapter:

- Exploring range operators
- Exploring loops

Technical requirements

The Xcode playground for this chapter is in the `Chapter04` folder of the code bundle for this book, which can be downloaded here:

`https://github.com/PacktPublishing/iOS-18-Programming-for-Beginners-Ninth-Edition`.

Check out the following video to see the code in action:

`https://youtu.be/swoigirsG_s`

If you wish to start from scratch, create a new playground and name it `RangeOperatorsAndLoops`.

You can type in and run all the code in this chapter as you go along. Let's start with specifying a range of numbers using range operators.

Exploring range operators

Imagine you need to write a program for a department store that automatically sends a discount voucher to customers between the ages of 18 and 30. It would be very cumbersome if you needed to set up an `if` or `switch` statement for each age. It's much more convenient to use a range operator in this case.

Range operators allow you to represent a range of values. Let's say you want to represent a sequence of numbers starting with `firstNumber` and ending with `lastNumber`. You don't need to specify every value; you can just specify the range in this way:

```
firstNumber...lastNumber
```

For more information on range operators, visit `https://docs.swift.org/swift-book/documentation/the-swift-programming-language/basicoperators`.

Let's try this out in the playground. Follow these steps:

1. Add the following code to your playground and run it:

   ```
   let myRange = 10...20
   ```

   This will assign a number sequence that starts with `10` and ends with `20`, including both numbers, to the `myRange` constant. This is known as a **closed-range operator**. The start and end values for `myRange` will be displayed in the Results area.

2. The result displayed in the Results area may be truncated. Click the square icon to the right of the result. It will be displayed inline in the Editor area:

Figure 4.1: Editor area displaying inline result

You can now see the complete result in a box under the line of code. You can drag the right edge of the box to make it bigger if you wish.

Remember you can drag the border between the Results and Editor areas to increase the size of the Results area.

3. Replace the `...` with `..<` if you don't want to include the last number of the sequence in the range. Type in and run the following statement on the next line:

   ```
   let myRange2 = 10..<20
   ```

   This will store the sequence starting with `10` and ending with `19` in the `myRange2` constant and is known as a **half-open range operator**.

There is one more type of range operator, the **one-sided range operator**, and you will learn about that in the next chapter.

Now that you know how to create and use ranges, you will learn about loops, the different loop types, and how to use them in the next section.

Exploring loops

In programming, you frequently need to do the same thing repeatedly. For example, each month, a company will need to generate payroll slips for each employee. If the company has 10,000 employees, it would be inefficient to write 10,000 instructions to create the payroll slips. Repeating a single instruction 10,000 times would be better, and loops are used for this.

There are three loop types: the `for-in` loop, the `while` loop, and the `repeat-while` loop. The `for-in` loop will repeat a known number of times, and the `while` and `repeat-while` loops will repeat if the loop condition is `true`.

For more information on loops, visit `https://docs.swift.org/swift-book/documentation/the-swift-programming-language/controlflow`.

Let's look at each type in turn, starting with the `for-in` loop, which is used when you know how many times a loop should be repeated.

The for-in loop

The `for-in` loop steps through every value in a sequence, and a set of statements in curly braces, known as the **loop body**, are executed each time. Each value is assigned to a temporary variable in turn, and the temporary variable can be used within the loop body. This is what it looks like:

```
for item in sequence {
    code
}
```

The number of times the loop repeats is dictated by the number of items in the sequence. Let's begin by creating a for-in loop to display all the numbers in myRange. Follow these steps:

1. Add the following code to your playground and run it:

```
for number in myRange {
    print(number)
}
```

   You should see each number in the sequence displayed in the Debug area. Note that the statements inside the loop are executed 11 times since myRange includes the last number in the range.

2. Let's try the same program, but this time with myRange2. Modify the code as follows and run it:

```
for number in myRange2 {
    print(number)
}
```

   The statements inside the loop are executed 10 times, and the last value printed in the Debug area is **19**.

3. You can even use a range operator directly after the in keyword. Type and run the following code:

```
for number in 0...5 {
    print(number)
}
```

   Each number from 0 to 5 is displayed in the Debug area.

4. If you want the sequence to be reversed, use the reversed() function. Modify the code as follows and run it:

```
for number in (0...5).reversed() {
    print(number)
}
```

   Each number from 5 to 0 is displayed in the Debug area.

Great job! Let's check out while loops in the next section, which are used when a loop sequence should be repeated if a condition is true.

The while loop

A while loop contains a condition and a set of statements in curly braces, known as the loop body. The condition is checked first; if true, the loop body is executed, and the loop repeats until the condition is false. Here is an example of what a while loop looks like:

```
while condition == true {
    code
}
```

Add and run the following code to create a variable, increment it by 5, and keep on doing it as long as the variable's value is less than 50:

```
var x = 0
while x < 50 {
   x += 5
   print("x is \(x)")
}
```

Let's walk through the code. Initially, `x` is set to `0`. The `x < 50` condition is checked and returns `true`, so the loop body is executed. The value of `x` is incremented by `5`, and **x is 5** is printed in the Debug area. The loop repeats, and `x < 50` is checked again. Since `x` is now `5` and `5 < 50` still returns `true`, the loop body is executed again. This is repeated until the value of `x` is `50`, at which point `x < 50` returns `false` and the loop stops.

If the `while` loop's condition is `false` to begin with, the loop body will never be executed. Try changing the value of `x` to `100` to see this.

In the next section, you'll study `repeat-while` loops. These will execute the statements in the loop body first before checking the loop condition.

The repeat-while loop

Like a `while` loop, a `repeat-while` loop also contains a condition and a loop body, but the loop body is executed first before the condition is checked. If the condition is `true`, the loop repeats until the condition returns `false`. Here is an example of what a `repeat-while` loop looks like:

```
repeat {
   code
} while condition == true
```

Add and run the following code to create a variable, increment it by 5, and keep on doing it as long as the variable's value is less than 50:

```
var y = 0
repeat {
   y += 5
   print("y is \(y)")
} while y < 50
```

Let's walk through the code. Initially, `y` is set to `0`. The loop body is executed. The value of `y` is incremented by `5`, so now `y` contains `5`, and **y is 5** is printed to the Debug area. The `y < 50` condition is checked, and since it returns `true`, the loop is repeated. The value of `y` is incremented by `5`, so now `y` contains `10`, and **y is 10** is printed to the Debug area. The loop is repeated until `y` contains `50`, at which point `x < 50` returns `false` and the loop stops.

The loop body will be executed at least once, even if the condition is `false` to begin with. Try changing the value of `y` to `100` to see this.

You now know how to create and use different loop types. Awesome!

Summary

In this chapter, you looked at closed and half-open range operators, which allow you to specify a range of numbers rather than specifying every individual number discretely.

You also learned about the three different loop types: the `for-in` loop, the `while` loop, and the `repeat-while` loop. The `for-in` loop allows you to repeat a set of statements a fixed number of times, and the `while` and `repeat-while` loops allow you to repeat a set of statements if a condition is true. Great job!

In the next chapter, you will study collection types, which allow you to store a collection of data referenced by an index, a collection of key-value pairs, and an unstructured collection of data.

Join us on Discord!

Read this book alongside other users, experts, and the author himself. Ask questions, provide solutions to other readers, chat with the author via Ask Me Anything sessions, and much more. Scan the QR code or visit the link to join the community.

`https://packt.link/ios-Swift`

5

Collection Types

You've learned quite a lot at this point! You can now create a program that stores data in constants or variables and performs operations on them, and you can control the flow using conditionals and loops. But so far, you've mostly been storing single values.

In this chapter, you will learn ways to store collections of values. Swift has three collection types: `arrays`, which store an ordered list of values; `dictionaries`, which store an unordered list of key-value pairs; and `sets`, which store an unordered list of values.

By the end of this chapter, you'll have learned how to create arrays, dictionaries, and sets, and how to perform operations on them.

The following topics will be covered in this chapter:

- Exploring arrays
- Exploring dictionaries
- Exploring sets

Technical requirements

The Xcode playground for this chapter is in the `Chapter05` folder of the code bundle for this book, which can be downloaded here:

`https://github.com/PacktPublishing/iOS-18-Programming-for-Beginners-Eighth-Edition`

Check out the following video to see the code in action:

`https://youtu.be/W13l9tOMFx8`

If you wish to start from scratch, create a new playground and name it `CollectionTypes`. You can type in and run all the code in this chapter as you go along.

To find out more about arrays, dictionaries, and sets, visit `https://docs.swift.org/swift-book/documentation/the-swift-programming-language/collectiontypes`.

The first collection type you will learn about is arrays, which let you store information in an ordered list.

Exploring arrays

Let's say you want to store the following:

- List of items to buy at a convenience store
- Chores that you must do every month

Arrays would be suitable for this. An array stores values in an ordered list. Here's what it looks like:

Index	Value
0	value1
1	value2
2	value3

Figure 5.1: Representation of an array

Values must be of the same type. You can access any value in an array by using the array index, which starts with 0.

If you create an array using the `let` keyword, its contents can't be changed after it has been created. If you want to change an array's contents after creation, use the `var` keyword.

Let's see how to work with arrays. You'll create an array by assigning a value to it in the next section.

Creating an array

In previous chapters, you created a constant or variable by declaring it and assigning an initial value to it. You can create an array the same way.

Imagine that your spouse has asked you to get some items from a convenience store. Let's implement a shopping list using an array. Add the following code to your playground and run it:

```
var shoppingList = ["Eggs", "Milk"]
```

This instruction creates an array variable named `shoppingList`. The assigned value, `["Eggs", "Milk"]`, is an **array literal**. It represents an array with two elements of type `String`, with `"Eggs"` at index `0` and `"Milk"` at index `1`.

Using the `var` keyword here means that the array's contents can be modified. Since Swift uses type inference, this array's elements will be of type `String`.

Imagine that you need to check how many items you need to get at the store. In the next section, you'll learn how to determine the number of elements in an array.

Checking the number of elements in an array

To find out how many elements there are in an array, use `count`. Type in and run the following code:

```
shoppingList.count
```

As the `shoppingList` array contains two elements, 2 is displayed in the Results area.

You can check to see if an array is empty by using `isEmpty`. Type in and run the following code:

```
shoppingList.isEmpty
```

As the `shoppingList` array contains two elements, **false** is displayed in the Results area.

It is also possible to see if an array is empty by using `shoppingList.count == 0` but using `shoppingList.isEmpty` offers better performance.

Imagine that your spouse called and asked you if you could get cooking oil, fish and chicken while you're at the store. In the next section, you'll see how to add elements to the end of an array, and at a specified array index.

Adding a new element to an array

You can add a new element to the end of an array by using `append(_:)`. Type in and run the following code:

```
shoppingList.append("Cooking Oil")
```

`"Cooking Oil"` has been added to the end of the `shoppingList` array, which now contains three elements – `"Eggs"`, `"Milk"`, and `"Cooking Oil"`, and **["Eggs", "Milk", "Cooking Oil"]** is displayed in the Results area.

You can also add an array to another array with the + operator, using the following code:

```
shoppingList = shoppingList + ["Fish"]
```

You can add a new item at a specified index using `insert(_:at:)`. Type and run the following code:

```
shoppingList.insert("Chicken", at: 1)
```

This inserts "Chicken" at index 1, so now the shoppingList array contains "Eggs", "Chicken", "Milk", "Cooking Oil", and "Fish". Note that "Chicken" is the second element in the array as the first element is at index 0. This can be seen when you click the Quick Look icon:

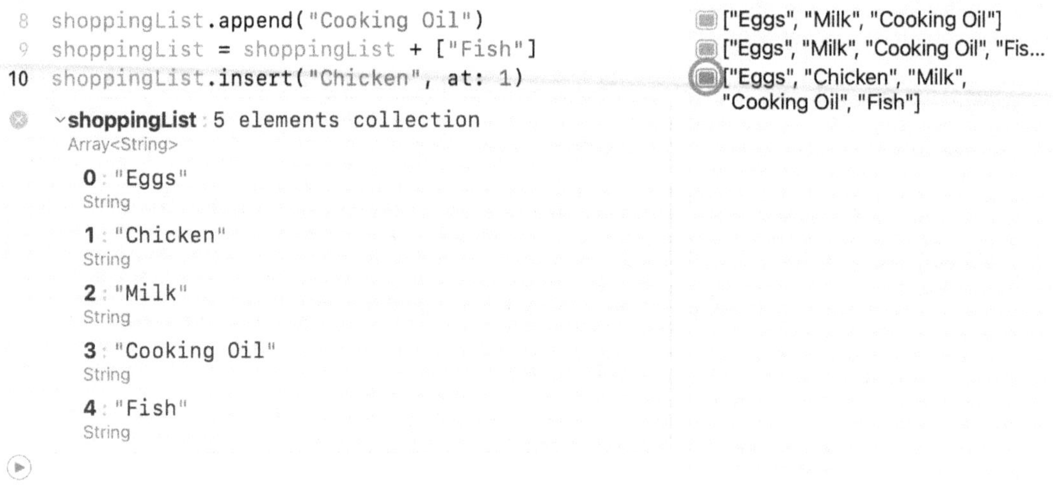

```
shoppingList.append("Cooking Oil")                    ["Eggs", "Milk", "Cooking Oil"]
shoppingList = shoppingList + ["Fish"]                ["Eggs", "Milk", "Cooking Oil", "Fis...
shoppingList.insert("Chicken", at: 1)                 ["Eggs", "Chicken", "Milk",
                                                      "Cooking Oil", "Fish"]
```

```
shoppingList 5 elements collection
Array<String>
  0 "Eggs"
  String
  1 "Chicken"
  String
  2 "Milk"
  String
  3 "Cooking Oil"
  String
  4 "Fish"
  String
```

Figure 5.2: Array contents displayed inline in the playground

Imagine that you've got the first item on your shopping list, and now you need to know the next item on the list. In the next section, you'll see how to access a specific array element using the array index.

Accessing an array element

You can specify an array index to access a particular element. Type in and run the following code:

```
shoppingList[2]
```

This returns the array element stored at index 2, and "Milk" is displayed in the Results area.

Imagine that your spouse called and asked you to get soy milk instead of cow's milk. As this array was declared using the var keyword, you can modify the values stored in it. You'll learn how in the next section.

Assigning a new value to a specified index

You can replace an existing array element by specifying the index and assigning a new value to it. Type in and run the following code:

```
shoppingList[2] = "Soy Milk"
shoppingList
```

This replaces the value stored at index 2, "Milk", with "Soy Milk". The shoppingList array now contains "Eggs", "Chicken", "Soy Milk", "Cooking Oil", and "Fish", as shown in the Results area.

Note that the index used must be valid. For instance, you can't use index 5 as the only valid indexes here are 0, 1, 2, 3, and 4. Doing so would cause the program to crash.

Imagine that your spouse called and told you that there was chicken and fish in the fridge, so you no longer have to get them. In the next section, you'll see two ways to remove elements from an array.

Removing an element from an array

You can remove an element from an array by using `remove(at:)`. Type in and run the following code:

```
let oldArrayValue = shoppingList.remove(at: 1)
oldArrayValue
shoppingList
```

This removes the item at index `1`, `"Chicken"`, from the `shoppingList` array, so now it contains `"Eggs"`, `"Soy Milk"`, `"Cooking Oil"`, and `"Fish"`. The item that has been removed is stored in `removedValue`. You can see this in the Results area.

You can also choose not to keep the removed values. Type in and run the following code:

```
shoppingList.remove(at:  3)
shoppingList
```

This removes the item at index `3`, `"Fish"`, from the `shoppingList` array, so now it contains `"Eggs"`, `"Soy Milk"`, and `"Cooking Oil"`.

If you're removing the last item from the array, you can use `removeLast()` instead, and optionally assign the removed value to a constant or variable as well.

Imagine that you've obtained every item in the list, and you would like to go through your list again to make sure. You'll need to access every array element in turn and perform operations on each element. You'll see how to do this in the next section.

Iterating over an array

Remember the `for-in` loop you studied in the previous chapter? You can use it to iterate over every element in an array. Type in and run the following code:

```
for shoppingListItem in shoppingList {
   print(shoppingListItem)
}
```

This prints out every element in the array to the Debug area.

You can also use **one-sided range operators**. These are range operators with only the starting value, for example, `1...`. Type in and run the following code:

```
for shoppingListItem in shoppingList[1...] {
   print(shoppingListItem)
}
```

This prints out the elements in the array, starting from the element at index `1` to the Debug area.

You now know how to use an array to create an ordered list, such as a shopping list, and how to perform array operations such as accessing, adding, and removing elements. In the next section, let's look at how to store an unordered list of key-value pairs using a dictionary.

Exploring dictionaries

Let's say you're writing a *Contacts* app. You would need to store a list of names and their corresponding contact numbers. A dictionary would be perfect for this, as it allows you to associate a phone number with a contact name.

A dictionary stores key-value pairs in an unordered list. Here's what it looks like:

Key	Value
key1	value1
key2	value2
key3	value3

Figure 5.3: Representation of a dictionary

All keys must be of the same type and must be unique. All values must be of the same type, but must not necessarily be unique. Keys and values don't have to be of the same type as each other. You use the key to get the corresponding value.

If you create a dictionary using the `let` keyword, its contents can't be changed after it has been created. If you want to change the contents after creation, use the `var` keyword.

Let's look at how to work with dictionaries. You'll create a dictionary by assigning a value to it in the next section.

Creating a dictionary

Imagine that you're creating a *Contacts* app. For this app, you'll use a dictionary to store your contacts. Just like an array, you can create a new dictionary by declaring it and assigning an initial value to it. Add the following code to your playground and run it:

```
var contactList = ["Shah" : "+60123456789", "Aamir" : "+0223456789"]
```

This instruction creates a dictionary variable named `contactList`. The assigned value, `["Shah" : "+60123456789", "Aamir" : "+0223456789"]`, is a **dictionary literal**. It represents a dictionary with two elements. Each element is a key-value pair, with the contact name as the key and the contact number as the value. Note that since the contact name is the key field, it should be unique.

Since the `contactList` dictionary is a variable, you can change the contents of the dictionary after it has been created. Both key and value are of type `String` due to type inference.

Imagine that your app must display the total number of contacts. In the next section, you'll learn how to determine the number of elements in a dictionary.

Checking the number of elements in a dictionary

To find out how many elements there are in a dictionary, use `count`. Type in and run the following code:

```
contactList.count
```

As there are two elements in the `contactList` dictionary, **2** is displayed in the Results area.

You can check whether a dictionary is empty by using `isEmpty`. Type in and run the following code:

```
contactList.isEmpty
```

Since the `contactList` dictionary has two elements, **false** is displayed in the Results area.

It is also possible to see if a dictionary is empty by using `contactlist.count == 0` but using `contactList.isEmpty` offers better performance.

Imagine that you just finished a meeting, and want to add a new contact to your app. As this dictionary was declared using the `var` keyword, you can add key-value pairs to it. You'll learn how in the next section.

Adding a new element to a dictionary

To add a new element to a dictionary, provide a key and assign a value to it. Type in and run the following code:

```
contactList["Meena"] = "+0229876543"
contactList
```

This adds a new key-value pair with the key `"Meena"` and the value `"+0229876543"` to the `contactList` dictionary. It now consists of `"Shah" : "+60126789345"`, `"Aamir" : "+0223456789"`, and `"Meena" : "+0229876543"`. You can see this in the Results area.

Imagine that you want to call one of your contacts, and you want the phone number for that contact. In the next section, you'll see how to access dictionary elements by specifying the key for the desired value.

Accessing a dictionary element

You can specify a dictionary key to access its corresponding value. Type in and run the following code:

```
contactList["Shah"]
```

This returns the value for the key `"Shah"`, and **+60123456789** is displayed in the Results area.

Imagine that one of your contacts has a new phone, so you must update the phone number for that contact. You can modify the key-value pairs stored in a dictionary. You'll learn how in the next section.

Assigning a new value to an existing key

You can assign a new value to an existing key. Type and run the following code:

```
contactList["Shah"] = "+60126789345"
contactList
```

This assigns a new value to the key `"Shah"`. The `contactList` dictionary now contains `"Shah" : "+60126789345"`, `"Aamir" : "+0223456789"`, and `"Meena" : "+0229876543"`. You can see this in the Results area.

Imagine that you have to remove a contact from your app. Let's see how you can remove elements from a dictionary in the next section.

Removing an element from a dictionary

To remove an element from a dictionary, assign `nil` to an existing key. Type in and run the following code:

```
contactList["Aamir"] = nil
contactList
```

This removes the element with the key `"Aamir"` from the `contactList` dictionary, and it now contains `"Shah" : "+60126789345"` and `"Meena" : "+0229876543"`. You can see this in the Results area.

If you want to retain the value you are removing, use `removeValue(for:Key)` instead. Type in and run the following code:

```
var oldDictValue = contactList.removeValue(forKey: "Meena")
oldDictValue
contactList
```

This removes the element with the key `"Meena"` from the `contactList` dictionary and assigns its value to `oldDictValue`. `oldDictValue` now contains `"+0229876543"` and the `contactList` dictionary contains `"Shah" : "+60126789345"`.

You can also choose to just remove the value without having to assign it to a constant or variable, like this:

```
contactList.removeValue(forKey: "Meena")
```

Imagine that you would like to call each contact to wish them a happy New Year. You'll have to access every dictionary element in turn and perform operations on each element. You'll see how to do this in the next section.

Iterating over a dictionary

Just like arrays, you can use a `for-in` loop to iterate over every element in a dictionary. Type in and run the following code:

```
for (name, contactNumber) in contactList {
   print("\(name) : \(contactNumber)")
}
```

This prints every element in the dictionary to the Debug area. Since dictionaries are unordered, you may get the results in a different order when you run this code again.

You now know how to use a dictionary to create an unordered list of key-value pairs, such as a contact list, and how to perform dictionary operations. In the next section, let's see how to store an unordered list of values in a set.

Exploring sets

Let's say you're writing a *Movies* app and you want to store a list of movie genres. You could do this with a set.

A set stores values in an unordered list. Here's what it looks like:

Value
value1
value2
value3

Figure 5.4: Representation of a set

All values are of the same type.

If you create a set using the `let` keyword, its contents can't be changed after it has been created. If you want to change the contents after creation, use the `var` keyword.

Let's look at how to work with sets. You'll create a set by assigning a value to it in the next section.

Creating a set

Imagine that you are creating a *Movies* app and you would like to store movie genres in your app. The app will store movie genres that you like and it can check to see if a movie that you're thinking of seeing is among them. Compared to arrays, sets are unordered and only contain unique values, whereas arrays are ordered by index and can contain duplicates. Since you do not need to store movie genres in order and each genre is unique, you will use a set for this purpose.

As you have seen for arrays and dictionaries, you can create a set by declaring it and assigning a new value to it. Add the following code to your playground and run it:

```
var movieGenres: Set = ["Horror", "Action", "Romantic Comedy" ]
```

This instruction creates a set variable named `movieGenres`. Note that the **set literal** assigned to it, `["Horror", "Action", "Romantic Comedy" ]`, has the same format as an array literal, so you use type annotation to set the type of `movieGenres` to `Set`. Otherwise, Swift's type inference will create an array variable and not a set variable.

Using the `var` keyword here means that the set's contents can be modified. This set's elements will be of type `String` due to type inference.

Imagine that you need to show the total number of genres in your app. Let's see how to find the number of elements there are in a set in the next section.

Checking the number of elements in a set

To find out how many elements there are in a set, use `count`. Type in and run the following code:

```
movieGenres.count
```

Since the `movieGenres` set contains three elements, **3** is displayed in the Results area.

You can check whether a set is empty by using `isEmpty`. Type in and run the following code:

```
movieGenres.isEmpty
```

As `movieGenres` contains three elements, **false** is displayed in the Results area.

It is also possible to see if a set is empty by using `movieGenres.count == 0` but using `movieGenres.isEmpty` offers better performance.

Imagine that users of your app can add more genres to it. As this set was declared using the `var` keyword, you can add elements to it. You'll learn how in the next section.

Adding a new element to a set

You can add a new element to a set by using `insert(_:)`. Type in and run the following code:

```
movieGenres.insert("War")
movieGenres
```

This adds a new item, `"War"`, to the `movieGenres` set, which now contains `"Horror"`, `"Romantic Comedy"`, `"War"`, and `"Action"`. This is visible when you click the Quick Look icon.

Imagine that a user would like to know if a certain genre is available in your app. In the next section, you'll learn how to check if an element is in a set.

Checking whether a set contains an element

To check whether a set contains an element, use `contains(_:)`. Type in and run the following code:

```
movieGenres.contains("War")
```

As `"War"` is one of the elements inside the `movieGenres` set, **true** is displayed in the Results area.

Imagine that a user wants to remove a genre from their list of genres. Let's see how to remove items from a set that are no longer needed in the next section.

Removing an item from a set

To remove an item from a set, use `remove(_:)`. The value you are removing can be discarded or assigned to a variable or a constant. If the value doesn't exist in the set, `nil` will be returned. Type in and run the following code:

```
var oldSetValue = movieGenres.remove("Action")
oldSetValue
movieGenres
```

`"Action"` is removed from the `movieGenres` set and assigned to `oldSetValue`, and the `movieGenres` set now contains `"Horror"`, `"Romantic Comedy"`, and `"War"`.

To remove all the elements from a set, use `removeAll()`.

Imagine that you would like to display all the genres your app has as recommendations for your app's users. You can iterate over and perform operations on each set element. Let's see how to do so in the next section.

Iterating over a set

As with arrays and dictionaries, you can use a `for-in` loop to iterate over every element in a set. Type in and run the following code:

```
for genre in movieGenres {
   print(genre)
}
```

You should see each set element in the Debug area. Since sets are unsorted, you may get the results in a different order when you run this code again.

Imagine that you want your app to perform operations on the genres you like with the genres that another person likes. In the next section, you will learn about the various operations that you can do with sets in Swift.

Performing set operations

It's easy to perform set operations such as **union, intersection, subtracting,** and **symmetric difference.** Type in and run the following code:

```
let movieGenres2: Set = ["Science Fiction", "War", "Fantasy"]
movieGenres.union(movieGenres2)
movieGenres.intersection(movieGenres2)
movieGenres.subtracting(movieGenres2)
movieGenres.symmetricDifference(movieGenres2)
```

Here, you are performing set operations on two sets, `movieGenres` and `movieGenres2`. Let's see the results of each set operation:

- `union(_:)` returns a new set containing all the values in both sets, so **"Horror"**, **"Romantic Comedy"**, **"War"**, **"Science Fiction"** and **"Fantasy"** will be displayed when you click the Quick Look icon.
- `intersection(_:)` returns a new set containing only the values common to both sets, so **"War"** will be displayed when you click the Quick Look icon.
- `subtracting(_:)` returns a new set without the values in the specified set, so **"Horror"** and **"Romantic Comedy"** will be displayed when you click the Quick Look icon.
- `symmetricDifference(_:)` returns a new set without the values common to both sets, so **"Horror"**, **"Romantic Comedy"**, **"Science Fiction"** and **"Fantasy"** will be displayed when you click the Quick Look icon.

Imagine that you want your app to compare the genres you like to the genres that another person likes. In the next section, you'll learn how to check if a set is equal to another set, is part of another set, or has nothing in common with another set.

Understanding set membership and equality

It's easy to check if a set is equal to a **subset**, a **superset**, or a **disjoint** of another set. Type in and run the following code:

```
let movieGenresSubset: Set = ["Horror", "Romantic Comedy"]
let movieGenresSuperset: Set = ["Horror", "Romantic Comedy", "War", "Science
Fiction", "Fantasy"]
let movieGenresDisjoint: Set = ["Bollywood"]
movieGenres == movieGenres2
movieGenresSubset.isSubset(of: movieGenres)
movieGenresSuperset.isSuperset(of: movieGenres)
movieGenresDisjoint.isDisjoint(with: movieGenres)
```

Let's see how this code works:

- The == operator checks whether all the members of one set are the same as those of another set. Since not all the members of the `movieGenres` set are the same as those in the `movieGenres2` set, **false** will be displayed in the Results area.
- `isSubset(of:)` checks whether a set is a subset of another set. Since all the members of the `movieGenresSubset` set are in the `movieGenres` set, **true** will be displayed in the Results area.
- `isSuperset(of:)` checks whether a set is a superset of another set. Since all the members of the `movieGenres` set are in the `movieGenresSuperset` set, **true** will be displayed in the Results area.
- `isDisjoint(with:)` checks whether a set has no values in common with another set. Since the `movieGenresDisjoint` set has no members in common with the `movieGenres` set, **true** will be displayed in the Results area.

You now know how to use a set to create an unordered list of values, such as a list of movie genres, and how to perform set operations. This concludes the chapter on collection types. Well done!

Summary

In this chapter, you looked at collection types in Swift. First, you learned about arrays. These allow you to use an ordered list of values to represent an item like a shopping list and perform operations on it.

Next, you learned about dictionaries. These allow you to use an unordered list of key-value pairs to represent an item like a contact list and perform operations on it.

Finally, you learned about sets. These allow you to use an unordered list of values to represent an item like a movie genre list and perform operations on it. You also learned why it may be more appropriate to use a set instead of an array in this instance.

In the next chapter, you will study how to group a set of instructions together using functions. This is handy when you want to execute a set of instructions multiple times in your program.

Join us on Discord!

6

Functions and Closures

At this point, you can write reasonably complex programs that can make decisions and repeat instruction sequences. You can also store data for your programs using collection types. As the programs you write grow in size and complexity, it will become harder to comprehend what they do.

To make large programs easier to understand, Swift allows you to create **functions**, which let you combine several instructions and execute them by calling a single name. You can also create **closures**, which let you combine several instructions without a name and assign them to a constant or variable.

By the end of this chapter, you'll have learned about functions, nested functions, functions as return types, functions as arguments, and the `guard` statement. You'll also have learned how to create and use closures.

The following topics will be covered in this chapter:

- Exploring functions
- Exploring closures

Technical requirements

The Xcode playground for this chapter is in the `Chapter06` folder of the code bundle for this book, which can be downloaded here:

`https://github.com/PacktPublishing/iOS-18-Programming-for-Beginners-Ninth-Edition`

Check out the following video to see the code in action:

`https://youtu.be/lgPDkddf_tc`

If you wish to start from scratch, create a new playground and name it `FunctionsAndClosures`.

You can type in and run all the code in this chapter as you go along. Let's start by learning about functions.

Exploring functions

Functions are useful for encapsulating several instructions that collectively perform a specific task, such as the following:

- Calculating the 10% service charge for a meal at a restaurant
- Calculating the monthly payment for a car that you wish to purchase

Here's what a function looks like:

```
func functionName(parameter1: ParameterType, ...) -> ReturnType {
  code
}
```

Every function has a descriptive name. You can define one or more values that the function takes as input, known as **parameters**. You can also define what the function will output when done, known as its **return type**. Both parameters and return types are optional.

You "call" a function's name to execute it. This is what a function call looks like:

```
functionName(parameter1: argument1, …)
```

You provide input values (known as **arguments**) that match the type of the function's parameters.

To learn more about functions, visit `https://docs.swift.org/swift-book/documentation/the-swift-programming-language/functions/`.

Let's see how you can create a function to calculate a service charge in the next section.

Creating a function

In its simplest form, a function just executes some instructions and does not have any parameters or return types. You'll see how this works by writing a function to calculate the service charge for a meal. The service charge should be 10% of the meal cost.

Add the following code to your playground to create and call this function and run it:

```
func serviceCharge() {
  let mealCost = 50
  let serviceCharge = mealCost / 10
  print("Service charge is \(serviceCharge)")
}
serviceCharge()
```

You've just created a very simple function named `serviceCharge()`. All it does is calculate the 10% service charge for a meal costing $50, which is `50 / 10`, returning `5`. You then call this function using its name. You'll see **Service charge is 5** displayed in the Debug area.

This function is not very useful because `mealCost` is always `50` every time you call this function. Also, the result is only printed in the Debug area and can't be used elsewhere in your program. Let's add some parameters and a return type to this function to make it more useful.

Modify your code as shown:

```
func serviceCharge(mealCost: Int) -> Int {
  return mealCost / 10
}
let serviceChargeAmount = serviceCharge(mealCost: 50)
print(serviceChargeAmount)
```

This is much better. Now, you can set the meal cost when you call the `serviceCharge(mealCost:)` function, and the result can be assigned to a variable or constant. It looks a bit awkward, though. You should try to make function signatures in Swift read like an English sentence, as this is considered a best practice. Let's see how to do that in the next section, where you'll use **custom labels** to make your function more English-like and easier to understand.

Using custom argument labels

Note that the `serviceCharge(mealCost:)` function is not very English-like. You can add a custom label to the parameter to make the function easier to understand.

Modify your code as shown:

```
func serviceCharge(forMealPrice mealCost: Int) -> Int {
  mealCost / 10
}
let serviceChargeAmount = serviceCharge(forMealPrice: 50)
print(serviceChargeAmount)
```

The function works the same as before, but to call it, you use `serviceCharge(forMealPrice:)`. This sounds more like English and makes it easier to figure out what the function does. Also, note that if your function body only consists of a single statement, the `return` keyword is optional.

In the next section, you'll learn how to use several smaller functions within the bodies of other functions, and these are known as **nested functions**.

Using nested functions

It's possible to have a function within the body of another function, and these are called nested functions. This allows you to keep a number of related functions together in one place and makes the enclosing function easier to understand.

A nested function can use the variables of the enclosing function. Let's see how nested functions work by writing a function to calculate monthly payments for a loan.

Type in and run the following code:

```
func calculateMonthlyPayments(carPrice: Double, downPayment: Double,
interestRate: Double, paymentTerm: Double) -> Double {
  func loanAmount() -> Double {
    carPrice - downPayment
  }
  func totalInterest() -> Double {
    interestRate * paymentTerm
  }
  func numberOfMonths() -> Double {
    paymentTerm * 12
  }
  return ((loanAmount() + (loanAmount() *
  totalInterest() / 100 )) / numberOfMonths())
}
calculateMonthlyPayments(carPrice: 50000, downPayment: 5000, interestRate: 3.5,
paymentTerm: 7.0)
```

Here, there are three functions within `calculateMonthlyPayments(carPrice:downPayment:interestRate:paymentTerm:)`. Let's look at them:

- The first nested function, `loanAmount()`, calculates the total loan amount by subtracting `downPayment` from `carPrice`. It returns `50000 - 5000` = `45000`.
- The second nested function, `totalInterest()`, calculates the total interest amount incurred for the payment term by multiplying `interestRate` by `paymentTerm`. It returns `3.5 * 7` = `24.5`.
- The third nested function, `numberOfMonths()`, calculates the total number of months in the payment term by multiplying `paymentTerm` by `12`. It returns `7 * 12` = `84`.

Note that the three nested functions all use the variables of the enclosing function. The value returned is `( 45000 + ( 45000 * 24.5 / 100 ) ) / 84` = `666.96`, which is the amount you must pay monthly for seven years to buy this car.

As you have seen, functions in Swift are like functions in other languages, but they have a cool feature. Functions are **first-class types** in Swift, so they can be used as parameters and return types. Let's see how that is done in the next section.

Using functions as return types

A function can return another function as its return type. Type in and run the following code to create a function that generates the value of pi using one of two possible ways:

```
func approximateValueOfPi1() -> Double {
  3.14159
```

```
}
func approximateValueOfPi2() -> Double {
  22.0 / 7.0
}
func pi() -> (() -> Double) {
  approximateValueOfPi1
  // approximateValueOfPi2
}
pi()()
```

Both `approximateValueOfPi1()` and `approximateValueOfPi2()` are functions that have no parameters and return the approximate value of pi. The `pi()` function's return type is a function that has no parameters and returns a `Double`. This means that it can either return `approximateValueOfPi1` (as shown here) or `approximateValueOfPi2`, since both functions match the expected return type.

`pi()()` calls the function `approximateValueOfPi1`, which returns `3.14159`. **3.14159** is displayed in the Results area.

Let's see how a function can be used as a parameter for another function in the next section.

Using functions as parameters

A function can take another function as a parameter. Type in and run the following code to create a function that determines whether a number meeting a certain condition exists within a list of numbers:

```
func isThereAMatch(listOfNumbers: [Int], condition: (Int) -> Bool) -> Bool {
  for number in listOfNumbers {
    if condition(number) {
      return true
    }
  }
  return false
}
func numberIsOdd(number: Int) -> Bool {
  (number % 2) > 0
}
func numberIsEven(number: Int) -> Bool {
  (number % 2) == 0
}
let numbersList = [1, 3, 5, 7]
isThereAMatch(listOfNumbers: numbersList, condition: numberIsOdd)
```

`isThereAMatch(listOfNumbers:condition:)` has two parameters: an array of integers and a function. The function provided as an argument must take an integer value and return a Boolean value.

Both `numberIsOdd(number:)` and `numberIsEven(_:)` take an integer and return a Boolean value, which means either function can be an argument for the second parameter. `numbersList`, an array containing odd numbers, is used as the argument for the first parameter. When `numberIsOdd` is used as an argument for the second parameter, `isThereAMatch(listOfNumbers:condition:)` will return `true` when called. Try using `numberisEven` as an argument for the second parameter as well.

Functions as parameters and return types can be difficult to understand, but are relatively rare at this point in your learning journey, so don't worry if you don't get it at first. As you gain experience, it will become clearer to you.

In the next section, you'll see how you can perform an early exit on a function if the arguments used are not suitable.

Using a guard statement to exit a function early

Let's say you need a function to be used in an online purchasing terminal. This function will calculate the remaining balance of a debit or credit card when you buy something. The price of the item that you want to buy is entered in a text field.

The value in the text field is converted into an integer so that you can calculate the remaining card balance. If there is something wrong with the input data, it is useful to be able to exit a function early.

Type in and run the following code:

```
func buySomething(itemValueEntered itemValueField: String, cardBalance: Int) ->
Int {
  guard let itemValue = Int(itemValueField) else {
    print("Error in item value")
    return cardBalance
  }
  let remainingBalance = cardBalance - itemValue
  return remainingBalance
}
print(buySomething(itemValueEntered: "10", cardBalance: 50))
print(buySomething(itemValueEntered: "blue", cardBalance: 50))
```

You should see this result in the Debug area:

```
40
Error in item value
50
```

Let's see how this function works. The first line in the function body is a `guard` statement. This checks to see whether a condition is `true`; if not, it exits the function. Here, it is used to check and see whether the user entered a valid price in the online purchasing terminal.

If so, the value can be converted successfully into an integer, and you can calculate the remaining card balance. Otherwise, the `else` clause in the `guard` statement is executed. An error message is printed to the Debug area and the unchanged card balance is returned.

For `print(buySomething(itemValueEntered: "10", cardBalance: 50))`, the item price is deducted successfully from the card balance, and `40` is returned.

For `print(buySomething(itemValueEntered: "blue", cardBalance: 50))`, the `guard` statement's condition fails and its `else` clause is executed, resulting in an error message being printed to the Debug area and `50` being returned.

You now know how to create and use functions. You have also seen how to use custom argument labels, nested functions, functions as parameters or return types, and the `guard` statement.

Now, let's look at closures. Like functions, closures allow you to combine several instructions, but closures do not have names and can be assigned to a constant or a variable. You'll see how they work in the next section.

Exploring closures

A closure, like a function, contains a sequence of instructions and can take arguments and return values. However, closures don't have names. The sequence of instructions in a closure is surrounded by curly braces (`{ }`), and the `in` keyword separates the arguments and return type from the closure body.

Closures can be assigned to a constant or variable, so they're handy if you need to pass them around inside your program. For instance, let's say you have an app that downloads a file from the internet, and you need to do something to the file once it has finished downloading. You can put a list of instructions to process the file inside a closure and have your program execute it once the file finishes downloading.

To learn more about closures, visit `https://docs.swift.org/swift-book/documentation/the-swift-programming-language/closures/`.

You'll now write a closure that applies a calculation to each element of an array of numbers. Add the following code to your playground and click the **Run** button to run it:

```
var numbersArray = [2, 4, 6, 7]
let myClosure = { (number: Int) -> Int in
   let result = number * number
   return result
}
let mappedNumbers = numbersArray.map(myClosure)
```

This assigns a closure that calculates a number's power of two to `myClosure`. The `map()` function then applies this closure to every element in `numbersArray`. Each element is multiplied by itself, and `[4, 16, 36, 49]` appears in the Results area.

It's possible to write closures in a more concise fashion, and you'll see how to do that in the next section.

Simplifying closures

One of the things that new developers have trouble with is the very concise method used by experienced Swift programmers to write closures. Consider the code shown in the following example:

```
var testNumbers = [2, 4, 6, 7]
let mappedTestNumbers = testNumbers.map({
  (number: Int) -> Int in
  let result = number * number
  return result
})
print(mappedTestNumbers)
```

Here, you have `testNumbers`, an array of numbers, and you use the `map(_:)` function to map a closure to each element of the array in turn. The code in the closure multiplies the number by itself, generating the square of that number. The result, **[4, 16, 36, 49]**, is then printed to the Debug area. As you will see, the closure code can be written more concisely.

When a closure's type is already known, you can remove the parameter type, return type, or both. Single-statement closures implicitly return the value of their only statement, which means you can remove the `return` statement as well. So, you can write the closure as follows:

```
let mappedTestNumbers = testNumbers.map({ number in
  number * number
})
```

When a closure is the only argument to a function, you can omit the parentheses enclosing the closure, as follows:

```
let mappedTestNumbers = testNumbers.map { number in
  number * number
}
```

You can refer to parameters by a number expressing their relative position in the list of arguments instead of by name, as follows:

```
let mappedTestNumbers = testNumbers.map { $0 * $0 }
```

So, the closure is now very concise indeed, but will be challenging for new developers to understand. Feel free to write closures in a way that you are comfortable with.

You now know how to create and use closures, and how to write them more concisely. Great!

Summary

In this chapter, you studied how to group statements into functions. You learned how to use custom argument labels, functions inside other functions, functions as return types, functions as parameters, and the `guard` statement. This will be useful later when you need to accomplish the same task at different points in your program.

You also learned how to create closures. This will be useful when you need to pass around blocks of code within your program.

In the next chapter, we will study classes, structures, and enumerations. Classes and structures allow for the creation of complex objects that can store state and behavior, and enumerations can be used to limit the values that can be assigned to a variable or constant, reducing the chances of error.

Join us on Discord!

Read this book alongside other users, experts, and the author himself. Ask questions, provide solutions to other readers, chat with the author via Ask Me Anything sessions, and much more. Scan the QR code or visit the link to join the community.

`https://packt.link/ios-Swift`

7

Classes, Structures, and Enumerations

In the previous chapter, you learned how to group instruction sequences using functions and closures.

It's time to think about how to represent complex objects in your code. For example, think about a car. You could use a `String` constant to store a car name and a `Double` variable to store a car price, but they are not associated with one another. You've seen that you can group instructions to make functions and closures. In this chapter, you'll learn how to group constants and variables in a single entity using **classes** and **structures**, and how to manipulate them. You'll also learn how to use **enumerations** to group a set of related values.

By the end of this chapter, you'll have learned how to create and initialize a class, create a subclass from an existing class, create and initialize a structure, differentiate between classes and structures, and create an enumeration.

The following topics will be covered in this chapter:

- Understanding classes
- Understanding structures
- Understanding enumerations

Technical requirements

The Xcode playground for this chapter is in the `Chapter07` folder of the code bundle for this book, which can be downloaded here:

`https://github.com/PacktPublishing/iOS-18-Programming-for-Beginners-Eighth-Edition`

Check out the following video to see the code in action:

`https://youtu.be/Yl9UuzSR_oE`

If you wish to start from scratch, create a new playground and name it `ClassesStructuresAndEnumerations`. You can type in and run all the code in this chapter as you go along. Let's start with learning what a class is and how to declare and define it.

Understanding classes

Classes are useful for representing complex objects, for example:

- Individual employee information for a company
- Items for sale at an e-commerce site
- Items you have in your house for insurance purposes

Here's what a class declaration and definition look like:

```
class ClassName {
  property1
  property2
  property3
  method1() {
    code
  }
  method2() {
    code
  }
}
```

Every class has a descriptive name, and it contains variables or constants used to represent an object. Variables or constants associated with a class are called **properties**.

A class can also contain functions that perform specific tasks. The functions associated with a class are called **methods**.

Once you have declared and defined a class, you can create **instances** of that class. Imagine you are creating an app for a zoo. If you have an `Animal` class, you can use instances of that class to represent different animals at the zoo. Each of these instances will have different values for their properties.

To learn more about classes, visit `https://docs.swift.org/swift-book/documentation/the-swift-programming-language/classesandstructures`.

Let's look at how to work with classes. You'll learn how to declare and define classes, create instances based on the class declaration, and manipulate those instances. You'll start by creating a class to represent animals in the next section.

Creating a class declaration

Let's declare and define a class that can store details about animals. Add the following code to your playground:

```
class Animal {
  var name: String = ""
  var sound: String = ""
  var numberOfLegs: Int = 0
  var breathesOxygen: Bool = true
  func makeSound() {
    print(sound)
  }
}
```

You've just declared a very simple class named `Animal`. Convention dictates that class names start with a capital letter. This class has properties to store the name of the animal, the sound it makes, the number of legs it has, and whether it breathes oxygen or not. This class also has a method, `makeSound()`, that prints the noise it makes to the Debug area.

Now that you have an `Animal` class, let's use it to create an instance that represents an animal in the next section.

Making an instance of the class

Once you have declared and defined a class, you can create instances of that class. You will now create an instance of the `Animal` class that represents a cat. Follow these steps:

1. To create an instance of the `Animal` class, list all its properties and call its `makeSound()` method; type the following code after your class declaration and run it:

   ```
   let cat = Animal()
   print(cat.name)
   print(cat.sound)
   print(cat.numberOfLegs)
   print(cat.breathesOxygen)
   cat.makeSound()
   ```

   You can access instance properties and methods by typing a dot after the instance name, followed by the property or method you want. You'll see the values for the instance properties and method calls listed in the Debug area. Since the values are the default values assigned when the class was created, `name` and `sound` contain empty strings, `numberOfLegs` contains `0`, `breathesOxygen` contains `true`, and the `makeSound()` method prints an empty string.

2. Let's assign some values to this instance's properties. Modify your code as shown:

   ```
   let cat = Animal()
   cat.name = "Cat"
   ```

```
cat.sound = "Mew"
cat.numberOfLegs = 4
cat.breathesOxygen = true
print(cat.name)
```

Now, when you run the program, the following is displayed in the Debug area:

```
Cat
Mew
4
true
Mew
```

The values for all the instance properties and the result of the `makeSound()` method are printed to the Debug area.

Note that here, you create the instance first, and then assign values to that instance. It is also possible to assign the values when the instance is being created, and you can do this by implementing an **initializer** in your class declaration.

3. An initializer is responsible for ensuring all the instance properties have valid values when a class is created. Let's add an initializer for the `Animal` class. Modify your class definition as shown:

```
class Animal {
  var name: String
  var sound: String
  var numberOfLegs: Int
  var breathesOxygen: Bool
  init(name: String, sound: String, numberOfLegs:
  Int, breathesOxygen: Bool) {
    self.name = name
    self.sound = sound
    self.numberOfLegs = numberOfLegs
    self.breathesOxygen = breathesOxygen
  }
  func makeSound() {
    print(sound)
  }
}
```

As you can see, the initializer uses the `init` keyword and has a list of parameters that will be used to set the property values. Note that the `self` keyword distinguishes the property names from the parameters. For example, `self.name` refers to the property and `name` refers to the parameter.

At the end of the initialization process, every property in the class should have a valid value.

4. You'll see some errors in your code at this point as the function call does not have any parameters. You will need to update your function call to address this. Modify your code as shown and run it:

```
    func makeSound() {
      print(sound)
    }
  }
  let cat = Animal(name: "Cat", sound: "Mew", numberOfLegs: 4,
  breathesOxygen: true)
  print(cat.name)
```

   The results are the same as those in *Step 2*, but you created the instance and set its properties in a single instruction. Excellent!

Now there are different types of animals, such as mammals, birds, reptiles, and fish. You could create a class for each type, but you could also create a **subclass** based on an existing class. Let's see how to do that in the next section.

Making a subclass

A subclass of a class inherits all the methods and properties of an existing class. You can also add additional properties and methods to it if you wish. For instance, for an IT company, you could have `CustomerSupportAgent` as a subclass of `Employee`. This class would have all the properties of the `Employee` class, as well as additional properties required for the customer support role.

You'll now create `Mammal`, a subclass of the `Animal` class. Follow these steps:

1. To declare the `Mammal` class, type in the following code after the `Animal` class declaration:

```
  class Mammal: Animal {
    let hasFurOrHair: Bool = true
  }
```

   Typing `: Animal` after the class name makes the `Mammal` class a subclass of the `Animal` class. It has all the properties and methods declared in the `Animal` class, and one additional property: `hasFurOrHair`. Since the `Animal` class is the parent of the `Mammal` class, you can refer to it as the superclass of the `Mammal` class.

2. Modify your code that creates an instance of your class, as shown, and run it:

```
  let cat = Mammal(name: "Cat", sound: "Mew", numberOfLegs: 4,
  breathesOxygen: true)
```

   `cat` is now an instance of the `Mammal` class instead of the `Animal` class. As you can see, the results displayed in the Debug area are the same as before, and there are no errors. The value for `hasFurOrHair` has not been displayed, though. Let's fix that.

3. Type in the following code after all the other code in your playground to display the contents of the `hasFurOrHair` property and run it:

```
print(cat.hasFurOrHair)
```

Since the initializer for the `Animal` class does not have a parameter to assign a value to `hasFurOrHair`, the default value is used, and `true` will be displayed in the Debug area.

You have seen that a subclass can have additional properties. A subclass can also have additional methods, and method implementation in a subclass can differ from the superclass implementation. Let's see how to do that in the next section.

Overriding a superclass method

So far, you've been using multiple `print()` statements to display the values of the class instance. You'll implement a `description()` method to display all the instance properties in the Debug area, so multiple `print()` statements will no longer be required. Follow these steps:

1. Modify your `Animal` class declaration to implement a `description()` method, as shown:

```
class Animal {
  var name: String
  var sound: String
  var numberOfLegs: Int
  var breathesOxygen: Bool = true
  init(name: String, sound: String, numberOfLegs:
  Int, breathesOxygen: Bool) {
    self.name = name
    self.sound = sound
    self.numberOfLegs = numberOfLegs
    self.breathesOxygen = breathesOxygen
  }
  func makeSound() {
    print(sound)
  }
  func description() -> String {
    "name: \(name) sound: \(sound)
    numberOfLegs: \(numberOfLegs)
    breathesOxygen: \(breathesOxygen)"
  }
}
```

2. Modify your code as shown to use the description() method in place of the multiple print() statements, and run the program:

```
let cat = Mammal(name: "Cat", sound: "Mew",
numberOfLegs: 4, breathesOxygen: true)
print(cat.description())
cat.makeSound()
```

You will see the following in the Debug area:

```
name: Cat sound: Mew numberOfLegs: 4 breathesOxygen: true
Mew
```

As you can see, even though the description() method is not implemented in the Mammal class, it is implemented in the Animal class. This means it will be inherited by the Mammal class, and the instance properties will be printed to the Debug area. Note that the value for the hasFurOrHair property is missing, and you can't put it in the description() method because the hasFurOrHair property does not exist for the Animal class.

3. You can change the implementation of the description() method in the Mammal class to display the hasFurOrHair property's value. Add the following code to your Mammal class definition and run it:

```
class Mammal: Animal {
  let hasFurOrHair: Bool = true
  override func description() -> String {
    super.description() + " hasFurOrHair:
    \(hasFurOrHair)"
  }
}
```

The override keyword is used here to specify that the description() method implemented is to be used in place of the superclass implementation. The super keyword is used to call the superclass implementation of description(). The value in hasFurOrHair is then added to the string returned by super.description().

You will see the following in the Debug area:

```
name: Cat sound: Mew numberOfLegs: 4 breathesOxygen: true hasFurOrHair:
true
Mew
```

The hasFurOrHair property's value is displayed in the Debug area, showing that you are using the Mammal subclass implementation of the description() method.

You've created class and subclass declarations and made instances of both. You've also added initializers and methods to both. Cool! Let's look at how to declare and use structures in the next section.

Understanding structures

Like classes, structures also group together properties and methods used to represent an object and do specific tasks. Remember the `Animal` class you created? You can also use a structure to accomplish the same thing. There are differences between classes and structures though, and you will learn more about those later in this chapter.

Here's what a structure declaration and definition look like:

```
struct StructName {
  property1
  property2
  property3
  method1() {
    code
  }
  method2(){
    code
  }
}
```

As you can see, a structure is very similar to a class. It also has a descriptive name and can contain properties and methods. You can also create instances of a structure.

To learn more about structures, visit `https://docs.swift.org/swift-book/documentation/the-swift-programming-language/classesandstructures`.

Let's look at how to work with structures. You'll learn how to declare and define structures, create instances based on the structure, and manipulate them. You'll start by creating a structure to represent reptiles in the next section.

Creating a structure declaration

Continuing with the animal theme, let's declare and define a structure that can store details about reptiles. Add the following code after all the other code in your playground:

```
struct Reptile {
  var name: String
  var sound: String
  var numberOfLegs: Int
  var breathesOxygen: Bool
  let hasFurOrHair: Bool = false
  func makeSound() {
```

```
        print(sound)
      }
      func description() -> String {
        "Structure: Reptile name: \(name)
        sound: \(sound)
        numberOfLegs: \(numberOfLegs)
        breathesOxygen: \(breathesOxygen)
        hasFurOrHair: \(hasFurOrHair)"
      }
    }
```

As you can see, this is almost the same as the `Animal` class declaration you did earlier. Structure names should also start with a capital letter, and this structure has properties to store the name of the animal, the sound it makes, how many legs it has, whether it breathes oxygen, and whether it has fur or hair. This structure also has a method, `makeSound()`, that prints the sound it makes to the Debug area.

Now that you have a `Reptile` structure declaration, let's use it to create an instance representing a snake in the next section.

Making an instance of the structure

As with classes, you can create instances from a structure declaration. You will now create an instance of the `Reptile` structure that represents a snake, print out the property values of that instance, and call the `makeSound()` method. Type the following after all the other code in your playground and run it:

```
var snake = Reptile(name: "Snake", sound: "Hiss",
numberOfLegs: 0, breathesOxygen: true)
print(snake.description())
snake.makeSound()
```

Note that you did not need to implement an initializer; structures automatically get an initializer for all their properties called the **memberwise initializer**. Neat! The following will be displayed in the Debug area:

```
Structure: Reptile name: Snake sound: Hiss numberOfLegs: 0 breathesOxygen: true
hasFurOrHair: false
Hiss
```

Even though the structure declaration is very similar to the class declaration, there are two differences between a class and a structure:

- Structures cannot inherit from another structure
- Classes are **reference types**, while structures are **value types**

Let's look at the difference between value types and reference types in the next section.

Comparing value types and reference types

Classes are reference types. This means when you assign a class instance to a variable, you are storing the memory location of the original instance in the variable instead of the instance itself.

Structures are value types. This means when you assign a structure instance to a variable, that instance is copied, and whatever changes you make to the original instance do not affect the copy.

Now, you will create an instance of a class and a structure and observe the differences between them. Follow these steps:

1. You'll start by creating a variable containing a structure instance and assigning it to a second variable, then change the value of a property in the second variable. Type in the following code and run it:

    ```
    struct SampleValueType {
      var sampleProperty = 10
    }
    var a = SampleValueType()
    var b = a
    b.sampleProperty = 20
    print(a.sampleProperty)
    print(b.sampleProperty)
    ```

    In this example, you declared a structure, `SampleValueType`, that contains one property, `sampleProperty`. Next, you created an instance of that structure and assigned it to a variable, `a`. After that, you assigned `a` to a new variable, `b`. Then, you changed the `sampleProperty` value of `b` to `20`.

    When you print out the `sampleProperty` value of `a`, `10` is printed in the Debug area, showing that any changes made to the `sampleProperty` value of `b` do not affect the `sampleProperty` value of `a`. This is because when you assigned `a` to `b`, a copy of `a` was assigned to `b`, so they are separate instances that don't affect one another.

2. Next, you'll create a variable containing a class instance and assign it to a second variable, then change the value of a property in the second variable. Type in the following code and run it:

    ```
    class SampleReferenceType {
      var sampleProperty = 10
    }
    var c = SampleReferenceType()
    var d = c
    c.sampleProperty = 20
    print(c.sampleProperty)
    print(d.sampleProperty)
    ```

In this example, you declared a class, `SampleReferenceType`, that contains one property, `sampleProperty`. Then, you created an instance of that class and assigned it to a variable, `c`. After that, you assigned `c` to a new variable, `d`. Next, you changed the `sampleProperty` value of `d` to `20`.

When you print out the `sampleProperty` value of `c`, `20` is printed in the Debug area, showing that any changes made to `c` or `d` affect the same `SampleReferenceType` instance.

Now, the question is, which should you use, classes or structures? Let's explore that in the next section.

Deciding between classes and structures

You've seen that you can use either a class or a structure to represent a complex object. So, which should you use?

It is recommended to use structures unless you need something that requires classes, such as subclasses. This helps prevent some subtle errors that may occur due to classes being reference types.

Fantastic! Now that you have learned about classes and structures, let's look at enumerations, which allow you to group related values, in the next section.

Understanding enumerations

Enumerations allow you to group related values, such as the following:

- Compass directions
- Traffic light colors
- The colors of a rainbow

To understand why enumerations would be ideal for this purpose, let's consider the following example.

Imagine you're programming a traffic light. You can use an integer variable to represent different traffic light colors where `0` is red, `1` is yellow, and `2` is green, like this:

```
var trafficLightColor = 2
```

Although this is a possible way to represent a traffic light, what happens when you assign `3` to `trafficLightColor`? This is an issue as `3` does not represent a valid traffic light color. So, it would be better if we could limit the possible values of `trafficLightColor` to the colors it can display.

Here's what an enumeration declaration and definition look like:

```
enum EnumName {
  case value1
  case value2
  case value3
}
```

Every enumeration has a descriptive name, and the body contains the associated values for that enumeration.

To learn more about enumerations, visit https://docs.swift.org/swift-book/documentation/the-swift-programming-language/enumerations.

Let's look at how to work with enumerations. You'll learn how to create and manipulate them. You'll start by creating one to represent a traffic light color in the next section.

Creating an enumeration

Let's create an enumeration to represent a traffic light. Follow these steps:

1. Add the following code to your playground and run it:

   ```
   enum TrafficLightColor {
     case red
     case yellow
     case green
   }
   var trafficLightColor = TrafficLightColor.red
   ```

   This creates an enumeration named `TrafficLightColor`, which groups together the red, yellow, and green values. The value for the `trafficLightColor` variable is limited to `red`, `yellow`, and `green`; setting any other value will generate an error.

2. Just like classes and structures, enumerations can contain methods. Let's add a method to `TrafficLightColor`. Modify your code as shown to make `TrafficLightColor` return a string representing the traffic light color and run it:

   ```
   enum TrafficLightColor {
     case red
     case yellow
     case green
     func description() -> String {
       switch self {
       case .red:
         "red"
       case .yellow:
         "yellow"
       case .green:
         "green"
       }
     }
   }
   ```

```
var trafficLightColor = TrafficLightColor.red
print(trafficLightColor.description())
```

The `description()` method returns a string depending on the value of `trafficLightColor`. Since the value of `trafficLightColor` is `TrafficLightColor.red`, **red** will appear in the Debug area.

You've learned how to create and use enumerations to store grouped values, and how to add methods to them. Good job!

Summary

In this chapter, you learned how to declare complex objects using a class, create instances of a class, create a subclass, and override a class method. You also learned how to declare a structure, create instances of a structure, and understand the difference between reference and value types. Finally, you learned how to use enumerations to represent a specific set of values.

You now know how to use classes and structures to represent complex objects, and how to use enumerations to group related values together in your own programs.

In the next chapter, you will study how to specify common traits in classes and structures using protocols, extend the capability of built-in classes using extensions, and handle errors in your programs.

Join us on Discord!

Read this book alongside other users, experts, and the author himself. Ask questions, provide solutions to other readers, chat with the author via Ask Me Anything sessions, and much more. Scan the QR code or visit the link to join the community.

`https://packt.link/ios-Swift`

8

Protocols, Extensions, and Error Handling

In the previous chapter, you learned how to represent complex objects using classes or structures and how to use enumerations to group related values together.

In this chapter, you'll learn about **protocols**, **extensions**, and **error handling**. Protocols define a blueprint of methods, properties, and other requirements that can be adopted by a class, structure, or enumeration. Extensions enable you to provide new functionality for an existing class, structure, or enumeration. Error handling covers how to respond to and recover from errors in your program.

By the end of this chapter, you'll be able to write your own protocols to meet the requirements of your apps, use extensions to add new capabilities to existing types, and handle error conditions in your apps without crashing.

The following topics will be covered in this chapter:

- Exploring protocols
- Exploring extensions
- Exploring error handling

Technical requirements

The Xcode playground for this chapter is in the `Chapter08` folder of the code bundle for this book, which can be downloaded here:

`https://github.com/PacktPublishing/iOS-18-Programming-for-Beginners-Ninth-Edition`

Check out the following video to see the code in action:

`https://youtu.be/fV6VNlDyyG0`

If you wish to start from scratch, create a new playground and name it `ProtocolsExtensionsAndErrorHandling`. You can type in and run all the code in this chapter as you go along. Let's start with protocols, which is a way of specifying the properties and methods that a class, structure, or enumeration should have.

Exploring protocols

Protocols are like blueprints that determine what properties or methods an object must have. After you've declared a protocol, classes, structures, and enumerations can adopt it and provide their own implementation for the required properties and methods.

Here's what a protocol declaration looks like:

```
protocol ProtocolName {
  var readWriteProperty1 {get set}
  var readOnlyProperty2 {get}
  func methodName1()
  func methodName2()
}
```

Just like classes and structures, protocol names start with an uppercase letter. Properties are declared using the `var` keyword. You use `{get set}` if you want a property that can be read from or written to, and you use `{get}` if you want a read-only property. Note that you just specify property and method names; the implementation is done within the adopting class, structure, or enumeration.

For more information on protocols, visit `https://docs.swift.org/swift-book/documentation/the-swift-programming-language/protocols`.

To help you understand protocols, imagine an app used by a fast-food restaurant. The management has decided to show calorie counts for the meals being served. The app currently has the following class, structure, and enumeration, and none of them have calorie counts implemented:

- A `Burger` class
- A `Fries` structure
- A `Sauce` enumeration

Add the following code to your playground to declare the `Burger` class, the `Fries` structure, and the `Sauce` enumeration:

```
class Burger {
}
struct Fries {
}
enum Sauce {
```

```
    case chili
    case tomato
}
```

These represent the existing class, structure, and enumeration in the app. Don't worry about the empty definitions, as they are not required for this lesson. As you can see, none of them have calorie counts at present. Let's learn how to create a protocol that specifies the properties and methods needed to implement calorie counts. You'll start by declaring this protocol in the next section.

Creating a protocol declaration

Let's create a protocol that specifies a required property, `calories`, and a method, `description()`. Type the following into your playground before the class, structure, and enumeration declarations:

```
protocol CalorieCountable {
  var calories: Int { get }
  func description() -> String
}
```

This protocol is named `CalorieCountable`. It specifies that any object that adopts it must have a property, `calories`, that holds the calorie count, and a method, `description()`, that returns a string. `{ get }` means that you only need to be able to read the value stored in `calories`, and you don't have to write to it. Note that the definition of the `description()` method is not specified, as that will be done in the class, structure, or enumeration. All you need to do to adopt a protocol is type a colon after the class name, followed by the protocol name, and implement the required properties and methods.

To make the `Burger` class conform to this protocol, modify your code as follows:

```
class Burger: CalorieCountable {
  let calories = 800
  func description() -> String {
    "This burger has \(calories) calories"
  }
}
```

As you can see, the `calories` property and the `description()` method have been added to the `Burger` class. Even though the protocol specifies a variable, you can use a constant here because the protocol only requires that you get the value for `calories`, not set it.

Let's make the `Fries` structure adopt this protocol as well. Modify your code for the `Fries` structure as follows:

```
struct Fries: CalorieCountable {
  let calories = 500
  func description() -> String {
    "These fries have \(calories) calories"
  }
}
```

The code added to the `Fries` structure is similar to that added to the `Burger` class, and it now conforms to the `CalorieCountable` protocol as well.

You could modify the `Sauce` enumeration in the same way, but let's do it using extensions instead. Extensions extend the capabilities of an existing class, structure, or enumeration. You'll add the `CalorieCountable` protocol to the `Sauce` enumeration using an extension in the next section.

Exploring extensions

Extensions allow you to provide extra capabilities to an object without modifying the original object definition. You can use them on Apple-provided objects (where you don't have access to the object definition) or when you wish to segregate your code for readability and ease of maintenance. Here's what an extension looks like:

```
class ExistingType {
  property1
  method1()
}
extension ExistingType : ProtocolName {
  property2
  method2()
}
```

Here, an extension is used to provide an additional property and method to an existing class.

For more information on extensions, visit `https://docs.swift.org/swift-book/documentation/the-swift-programming-language/extensions`.

Let's look at how to use extensions. You'll start by making the `Sauce` enumeration conform to the `CalorieCountable` protocol using an extension in the next section.

Adopting a protocol via an extension

At present, the `Sauce` enumeration does not conform to the `CalorieCountable` protocol. You'll use an extension to add the properties and methods required to make it conform. Type in the following code after the declaration for the `Sauce` enumeration:

```
enum Sauce {
  case chili
  case tomato
}
extension Sauce: CalorieCountable {
  var calories: Int {
    switch self {
```

```
      case .chili:
        20
      case .tomato:
        15
      }
    }
    func description() -> String {
      "This sauce has \(calories) calories"
    }
  }
```

As you can see, no changes were made to the original definition for the `Sauce` enumeration. This is also useful if you want to extend the capabilities of existing Swift standard types, such as `String` and `Int`.

Enumeration instances can't store values in properties the way structures and classes can, so a `switch` statement is used to return the number of calories based on the enumeration's value. The `description()` method is the same as the one in the `Burger` class and the `Fries` structure.

All three objects have a `calories` property and a `description()` method. Great!

Let's see how you can put them in an array and perform an operation to get the total calorie count for a meal in the next section.

Creating an array of different types of objects

Ordinarily, an array's elements must be of the same type. However, since the `Burger` class, the `Fries` structure, and the `Sauce` enumeration all conform to the `CalorieCountable` protocol, you can make an array that contains elements conforming to this protocol. Follow these steps:

1. To add instances of the `Burger` class, the `Fries` structure, and the `Sauce` enumeration to an array, type in the following code after all other code in the file:

   ```
   let burger = Burger()
   let fries = Fries()
   let sauce = Sauce.tomato
   let foodArray: [CalorieCountable] = [burger, fries, sauce]
   ```

2. To get the total calorie count, add the following code after the line where you created the `foodArray` constant:

   ```
   let totalCalories = foodArray.reduce(0, {$0 + $1.calories})
   print(totalCalories)
   ```

   The `reduce` method is used to produce a single value from the elements of the `foodArray` array. The first parameter of this method is the initial value, and it is set to `0`. The second parameter is a closure that combines the initial value with the value stored in an element's `calories` property. This is repeated for each element in the `foodArray` array and the result is assigned to `totalCalories`. The total amount, **1315**, will be displayed in the Debug area.

You have learned how to create a protocol and make a class, structure, or enumeration conform to it, either within the class definition or via extensions. Let's look at error handling next, and see how to respond to or recover from errors in your program.

Exploring error handling

When you write apps, bear in mind that error conditions may happen, and error handling is how your app responds to and recovers from such conditions.

First, you create a type that conforms to Swift's `Error` protocol, which lets this type be used for error handling. Enumerations are normally used, as you can specify associated values for different kinds of errors. When something unexpected happens, you can stop program execution by throwing an error. You use the `throw` statement for this and provide an instance of the type conforming to the `Error` protocol, with the appropriate value. This allows you to see what went wrong.

Of course, it would be better if you could respond to an error without stopping your program. For this, you can use a `do-catch` block, which looks like this:

```
do {
  try expression1
  statement1
} catch {
  statement2
}
```

Here, you attempt to execute code in the `do` block using the `try` keyword. If an error is thrown, the statements in the `catch` block are executed. You can have multiple `catch` blocks to handle different error types.

For more information on error handling, visit `https://docs.swift.org/swift-book/documentation/the-swift-programming-language/errorhandling`.

As an example, let's say you have an app that needs to access a web page. However, if the server where that web page is located is down, it is up to you to write the code to handle the error, such as trying an alternative web server or informing the user that the server is down.

Let's create an enumeration that conforms to the `Error` protocol, use a `throw` statement to stop program execution when an error occurs, and use a `do-catch` block to handle an error. Follow these steps:

1. Type the following code into your playground:

```
enum WebsiteError: Error {
  case noInternetConnection
  case siteDown
```

```
    case wrongURL
}
```

This declares an enumeration, `WebsiteError`, that adopts the `Error` protocol. It covers three possible error conditions: there is no internet connection, the website is down, or the URL cannot be resolved.

2. Type the following code after the `WebsiteError` definition to declare a function that checks if a website is up after the `WebpageError` declaration:

```
func checkWebsite(siteUp: Bool) throws -> String {
  if !siteUp {
    throw WebsiteError.siteDown
  }
  return "Site is up"
}
```

If `siteUp` is `true`, `"Site is up"` is returned. If `siteUp` is `false`, the program will stop executing and throw an error.

3. Type the following code after the `checkWebsite(siteUp:)` function definition and run your program:

```
let siteStatus = true
try checkWebsite(siteUp: siteStatus)
```

Since `siteStatus` is `true`, **Site is up** will appear in the Results area.

4. Change the value of `siteStatus` to `false` and run your program. Your program crashes and the following error message is displayed in the Debug area:

> `Playground execution terminated: An error was thrown and was not caught:`
> `error: error:`

5. Of course, it is always better if you can handle errors without making your program crash. You can do this by using a `do-catch` block. Modify your code as shown and run it:

```
let siteStatus = false
do {
  print(try checkWebsite(siteUp: siteStatus))
} catch {
  print(error)
}
```

The do block tries to execute the `checkWebsite(siteUp:)` function and prints the status if successful. If there is an error, instead of crashing, the statements in the `catch` block are executed, and the error message `siteDown` appears in the Debug area.

You can make your program handle different error conditions by implementing multiple `catch` blocks. See this link for details: `https://docs.swift.org/swift-book/documentation/the-swift-programming-language/errorhandling`.

You have learned how to handle errors in your app without making it crash. Great!

Summary

In this chapter, you learned how to write protocols and how to make classes, structures, and enumerations conform to them. You also learned how to extend the capabilities of a class by using an extension. Finally, you learned how to handle errors using the `do-catch` block.

These may seem rather abstract and hard to understand now, but in *Part 3* of this book, you will see how to use protocols to implement common functionalities in different parts of your program instead of writing the same program over and over. You will see how useful extensions are in organizing your code, which makes it easy to maintain. Finally, you'll see how good error handling makes it easy to pinpoint the mistakes you made while coding your app.

In the next chapter, you will learn about **Swift concurrency**, a new way to handle asynchronous operations in Swift.

Join us on Discord!

Read this book alongside other users, experts, and the author himself. Ask questions, provide solutions to other readers, chat with the author via Ask Me Anything sessions, and much more. Scan the QR code or visit the link to join the community.

`https://packt.link/ios-Swift`

Swift Concurrency

Apple introduced **Swift concurrency**, which adds support for structured asynchronous and parallel programming to Swift 5.5, during WWDC21. It allows you to write concurrent code, which is more readable and easier to understand. During WWDC24, Apple introduced **Swift 6**, which makes concurrency programming easier by diagnosing **data races** at compile time.

At the present time, it is not recommended to turn strict concurrency on for large existing projects, as it is likely to generate multiple errors and warnings. However, as this is Apple's direction going forward, you will be turning it on for the project in this chapter and in *Part 3* of this book so you may learn and gain experience with it.

In this chapter, you will learn the basic concepts of Swift concurrency. Next, you will examine an app without concurrency and explore its issues. After that, you will use `async/await` to implement concurrency in the app. Finally, you'll make your app more efficient by using `async-let`.

By the end of this chapter, you'll have learned the basics of how Swift concurrency works and how to update your own apps to use it.

The following topics will be covered:

- Understanding Swift concurrency
- Examining an app without concurrency
- Updating the app using `async/await`
- Improving efficiency using `async-let`

Technical requirements

We will use an example app, *BreakfastMaker*, to understand the concepts of Swift concurrency.

The completed Xcode project for this chapter is in the `Chapter09` folder of the code bundle for this book, which can be downloaded here:

`https://github.com/PacktPublishing/iOS-18-Programming-for-Beginners-Ninth-Edition`

Check out the following video to see the code in action:

`https://youtu.be/uEckcWHFeiE`

Let's start by learning about Swift concurrency in the next section.

Understanding Swift concurrency

In Swift 5.5, Apple added support for writing **asynchronous** and **parallel** code in a structured way.

Asynchronous code allows your app to suspend and resume code. Parallel code allows your app to run multiple pieces of code simultaneously. This allows your app to do things like update the user interface while still performing operations like downloading data from the internet.

You can find links to all of Apple's Swift concurrency videos during WWDC21 at `https://developer.apple.com/news/?id=2o3euotz`.

You can read Apple's Swift concurrency documentation at `https://developer.apple.com/news/?id=2o3euotz`.

During WWDC24, Apple released Swift 6. With the Swift 6 language mode, the compiler can now guarantee that concurrent programs are free of data races. This means that code from one part of your app can no longer access the same area of memory that is being modified by code from another part of your app. However, when you create a new Xcode project, it defaults to the Swift 5 language mode, and you have to turn on the Swift 6 language mode to enable this feature.

To view Apple's WWDC24 video on migrating your app to Swift 6, click this link: `https://developer.apple.com/videos/play/wwdc2024/10169/`

To view Apple's documentation on migrating your app to Swift 6, click this link: `https://www.swift.org/migration/documentation/migrationguide/`

To give you an idea of how Swift concurrency works, imagine that you are making soft-boiled eggs and toast for breakfast. Here is one way of doing it:

1. Put two slices of bread into the toaster.
2. Wait two minutes until the bread is toasted.
3. Put two eggs in a pan containing boiling water, and cover them.
4. Wait seven minutes until the eggs are cooked.
5. Plate and serve your breakfast.

This takes nine minutes in total. Now, think about this sequence of events. Do you spend that time just staring at the toaster and the pan? You'll probably be using your phone while the bread is in the toaster and the eggs are in the pan. In other words, you can do other things while the toast and eggs are being prepared. So, the sequence of events would be more accurately described as follows:

1. Put two slices of bread into the toaster.
2. Use your phone for two minutes until the bread is toasted.
3. Put two eggs in a pan containing boiling water, and cover them.
4. Use your phone for seven minutes until the eggs are cooked.
5. Plate and serve your breakfast.

Here, you can see that your interaction with the toaster and pan can be suspended and then resumed, which means these operations are asynchronous. The operation still takes nine minutes, but you were able to do other things during that time.

There is another factor to consider. You don't need to wait for the bread to finish toasting before you put the eggs in the pan. This means you could modify the sequence of steps as follows:

1. Put two slices of bread into the toaster.
2. While the bread is toasting, put two eggs in a pan containing boiling water, and cover them.
3. Use your phone for seven minutes. During that time, the bread will be toasted and the eggs will be cooked.
4. Plate and serve your breakfast.

Toasting the bread and boiling the eggs are now carried out in parallel, which saves you two minutes. Great! However, do note that you have more things to keep track of.

Now that you understand the concepts of asynchronous and parallel operations, let's study the issues that an app without concurrency has in the next section.

Examining an app without concurrency

You've seen how asynchronous and parallel operations can help you prepare breakfast faster and allow you to use your phone while you're doing it. Now, let's look at a sample app that simulates the process of preparing breakfast. Initially, this app does not have concurrency implemented so you can see how that affects the app. Follow these steps:

1. If you have not already done so, download the `Chapter09` folder of the code bundle for this book at this link: `https://github.com/PacktPublishing/iOS-18-Programming-for-Beginners-Eighth-Edition`.
2. Open the `Chapter09` folder, and you'll see two folders, `BreakfastMaker-start` and `BreakfastMaker-complete`. The first folder contains the app that you will be modifying in this chapter, and the second contains the completed app.

3. Open the `BreakfastMaker-start` folder and then the `BreakfastMaker` Xcode project. Click on the **Main** storyboard file in the Project navigator. You should see four labels and a button in the **View Controller Scene**, as shown here:

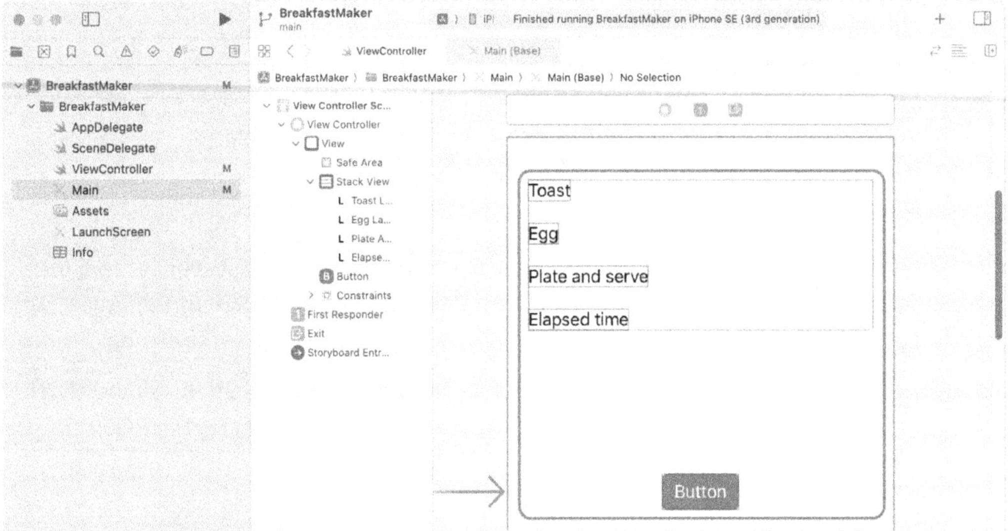

Figure 9.1: Main storyboard file showing the View Controller Scene

The app will display a screen that shows the status of the toast and eggs, and the time taken to plate and serve your breakfast. The app will also display a button that you can use to test the responsiveness of the user interface.

Don't worry if some of these concepts are not familiar to you. You will learn how to build user interfaces using storyboards for your apps in the next chapter, *Chapter 10, Setting Up the User Interface*.

4. Click the **ViewController** file in the Project navigator. You should see the following code in the Editor area:

```
import UIKit
class ViewController: UIViewController {
   @IBOutlet var toastLabel: UILabel!
   @IBOutlet var eggLabel: UILabel!
   @IBOutlet var plateAndServeLabel: UILabel!
   @IBOutlet var elapsedTimeLabel: UILabel!
   override func viewDidAppear(_ animated: Bool) {
      super.viewDidAppear(animated)
      let startTime = Date().timeIntervalSince1970
      toastLabel.text = "Making toast..."
      toastLabel.text = makeToast()
```

```
        eggLabel.text = "Boiling eggs..."
        eggLabel.text = boilEggs()
        plateAndServeLabel.text = plateAndServe()
        let endTime = Date().timeIntervalSince1970
        elapsedTimeLabel.text = "Elapsed time is
        \(((endTime - startTime) * 100).rounded()
        / 100) seconds"
    }
    func makeToast() -> String {
        sleep(2)
        return "Toast done"
    }
    func boilEggs() -> String {
        sleep(7)
        return "Eggs done"
    }
    func plateAndServe() -> String {
        return "Plating and serving done"
    }
    @IBAction func testButton(_ sender: UIButton) {
        print("Button tapped")
    }
}
```

As you can see, this code simulates the process of making breakfast that was described in the previous section. Let's break it down:

```
@IBOutlet var toastLabel: UILabel!
@IBOutlet var eggLabel: UILabel!
@IBOutlet var plateAndServeLabel: UILabel!
@IBOutlet var elapsedTimeLabel: UILabel!
```

These outlets are linked to four labels in the `Main` storyboard file. When you run the app, these labels will display the status of the toast and eggs, plating, and serving, as well as the time taken to complete the process.

```
override  func viewDidAppear(_ animated: Bool) {
```

This statement method is called when the view controller's view appears onscreen.

```
let startTime = Date().timeIntervalSince1970
```

This statement sets startTime to the current time, so the app can later calculate how long it takes to make the meal.

```
toastLabel.text = "Making toast..."
```

This statement makes toastLabel display the text Making toast....

```
toastLabel.text = makeToast()
```

This statement calls the makeToast() method, which waits for two seconds to simulate the time taken to make toast, and then returns the text Toast done, which will be displayed by toastLabel.

```
eggLabel.text = "Boiling eggs..."
```

This statement makes eggLabel display the text Boiling eggs....

```
eggLabel.text = boilEggs()
```

This statement calls the boilEggs() method, which waits for seven seconds to simulate the time taken to boil two eggs, and then returns the text Eggs done, which will be displayed by eggLabel.

```
plateAndServeLabel.text = plateAndServe()
```

This statement calls the plateAndServe() method, which returns the text Plating and serving done, which will be displayed by `plateAndServeLabel`.

```
let endTime = Date().timeIntervalSince1970
```

This statement sets `endTime` to the current time.

```
elapsedTimeLabel.text = "Elapsed time is
\(((endTime - startTime) * 100).rounded()
/ 100) seconds"
```

This statement calculates the elapsed time (approximately eight seconds), which will be displayed by elapsedTime`Label`.

```
@IBAction func testButton(_ sender: UIButton) {
    print("Button tapped")
}
```

This method displays **Button tapped** in the Debug area each time the button onscreen is tapped.

Build and run the app, and tap the button the moment the user interface appears:

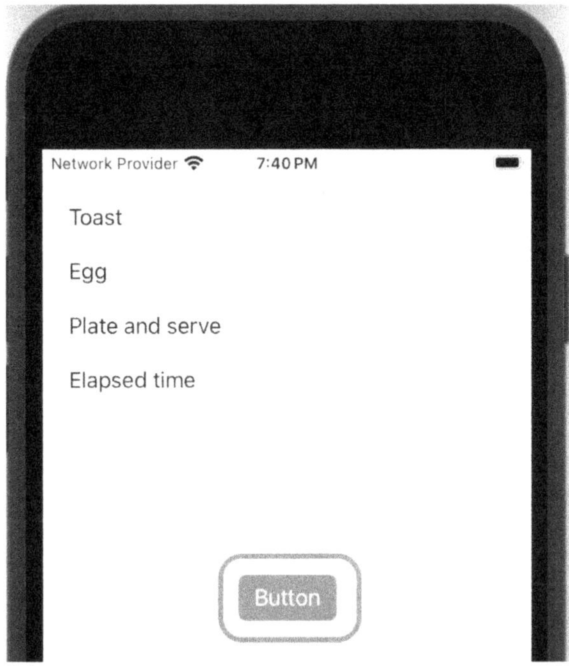

Figure 9.2: iOS Simulator running the BreakfastMaker app, showing the button to be tapped

You should notice the following issues:

- Tapping the button has no effect initially, and you'll only see **Button tapped** in the Debug area after approximately nine seconds.
- **Making toast...** and **Boiling eggs...** are never displayed, and **Toast done** and **Eggs done** only appear after approximately nine seconds.

The reason why this happens is that your app's code did not update the user interface while the `makeToast()` and `boilEggs()` methods were running. Your app did register the button taps but was only able to process them and update the labels after `makeToast()` and `boilEggs()` had completed their execution. These issues do not offer a good user experience with your app.

You have now experienced the issues presented by an app that does not have concurrency implemented. In the next section, you'll modify the app using `async/await` so that it can update the user interface while the `makeToast()` and `boilEggs()` methods are running.

Updating the app using async/await

As you saw previously, the app is unresponsive when the `makeToast()` and `poachEgg()` methods are running. To resolve this, you will use async/await in the app.

Writing the `async` keyword in the method declaration indicates that the method is asynchronous. This is what it looks like:

```
func methodName() async -> returnType {
```

Writing the `await` keyword in front of a method call marks a point where execution may be suspended, thus allowing other operations to run. This is what it looks like:

```
await methodName()
```

You can watch Apple's WWDC21 video discussing async/await at `https://developer.apple.com/videos/play/wwdc2021/10132/`.

You will modify your app to use async/await. This will enable it to suspend the `makeToast()` and `poachEgg()` methods to process button taps, update the user interface, and then resume execution of both methods afterward. You will also enable strict concurrency checking for your app by turning on the Swift 6 language mode. Follow these steps:

1. In the Project navigator, click the **BreakfastMaker** icon at the top and then the **BreakfastMaker** target. In the **Build Settings** tab, change **Swift Language Version** to **Swift 6**:

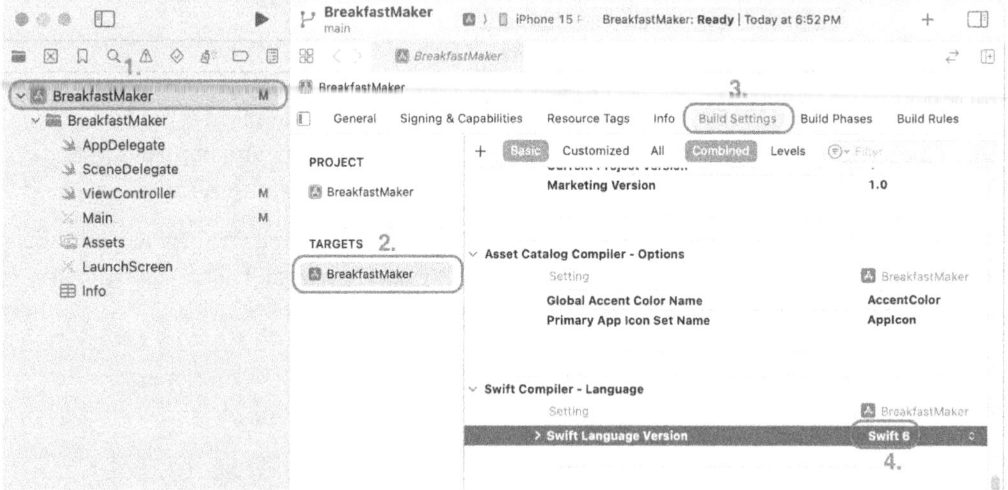

Figure 9.3: BreakfastMaker project with Swift Language Version set to Swift 6

This enables strict concurrency checking for your app.

2. Click the **ViewController** file in the Project navigator. Modify the `makeToast()` and `boilEggs()` methods, as shown here, to make the code in their bodies asynchronous:

```
func makeToast() -> String {
   try? await Task.sleep(for: .seconds(2))
   return "Toast done"
}
func boilEggs() -> String {
   try? await Task.sleep(for: .seconds(7))
   return "Eggs done"
}
```

`Task` represents a unit of asynchronous work. It has a static method, `sleep(for:)`, which pauses execution for a specified duration, measured in seconds. Since this method is a throwing method, you'll use the `try?` keyword to call it without having to implement a `do-catch` block. The `await` keyword indicates that this code can be suspended to allow other code to run.

Using `try?` will result in any errors being suppressed or ignored. This is acceptable in this case because sleeping for 2 or 7 seconds is unlikely to generate an error. This may not be acceptable in other situations, where a `do-catch` block is a better solution. You may wish to reread *Chapter 8, Protocols, Extensions, and Error Handling* for information on how to implement a `do-catch` block.

3. Errors will appear for both `makeToast()` and `boilEggs()`. Click either error icon to display the error message:

'async' call in a function that does not support concurrency

Add 'async' to function 'makeToast()' to make it asynchronous   Fix

Figure 9.4: Error message when the error icon is clicked

The error is displayed because you're calling an asynchronous method inside a method that does not support concurrency. You will need to add the `async` keyword to the method declaration to indicate that it is asynchronous.

4. For each method, click the **Fix** button to add the `async` keyword to the method declaration.
5. Verify that your code looks like this after you're done:

```
func makeToast() async -> String {
   try? await Task.sleep(for: .seconds(2))
   return "Toast done"
}
func boilEggs() async -> String {
   try? await Task.sleep(for: .seconds(2))
```

```
        return "Eggs done"
    }
```

6. The errors in the `makeToast()` and `poachEgg()` methods should be gone, but new errors will appear in the `viewDidAppear()` method. Click one of the error icons to see the error message, which will be the same as the message you saw in *step 2*. This is because you're calling an asynchronous method inside a method that does not support concurrency.
7. Click the **Fix** button, and more errors will appear.
8. Ignore the error in the method declaration for now, and click the one next to the `makeToast()` method call to see the error message:

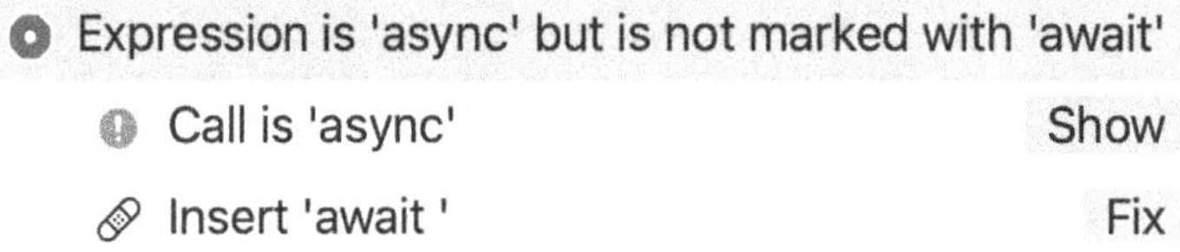

Figure 9.5: Error message when the error icon is clicked

This error message is displayed because you did not use `await` when calling an asynchronous function.

9. Click the **Fix** button to insert the `await` keyword before the method call.
10. Repeat *step 7* and *step 8* for the error next to the `boilEggs()` method call. The `await` keyword will be inserted for the `boilEggs()` method call as well.
11. Click the error icon in the `viewDidAppear()` method declaration to see the error message:

```
    override func viewDidAppear(_ animated: Bool) async {  M...
        super.viewDidAppear(animated)
        let startTime = Date().timeIntervalSince1970
        toastLabel.text = "Making toast..."
        toastLabel.text = await makeToast()
        eggLabel.text = "Boiling eggs..."
        eggLabel.text = await boilEggs()
```

Figure 9.6: Error with the error icon highlighted

This error is displayed because you can't use the `async` keyword to make the `viewDidAppear()` method asynchronous, as this capability is not present in the superclass.

12. To resolve this issue, you'll remove the `async` keyword and enclose all the code after `super.viewDidAppear()` in a `Task` block, which will allow it to execute asynchronously in a synchronous method. Modify your code as follows:

```
override func viewDidAppear(_ animated: Bool) {
    super.viewDidAppear(animated)
    Task {
        let startTime = Date().timeIntervalSince1970
```

```
            toastLabel.text = "Making toast..."
            toastLabel.text = await makeToast()
            eggLabel.text = "Boiling eggs..."
            eggLabel.text = await boilEggs()
            plateAndServeLabel.text = plateAndServe()
            let endTime = Date().timeIntervalSince1970
            elapsedTimeLabel.text = "Elapsed time is
            \(((endTime - startTime) * 100).rounded()
            / 100) seconds"
        }
    }
```

Build and run the app, and tap the button as soon as you see the user interface. Note that **Button tapped** now appears immediately in the Debug area, and the labels update as they should. This is because the app is now able to suspend the `makeToast()` and `boilEggs()` methods to respond to taps, update the user interface, and resume method execution later. Awesome!

However, if you look at the elapsed time, you'll see that the app takes slightly longer to prepare breakfast than it did before:

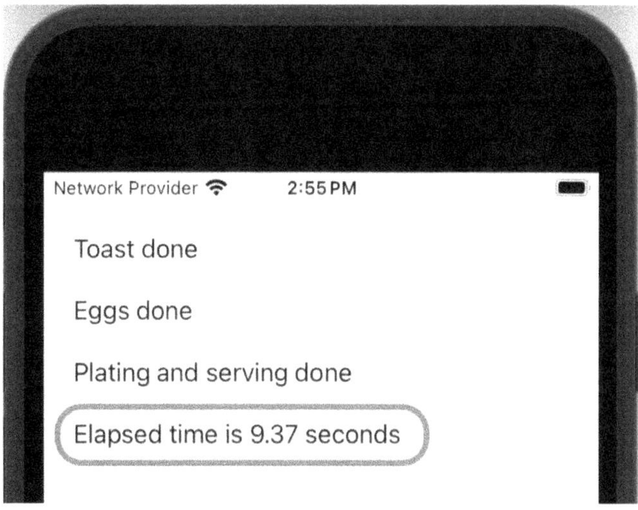

Figure 9.7: iOS Simulator running the BreakfastMaker app, showing the elapsed time

This is partly due to the additional processing required for the async/await suspending and resuming methods, but there is another factor involved. Even though the `makeToast()` and `boilEggs()` methods are now asynchronous, the `boilEggs()` method only starts execution after the `makeToast()` method has finished execution. In the next section, you'll see how you can use `async-let` to run the `makeToast()` and `boilEggs()` methods in parallel.

Improving efficiency using async-let

Even though your app is now responsive to button taps and can update the user interface while the `makeToast()` and `boilEggs()` methods are running, both methods still execute sequentially. The solution here is to use `async-let`.

Writing `async` in front of a `let` statement when you define a constant, and then writing `await` when you access the constant, allows the parallel execution of asynchronous methods, as shown here:

```
async let temporaryConstant1 = methodName1()
async let temporaryConstant2 = methodName2()
await variable1 = temporaryConstant1
await variable2 = temporaryConstant1
```

In this example, `methodName1()` and `methodName2()` will run in parallel.

You will modify your app to use `async-let` to enable the `makeToast()` and `poachEgg()` methods to run in parallel. In the `ViewController` file, modify the code in the `Task` block as follows:

```
Task {
    let startTime = Date().timeIntervalSince1970
    toastLabel.text = "Making toast..."
    async let tempToast = makeToast()
    eggLabel.text = "Boiling eggs..."
    async let tempEggs = boilEggs()
    await toastLabel.text = tempToast
    await eggLabel.text = tempEggs
    plateAndServeLabel.text = plateAndServe()
    let endTime = Date().timeIntervalSince1970
    elapsedTimeLabel.text = "Elapsed time is
    \(((endTime - startTime) * 100).rounded()
    / 100) seconds"
}
```

Build and run the app. You'll see that the elapsed time is now less than it was before:

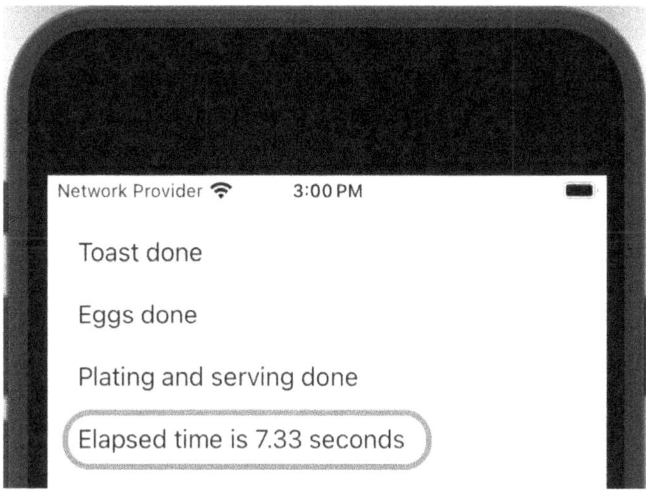

Figure 9.8: iOS Simulator running the BreakfastMaker app, showing the elapsed time

This is because using `async-let` allows both the `makeToast()` and `poachEgg()` methods to run in parallel, and the `poachEgg()` method no longer waits for the `makeToast()` method to complete before starting execution. Cool!

There is still lots more to learn about Swift concurrency, such as structured concurrency and actors, but that is beyond the scope of this chapter. You can learn more about structured concurrency at `https://developer.apple.com/videos/play/wwdc2021/10134/`, and you can learn more about actors at `https://developer.apple.com/videos/play/wwdc2021/10133/`.

You have successfully implemented asynchronous code in your app. Fantastic! There are still a lot of things to learn about Swift concurrency, such as structured concurrency and actors, but that is beyond the scope of this chapter.

Give yourself a pat on the back; you have completed the first part of this book!

Summary

In this chapter, you learned about Swift concurrency and how to implement it in the *BreakfastMaker* app.

You started by learning the basic concepts of Swift concurrency. Then, you examined an app without concurrency and explored its issues. After that, you turned on strict concurrency checking and implemented concurrency in the app, using `async/await`. Finally, you made your app more efficient by using `async let`.

You now understand the basics of Swift concurrency and will be able to use `async/await` and `async-let` in your own apps.

In the next chapter, you will start writing your first iOS application by creating the screens for it, using storyboards, which allow you to rapidly prototype an application without having to type a lot of code.

Leave a review!

Thank you for purchasing this book from Packt Publishing—we hope you enjoy it! Your feedback is invaluable and helps us improve and grow. Once you've completed reading it, please take a moment to leave an Amazon review; it will only take a minute, but it makes a big difference for readers like you. Scan the QR code below or visit the link to receive a free ebook of your choice.

`https://packt.link/NzOWQ`

Part 2

Design

Welcome to *Part 2* of this book. At this point, you're familiar with the Xcode user interface, and you have a solid foundation of using Swift. In this part, you'll start creating the user interface of a journal app, named *JRNL*. You will use Interface Builder to build the screens that your app will use, add elements such as buttons, labels, and fields to them, and connect them together using segues. As you will see, you can do this with a minimum of coding.

This part comprises the following chapters:

- *Chapter 10, Setting Up the User Interface*
- *Chapter 11, Building Your User Interface*
- *Chapter 12, Finishing Up Your User Interface*
- *Chapter 13, Modifying App Screens*

By the end of this part, you'll be able to navigate the various screens of your app in the iOS Simulator, and you will know how to prototype the user interface of your own apps. Let's get started!

10

Setting Up the User Interface

In *Part 1* of this book, you studied the Swift language and how it works. Now that you have a good working knowledge of the language, you can learn how to develop an iOS application. In this part, you will build the **user interface** (**UI**) of a journal app, *JRNL*. You will use Xcode's **Interface Builder** for this, and coding will be kept to a minimum.

You'll start this chapter by learning useful terms used in iOS app development, which are used extensively throughout this book. Next, you will take a tour of the screens used in the *JRNL* app and learn how a user would use the app. Finally, you will begin recreating the app's UI with Interface Builder, starting with the tab bar, which allows the user to select between the Journal List and Map screens. You'll add navigation bars to the top of both screens and configure the tab bar buttons.

By the end of this chapter, you'll have learned common terms used in iOS app development, what the flow of your app will look like, and how to use Interface Builder to add and configure UI elements.

The following topics will be covered in this chapter:

- Learning useful terms in iOS development
- A tour of the *JRNL* app
- Modifying your Xcode project
- Setting up a tab bar controller scene

Technical requirements

You will modify the `JRNL` Xcode project that you created in *Chapter 1*, *Exploring Xcode*.

The resource files and completed Xcode project for this chapter are in the `Chapter10` folder of the code bundle for this book, which can be downloaded here:

`https://github.com/PacktPublishing/iOS-18-Programming-for-Beginners-Ninth-Edition`

Check out the following video to see the code in action:

`https://youtu.be/lgyerQeTgN4`

Before you get started with the project, you'll learn some common terms used in iOS development.

Learning useful terms in iOS development

As you begin your journey into iOS app development, you will encounter special terms and definitions. Here are some of the most used terms and definitions. Just read through them for now. Even though you may not understand everything yet, things will become clearer as you go along:

- **View**: A view is an instance of the `UIView` class or one of its subclasses. Anything you see on your screen (buttons, text fields, labels, and so on) is a view. You will use views to build your UI.

Classes are covered in *Chapter 7, Classes, Structures, and Enumerations.*

- **Stack view**: A stack view is an instance of the `UIStackView` class, which is a subclass of `UIView`. It is used to group views together in a horizontal or vertical stack. This makes them easier to position on the screen using **Auto Layout**, which is discussed later in this section.
- **View controller**: A view controller is an instance of the `UIViewController` class. Every view controller has a `view` property, which contains a reference to a view. It determines what a view displays to a user and what happens when the user interacts with a view.

View controllers will be discussed in detail in *Chapter 14, Getting Started with MVC and Table Views.*

- **Table view controller**: A table view controller is an instance of the `UITableViewController` class, which is a subclass of the `UIViewController` class. Its `view` property has a reference to a `UITableView` instance (table view), which displays a single column of `UITableViewCell` instances (table view cells).

The **Settings** app displays your device settings in a table view:

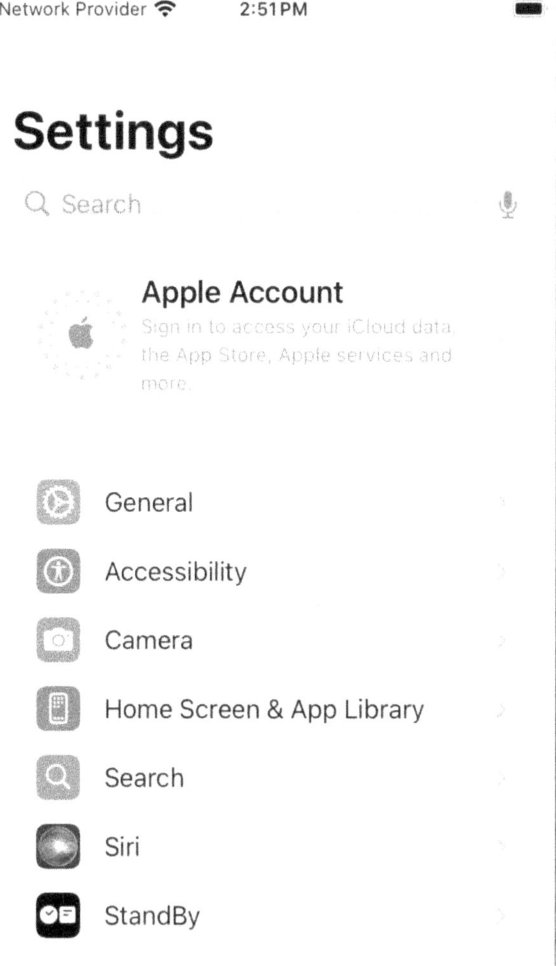

Figure 10.1: Settings app

As you can see, all the different settings (**General**, **Accessibility**, **Privacy**, and so on) are displayed in table view cells inside the table view.

- **Collection view controller**: A collection view controller is an instance of the `UICollectionViewController` class, which is a subclass of the `UIViewController` class. Its `view` property has a reference to a `UICollectionView` instance (collection view), which displays a grid of `UICollectionViewCell` instances (collection view cells).

The **Photos** app displays photos in a collection view:

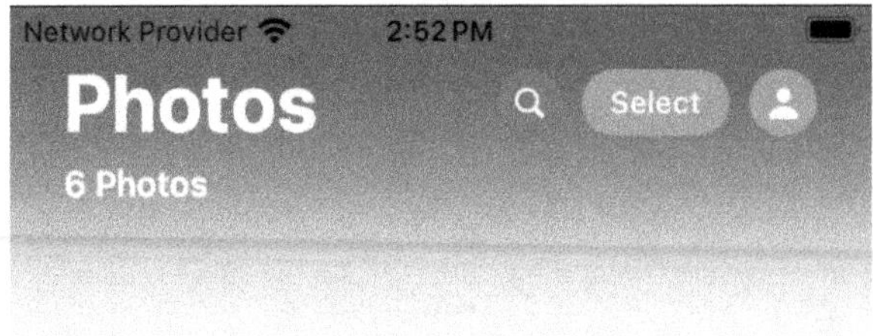

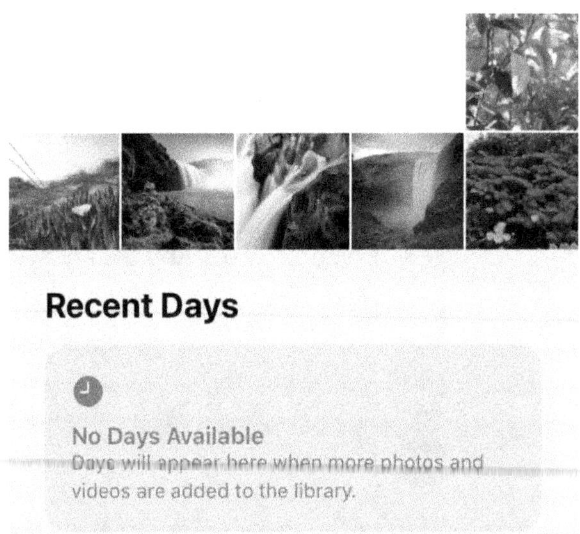

Figure 10.2: Photos app

As you can see, thumbnail pictures are displayed in collection view cells inside the collection view.

- **Navigation controller:** A navigation controller is an instance of the `UINavigationController` class, which is a subclass of the `UIViewController` class. It has a `viewControllers` property that holds an array of view controllers. The view of the last view controller in the array appears onscreen, along with a navigation bar at the top of the screen.

  The table view controller in the **Settings** app is embedded in a navigation controller, and you can see the navigation bar above the table view:

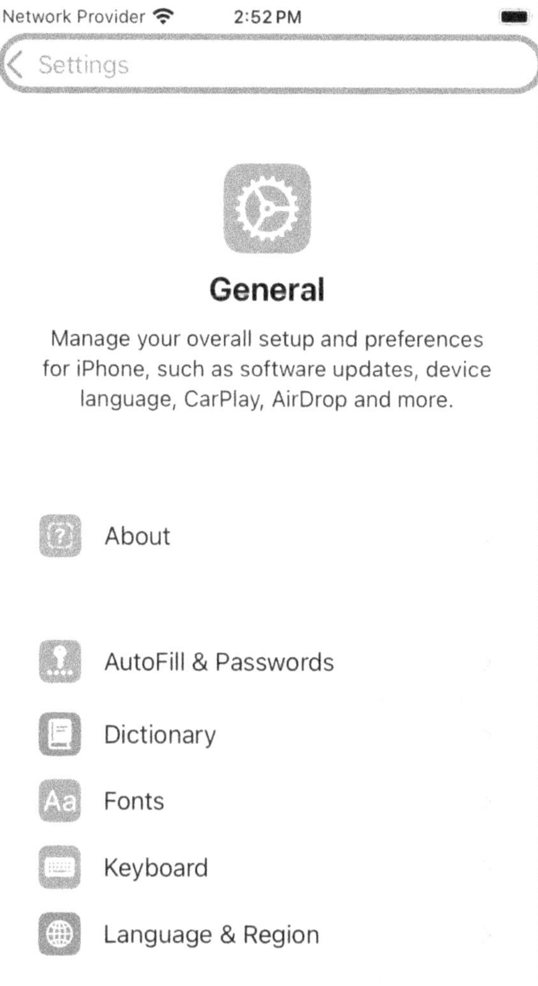

Figure 10.3: Navigation bar in the Settings app

When you tap on a setting, the view controller for that setting is added to the array of view controllers assigned to the `viewControllers` property. The user sees the view for that view controller slide in from the right. Note the navigation bar at the top of the screen, which can hold a title and buttons. A < **Settings** button appears on the top-left side of the navigation bar. Tapping this button returns you to the previous screen, and it removes the view controller for that setting from the array of view controllers assigned to the `viewControllers` property.

- **Tab bar controller**: A tab bar controller is an instance of the `UITabBarController` class, which is a subclass of the `UIViewController` class. It has a `viewControllers` property that holds an array of view controllers. The view of the first view controller in the array appears onscreen, along with a tab bar with buttons at the bottom. The button on the extreme left corresponds to the first view controller in the array and will already be selected. When you tap another button, the corresponding view controller is loaded, and its view appears on the screen.

The **Fitness** app uses a tab bar controller to navigate to different screens:

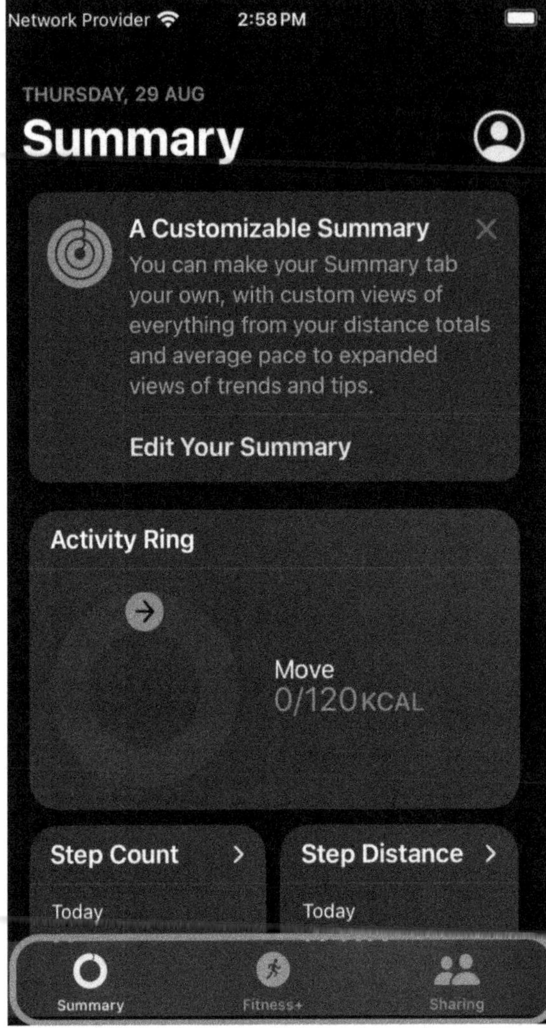

Figure 10.4: Tab bar in the Fitness app

As you can see, the different screens for this app (**All Photos**, **For You**, **Albums**, and **Search**) are accessed by tapping the corresponding tab bar button.

- **Model-View-Controller (MVC)**: This is a very common design pattern used in iOS app development. The user interacts with views onscreen. App data is stored in data model objects. Controllers manage the flow of information between views and data model objects.

MVC will be discussed in detail in *Chapter 14, Getting Started with MVC and Table Views*.

- **Storyboard file:** A storyboard file contains a visual representation of what a user sees. Each screen of an app is represented by a storyboard **scene**.

  Open the *JRNL* project that you created in *Chapter 1, Exploring Xcode,* and click the **Main** storyboard file.

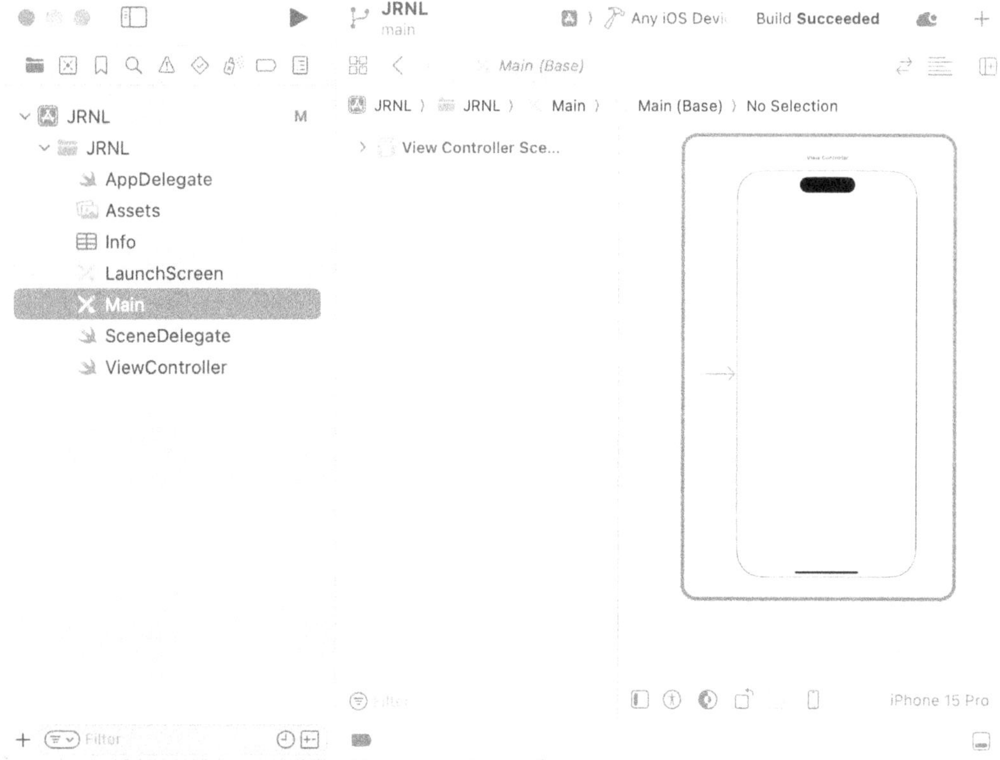

Figure 10.5: JRNL Xcode project showing the Main storyboard file

  You'll see one scene in it, and when you run your app in **Simulator**, the contents of this scene will be displayed on the screen. You can have more than one scene in a storyboard file.

- **Segue:** If you have more than one scene in an app, you use segues to move from one scene to another. The *JRNL* project does not have any segues, since there is just one scene in its storyboard file, but you will see them in a later part of this chapter.
- **Auto Layout:** As a developer, you must make sure that your app looks good on devices with different screen sizes. **Auto Layout** helps you lay out your UI based on the **constraints** you specify. For instance, you can set a constraint to make sure that a button is centered on the screen, regardless of screen size, or make a text field expand to the width of the screen when a device is rotated from portrait to landscape.

Now that you are familiar with the terms used in iOS app development, let's take a tour of the app you will build.

A tour of the JRNL app

Let's take a quick tour of the app that you will build. The *JRNL* app is a journal app that lets users write their own personal journal, with the option of storing a photo or a map location for each journal entry. Users can also view a map that shows the locations of entries that are close to a user's current location. You'll see all the screens used in the app and its overall flow in the next sections.

Using the Journal List screen

When the app is launched, you will see the Journal List screen:

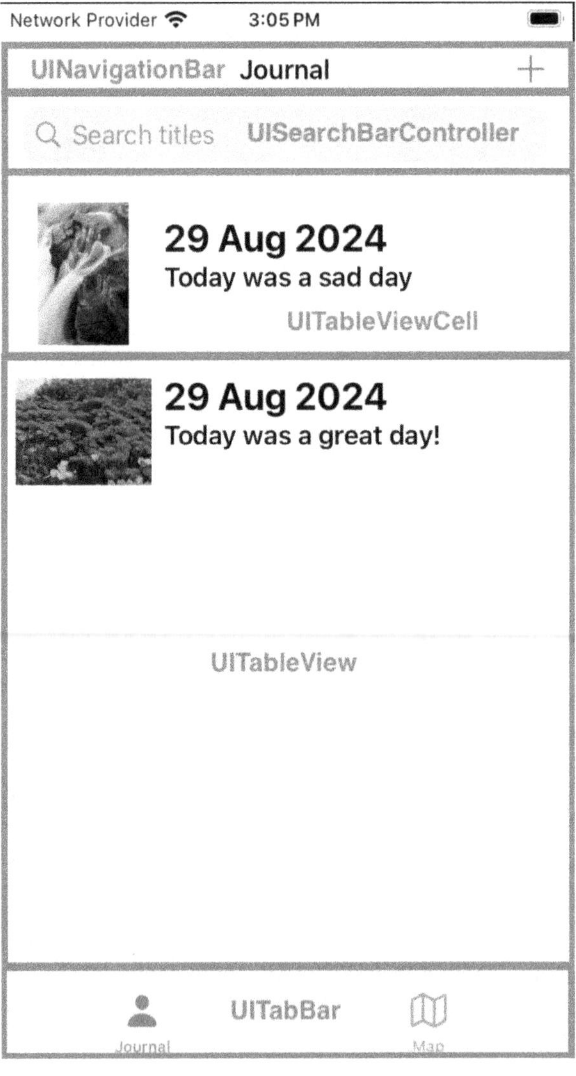

Figure 10.6: Journal List screen

Let's study the different parts of this screen.

A `UITabBar` instance (tab bar) at the bottom of the screen displays the **Journal** and **Map** buttons. The **Journal** button is selected, and you can see a table view displaying a list of journal entries in table view cells. A `UISearchController` instance displays a search bar at the top of the screen. This allows you to search for a particular journal entry.

To add a new journal entry, you tap the + button at the top of the screen. This displays the Add New Journal Entry screen.

Using the Add New Journal Entry screen

When you tap the + button at the top of the Journal List screen, you will see the Add New Journal Entry screen:

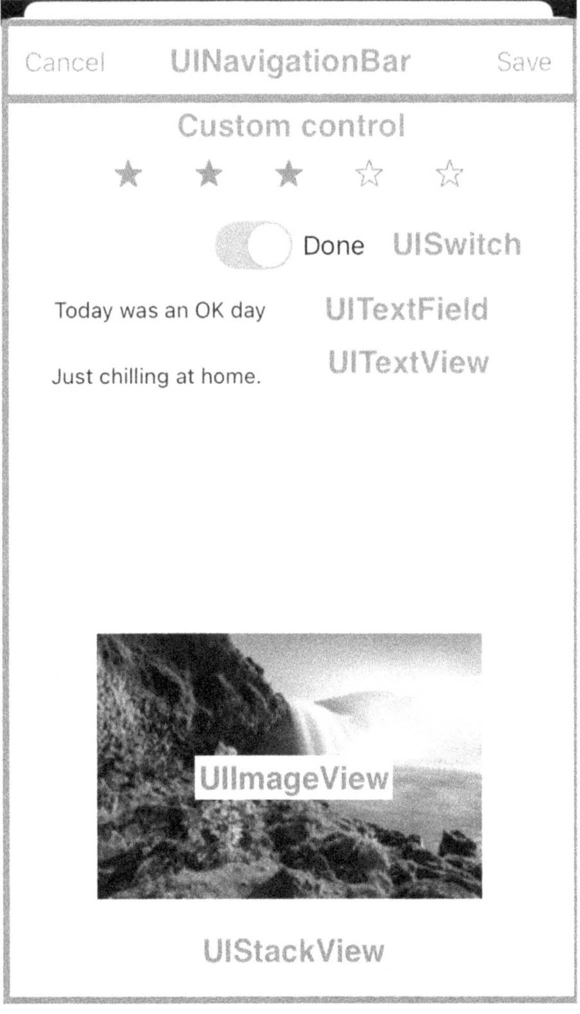

Figure 10.7: Add New Journal Entry screen

Let's study the different parts of this screen.

A navigation bar at the top of the screen contains the **Cancel** and **Save** buttons. A stack view displays a custom rating control, a switch, an entry title text field, a body text view, and a placeholder photo. Tapping the rating control allows you to assign 0 to 5 stars for this entry. Switching the switch on will obtain your current location.

You can enter the journal entry's title in the entry title text field, and the details in the body text view. You can also tap the placeholder photo to take a picture with your device camera. Once you tap **Save**, you are returned to the Journal List screen, and then the new entry will be visible in the table view. You can also tap **Cancel** to return to the Journal List screen without creating a new journal entry.

To see the details of a particular journal entry, tap the entry you want in the list, and then you will see the Journal Entry Detail screen.

Using the Journal Entry Detail screen

Tapping any one of the journal entries on the Journal List screen will display the corresponding Journal Entry Detail screen:

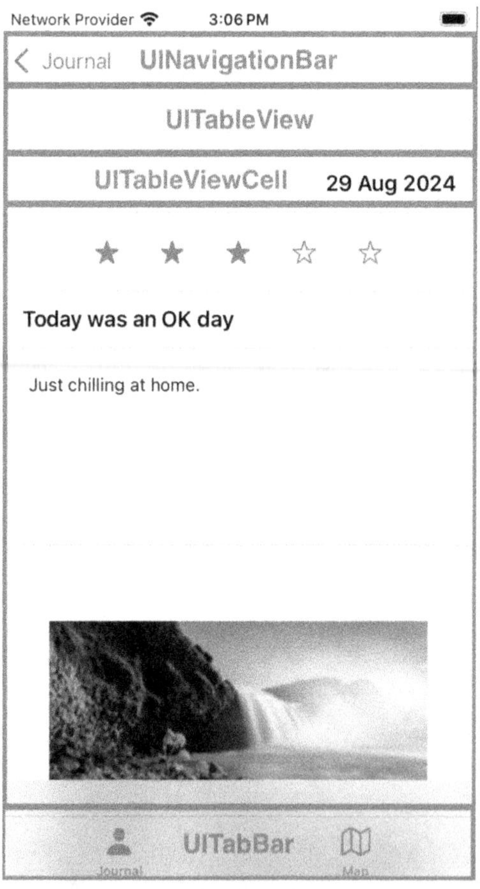

Figure 10.8: Journal Entry Detail screen

Let's study the different parts of this screen.

A navigation bar at the top of the screen contains a **Back** button. A table view displays the journal entry's date, rating, title text, body text, photo, and location map in table view cells.

You can tap the < **Journal** button to return to the Journal List screen.

Using the Map screen

Tapping the **Map** button in the tab bar displays the Map screen:

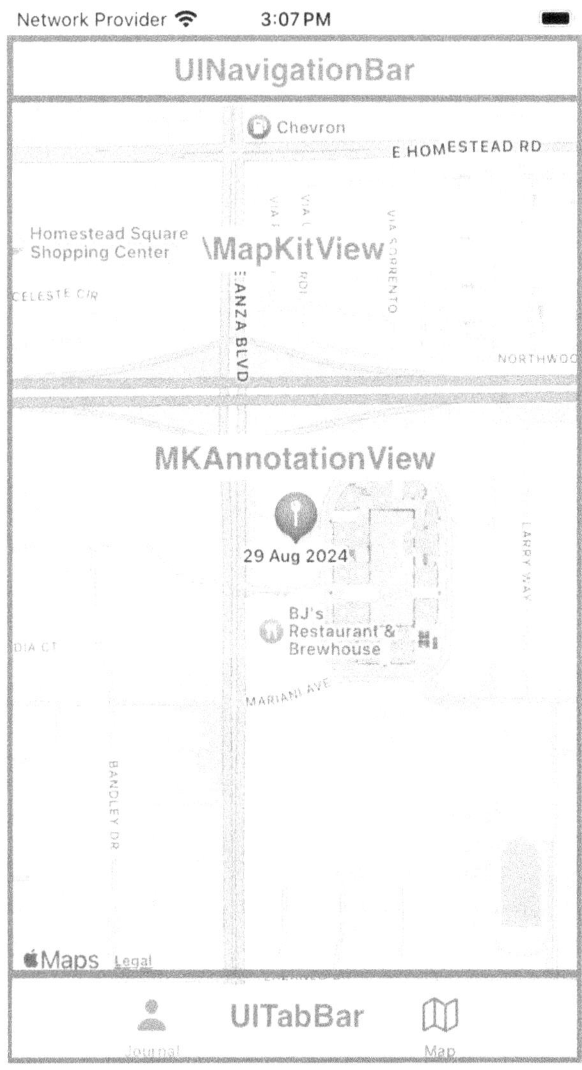

Figure 10.9: Map screen

Let's study the different parts of this screen.

A tab bar at the bottom of the screen displays the **Journal** and **Map** buttons. The **Map** button is selected, and you can see an `MKMapView` instance (map view) displaying a map on the screen, with pins indicating journal entries.

Tapping a pin will display an annotation, and tapping the button in the annotation will display the Journal Entry Detail screen for that journal entry.

This completes the tour of the app. Now, it's time to start building the UI for it!

Modifying your Xcode project

Now that you know what the screens of the app are going to look like, you can start building it. If you have not yet done so, open the `JRNL` project you created in *Chapter 1, Exploring Xcode*:

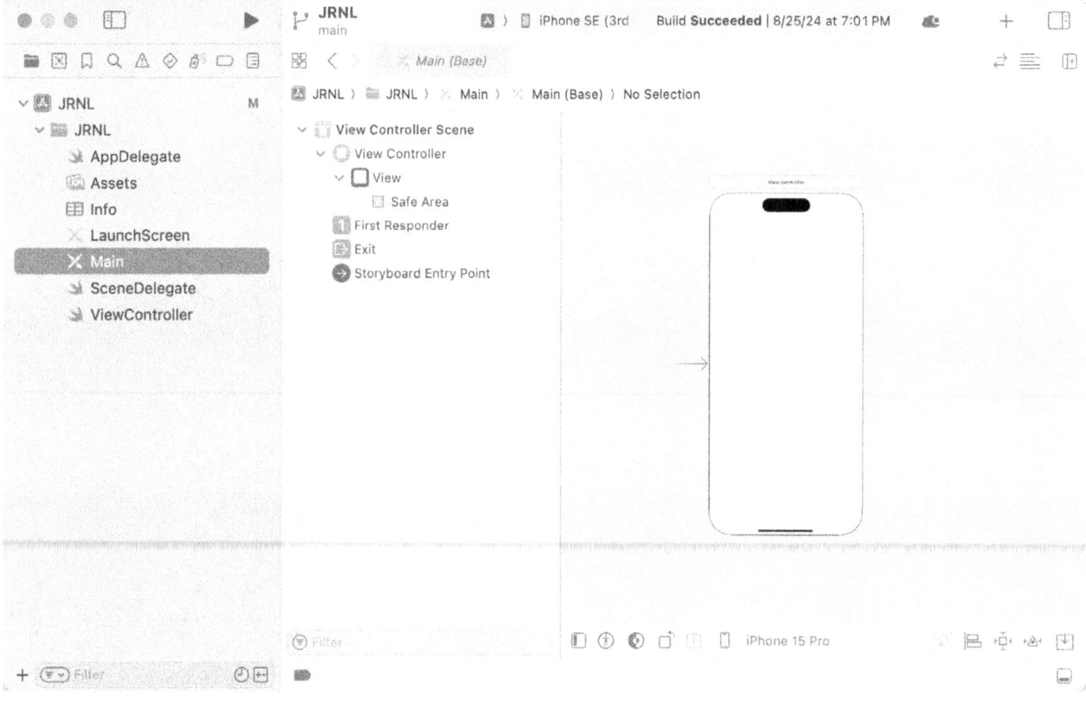

Figure 10.10: The JRNL project

Verify that **iPhone SE (3rd generation)** is selected from the Destination menu. Build and run your app. You will see a blank white screen. If you click the **Main** storyboard file in the Project navigator, you will see that it contains a single scene containing a blank view. This is why you only see a blank white screen when you run the app.

To configure the UI, you will modify the **Main** storyboard file using Interface Builder. Interface Builder allows you to add and configure scenes. Each scene represents a screen that a user will see. You can add UI objects such as views and buttons to a scene and configure them as required, using the Attributes inspector.

For more information on how to use Interface Builder, visit this link: `https://help.apple.com/xcode/mac/current/#/dev31645f17f`.

Now, you will embed the existing scene in a tab bar and add another scene to it. The tab bar scene will display a tab bar with two buttons at the bottom of the screen. Tapping a button will display the screen associated with it. These screens correspond to the Journal List and Map screens shown in the app tour. Let's see how to do this in the next section.

Setting up a tab bar controller scene

As you saw in the app tour, the *JRNL* app has a tab bar with two buttons at the bottom of the screen, which are used to display the Journal List and Map screens. You will embed the existing view controller scene in a tab bar and add a second view controller scene to the tab bar. Follow these steps:

1. Click the **Main** storyboard file in the Project navigator:

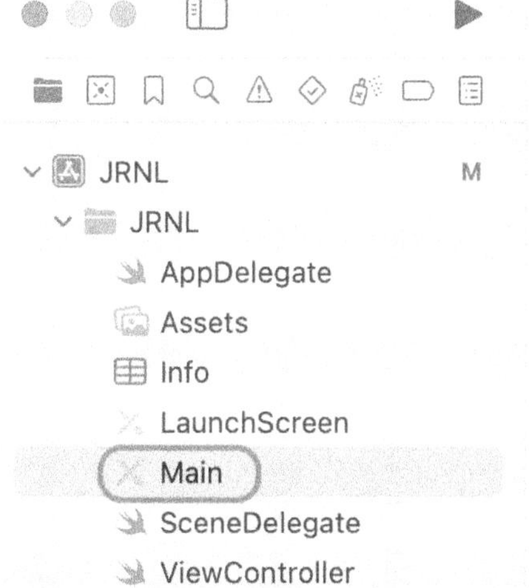

Figure 10.11: Project navigator with the Main storyboard file selected

The contents of the **Main** storyboard file appear in the Editor area.

2. Click the Document Outline button to display the document outline if it is not present:

Figure 10.12: Editor area with the Document Outline button shown

3. Select **View Controller** in the document outline:

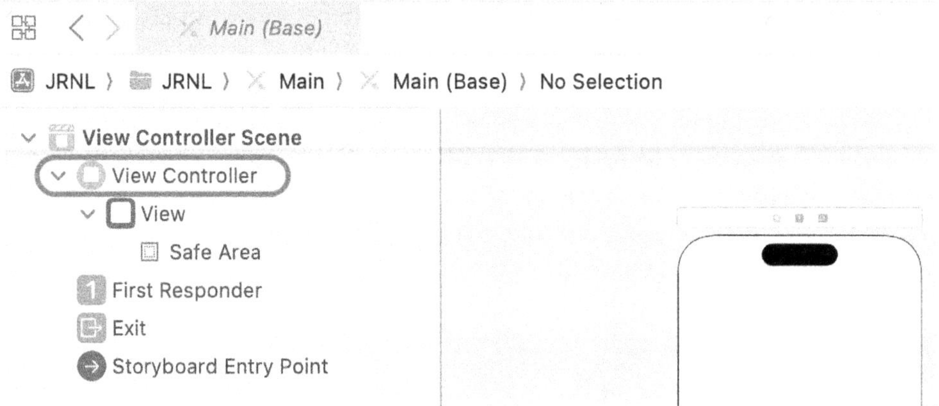

Figure 10.13: Document outline with View Controller selected

4. You'll embed the existing view controller scene in a tab bar controller scene. Choose **Embed In | Tab Bar Controller** from the **Editor** menu:

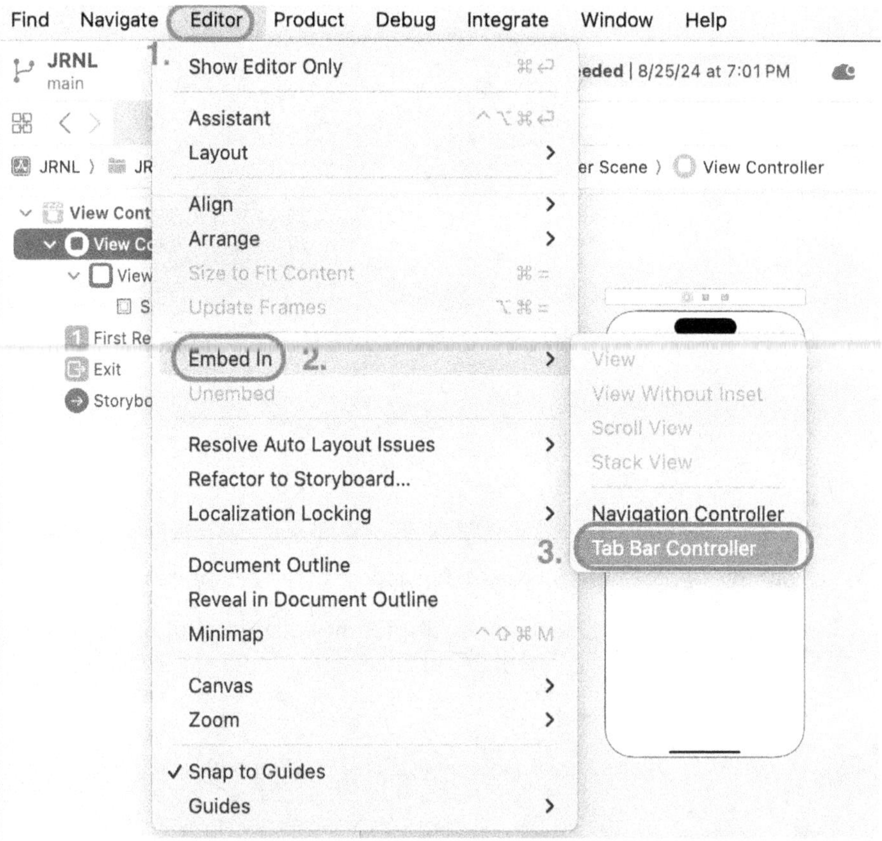

Figure 10.14: Editor menu with Embed In | Tab Bar Controller selected

You'll see a new tab bar controller scene appear in the Editor area.

5. Click the + button at the top-right side of the window to show the library:

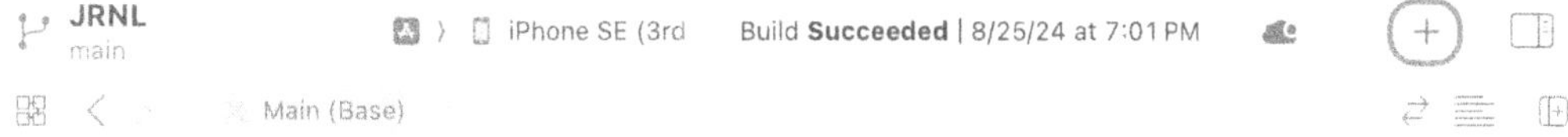

Figure 10.15: Toolbar with the + button shown

The library allows you to pick UI objects to be added to a scene.

6. Type `view con` in the library's filter field. A **View Controller** object will appear in the list of results:

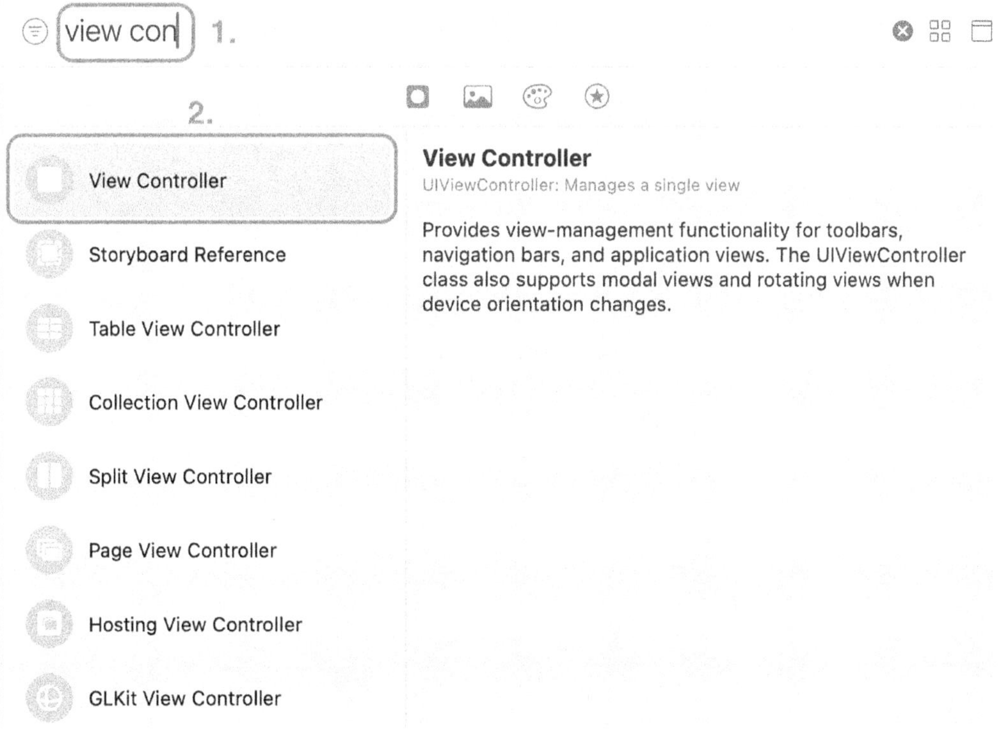

Figure 10.16: Library with the View Controller object selected

7. Drag the **View Controller** object to the storyboard to add a new view controller scene, and position it below the existing view controller scene:

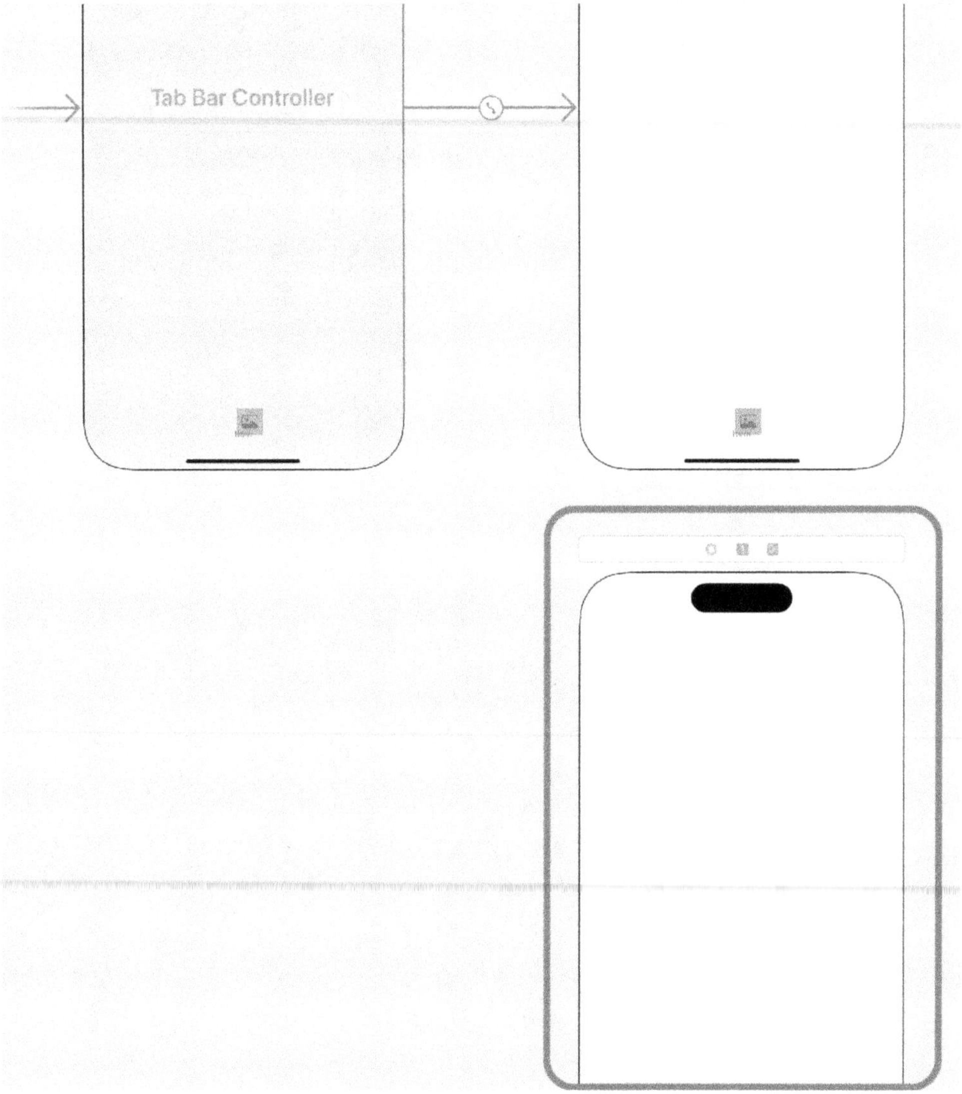

Figure 10.17: Main storyboard file with the view controller scene added

8. Click the - button to zoom out, and rearrange the scenes in the storyboard so that both the tab bar controller scene and the view controller scenes are visible:

Figure 10.18: Editor area with the zoom buttons shown

If the - and + buttons are not visible, try making the Xcode window larger. You could also try hiding the **Navigator** and **Inspector** areas using the **Navigator** and **Inspector** buttons.

9. Select **Tab Bar Controller** in the document outline. Press *Ctrl* and drag from **Tab Bar Controller** to the newly added view controller scene:

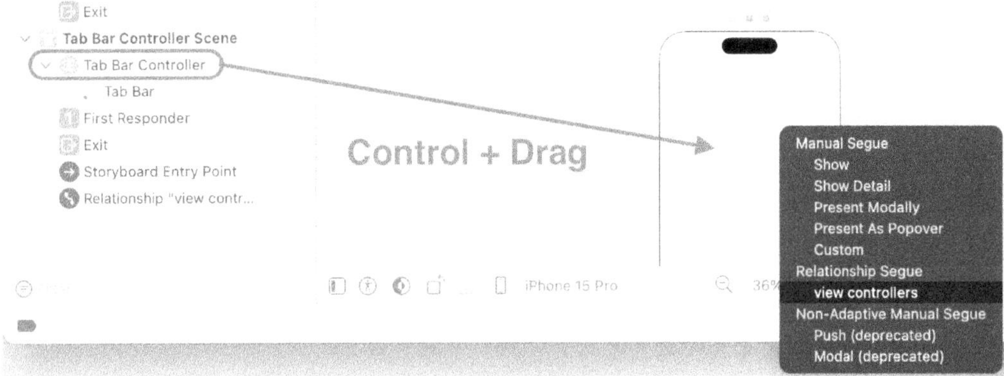

Figure 10.19: Editor area showing the drag destination

10. A segue pop-up menu will appear. Choose **view controllers** from this menu:

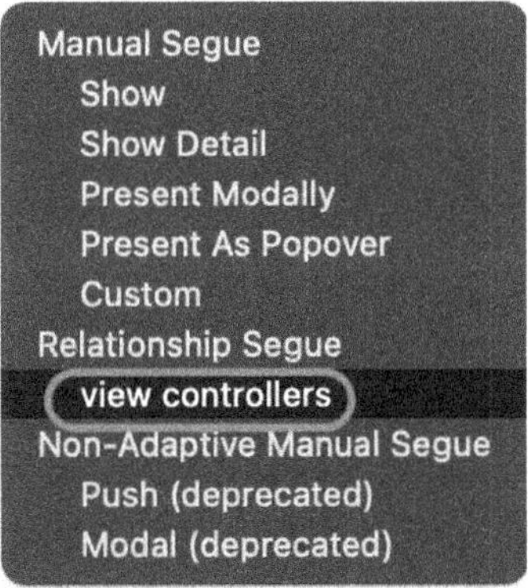

Figure 10.20: Segue pop-up menu

A segue connecting the tab bar controller scene to the view controller scene will appear.

11. Rearrange the scenes in the **Editor** area so that it looks like the screenshot below:

Figure 10.21: Editor area with rearranged scenes

12. Build and run your app in Simulator, and you'll see the tab bar with two buttons at the bottom of the screen:

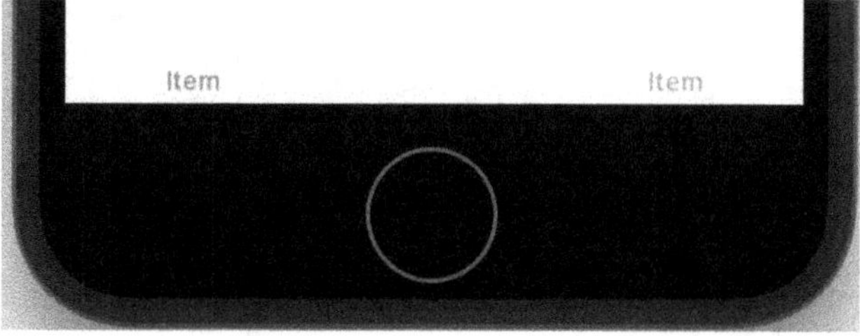

Figure 10.22: Simulator showing the tab bar with two buttons

You have successfully added a tab bar to your project, but as you can see, the button titles are currently both named **Item**. You will change them to **Journal** and **Map** in the next section.

Setting the tab bar button titles and icons

Your app now displays a tab bar at the bottom of the screen, but the button titles and icons do not match those shown in the app tour. To make them match, you will configure the button titles to read **Journal** and **Map** in the Attributes inspector and configure their icons as well. Follow these steps:

1. Click the **Main** storyboard file in the Project navigator. Click the Document Outline button to show the document outline if it is not present. Click the first **Item Scene** in the document outline:

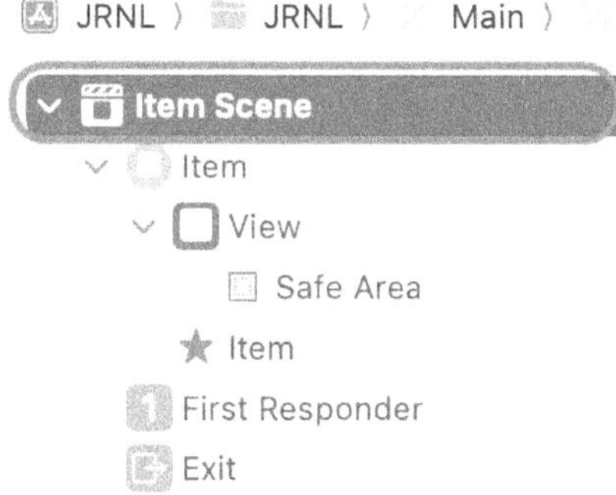

Figure 10.23: Document outline showing the first Item Scene selected

2. Click the **Item** button under **Item Scene**. Then, click the Attributes inspector button:

Figure 10.24: Attributes inspector selected

3. Under **Bar Item**, set **Title** to `Journal` and **Image** to `person.fill`:

Tab Bar Item

Badge Value

Default

System Item Custom

Selected Image Selected Image

Title Position Default Position

Drag and Drop Spring Loaded

Bar Item

Title Journal

Image person.fill System

Symbol Scale Unspecified

Render Mode Automatic

Value Variable

Landscape Image (iPhone)

Accessibility Large Content Size Image

Tag 0

Enabled

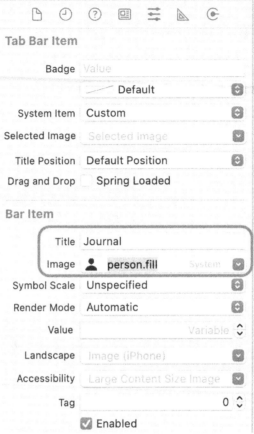

Figure 10.25: Attributes inspector with Title set to Journal and Image set to person.fill

4. Click the **Item** button in the second **Item** scene, and under **Bar Item**, set **Title** to `Map` and **Image** to `map`:

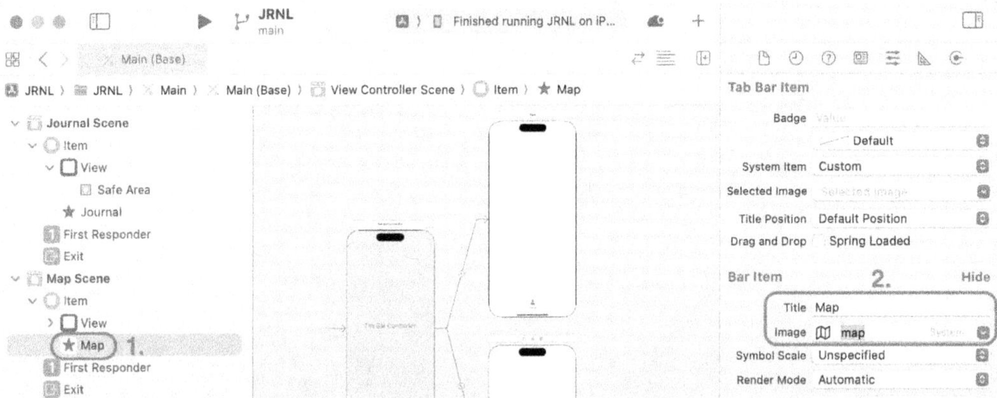

Figure 10.26: Attributes inspector with Title set to Map and Image set to map

5. Build and run your app in Simulator. You'll see that the titles for the buttons have changed to **Journal** and **Map**, respectively, and each button also has a custom icon:

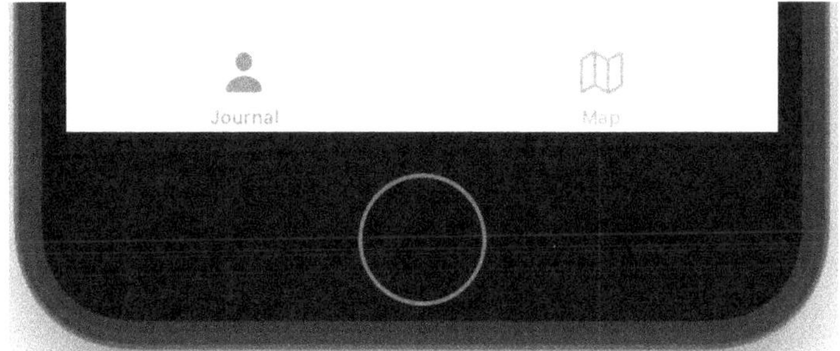

Figure 10.27: Simulator showing the tab bar with custom button titles and icons

Tapping the **Journal** and **Map** buttons will display the scenes for the Journal List and Map screens.

The `person.fill` and `map` icons are part of Apple's **SF Symbols** library. To learn more about it, visit this link: `https://developer.apple.com/design/human-interface-guidelines/sf-symbols`.

As you have seen in the app tour, some screens have titles and buttons in the navigation bar. In the next section, you will learn how to add navigation bars to your screens so that you can add buttons and titles to them later as required.

Embedding view controllers in navigation controllers

As you saw in the app tour, the Journal List and Map screens both have a navigation bar at the top of the screen. To add the navigation bars for both screens, you will embed the view controllers of the Journal and Map scenes in a navigation controller. This will make navigation bars appear at the top of the screen when the Journal List and Map screens are displayed. Follow these steps:

1. Click **Journal Scene** in the document outline:

Figure 10.28: Document outline with Journal Scene selected

2. Choose **Embed In | Navigation Controller** from the **Editor** menu:

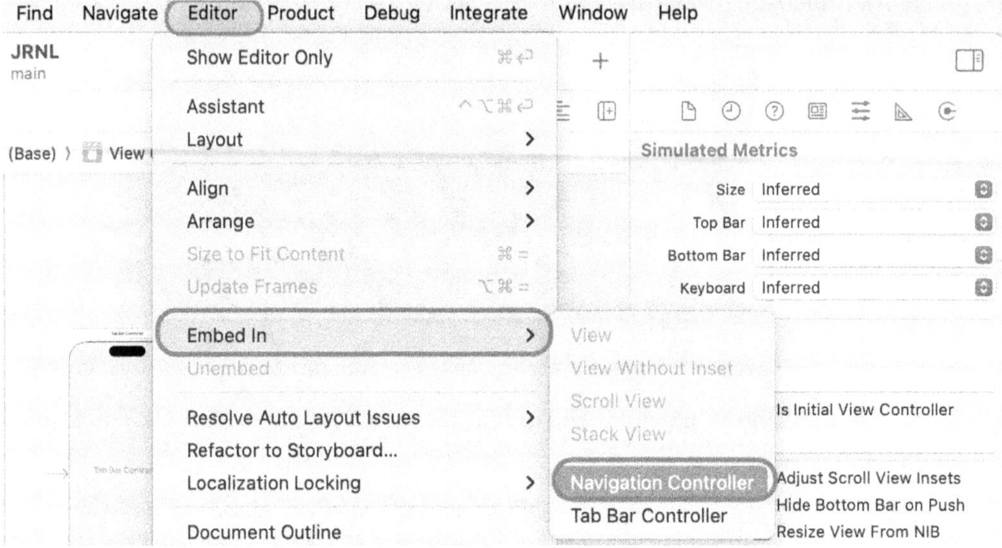

Figure 10.29: Editor menu with Embed In | Navigation Controller selected

3. Verify that a navigation controller scene has appeared between the tab bar controller scene and the journal scene:

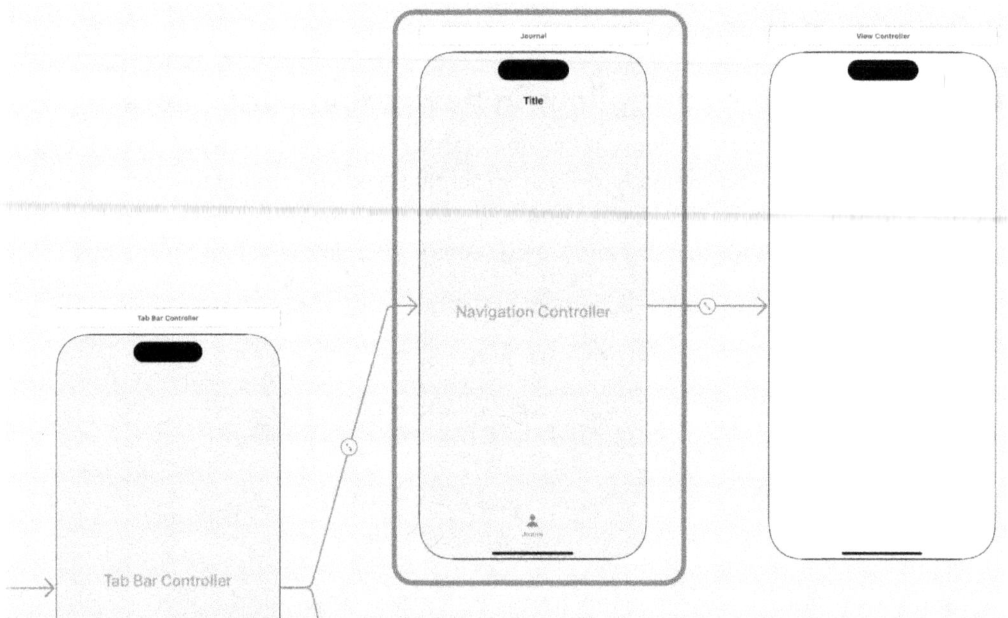

Figure 10.30: Editor area showing an added navigation controller scene

4. Click **Map Scene** in the document outline and repeat *step 2*.

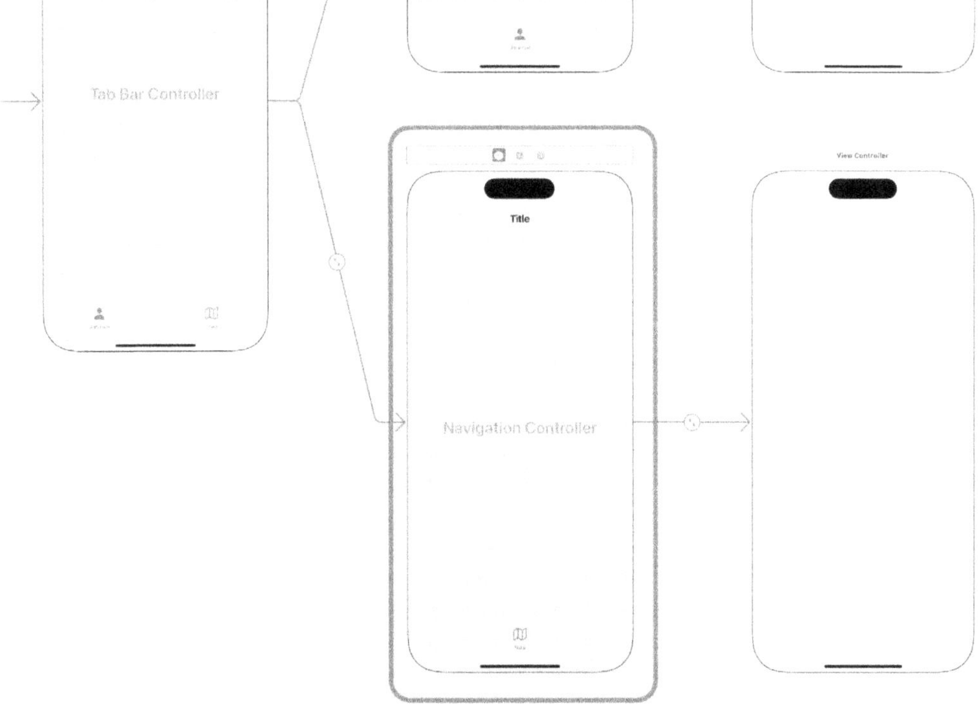

Figure 10.31: Editor area showing an added navigation controller scene

Both the **Journal List** screen and the **Map** screen now have navigation bars, but since they are the same color as the background, it is not apparent on the screen. You will set the titles for each scene's navigation item to distinguish between them.

5. Select the **Navigation Item** for the first **View Controller Scene** in the document outline. In the Attributes inspector, under **Navigation Item**, set **Title** to `Journal`:

Figure 10.32: Attributes inspector with Title set to Journal

6. Select the **Navigation Item** for the second **View Controller Scene** in the document outline. In the Attributes inspector, under **Navigation Item**, set **Title** to `Map`:

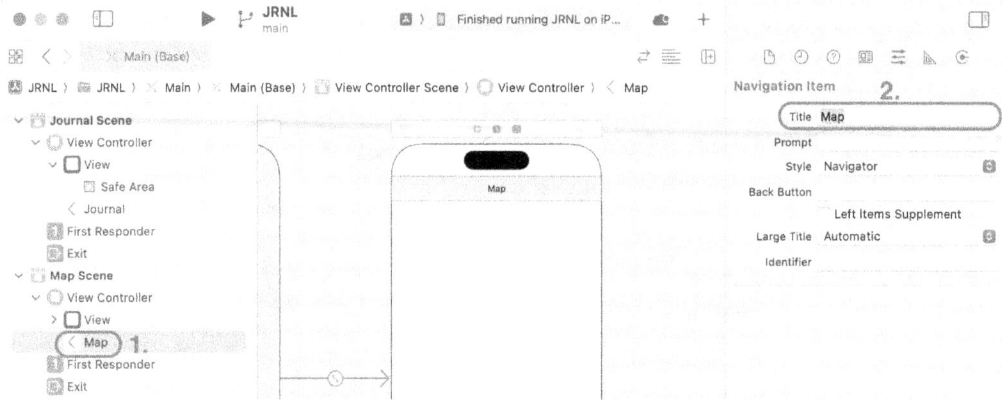

Figure 10.33: Attributes inspector with Title set to Map

7. Build and run your app, and tap each tab bar button to display the corresponding screen. Note that each screen displays a title in the navigation bar.

Embedding a view controller in a navigation controller adds that view controller to the navigation controller's `viewControllers` array. The navigation controller then displays the view controller's view on the screen. The navigation controller also displays a navigation bar with a title at the top of the screen.

Congratulations! You've just configured the tab bar and navigation controllers for your app!

You may have noticed that the screens represented in Interface Builder don't match the iPhone model you selected in the Destination menu, and you may find that the minimap display gets in the way of arranging screens in your app. Let's do some additional configuration of Interface Builder to fix that.

Configuring Xcode

Even though you have configured Simulator to use iPhone SE (3rd generation) for your app, the scenes shown in Interface Builder are for a different iPhone model. You may also wish to hide the minimap display. Let's configure the scenes in Interface Builder to use iPhone SE (3rd generation) and hide the minimap display. Follow these steps:

1. The **Main** storyboard file should still be selected. To configure the appearance of the scenes in Interface Builder, click the device configuration button:

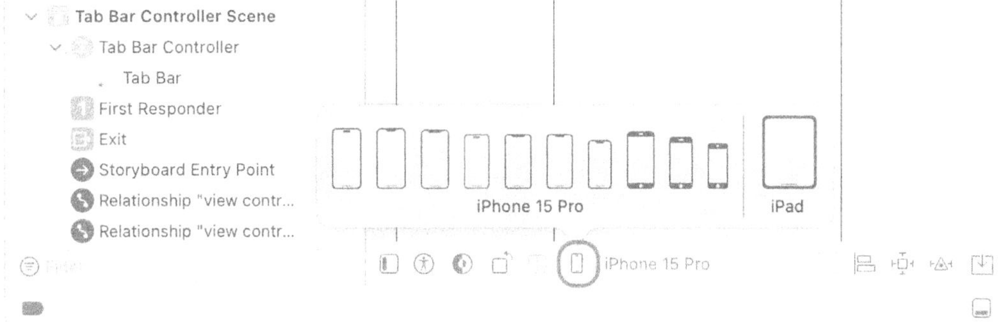

Figure 10.34: Editing area with the device configuration button shown

A pop-up window displaying different device screens will appear.

2. Choose **iPhone SE (3rd generation)** from this pop-up window, and click anywhere in the Editor area to dismiss it:

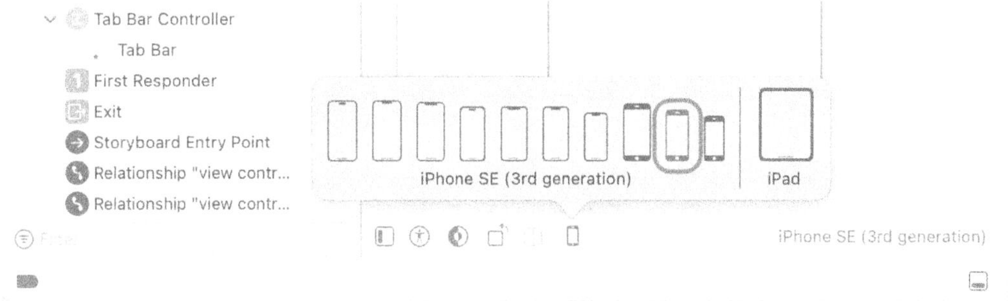

Figure 10.35: Device pop-up window with iPhone SE (3rd generation) selected

The appearance of the scenes in the storyboard will change to reflect the iPhone SE (3rd generation)'s screen.

3. If you wish to hide the minimap, choose **Minimap** from the **Editor** menu to deselect it.

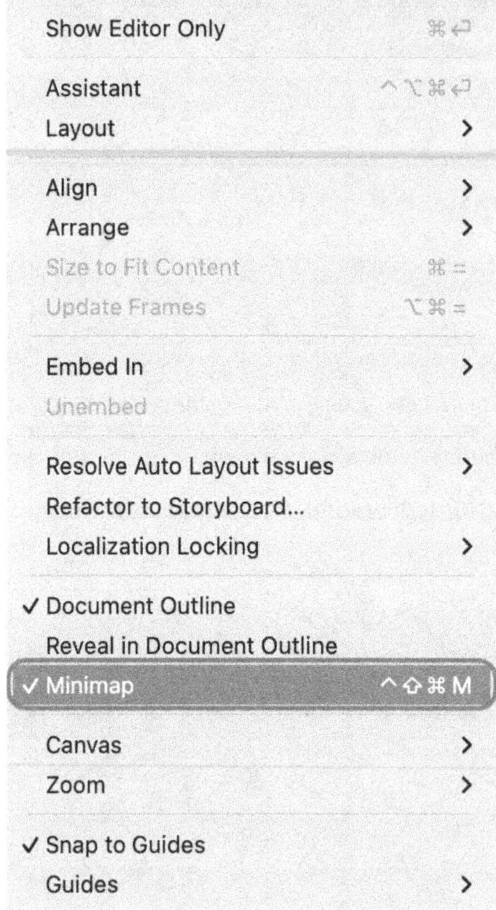

Figure 10.36: Editor menu with Minimap highlighted

4. Verify that you have the following scenes in the `Main` storyboard file:

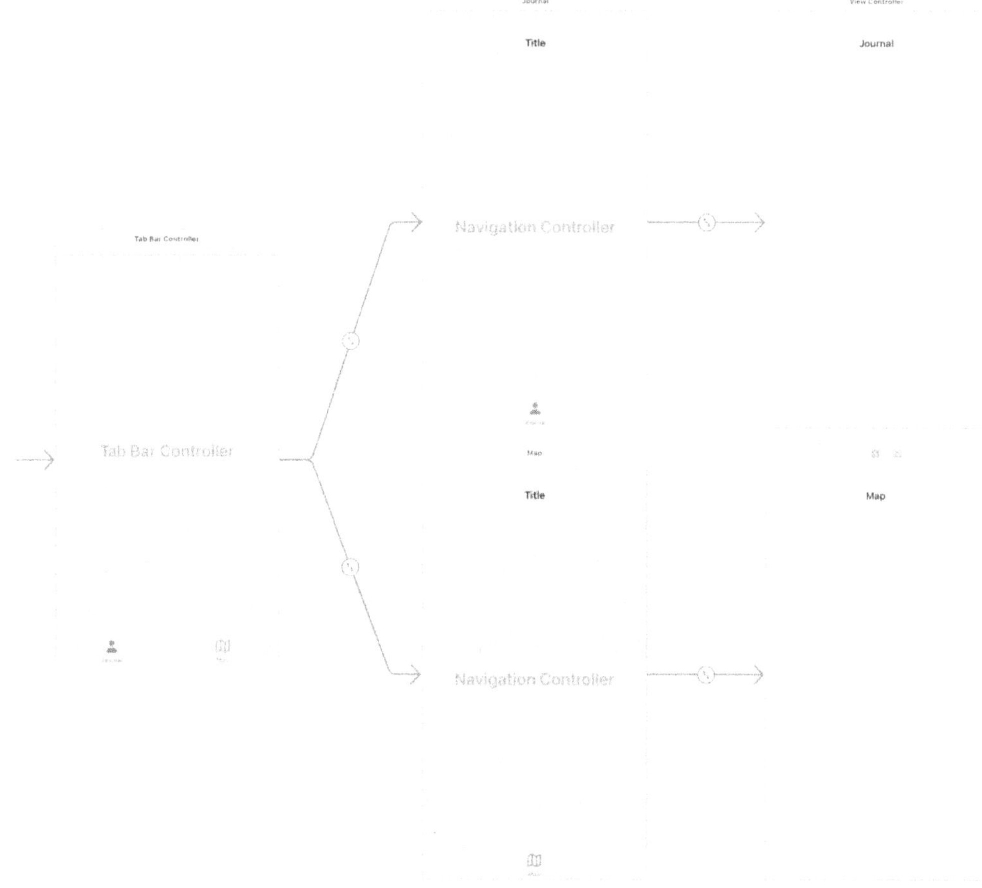

Figure 10.37: Editor area showing the completed Main storyboard file

5. Build and run your app. It should work just as it did before.

You have created the Journal List and Map screens for your app! Well done!

Summary

In this chapter, you learned some useful terms used in iOS app development. This will make it easier for you to understand the remainder of this book, as well as other books or online resources on the subject.

Then, you also learned about the different screens used in the *JRNL* app and how a user could use the app. As you recreate the app's UI from scratch, you're able to compare what you're doing to what the actual app looks like.

Finally, you learned how to use Interface Builder and storyboards to add a tab bar controller scene to your app, configure the button titles and icons, and add navigation controllers for the Journal List and Map screens. This will familiarize you with adding and configuring UI elements for your own apps.

In the next chapter, you will continue setting up your app's UI and become familiar with more UI elements. You will add and configure the remaining screens for your app.

Join us on Discord!

Read this book alongside other users, experts, and the author himself. Ask questions, provide solutions to other readers, chat with the author via Ask Me Anything sessions, and much more. Scan the QR code or visit the link to join the community.

`https://packt.link/ios-Swift`

11

Building Your User Interface

In the previous chapter, you modified an existing Xcode project, added a tab bar to your app that allowed the user to select between the Journal List and Map screens, and configured the tab bar button titles and icons. When your app is launched, the Journal List screen is displayed, but it is currently blank.

As you saw in the app tour in *Chapter 10, Setting Up the User Interface*, the Journal List screen should display a table view showing a list of journal entries in table view cells.

In this chapter, you will make the Journal List screen display a table view containing 10 empty table view cells, as well as a button that will display a view representing the Add New Journal Entry screen when tapped. You'll also configure a **Cancel** button to dismiss this view and return you to the Journal List screen.

You'll be adding a small amount of code to your app, but don't worry too much about this—you'll learn more about it in the next part of this book.

By the end of this chapter, you'll have learned how to add view controllers to a storyboard scene, link outlets in view controllers to scenes, set up table view cells, and present a view controller modally. This will be very useful when you're designing the user interface for your own apps.

The following topics will be covered in this chapter:

- Adding a table view to the Journal List screen
- Connecting storyboard elements to the view controller
- Configuring data source methods for the table view
- Presenting a view modally

Technical requirements

You will continue working on the `JRNL` Xcode project that you created in the previous chapter.

The completed Xcode project for this chapter is in the `Chapter11` folder of the code bundle for this book, which can be downloaded here:

`https://github.com/PacktPublishing/iOS-18-Programming-for-Beginners-Ninth-Edition`

Check out the following video to see the code in action:

https://youtu.be/EsDaVgrGLus

Let's start by adding a table view to the Journal List screen, which will eventually display the list of journal entries.

Adding a table view to the Journal List screen

As you saw in the app tour, the *JRNL* app displays journal entries in a table view. A table view is an instance of the `UITableView` class. It displays a column of cells. Each cell in a table view is a table view cell, which is an instance of the `UITableViewCell` class. In this section, you'll start by adding a table view to the view controller scene for the Journal List screen in the `Main` storyboard file, then you'll add Auto Layout constraints to make it fill the screen.

For more information on Auto Layout and how to use it, go to `https://developer.apple.com/library/archive/documentation/UserExperience/Conceptual/AutolayoutPG/`.

Open the *JRNL* project you created in the previous chapter and run the app to make sure everything still works as it should, then follow these steps:

1. Click the **Main** storyboard file in the Project navigator, select the view controller scene representing the Journal List screen, and click the Library button:

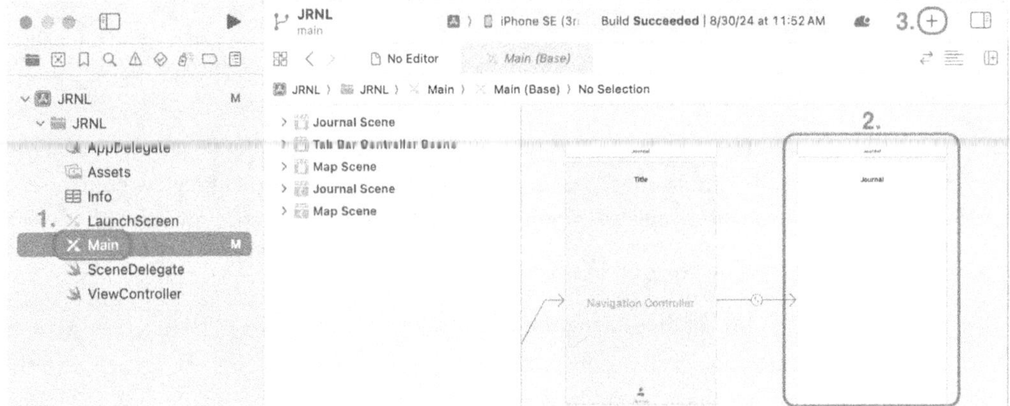

Figure 11.1: Toolbar with the Library button shown

2. The library will appear. Type `table` in the filter field. A **Table View** object will appear as one of the results. Drag it to the middle of the view of the view controller scene for the Journal List screen:

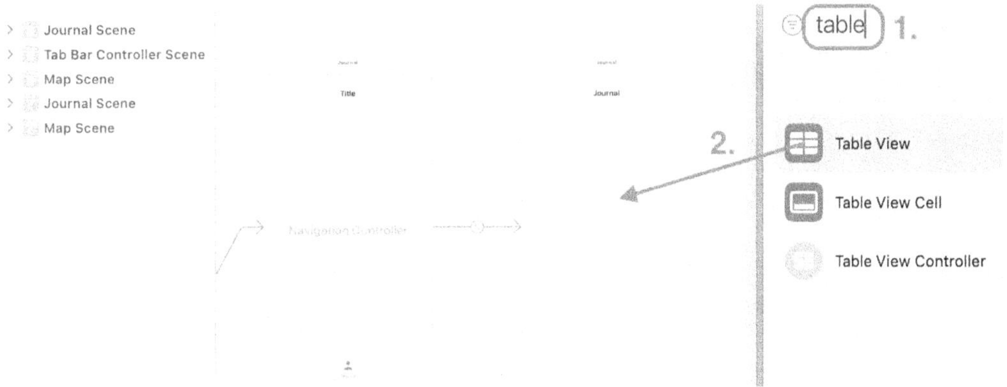

Figure 11.2: Library with Table View object selected

The table view has been added, but it only takes up a small part of the screen. As shown in the app tour in the previous chapter, it should fill the screen.

3. You will use the **Auto Layout Add New Constraints** button to bind the edges of the table view to the edges of its enclosing view. Make sure the table view is selected and click the Auto Layout Add New Constraints button:

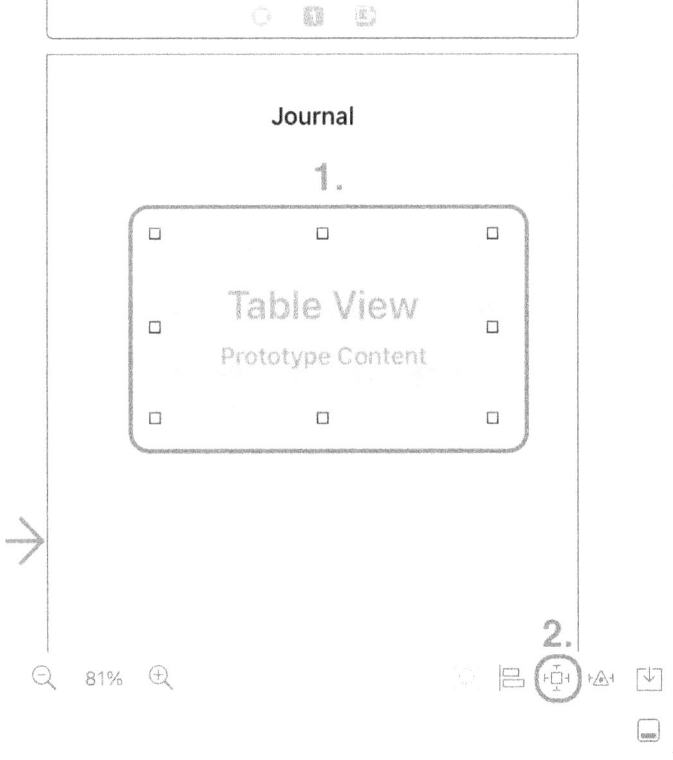

Figure 11.3: View controller scene with table view selected

4. Type `0` in the top, left, right, and bottom edge constraint fields and click all the pale red struts. Make sure all the struts have turned bright red. Click the **Add 4 Constraints** button:

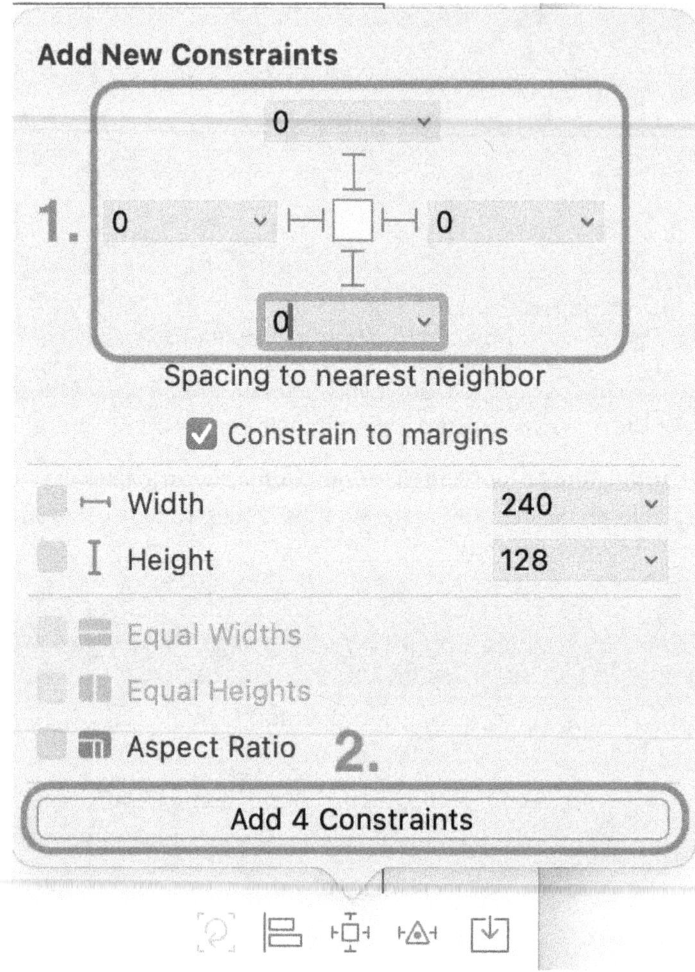

Figure 11.4: Auto Layout pop-up dialog box for adding new constraints

This sets the space between the edges of the table view and the edges of the enclosing view to 0, binding the table view's edges to those of the enclosing view. Now the table view will fill the screen, regardless of the device and orientation.

5. Verify that all four sides of the table view are now bound to the edges of the screen, as shown in the following screenshot:

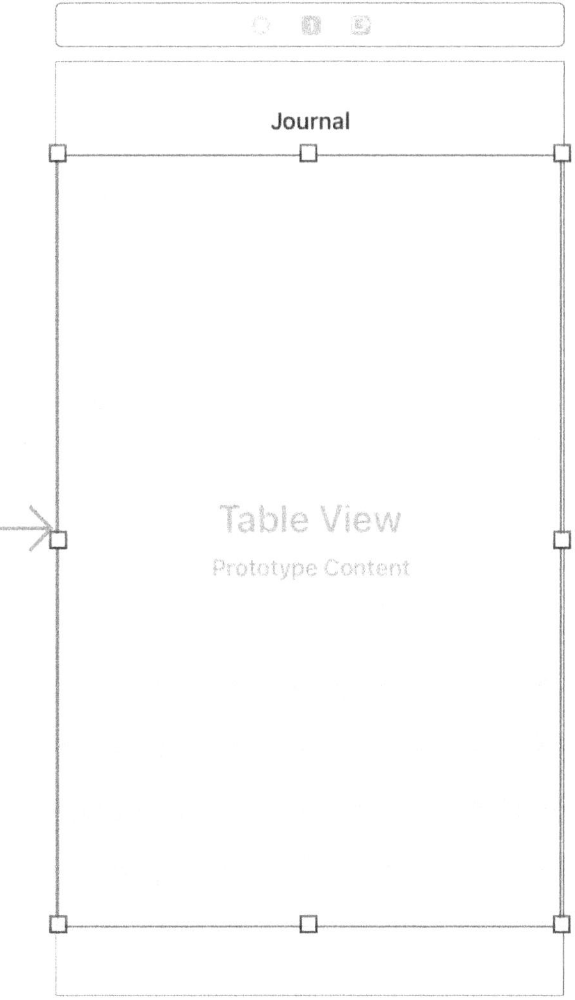

Figure 11.5: View controller scene with table view filling the screen

You have added a table view to the view of the view controller scene for the Journal List screen and used Auto Layout constraints to make it fill the screen, but the Journal List screen will still be blank when you build and run your app.

In the next section, you will implement the code for the `JournalListViewController` class, and you'll connect outlets in this class to the UI elements on the Journal List screen. This will enable an instance of the `JournalListViewController` class to control what is displayed by the Journal List screen.

Connecting storyboard elements to the view controller

You've added a table view to the Journal List screen, but it does not display anything yet. You'll need to modify the existing view controller to manage the table view in the Journal List screen. The `ViewController` file was automatically created by Xcode when you created the `JRNL` project.

It contains the declaration and definition of a `UIViewController` subclass named `ViewController`, and this class is currently set as the view controller for the Journal List screen. You'll change the name of the class in the `ViewController` file to `JournalListViewController` and create an outlet for the table view that you added to the view controller scene earlier. Follow these steps:

1. Click the **ViewController** file in the Project navigator. In the Editor area, right-click the class name and choose **Refactor** | **Rename...**

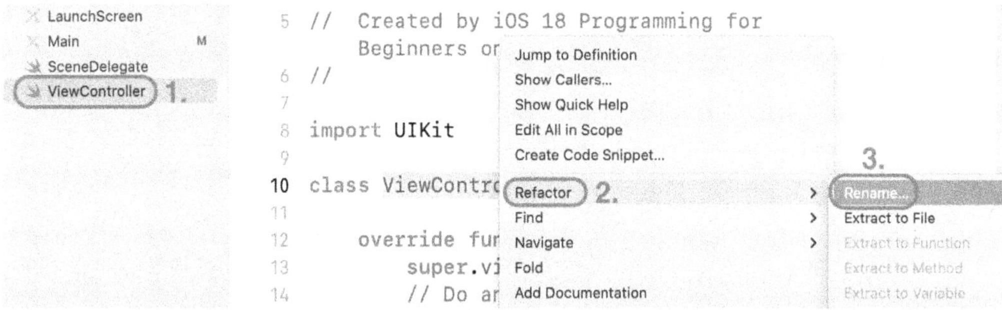

Figure 11.6: Editor area showing the pop-up menu with Rename...highlighted

2. Change the class name to `JournalListViewController` and click **Rename**:

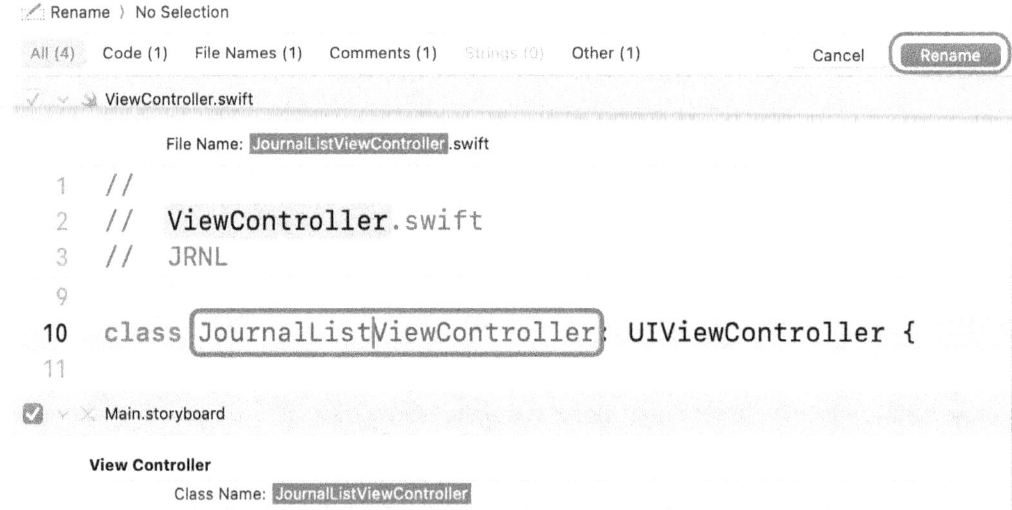

Figure 11.7: Editor area showing the new name for the ViewController class

3. Verify both the class name and the file name have been changed to **JournalListViewController**:

Figure 11.8: File name and class name both changed to JournalListViewController

4. Click on the **Main** storyboard file in the Project navigator and select the first **Journal Scene** (the one containing the table view) in the document outline.

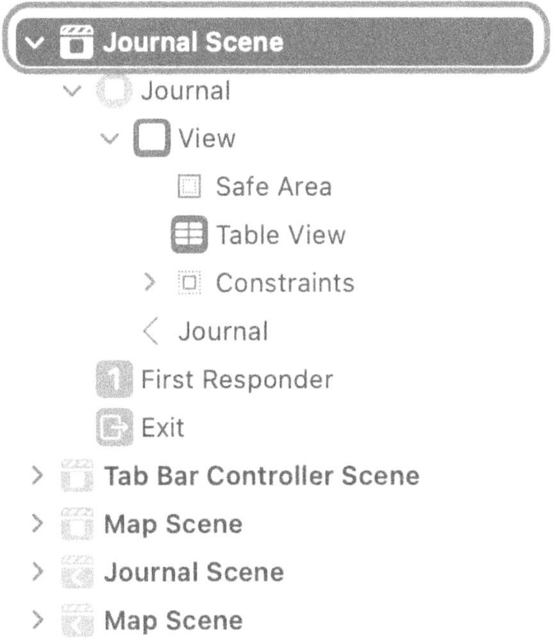

Figure 11.9: Document outline showing the first Journal Scene selected

5. Click on the Identity inspector button and verify that, under **Custom Class**, **Class** is set to **JournalListViewController**:

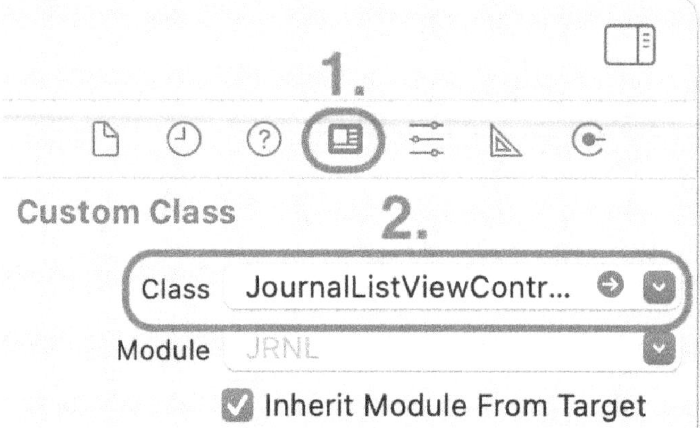

Figure 11.10: Identity inspector with Class set to JournalListViewController

This means that the content of the Journal List screen is being managed by an instance of the `JournalListViewController` class.

6. Click the Navigator and Inspector buttons to hide the Navigator and Inspector areas so you have more room to work:

Figure 11.11: Toolbar showing Navigator and Inspector buttons

7. Click the Adjust Editor Options button and choose **Assistant** from the pop-up menu:

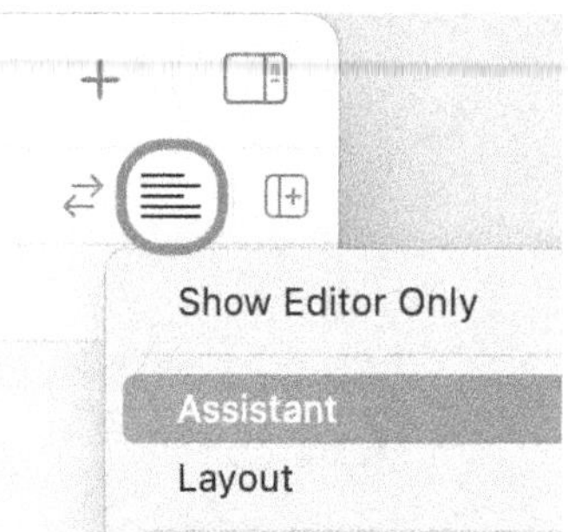

Figure 11.12: Adjust Editor Options menu with Assistant selected

This will display any Swift files associated with this scene in an assistant editor. As you can see, the **Main** storyboard file's content appears on the left side and the `JournalListViewController` class definition appears on the right side of the Editor area.

8. Look at the bar just above the code and verify that **JournalListViewController.swift** is selected:

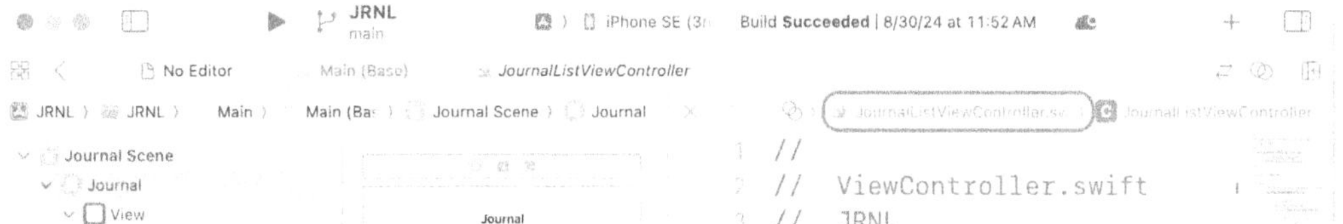

Figure 11.13: Bar showing JournalListViewController.swift selected

If you don't see **JournalListViewController.swift** selected, click the bar and select **JournalList-ViewController.swift** from the pop-up menu.

9. To connect the table view in the Journal scene to an outlet in the `JournalListViewController` class, *Ctrl* + *Drag* from the table view to the `JournalListViewController` file, just below the class name declaration:

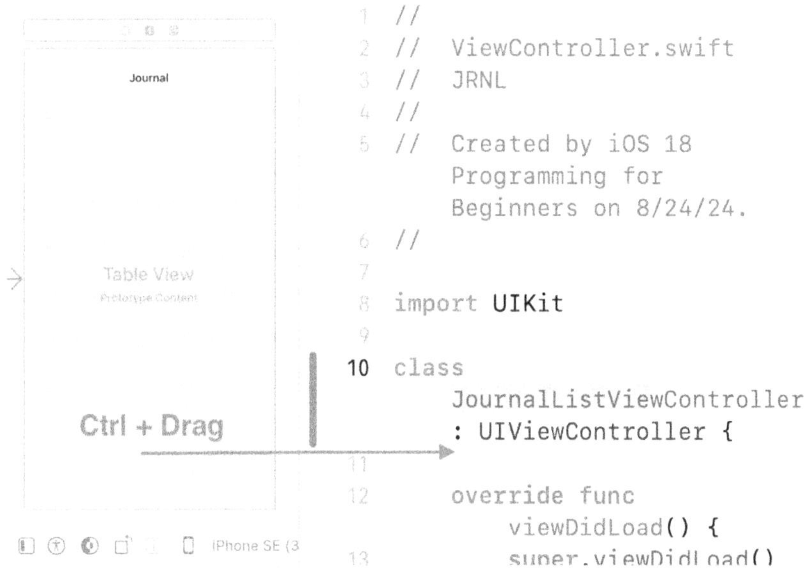

Figure 11.14: Editor area showing drag destination

You can also drag from the table view in the document outline.

10. A small pop-up dialog box will appear. Type the name of the outlet, `tableView`, into the **Name** text field, set **Storage** to **Strong**, and click **Connect**:

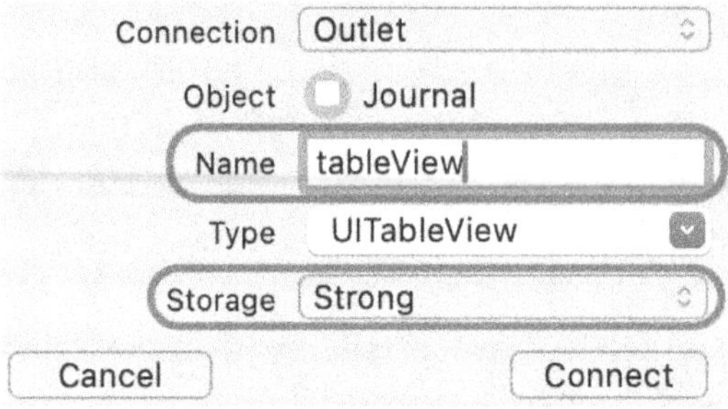

Figure 11.15: Pop-up dialog box for outlet creation

11. Verify that the `tableView` outlet declaration has been automatically added to the `JournalListViewController` class. After you have done so, click the **x** to close the assistant editor window:

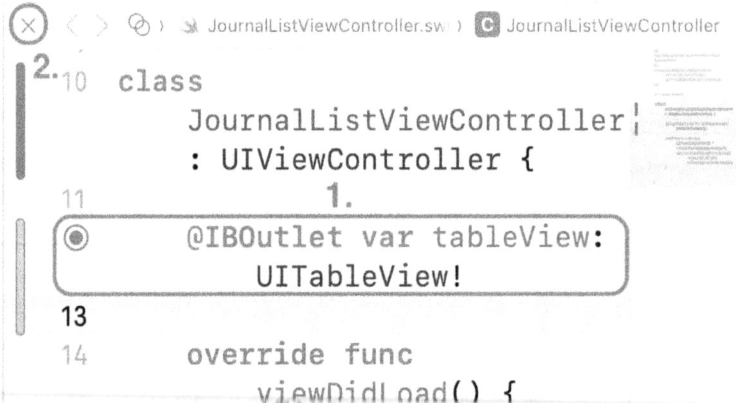

Figure 11.16: Editor area showing tableView outlet

The `JournalListViewController` class now has an outlet, `tableView`, for the table view in the Journal List screen. This means a `JournalListViewController` instance can manage what the table view displays.

It is common to make mistakes when using *Ctrl* + *Drag* to drag from an element in a storyboard scene to a file. If you make a mistake while doing so, this may cause a crash to occur when the app is launched. To check whether there are any errors in the connection between the table view and `JournalListViewController` class, follow these steps:

1. Click the Navigator and Inspector buttons to display the Navigator and Inspector areas.
2. With **Journal** in the **Journal Scene** selected, click the Connections inspector button:

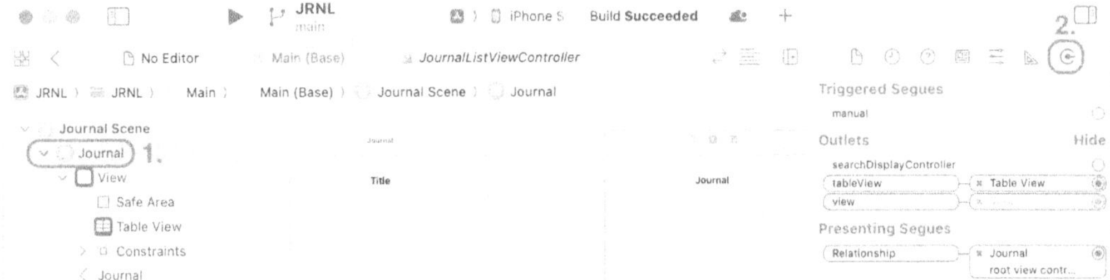

Figure 11.17: Connections inspector selected

The Connections inspector displays the links between your UI objects and your code. You will see the **tableView** outlet connected to the table view in the **Outlets** section.

3. If you see a tiny yellow warning icon, click on the **x** to break the connection:

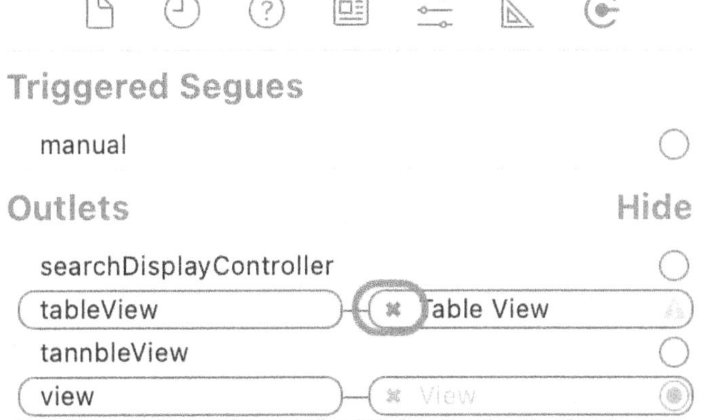

Figure 11.18: Connections inspector showing tableView outlet with yellow warning icon

4. Under **Outlets**, drag from the **tableView** outlet to the table view to re-establish the connection:

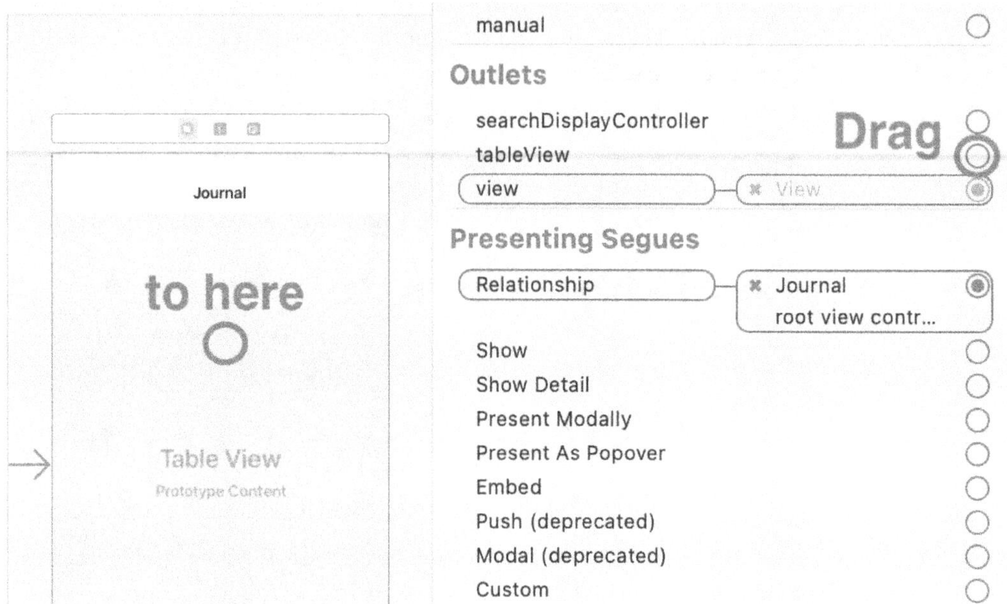

Figure 11.19: Editor area showing table view to be connected

If you need to change the name of an outlet in your code after it has been created, right-click the outlet name and choose **Refactor** | **Rename** from the pop-up menu instead of changing it manually to avoid errors.

You've successfully created an outlet in the `JournalListViewController` class for the table view. Great job!

To display table view cells onscreen, you will need to implement data source methods for the table view by adding some code to the `JournalListViewController` class. You will do this in the next section.

Configuring data source methods for the table view

When your app is running, an instance of the `JournalListViewController` class acts as the view controller for the Journal List screen. It is responsible for loading and displaying all the views on that screen, including the table view you added earlier. The table view needs to know how many table view cells to display and what to display in each cell. Normally, the view controller is responsible for providing this information. Apple has created a protocol, `UITableViewDataSource`, for this purpose. All you need to do is set the table view's `dataSource` property to the `JournalListViewController` class and implement the required methods of this protocol.

The table view also needs to know what to do if the user taps on a table view cell. Again, the view controller for the table view is responsible, and Apple has created the `UITableViewDelegate` protocol for this purpose. You will set the table view's `delegate` property to the `JournalListViewController` class, but you won't be implementing any methods from this protocol yet.

Protocols are covered in *Chapter 8, Protocols, Extensions, and Error Handling*.

You will need to type in a small amount of code in this chapter. Don't worry about what it means; you'll learn more about table view controllers and their associated protocols in *Part 3* of this book.

In the next section, you'll use the Connections inspector to assign the table view's `dataSource` and `delegate` properties to outlets in the `JournalListViewController` class.

Setting the delegate and data source properties of the table view

An instance of the `JournalListViewController` class will provide the data that the table view will display, as well as the methods that will be executed when the user interacts with the table view. To make this work, you'll connect the table view's `dataSource` and `delegate` properties to outlets in the `JournalListViewController` class. Follow these steps:

1. Click the Navigator and Inspector buttons to display the Navigator and Inspector areas again if you haven't done so already.
2. The **Main** storyboard file should still be selected. Select the **Table View** for the **Journal Scene** in the document outline and click the Connections inspector button. In the **Outlets** section, you will see two empty circles next to the **dataSource** and **delegate** outlets. Drag from each empty circle to the **Journal** icon in the document outline:

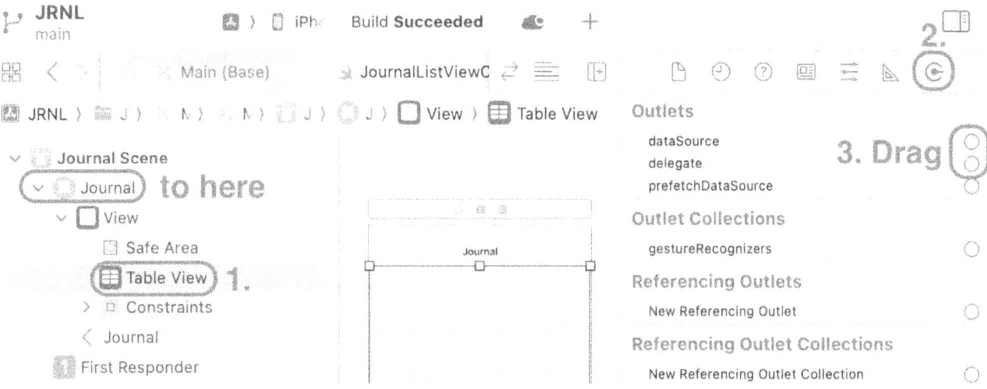

Figure 11.20: Connections inspector showing the dataSource and delegate outlets

3. Verify that the `dataSource` and `delegate` properties of the table view have been connected to outlets in the `JournalListViewController` class:

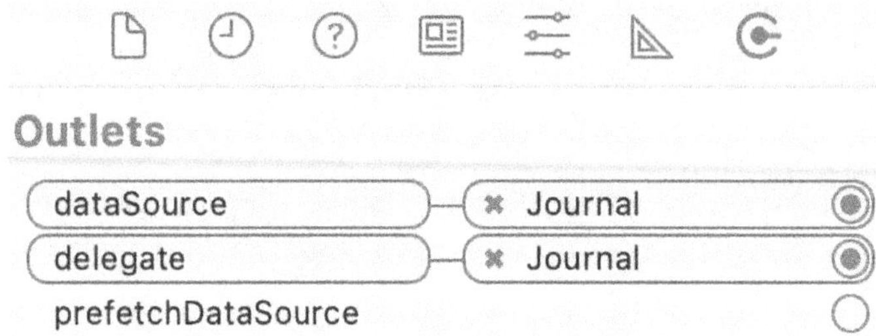

Figure 11.21: Connections inspector with the dataSource and delegate outlets set

In the next section, you will add some code to make the `JournalListViewController` class conform to the `UITableViewDataSource` protocol, and configure the table view to display 10 table view cells when you run your app.

Adopting the UITableViewDataSource and UITableViewDelegate protocols

So far, you've made the `JournalListViewController` class the data source and delegate for the table view. The next step is to make it adopt the `UITableViewDataSource` and `UITableViewDelegate` protocols and implement any required methods. You'll also change the color of the table view cells to make them visible onscreen. Follow these steps:

1. Click **Table View** in the document outline and click the Attributes inspector button. Under **Table View**, change the number of **Prototype Cells** to `1`:

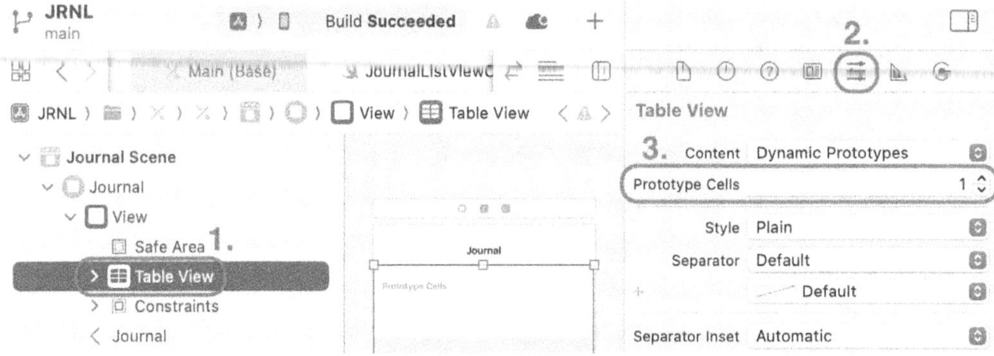

Figure 11.22: Attributes inspector showing Prototype Cells set to 1

2. Click the > button next to **Table View** in the document outline to display **Table View Cell**:

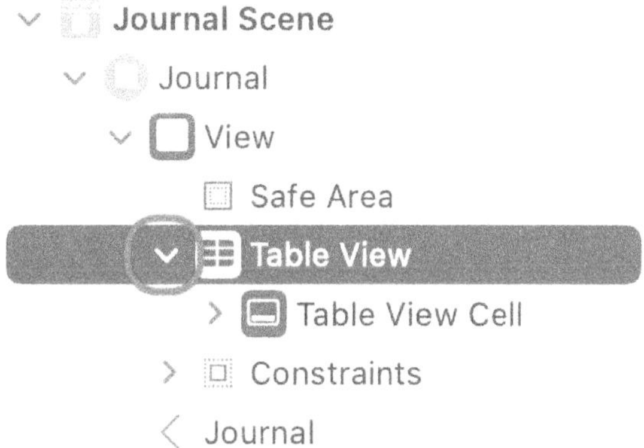

Figure 11.23: Document outline showing > button

This represents the table view cells that the table view will display.

3. Click **Table View Cell** in the document outline. In the **Attributes inspector** under **Table View Cell**, set **Identifier** to `journalCell` and press *Return*:

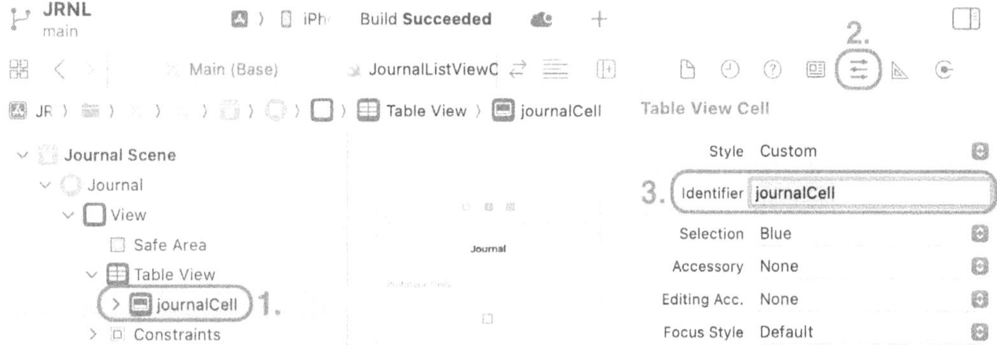

Figure 11.24: Attributes inspector with Identifier set

The name **Table View Cell** in the document outline will change to **journalCell**.

4. In the Attributes inspector under **View**, set **Background** to **System Cyan Color** so that you can see the table view cells easily when you run the app:

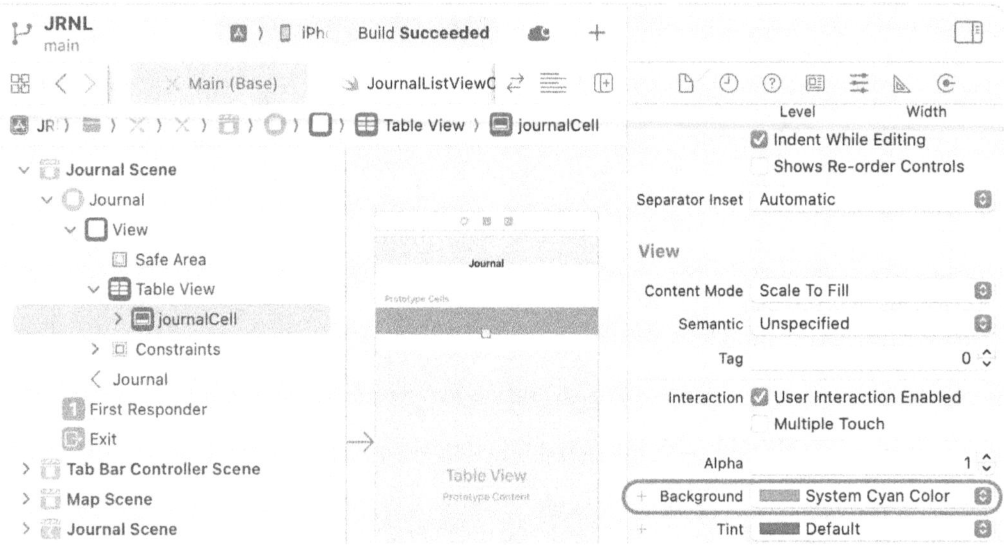

Figure 11.25: Attributes inspector with table view cell background color set

5. Click on the **JournalListViewController** file in the Project navigator. Type in the following code after the class declaration to make the `JournalListViewController` class adopt the `UITableViewDataSource` and `UITableViewDelegate` protocols:

```
class JournalListViewController: UIViewController, UITableViewDataSource,
UITableViewDelegate {
```

After a few seconds, an error will appear:

```
import UIKit

class JournalListViewController:
    UIViewController,
    UITableViewDataSource,
    UITableViewDelegate {

    @IBOutlet var tableView: UITableView!
```

Figure 11.26: Editor area showing error

6. Click on it to display an error message. The error message says **Type 'JournalListViewController' does not conform to protocol 'UITableViewDataSource'. Add stubs for conformance**:

```
class JournalListViewController:
    UIViewController,
    UITableViewDataSource,
    UITableViewDelegate {
```

Type 'JournalListViewController' does not conform to protocol 'UITableViewDataSource'

Add stubs for conformance   Fix

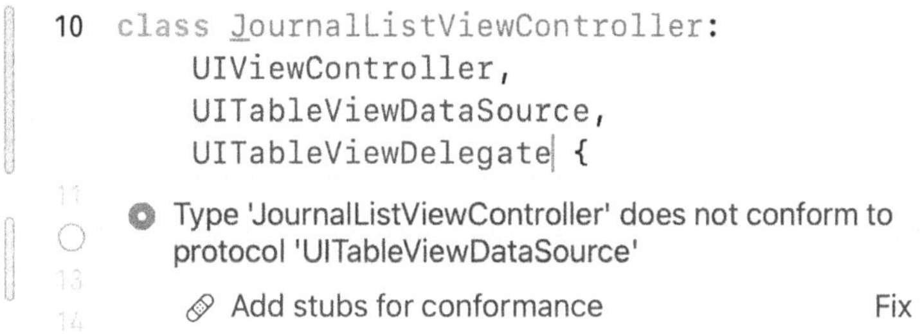

Figure 11.27: Editor area showing an error message

This means you need to implement the required methods for the `UITableViewDataSource` protocol to make `JournalListViewController` conform to it.

7. Click **Fix** to automatically add stubs for the required methods into the `JournalListViewController` class.
8. Verify the stubs for the two required methods for the `UITableViewDataSource` protocol have been automatically inserted into the `JournalListViewController` class, as shown here:

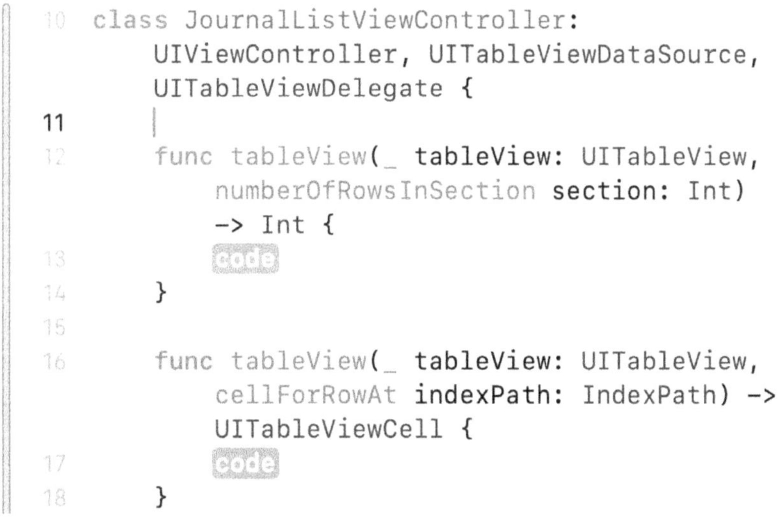

```
class JournalListViewController:
    UIViewController, UITableViewDataSource,
    UITableViewDelegate {

    func tableView(_ tableView: UITableView,
        numberOfRowsInSection section: Int)
        -> Int {
        code
    }

    func tableView(_ tableView: UITableView,
        cellForRowAt indexPath: IndexPath) ->
        UITableViewCell {
        code
    }
```

Figure 11.28: Editor area showing UITableViewDataSource method stubs

The first method tells the table view how many cells to display, while the second method tells the table view what to display in each table view cell.

9. Replace the placeholder text in the first method with `10` (the `return` keyword is optional if it's just a single line of code). This tells the table view to display 10 cells:

```
    func tableView(_ tableView: UITableView,
        numberOfRowsInSection section: Int)
        -> Int {
        10
    }
```

Figure 11.29: Editor area showing code to display 10 table view cells

10. Replace the placeholder text in the second method with the following code:

```
tableView.dequeueReusableCell(withIdentifier: "journalCell", for:
indexPath)
```

```
    func tableView(_ tableView: UITableView,
        cellForRowAt indexPath: IndexPath) ->
        UITableViewCell {
        tableView
            .dequeueReusableCell
            (withIdentifier: "journalCell",
            for: indexPath)
    }
```

Figure 11.30: Editor area showing code to display a table view cell for each row

Don't worry about what this means for now as you'll learn more about table views in *Part 3*

11. Build and run your app. Simulator will display a column of 10 cyan table view cells, as shown here:

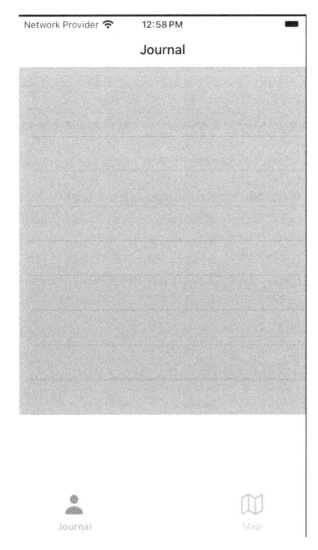

Figure 11.31: Simulator showing 10 table view cells

As you saw in the app tour in *Chapter 10, Setting Up the User Interface*, there should be a + button at the top right of this screen. You will add this button in the next section.

Presenting a view modally

The navigation bar for the Journal List screen can be configured to display a title and buttons. You have already configured the title in *Chapter 10, Setting Up the User Interface*. Now you will add and configure a bar button item to the navigation bar. When tapped, this button will display a view representing the Add New Journal Entry screen. This view will be from a new view controller scene embedded in a navigation controller, which you will add to the project. The view will be presented modally, which means you won't be able to do anything else until it is dismissed.

To dismiss it, you'll add a **Cancel** button to the view's navigation bar. You'll also add a **Save** button, but you'll only implement its functionality in *Chapter 16, Passing Data between View Controllers*. Let's start by adding a bar button item from the library to the navigation bar in the next section.

Adding a bar button to the navigation bar

As shown in the app tour in *Chapter 10, Setting Up the User Interface*, there is a + button in the top right-hand corner of the screen. To implement this, you'll add a bar button item to the Journal List screen's navigation bar. Follow these steps:

1. Click the **Main** storyboard file in the Project navigator. Make sure the first **Journal Scene** is selected in the document outline. Click the Library button to display the library:

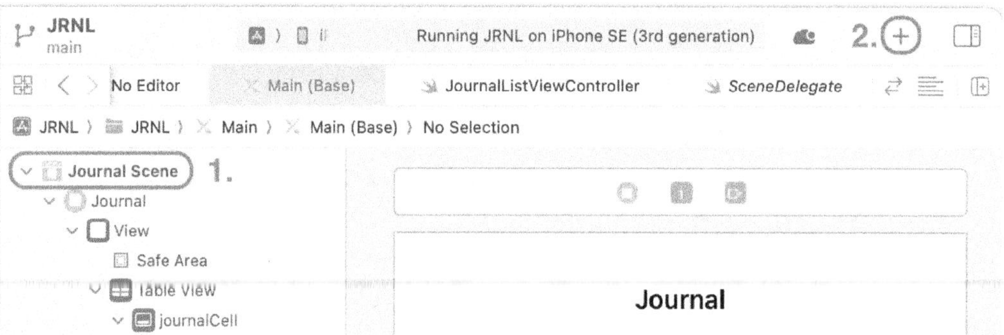

Figure 11.32: Toolbar with Library button shown

2. Type `bar b` in the filter field. A **Bar Button Item** object will appear in the results. Drag the bar button object to the right side of the navigation bar:

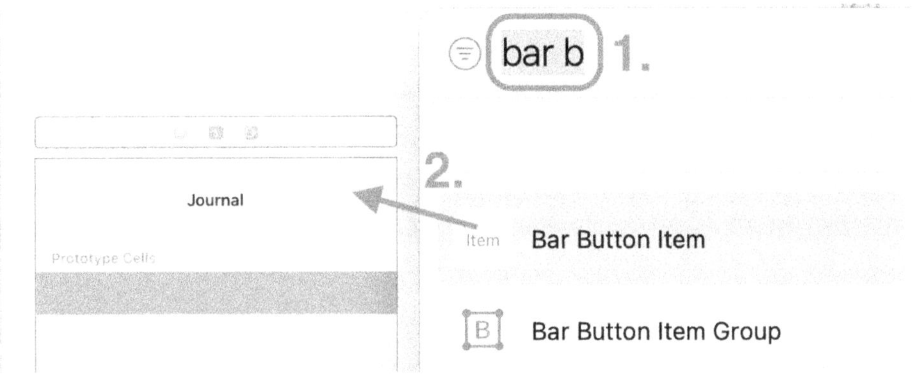

Figure 11.33: Library with Bar Button Item object selected

3. With the bar button selected, click the Attributes inspector button. Under **Bar Button Item**, set **System Item** to **Add**:

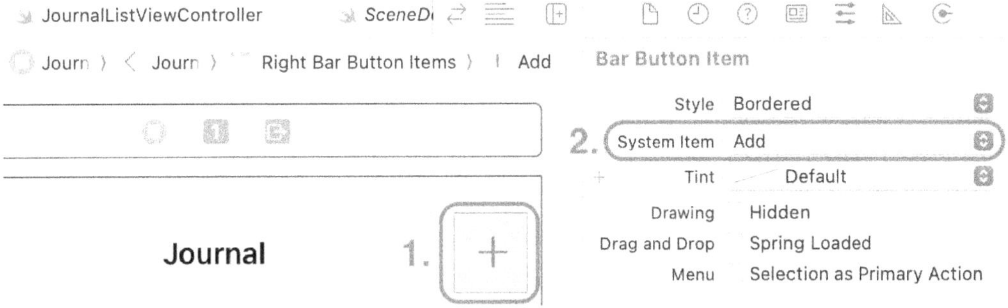

Figure 11.34: Attributes inspector with System Item set to Add

You now have a + button in your navigation bar. In the next section, you will add a view controller scene to represent the Add New Journal Entry screen that will appear when the button is tapped.

Adding a new view controller scene

As shown in the app tour in *Chapter 10, Setting Up the User Interface*, when you tap the + button in the navigation bar, the Add New Journal Entry screen will be displayed. You'll add a new view controller scene to your project to represent this screen. Follow these steps:

1. Click the Library button to display the library and type `view con` in the filter field. A **View Controller** object will be among the search results. Drag the **View Controller** object onto the storyboard:

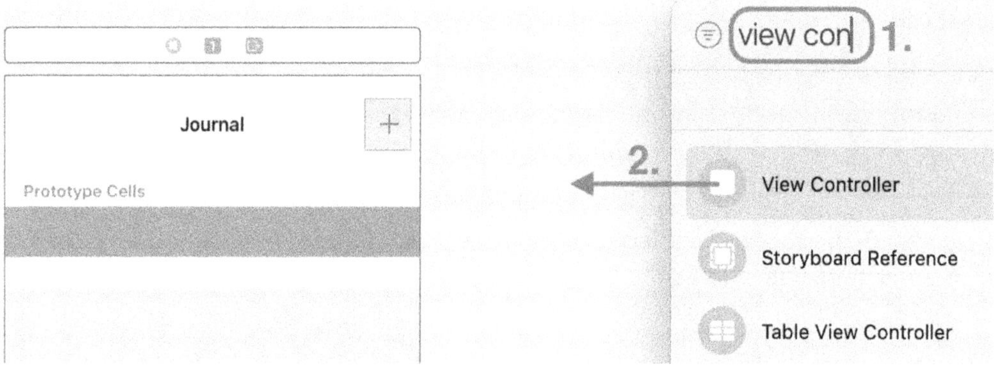

Figure 11.35: Library with the View Controller object selected

2. Position the view controller to the right of the **Journal** scene:

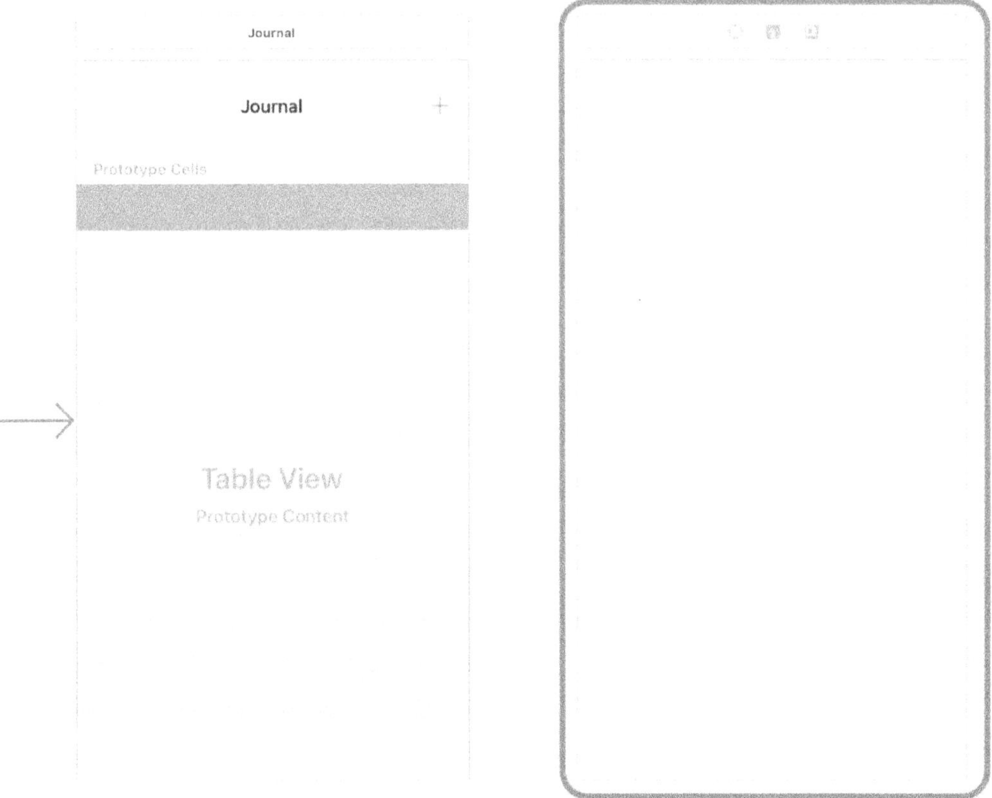

Figure 11.36: Editor area showing view controller scene next to Journal scene

3. Select the newly added view controller scene. In the document outline, click on the **View Controller** icon for this scene:

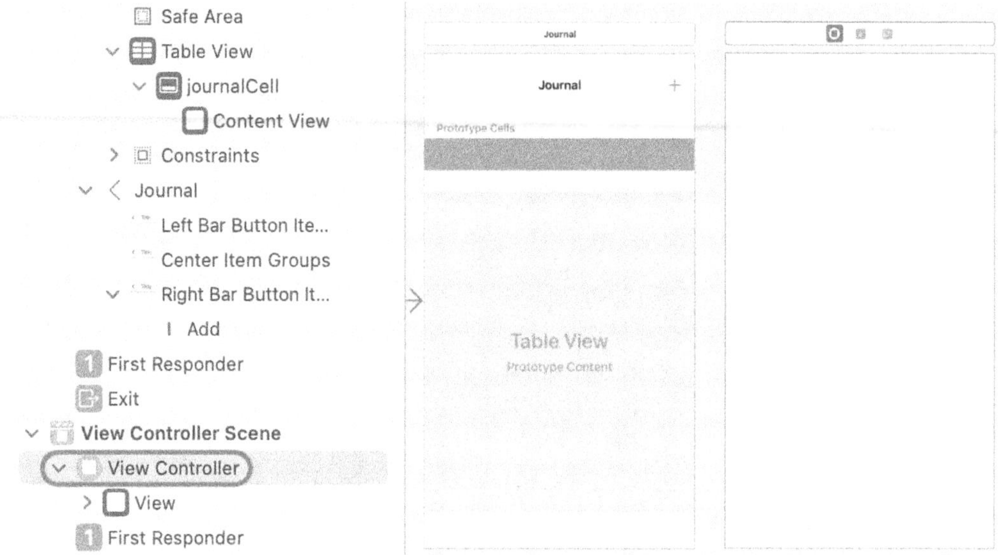

Figure 11.37: Document outline with View Controller selected

4. You will need space for the **Cancel** and **Save** buttons, so you will embed this view controller scene in a navigation controller to provide a navigation bar where the buttons can be placed. Choose **Embed In | Navigation Controller** from the **Editor** menu.
5. Verify that a navigation controller scene has appeared to the left of the view controller scene:

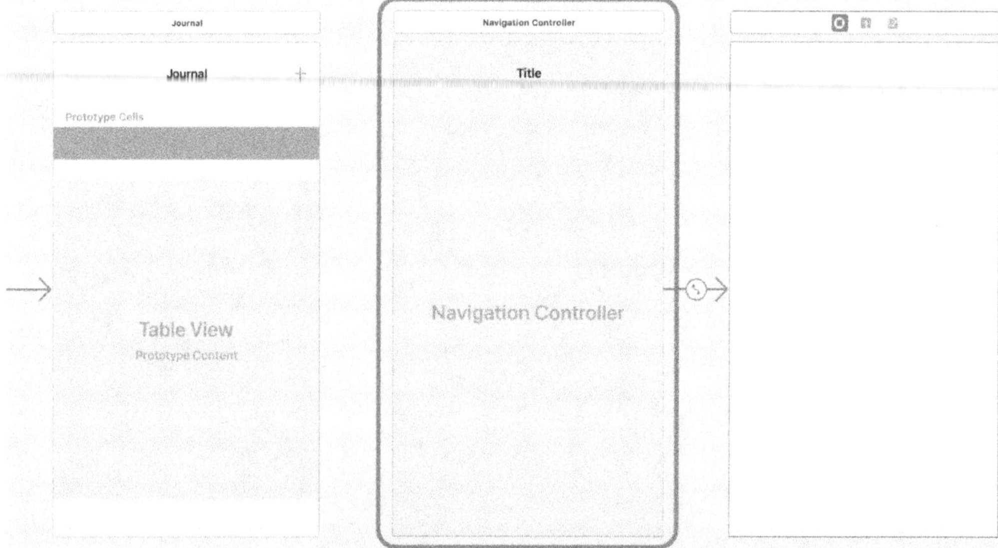

Figure 11.38: Editor area showing view controller scene embedded in a navigation controller

6. Click the **Navigation Item** for the new view controller scene in the document outline. In the Attributes inspector, under **Navigation Item**, set **Title** to `New Entry`:

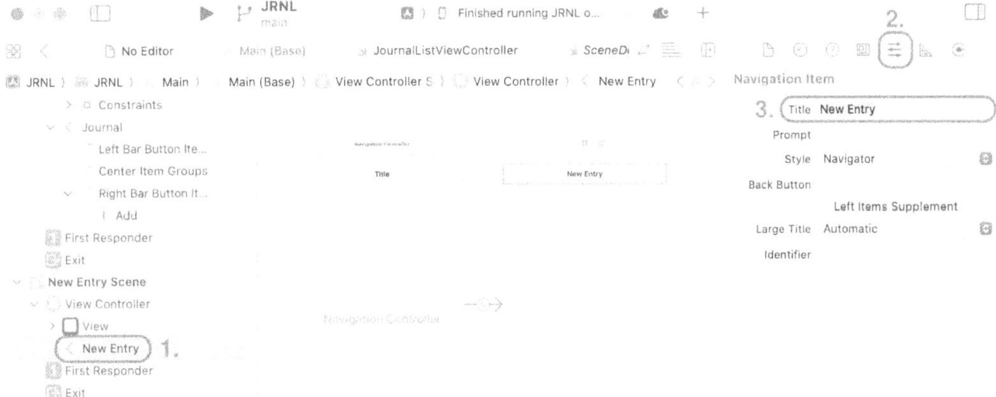

Figure 11.39: Attributes inspector with Title set to New Entry

The name of the navigation item will change to **New Entry**.

7. *Ctrl* + *Drag* from the button to the navigation controller scene:

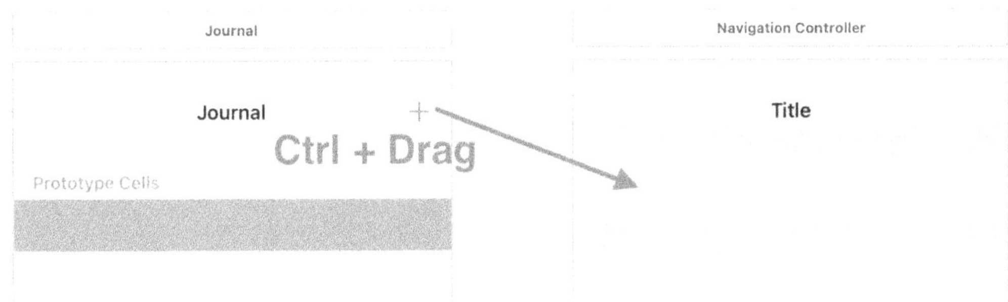

Figure 11.40: Editor area showing the drag destination

8. The segue pop-up menu will appear. Choose **Present Modally**:

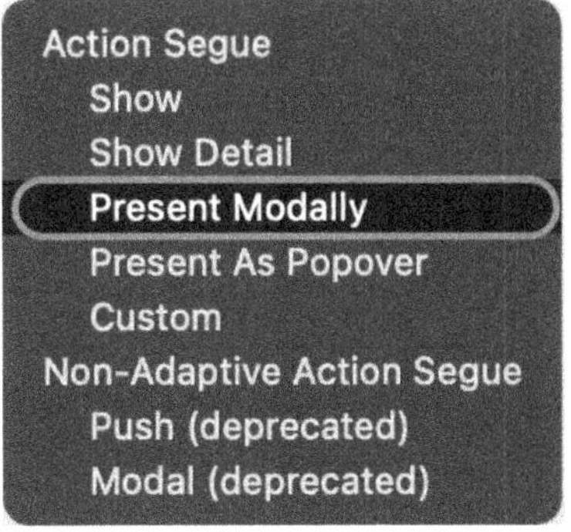

Figure 11.41: Segue pop-up menu with Present Modally selected

This makes the view controller's view slide up from the bottom of the screen when the button is tapped. You won't be able to interact with any other view until this view is dismissed.

9. Verify that a segue has linked the **Journal** scene and the **Navigation Controller** scene together:

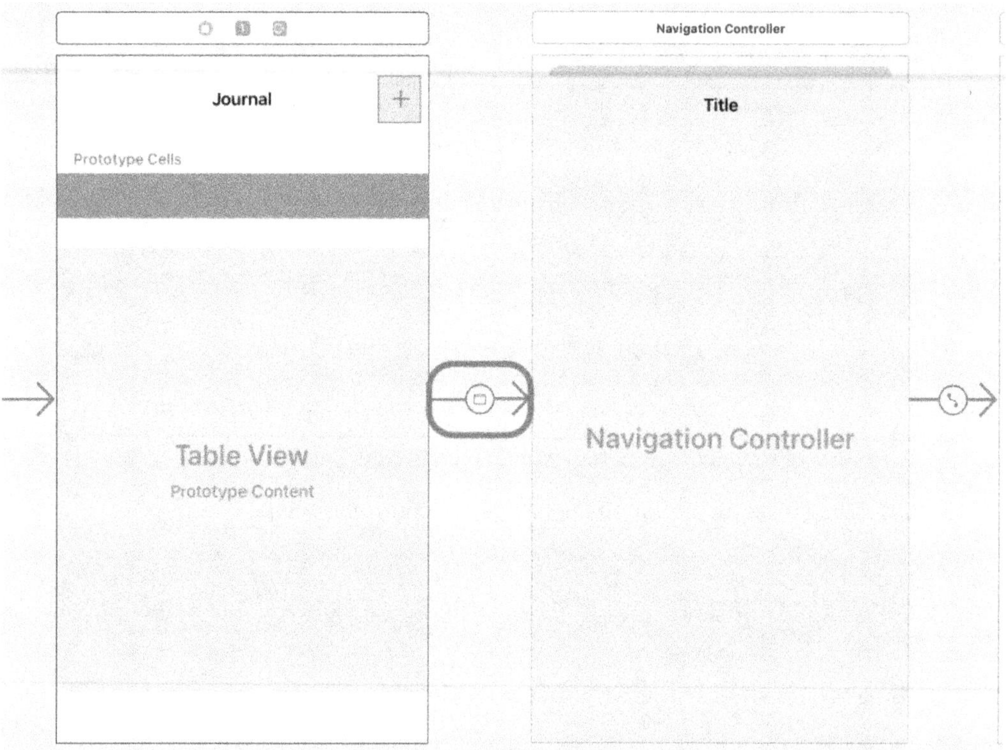

Figure 11.42: Editor area showing segue between the Journal scene and Navigation Controller scene

10. Build and run your app. Click the + button and the new view controller's view will slide up from the bottom of the screen:

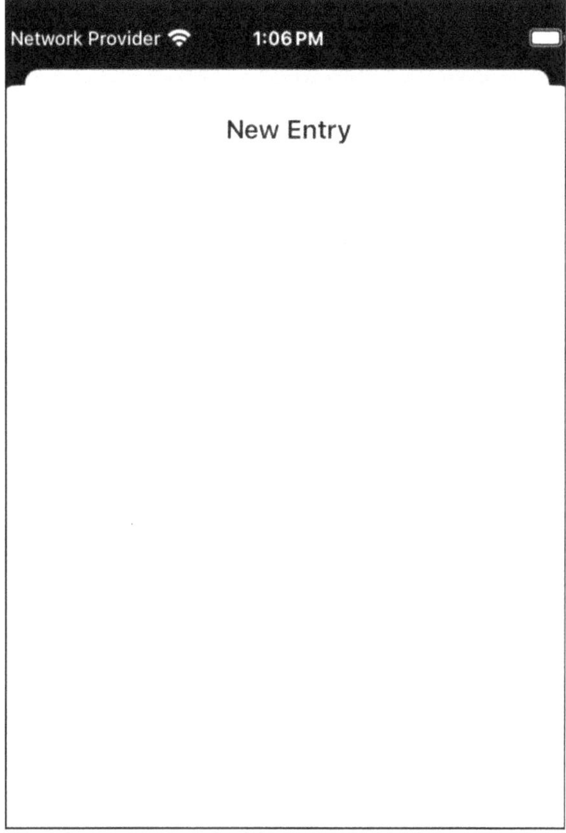

Figure 11.43: Simulator showing new view controller's view

You can only dismiss this view by dragging it downward at present. In the next section, you will add a **Cancel** button to the navigation bar and program it to dismiss the view. You'll also add a **Save** button, but you won't program it yet.

Adding Cancel and Save buttons to the navigation bar

As you have seen earlier, one of the benefits of embedding a view controller in a navigation controller is the navigation bar at the top of the screen. You can place buttons on its left and right sides. Follow these steps to add the **Cancel** and **Save** buttons to the navigation bar:

1. Click the **Navigation Item** for the **New Entry** scene in the document outline. Click the **Library** button:

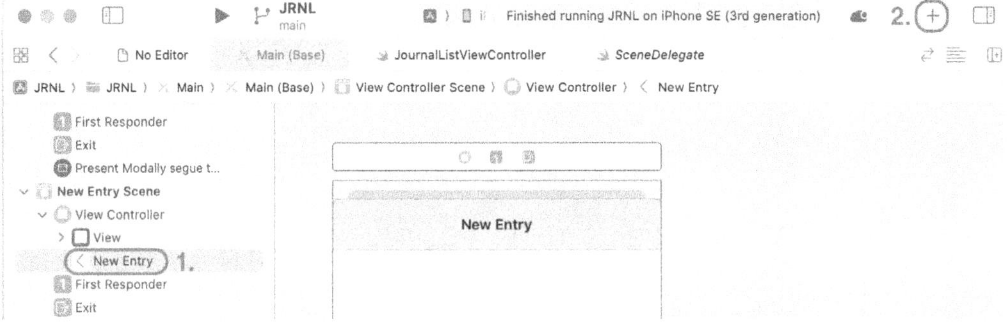

Figure 11.44: Toolbar with Library button shown

2. Type `bar b` into the filter field and drag a **Bar Button Item** object to each side of the navigation bar:

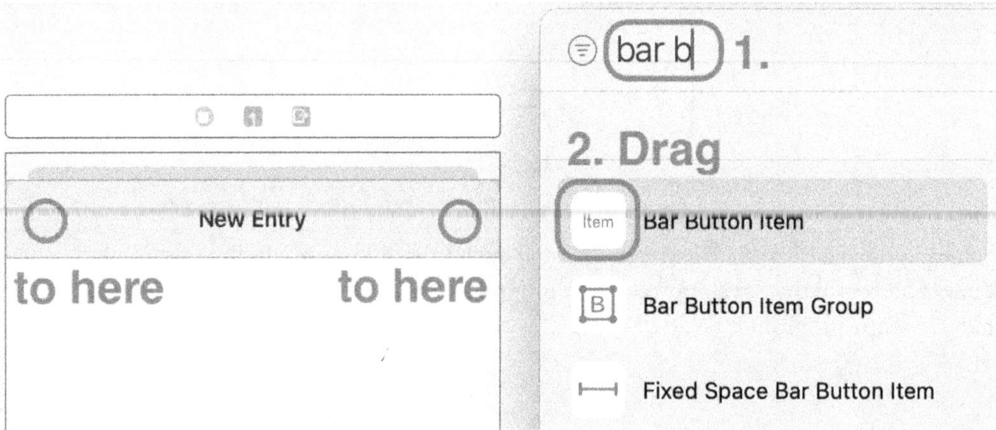

Figure 11.45: Library with Bar Button Item object selected

3. Click the right **Item** button. In the Attributes inspector under **Bar Button Item**, set **Style** to **Done** and set **System Item** to **Save**:

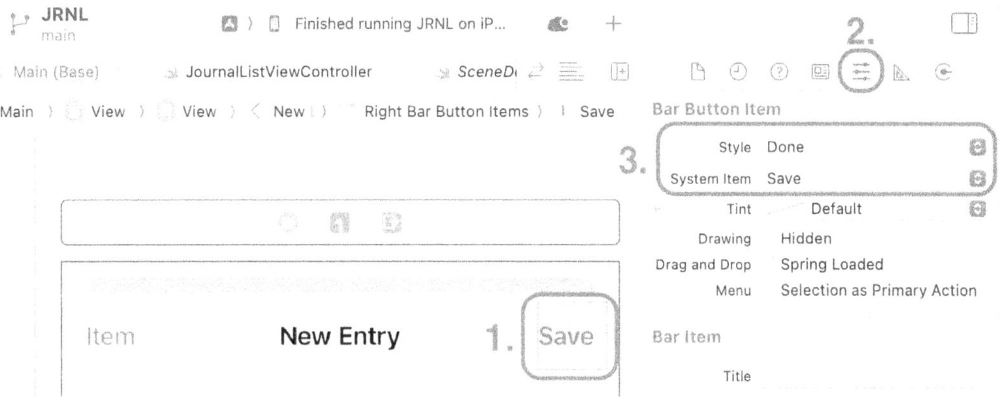

Figure 11.46: Attributes inspector with Style set to Done and System Item set to Save

4. Click the left **Item** button and set **System Item** to **Cancel**:

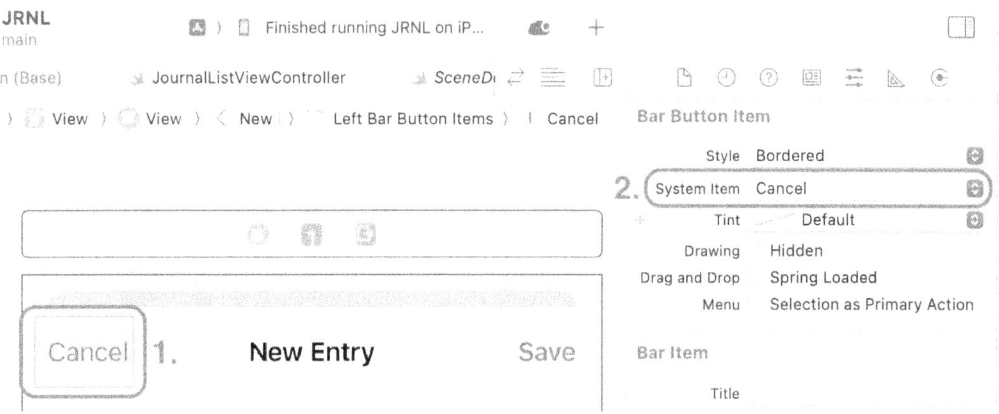

Figure 11.47: Attributes inspector with System Item set to Cancel

Remember that the navigation controller has a property, `viewControllers`, that holds an array of view controllers. When you click the + button on the Journal List screen, the new view controller is added to the `viewControllers` array and its view appears from the bottom of the screen, covering the Journal List screen, and the only way to dismiss the view is to drag it downward.

5. To enable the **Cancel** button to dismiss the view, you will link the **Cancel** button to the scene exit and implement a method in the `JournalListViewController` class that will be executed when the Journal List screen reappears. In the Project navigator, click the **JournalListViewController** file and add the following method at the bottom of the file before the final closing curly brace:

```
@IBAction func unwindNewEntryCancel(segue:
UIStoryboardSegue) {
}
```

6. Click on the **Main** storyboard file in the Project navigator, and click the **Cancel** button in the New Entry scene. In the document outline, *Ctrl* + *Drag* from the **Cancel** button to the scene exit icon and choose **unwindNewEntryCancelWithSegue:** from the pop-up menu:

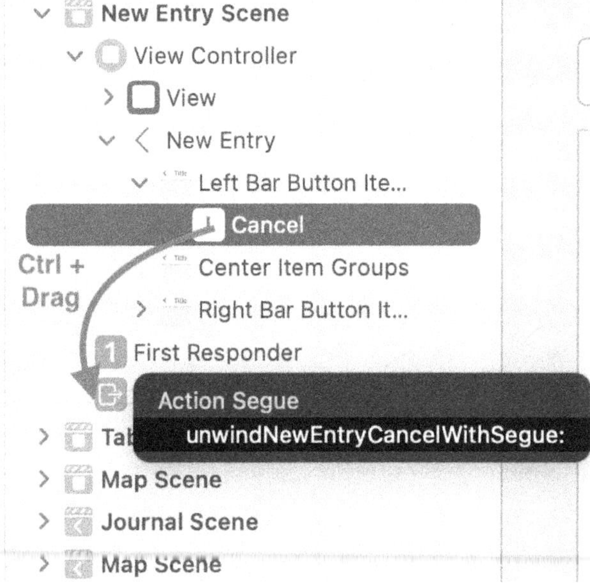

Figure 11.48: Document outline showing the Cancel button action being set

When your app is running, clicking the **Cancel** button will remove the view controller from the navigation controller's `viewControllers` array, dismiss the view that is presented modally, and execute the `unwindNewEntryCancel(segue:)` method. Note that this method doesn't do anything at present.

7. Build and run your app and click the + button in the navigation bar of the Journal List screen. The new view will appear. When you click the **Cancel** button, the new view disappears:

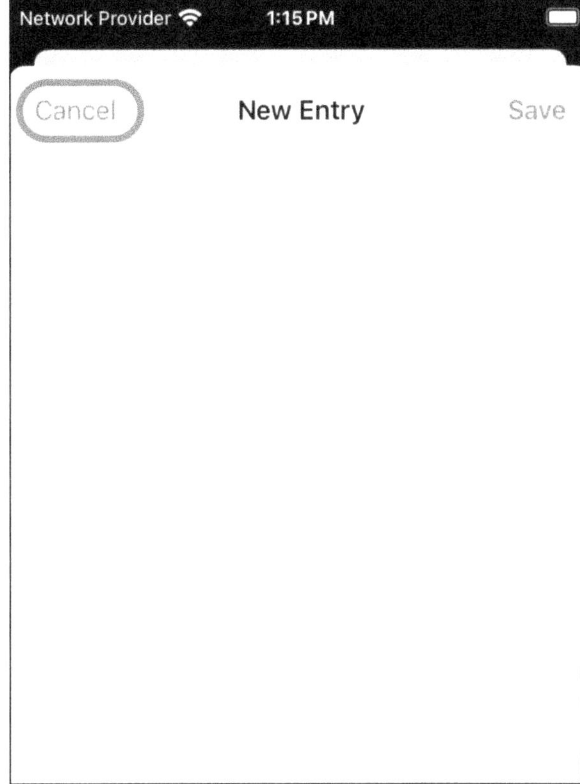

Figure 11.49: Simulator showing the Cancel button

Congratulations! You've completed the basic structure for the Journal List screen!

Summary

In this chapter, you added a table view to the Journal List screen in the Main storyboard file and modified the existing view controller class to implement the `JournalListViewController` class. Then, you modified the `JournalListViewController` class to have an outlet for the table view in the storyboard and made it the data source and delegate for the table view. Finally, you added a button to display a second view and configured a **Cancel** button to dismiss it.

At this point, you should be proficient in using Interface Builder to add views and view controllers to a storyboard scene, link view controller outlets to UI elements in storyboards, set up table views, and present views modally. This will be very useful when you're designing the UI for your own apps.

In the next chapter, you'll implement the Journal Entry Detail screen of your app and implement a map view for the Map screen.

Join us on Discord!

Read this book alongside other users, experts, and the author himself. Ask questions, provide solutions to other readers, chat with the author via Ask Me Anything sessions, and much more. Scan the QR code or visit the link to join the community.

`https://packt.link/ios-Swift`

12

Finishing Up Your User Interface

In the previous chapter, you configured the Journal List screen to display 10 empty table view cells in a table view, added a bar button item to the navigation bar to present a view representing the Add New Journal Entry screen modally, and added **Cancel** and **Save** buttons to it.

In this chapter, you'll add the remaining screens shown during the app tour in *Chapter 10, Setting Up the User Interface*. You'll add the Journal Entry Detail screen, which will be displayed when a table view cell in the Journal List screen is tapped. You'll configure this screen to display a table view with a fixed number of table view cells. You'll also make the Map screen display a map.

By the end of this chapter, you'll have learned how to add and configure a table view with a fixed number of cells to a storyboard scene, how to implement a segue that will display a screen when a cell in the Journal List screen is tapped, and how to add a map view to a scene. The basic user interface of your app will be complete, and you will be able to walk through all the screens in Simulator. None of the screens will be displaying data, but you will finish their implementation in *Part 3* of this book.

The following topics will be covered in this chapter:

- Implementing the Journal Entry Detail screen
- Adding a map view to the Map screen

Technical requirements

You will continue working on the `JRNL` project that you created in the previous chapter.

The completed Xcode project for this chapter is in the `Chapter12` folder of the code bundle for this book, which can be downloaded here:

https://github.com/PacktPublishing/iOS-18-Programming-for-Beginners-Ninth-Edition

Check out the following video to see the code in action:

https://youtu.be/TFaELYrzyxY

To start, you'll add a new table view controller scene to represent the Journal Entry Detail screen. This screen will be displayed when a cell in the Journal List screen is tapped. You'll do this in the next section.

Implementing the Journal Entry Detail screen

As shown in the app tour in *Chapter 10, Setting Up the User Interface*, when you tap a journal entry in the Journal List screen, a Journal Entry Detail screen containing the details of that journal entry will appear. In this section, you'll add a new table view controller scene to your storyboard to represent the Journal Entry Detail screen. Follow these steps:

1. Click the **Main** storyboard file in the Project navigator and click the **Library** button.
2. Type `table` in the filter field, and drag a **Table View Controller** object to the storyboard next to the **Map** scene:

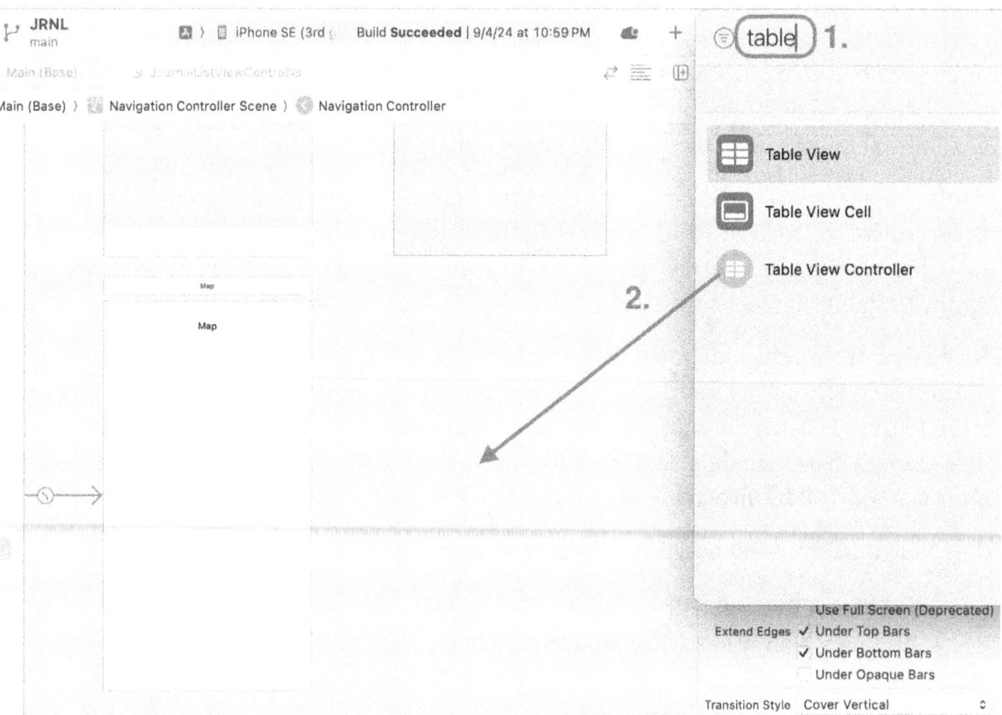

Figure 12.1: Library showing Table View Controller object

This will represent the Journal Entry Detail screen.

3. Verify that the **Table View Controller** scene has been added:

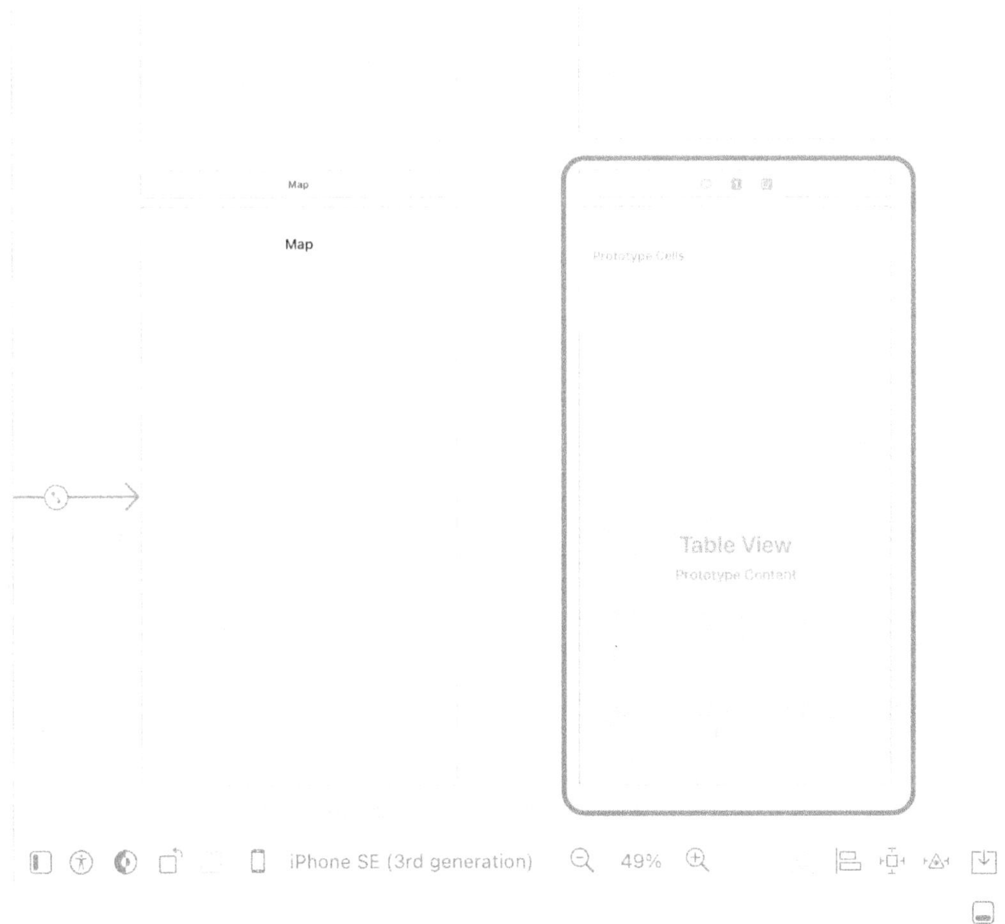

Figure 12.2: Editor area showing the Table View Controller scene next to the Map scene

Note that it already has a table view inside it, so you don't need to add a table view to the scene, like you did in the previous chapter.

4. To display the Journal Entry Detail screen when a table view cell in the Journal List screen is tapped, *Ctrl + Drag* from **journalCell** (in the document outline under **Journal Scene**) to the **Table View Controller** scene to add a segue between them:

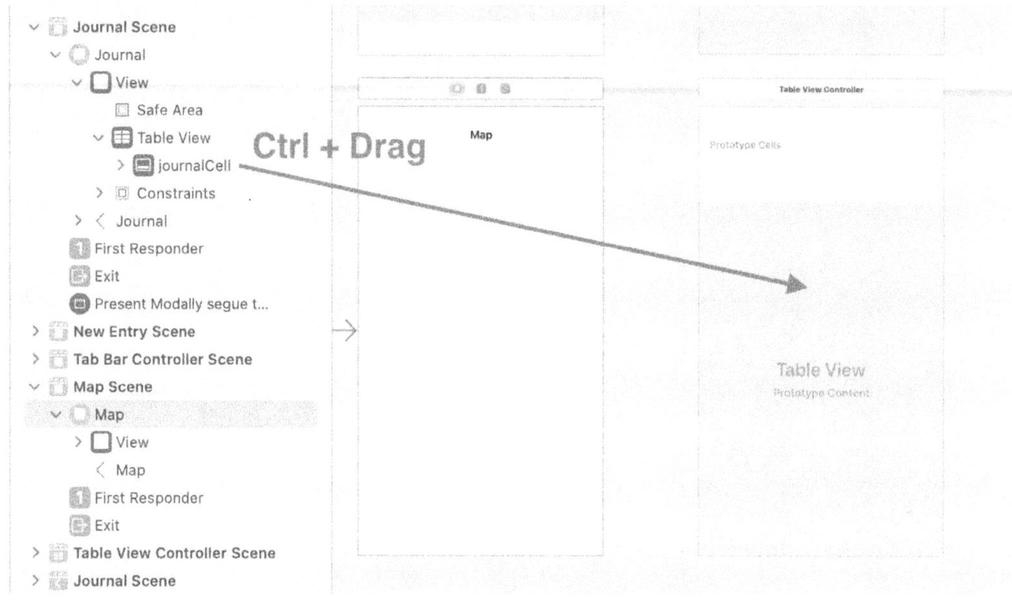

Figure 12.3: Document outline showing journalCell

5. In the pop-up menu, select **Show** under **Selection Segue**:

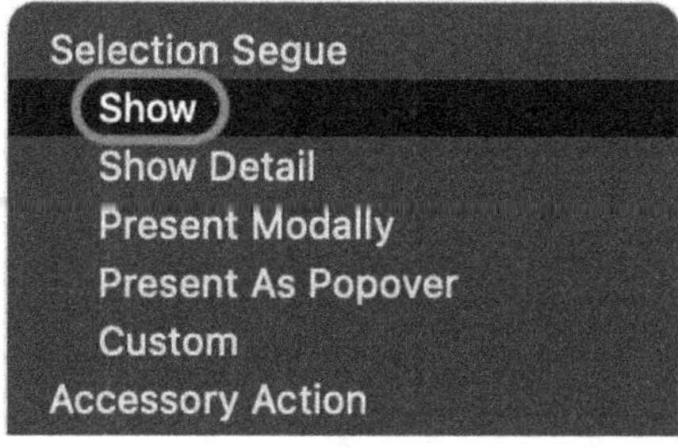

Figure 12.4: Pop-up menu with Show selected

This makes the Journal Entry Detail screen slide in from the right when a cell in the Journal List screen is tapped.

6. Verify that a segue has appeared between the two scenes:

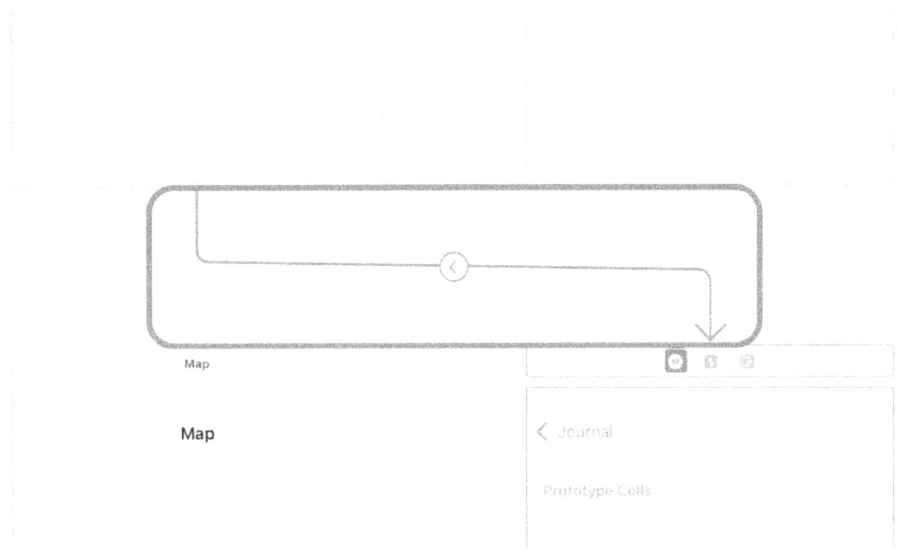

Figure 12.5: Editor area showing segue between the Journal scene and the Table View Controller scene

You can rearrange scenes in the storyboard to make the segues easier to see.

7. The Journal Entry Detail screen will always display a fixed number of cells. In the document outline, click **Table View** under the **Table View Controller Scene**:

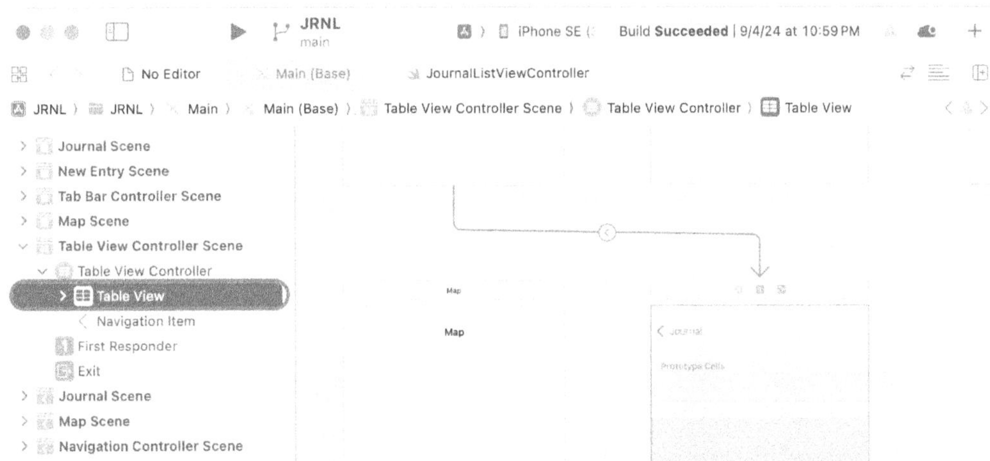

Figure 12.6: Table View in document outline selected

8. Click the Attributes inspector button and set **Content** to **Static Cells** to make the Journal Entry Detail screen display a fixed number of cells.

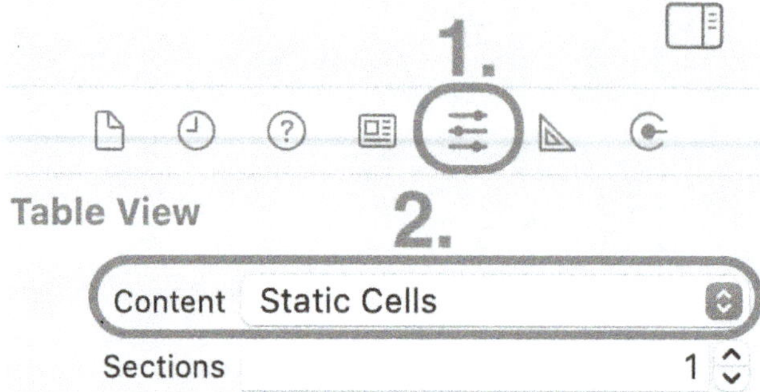

Figure 12.7: Attributes inspector with Content set to Static Cells

9. Build and run your app. Click on a cell in the Journal List screen to display the Journal Entry Detail screen:

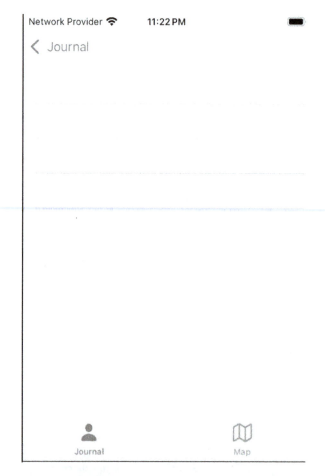

Figure 12.8: Simulator showing Journal Entry Detail screen

10. Click the < **Journal** button to go back to the Journal List screen.

You have successfully implemented the Journal Entry Detail screen! Great!

You can use this method if your app needs to show the details of an item on a list. Examples of this are the *Contacts* app and the *Settings* app on your iPhone.

In the next section, you will make the Map screen display a map.

Implementing the Map screen

When you launch the app, the Journal List screen is displayed. Tapping the Map button in the tab bar makes the Map screen appear, but it is blank. To make the Map screen display a map, you'll add a map view to the view in the view controller scene for the Map screen. Follow these steps:

1. Select the view controller scene for the Map screen in the Editor area, which will expand the corresponding **Map Scene** in the document outline:

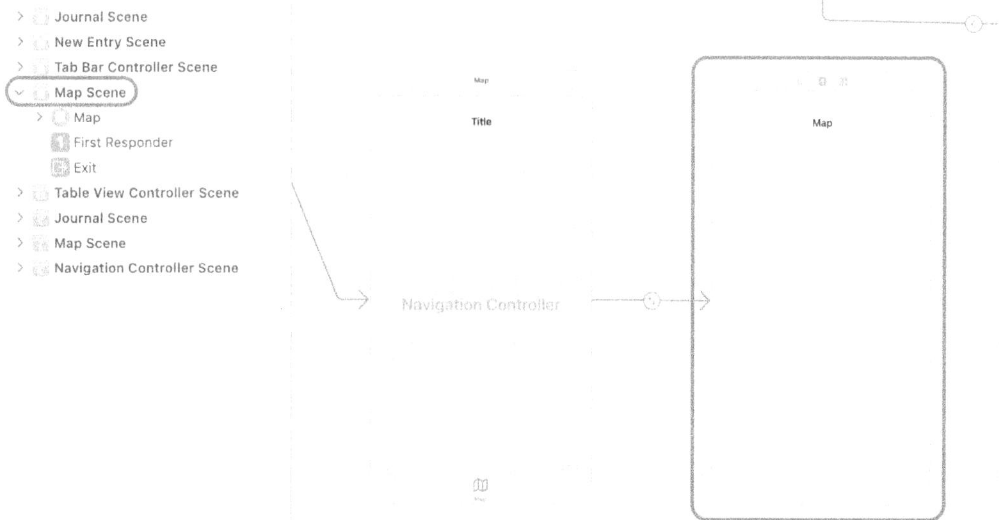

Figure 12.9: Editor area showing view controller scene for the Map scene

2. To make this scene display a map, click the Library button and type `map` in the filter field. A **Map Kit View** object appears as one of the results. Drag it to the view in the view controller scene:

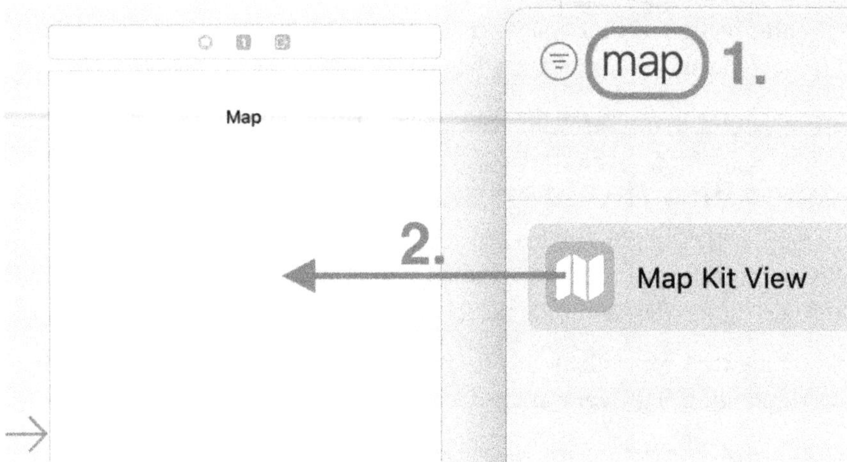

Figure 12.10: Library with Map Kit View object selected

3. To make the map view fill the whole screen, verify that it is selected and click the Add New Constraints button:

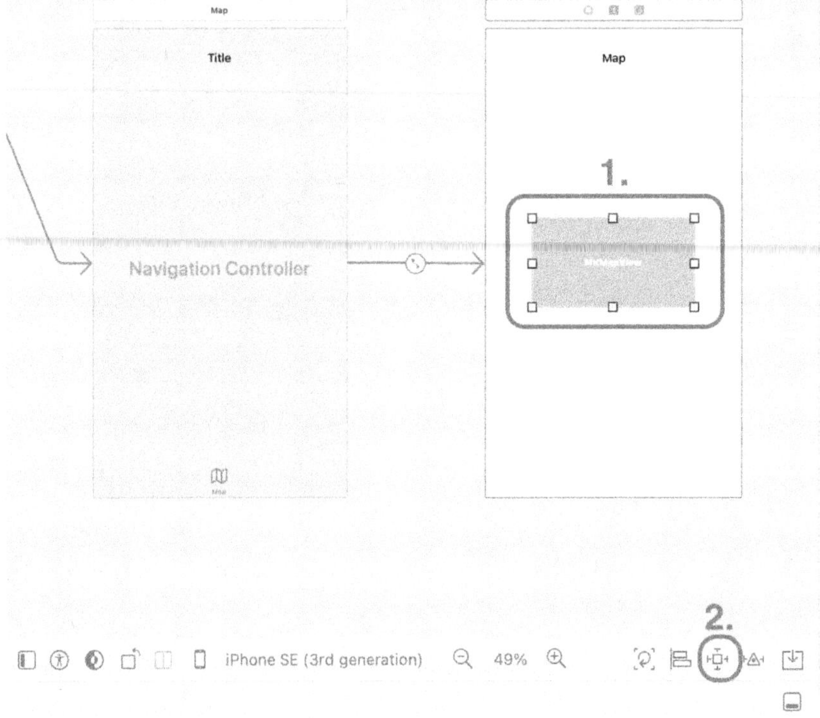

Figure 12.11: View controller scene with map view selected

4. Type `0` into all the **Spacing to nearest neighbor** fields and make sure that the pale red struts are selected (they will turn bright red). Click the **Add 4 Constraints** button:

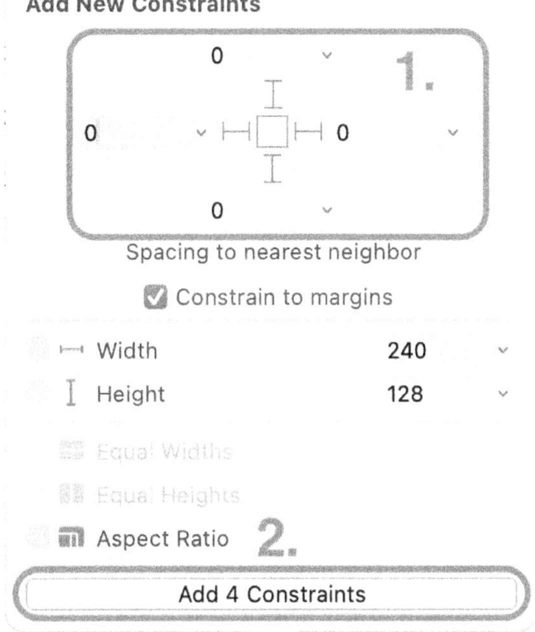

Figure 12.12: Auto Layout Add New Constraints pop-up dialog box

5. Verify that the map view fills the entire screen:

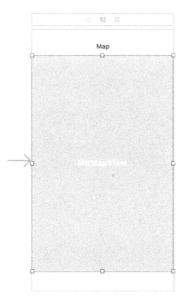

Figure 12.13: View controller scene with map view filling the screen

6. Build and run your app. Click the **Map** button. You should see a map like the one shown here:

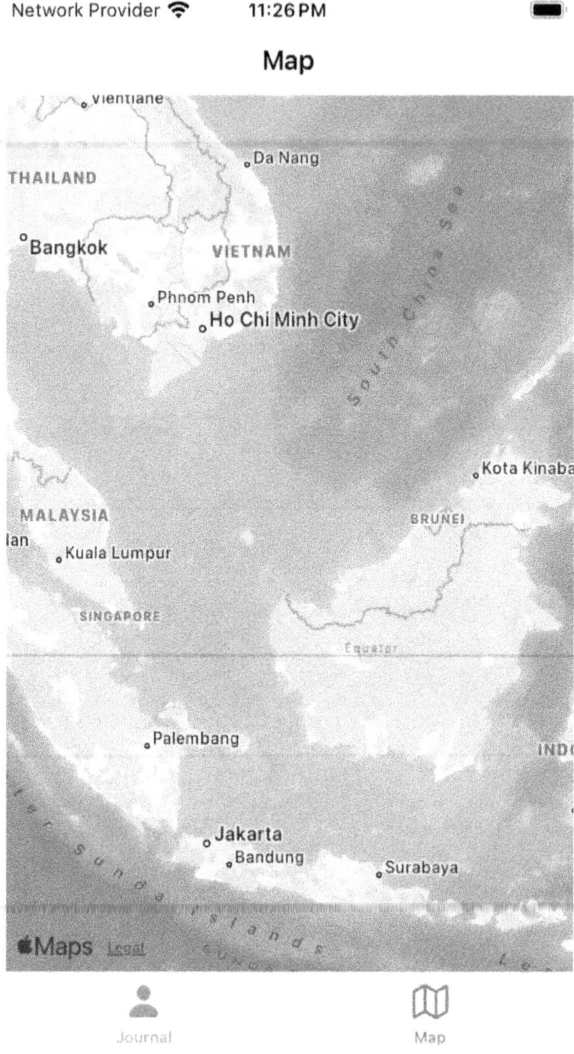

Figure 12.14: Simulator showing Map screen

7. Verify that all the screens required for the `JRNL` app have been created in the **Main** storyboard file:

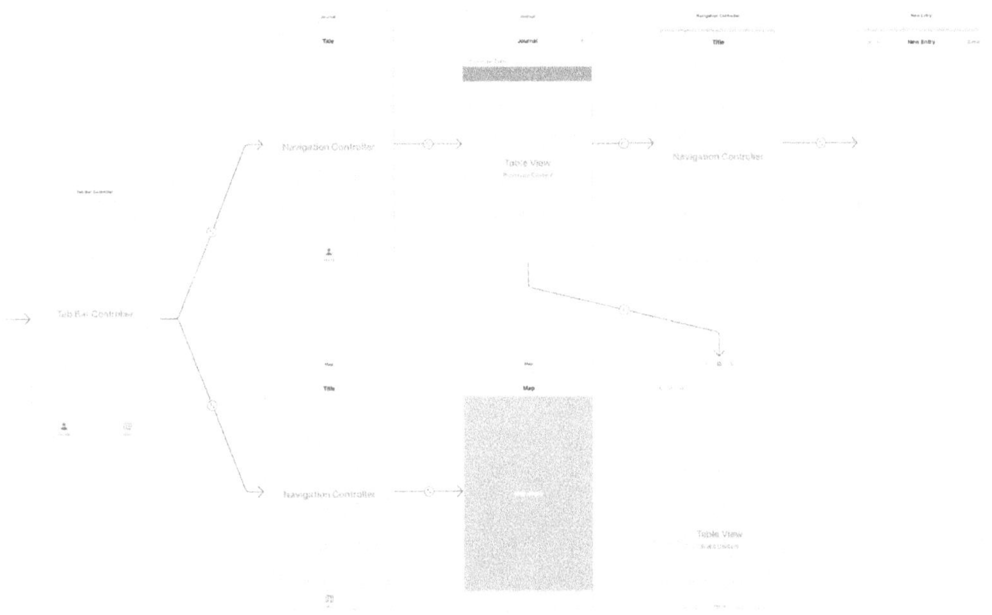

Figure 12.15: Editor area showing all the scenes in Main.storyboard

8. Verify that all the screens appear as they should when you run your app in Simulator.

Wonderful! You've now completed the basic user interface for your app!

Summary

In this chapter, you completed the basic structure of your app. You added a new table view controller scene to represent the Journal Entry Detail screen, configured a table view with static cells for this screen, and implemented a segue that will display this screen when a cell in the Journal List screen is tapped. You also added a map view to the view controller scene for the Map screen, and it now displays a map when the **Map** button is tapped.

You have successfully implemented all the screens required for your app, and you'll be able to test your app's flow when you run it in Simulator. You should also be more proficient with Interface Builder. Familiarity with using and positioning objects from the Library will be crucial when you're building user interfaces for your own apps.

In the next chapter, you'll modify the cells on the Journal List screen, the Add New Journal Entry screen, and the Journal Entry Detail screen so that they match the designs that were shown in the app tour.

Join us on Discord!

Read this book alongside other users, experts, and the author himself. Ask questions, provide solutions to other readers, chat with the author via Ask Me Anything sessions, and much more. Scan the QR code or visit the link to join the community.

`https://packt.link/ios-Swift`

13

Modifying App Screens

In *Chapter 11, Building Your User Interface*, you added some of the screens required for your app to match what was shown during the app tour. In *Chapter 12, Finishing Your User Interface*, you added the remaining screens required for your app. You're now able to navigate through all the screens of your app when you run it in Simulator, but the screens still lack the user interface elements required for data input and data display.

In this chapter, you'll add and configure the missing user interface elements to the Journal List, Add New Journal Entry, and Journal Entry Detail screens, to match the design shown in the app tour.

For the Journal List screen, you'll modify the `journalCell` table view cell by adding an image view and two labels to it, so that it can display the photo, date, and title of a journal entry. For the Add New Journal Entry screen, you'll modify it by adding a custom view, a switch, a text field, a text view, and an image view, so that you can enter the details of a new journal entry. You'll also configure the image view to show a default image. For the Journal Entry Detail screen, you'll add a text view, labels, and image views to it and configure the image views to show default images, so that the screen can display the details of an existing journal entry. With all the user interface elements in place, your app will be ready for code, which you will implement in *Part 3* of this book.

By the end of this chapter, you will be more proficient in adding and positioning user interface elements and will have gained more experience in how to use constraints to determine their position relative to one another. This will be useful to ensure compatibility with different screen sizes and orientations, enabling you to easily prototype the appearance and flow of your apps.

The following topics will be covered in this chapter:

- Modifying the Journal List screen
- Modifying the Add New Journal Entry screen
- Modifying the Journal Entry Detail screen

Technical requirements

You will continue working on the `JRNL` project that you modified in the previous chapter.

The completed Xcode project for this chapter is in the `Chapter13` folder of the code bundle for this book, which can be downloaded here:

`https://github.com/PacktPublishing/iOS-17-Programming-for-Beginners-Eighth-Edition`

Check out the following video to see the code in action:

`https://youtu.be/tgo2dT1LZeM`

Let's start by modifying the `journalCell` table view cell on the Journal List screen. In the next section, you'll add some user interface elements to make it match the table view cell shown in the app tour.

Modifying the Journal List screen

Let's see what the Journal List screen looked like in the app tour:

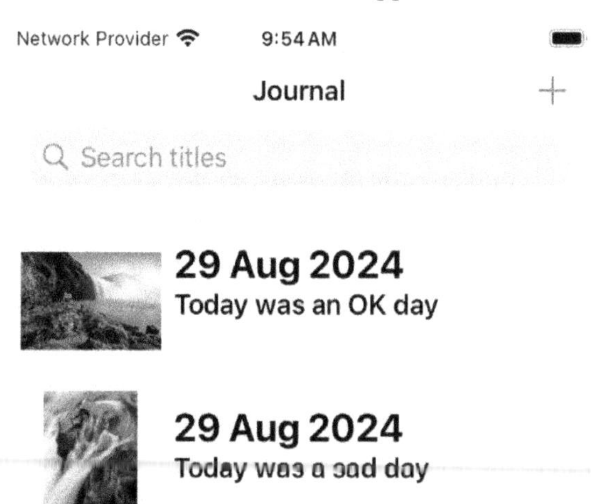

Figure 13.1: The Journal List screen for the completed JRNL app

As you can see, the table view cells on the Journal List screen have a photo, a date, and a journal entry title. In *Chapter 11, Building Your User Interface*, you set the background color for the `journalCell` table view cell to cyan and configured the table view to display a column of 10 cells. You'll now remove the background color and add user interface elements to the `journalCell` table view cell, matching the design shown in the app tour. You'll start by adding an image view to it in the next section.

Adding an image view to journalCell

An image view is an instance of the `UIImageView` class. It can display a single image or a sequence of animated images in your app. To add an image view to the `journalCell` table view cell, follow these steps:

1. To make the user interface elements easier to see when you add them to the storyboard, choose **Canvas | Bounds Rectangles** from the **Editor** menu:

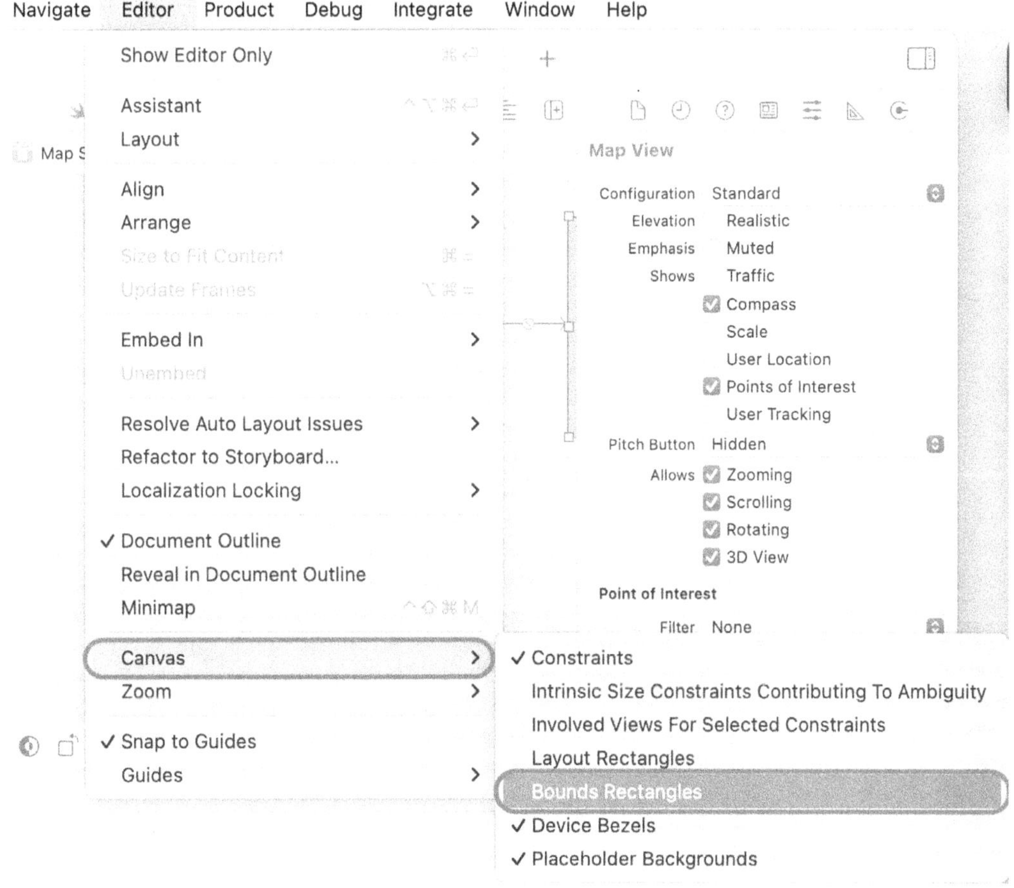

Figure 13.2: Editor menu with Canvas | Bounds Rectangles selected

This will apply a thin blue outline to the user interface elements in the storyboard.

2. Click the **Main** storyboard file in the Project navigator. Under the first **Journal Scene**, select **journalCell** in the document outline:

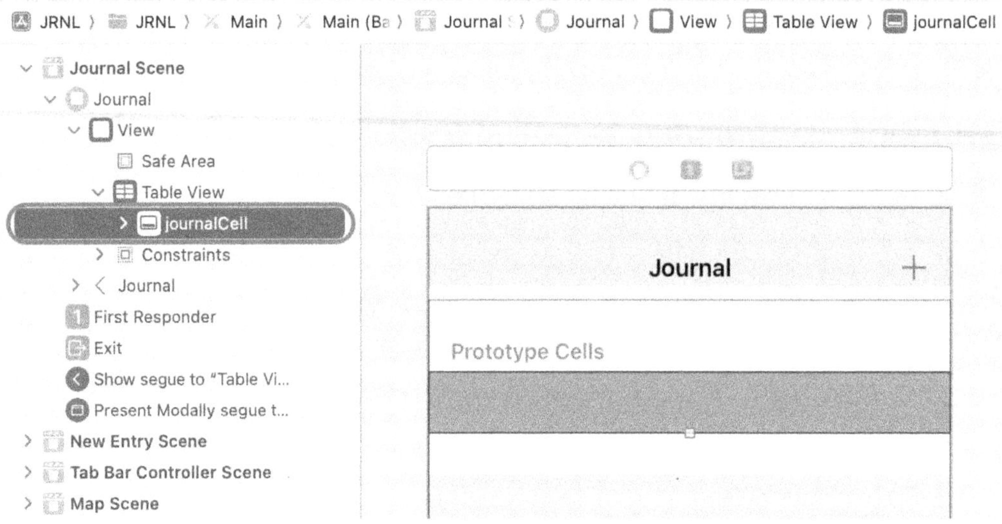

Figure 13.3: Document outline showing journalCell

3. You will need to remove the background color you set earlier prior to adding the image view. In the Attributes inspector, under **View**, set **Background** to **Default**:

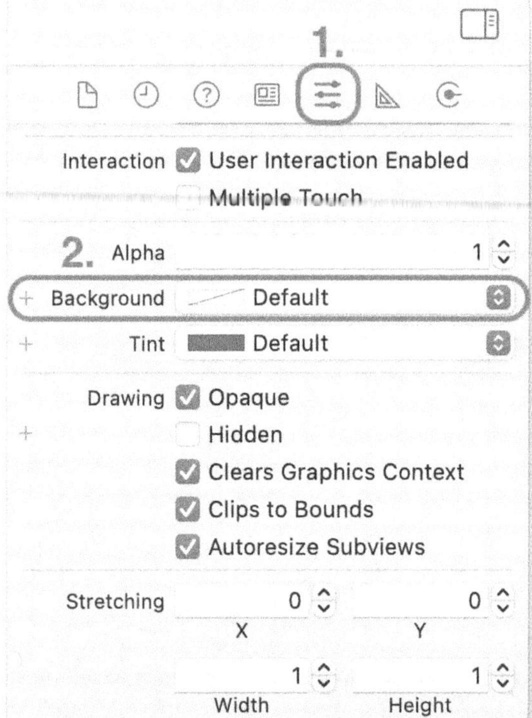

Figure 13.4: Attributes inspector settings for journalCell

4. To add an image view to the table view cell, click the **Library** button. Type `imag` into the filter field. An **Image View** object will appear in the results. Drag it into the prototype cell:

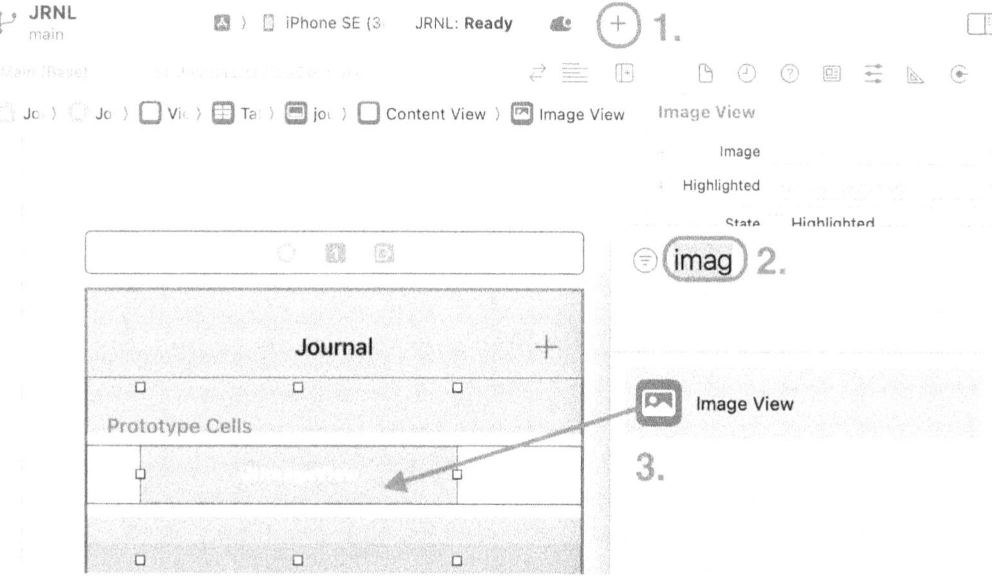

Figure 13.5: Prototype cell with image view added

5. To ensure the constraints for the newly added image view can be set properly, verify that it is a subview of the **journalCell** table view cell's **Content View** and is selected:

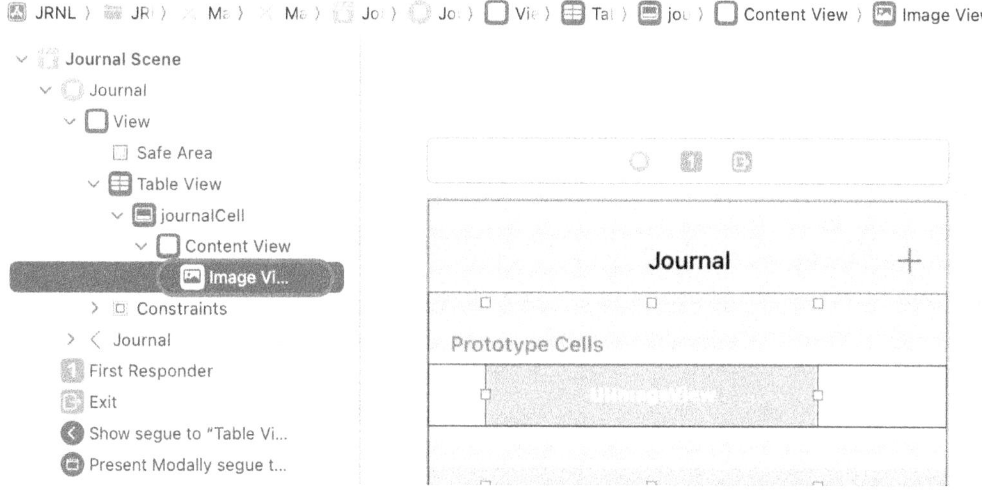

Figure 13.6: Document outline with Image View object selected

6. Click the Add New Constraints button, and enter the following values to set the constraints for the newly added image view:

    - Top: `0`

- Left: `0`
- Bottom: `0`
- Width: `90`
- Height: `90`

When done, click the **Add 5 Constraints** button.

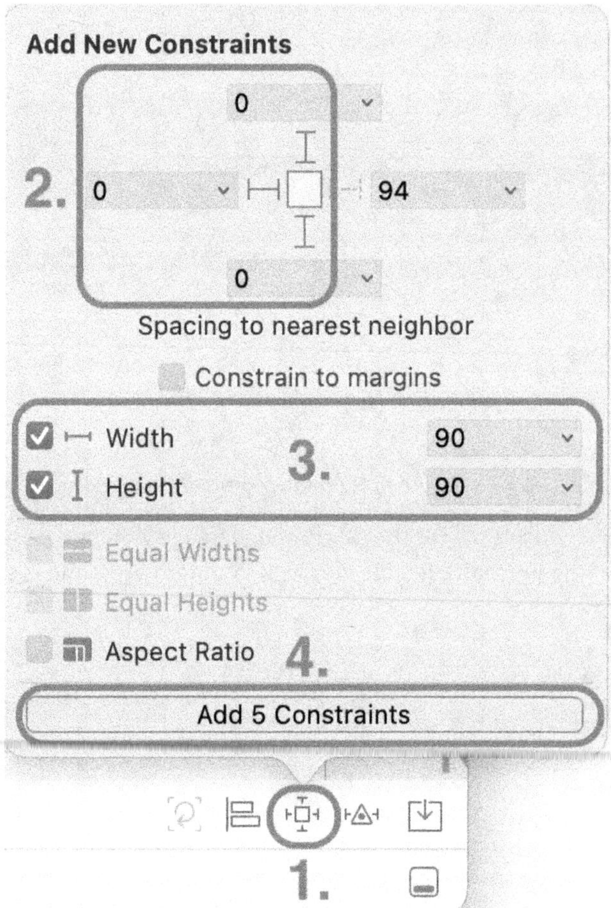

Figure 13.7: Add New Constraints dialog box

This binds the image view's top, left, and bottom edges to the corresponding edges of the `journalCell` table view cell, and sets its width and height to `90` points. It also implicitly sets the height of the table view cell to `90` points.

7. In the Attributes inspector, under **Image View**, set **Image** to `face.smiling`:

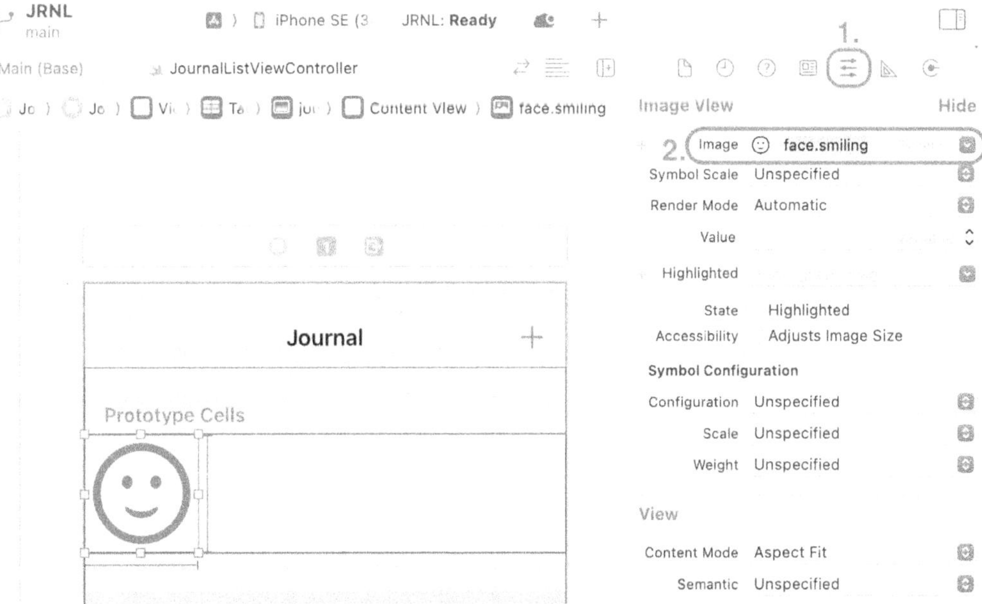

Figure 13.8: Image view with Image set to face.smiling

You have successfully added an image view to the table view cell, set its default image, and applied constraints to determine its position relative to its enclosing view. Cool!

In the next section, you'll add user interface elements that will be used to display the date and the title of the journal entry.

Adding labels to journalCell

You will use labels to display the date and the journal entry title in the `journalCell` table view cell. A label is an instance of the `UILabel` class. It can display one or more lines of text in your app.

To add labels to the `journalCell` table view cell, follow these steps:

1. First, you'll add a label to display the date. Click the Library button and drag a **Label** object to the space between the image view you just added and the right side of the prototype cell:

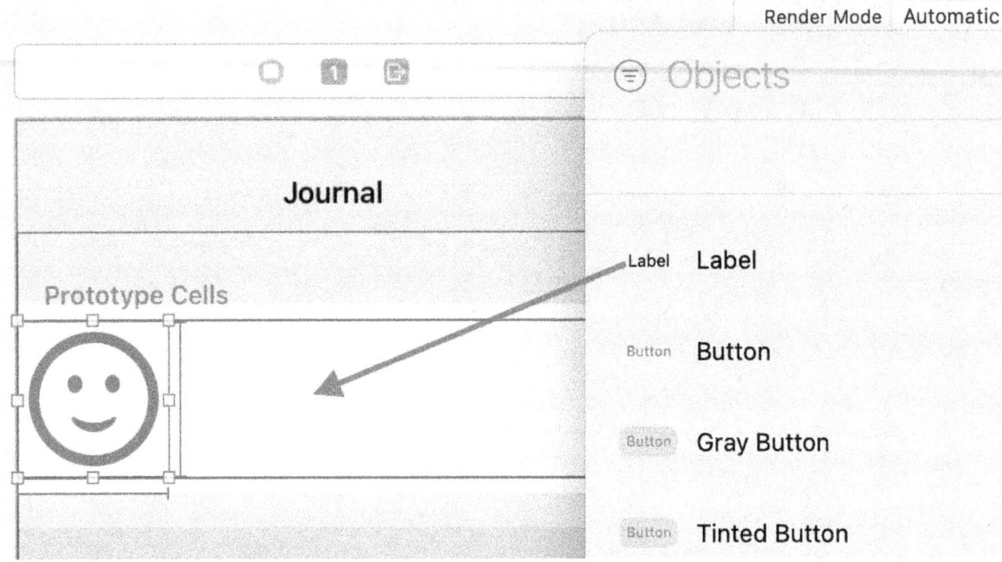

Figure 13.9: Library with the Label object selected

Note that **Label** appears in the document outline and is a subview of the **journalCell** table view cell's **Content View**.

2. In the Attributes inspector, under **Label**, set **Font** to **Title 1** using the **Font** menu:

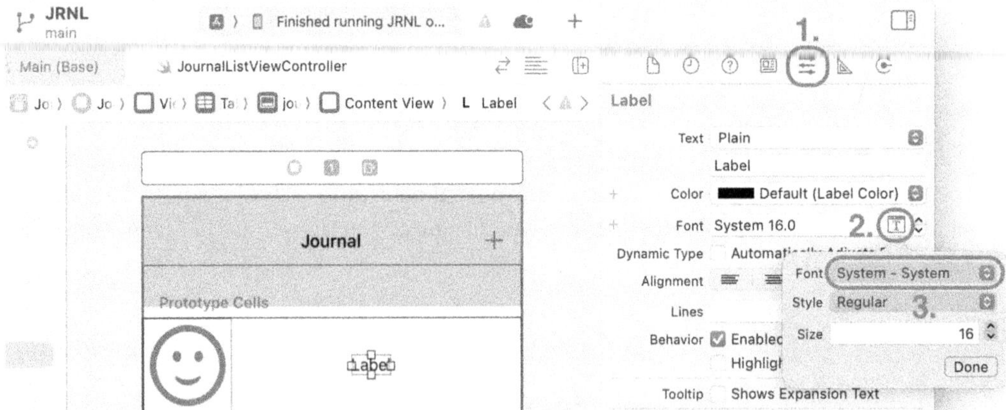

Figure 13.10: Attributes inspector for Label

3. Click the **Add New Constraints** button, and enter the following values to set the constraints for the label:

   - Top: `0`
   - Left: `8`
   - Right: `0`

   Constrain to Margins should already be checked, which sets a standard margin of `8` points as the space between the top and right sides of the label, and the top and right sides of the table view cell. When done, click the **Add 3 Constraints** button.

4. Verify that the position of the label appears as shown in the screenshot below and that the newly added constraints are in the document outline:

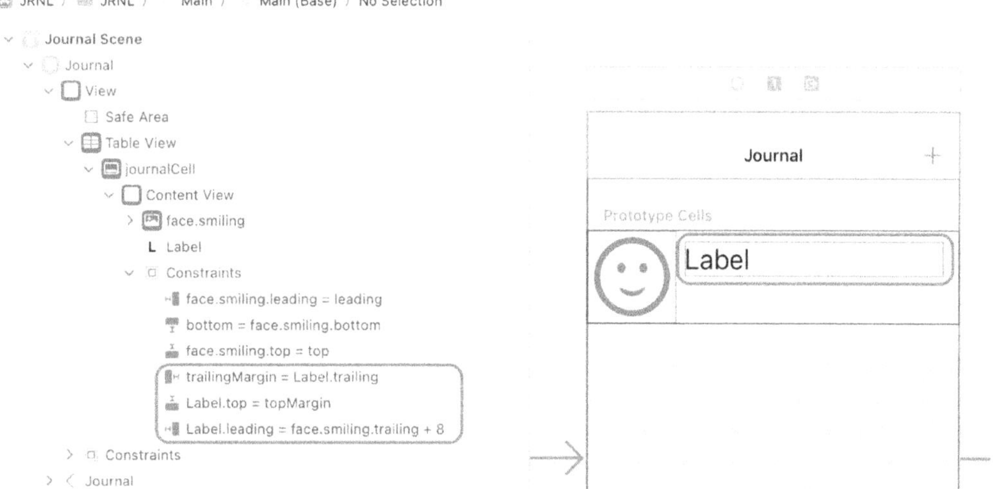

Figure 13.11: Label with constraints applied

The space between the top edge of the label and the top edge of the `journalCell` content view is set to `0 + 8` points. The space between the left edge of the label and the right edge of the image view is `8` points. The space between the right edge of the label and the right edge of the `journalCell` content view is `0 + 8` points. The position of the bottom edge of the label is automatically set by the text style you set earlier.

Next, you'll add a label to display the journal entry title. Follow these steps:

1. Click the Library button and drag a **Label** object to the space between the label you just added and the bottom of the prototype cell:

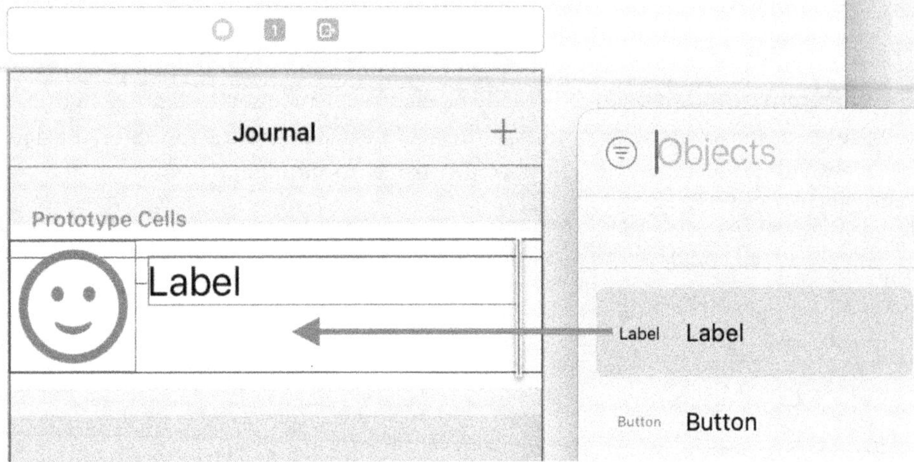

Figure 13.12: Library with the Label object selected

Note that **Label** appears in the document outline and is a subview of the **journalCell** table view cell's **Content View**.

2. In the Attributes inspector, under **Label**, set **Font** to **Body** using the **Font** menu, and set **Lines** to 2:

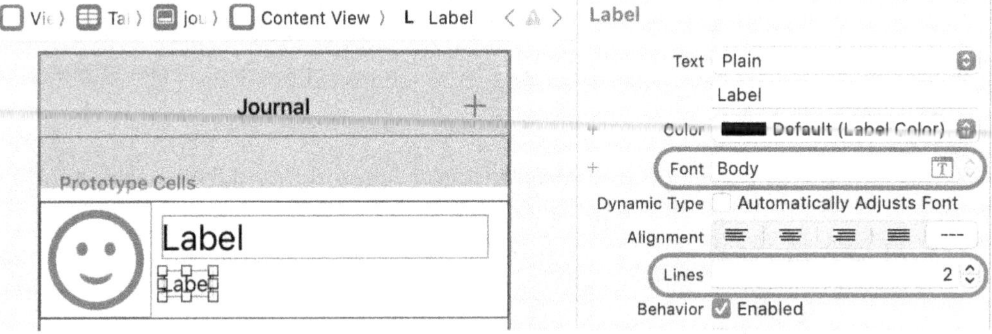

Figure 13.13: Attributes inspector for Label

Setting **Lines** to 2 will make the label display a maximum of two lines of text when the app is running.

3. Click the Add New Constraints button, and enter the following values to set the constraints for the label:

   - Top: `0`
   - Left: `8`
   - Right: `0`

   Constrain to Margins should already be checked, which sets a standard margin of 8 points as the space between the right side of the label and the right side of the table view cell. When done, click the **Add 3 Constraints** button.

4. Verify that the position of the label appears as shown in the screenshot below and that the newly added constraints are in the document outline:

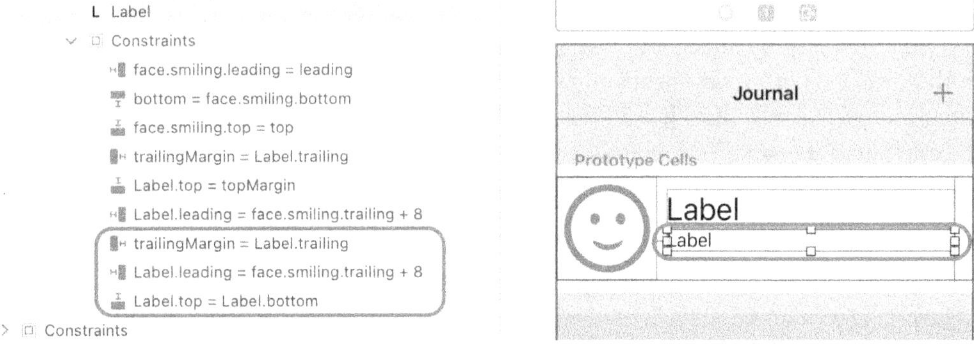

Figure 13.14: Label with constraints applied

The space between the top edge of the label and the bottom edge of the label you added earlier is set to `0` points. The space between the left edge of the label and the right edge of the image view is `8` points. The space between the right edge of the label and the right edge of the `journalCell` content view is `0 + 8` points. The position of the bottom edge of the label is automatically set by the text style and number of lines you set earlier.

You can click a constraint in the document outline and modify it in the Size inspector.

5. Build and run your app:

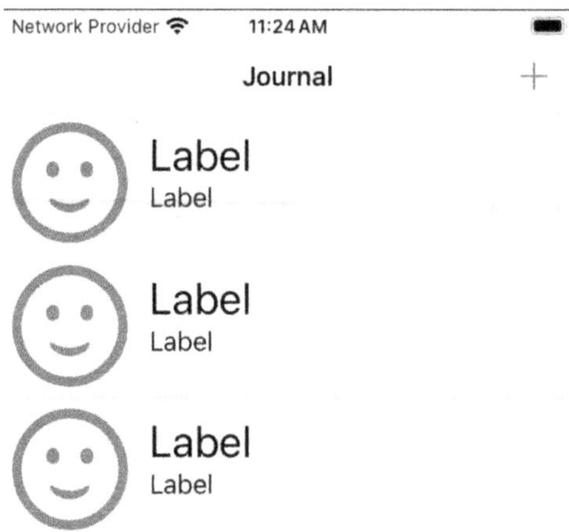

Figure 13.15: Simulator showing the completed journalCell table view cell

You have successfully added and configured labels to display the date and the journal entry title of the `journalCell` table view cell, and all the necessary constraints have been added. As you can see, the Journal List screen now has all the user interface elements required to display data as shown in the app tour. Fantastic!

In the next section, you'll add a stack view containing user interface elements to the Add New Journal Entry screen.

Modifying the Add New Journal Entry screen

Let's see what the Add New Journal Entry screen looks like in the app tour:

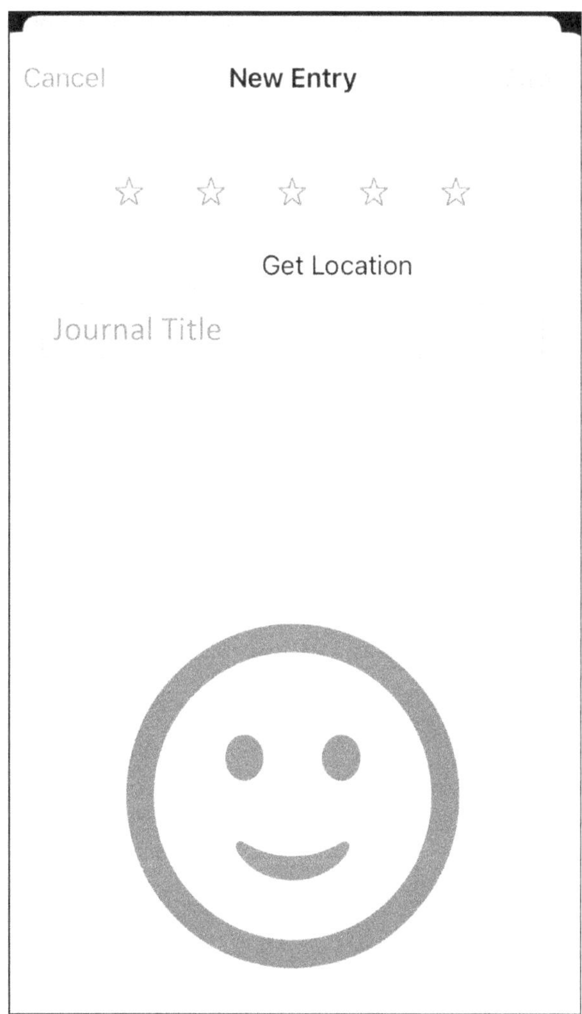

Figure 13.16: The Add New Journal Entry screen for the completed Journal app

Apple provides an extensive library of user interface elements that you can use in your own apps. This helps to give all iOS apps a consistent look and feel. As you can see, the Add New Journal Entry screen has the following elements:

- A custom view showing star ratings
- A switch that allows you to get your current location
- A text field for the journal entry title
- A text view for the journal entry body
- An image view for a photo that you will take with your phone's camera

You will now modify the screen to match the design shown in the app tour, beginning in the next section by adding a custom view that allows a user to set star ratings.

Adding a custom view to the New Entry scene

As you have seen in the app tour, the Add New Journal Entry screen has a custom view showing star ratings. This custom view is a subclass of a horizontal stack view. You'll add the horizontal stack view in this chapter and complete the implementation of the custom view in *Chapter 19, Getting Started with Custom Views.*

A stack view is an instance of the `UIStackView` class. It allows you to easily lay out a collection of views either in a column or a row. To add a stack view to the Add New Journal Entry screen, follow these steps:

1. In the **Main** storyboard file, click **New Entry Scene** in the document outline:

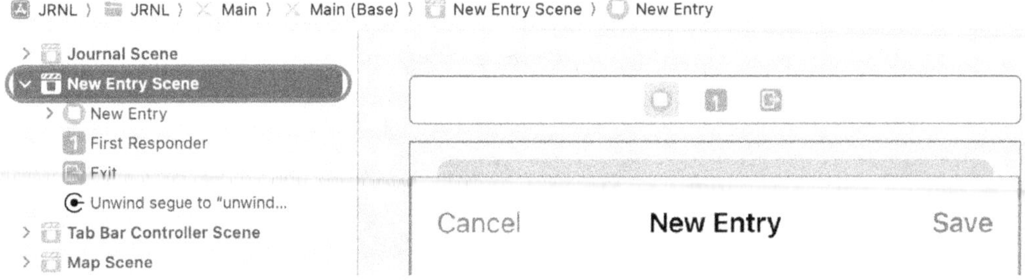

Figure 13.17: Editor area showing New Entry Scene

2. To add a horizontal stack view to the scene, click the Library button. Type `hori` into the filter field. A **Horizontal Stack View** object will appear in the results. Drag it to the view of the **New Entry** scene:

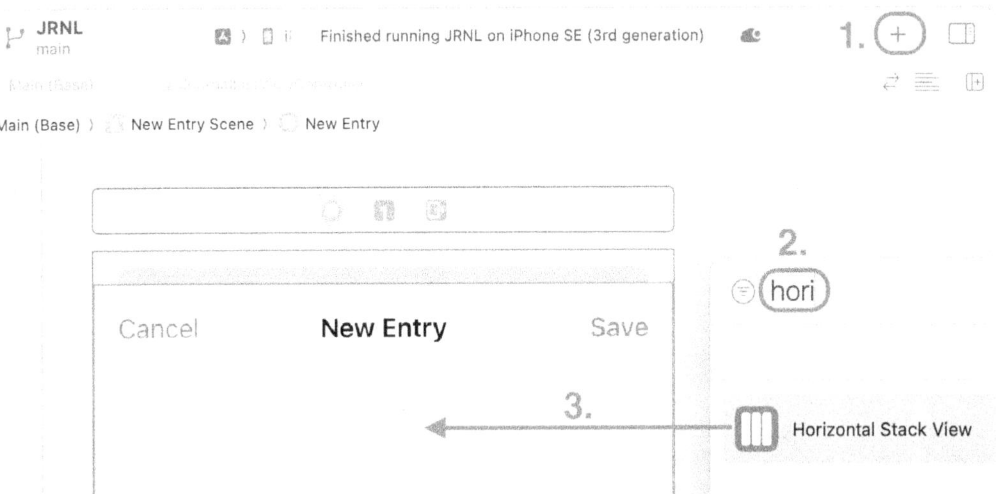

Figure 13.18: Library with the Horizontal Stack View object selected

Note that the stack view you just added appears in the document outline and is a subview of the view for **New Entry Scene**.

3. Click the Attributes inspector. Under **Stack View**, set **Spacing** to 8 if it was not already set, and under **View**, set **Background** to **System Cyan Color**:

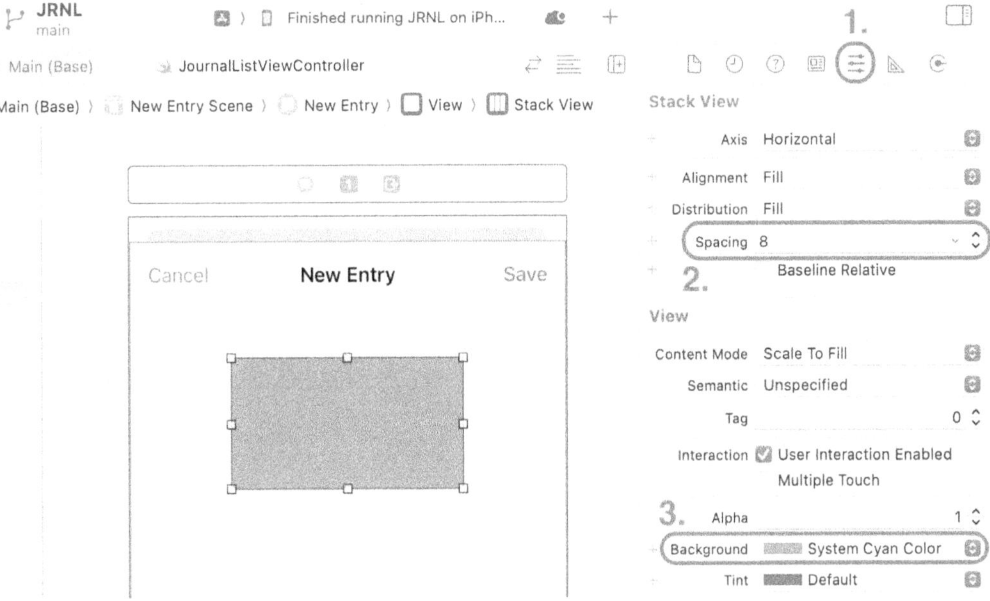

Figure 13.19: Attributes inspector showing the Spacing and Background settings

The **Spacing** value determines the spacing between elements in a stack view.

4. Click the Size inspector. Under **View**, set **Width** to 252 and **Height** to 44:

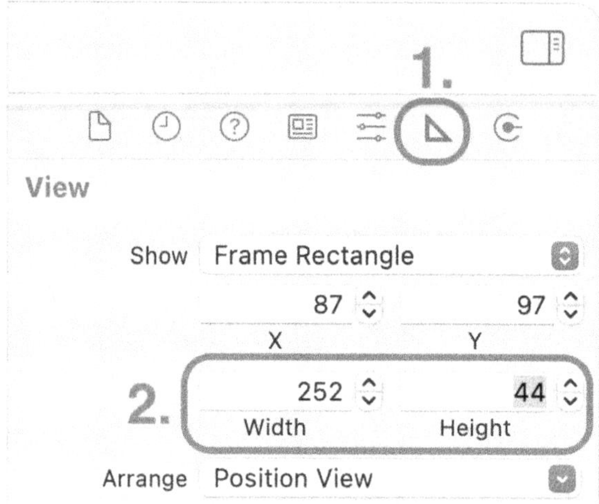

Figure 13.20: Size inspector showing the size of the stack view

The custom view showing star ratings will consist of five buttons. Each button has a height of 44 points and a width of 44 points, and the space between each button is 8 points. The total width of the custom view will be `5 x 44 + 4 x 8`, giving a total width of 252 points.

5. Click the Add New Constraints button, and enter the following values to set the constraints for the stack view:

   - Width: 252
   - Height: 44

   When done, click the **Add 2 Constraints** button.

   Setting the size of a UI element in the Size inspector makes it easier for you to add constraints later because the intended values will already be set in the **Add New Constraints** dialog box.

6. You'll see that the stack view is outlined in red. Click the little reddish-pink arrow in the document outline.

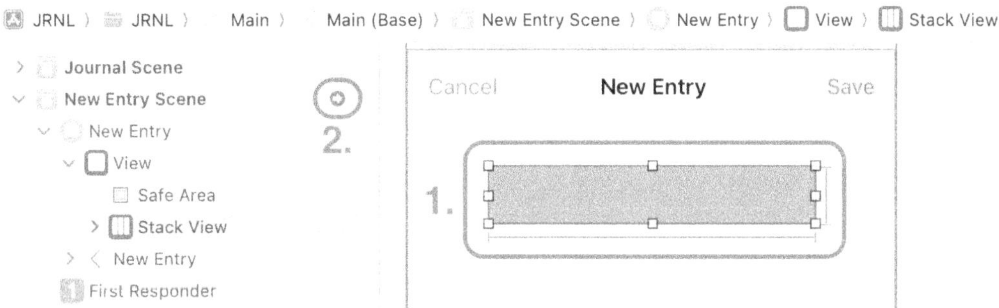

Figure 13.21: Arrow in the document outline

7. You'll see two **Missing Constraints** errors in the document outline:

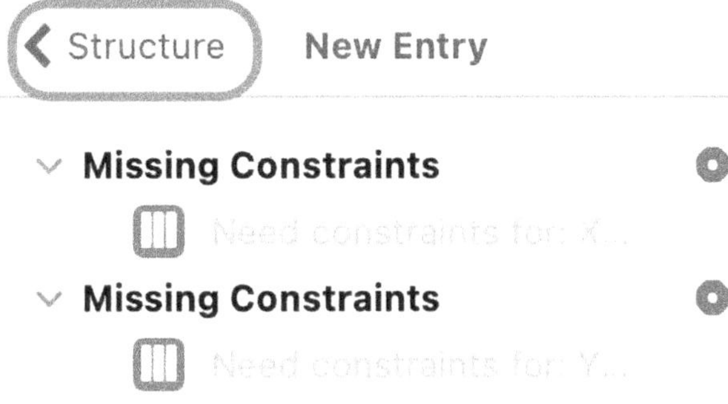

Figure 13.22: Missing constraints error displayed

The stack view is outlined in red because of the missing constraints errors. This means that the position of the stack view relative to its enclosing view is currently ambiguous. You will fix this when you embed this stack view in another stack view later.

8. Click the < **Structure** button to go back.

You have successfully added a horizontal stack view to the **New Entry** scene. In the next section, you'll add a UI element to it that allows you to get your current location.

Adding a switch to the New Entry scene

As shown in the app tour, you can toggle a switch to get your current location when creating a new journal entry. A switch is an instance of the `UISwitch` class. It displays a control that offers you a binary choice, such as on/off. You will also add a label to describe what the switch does and put both objects in a horizontal stack view.

Follow these steps:

1. To add a switch to the **New Entry** scene, click the Library button. Type `swi` into the filter field. A **Switch** object will appear in the results. Drag it to the view of the **New Entry** scene under the horizontal stack view:

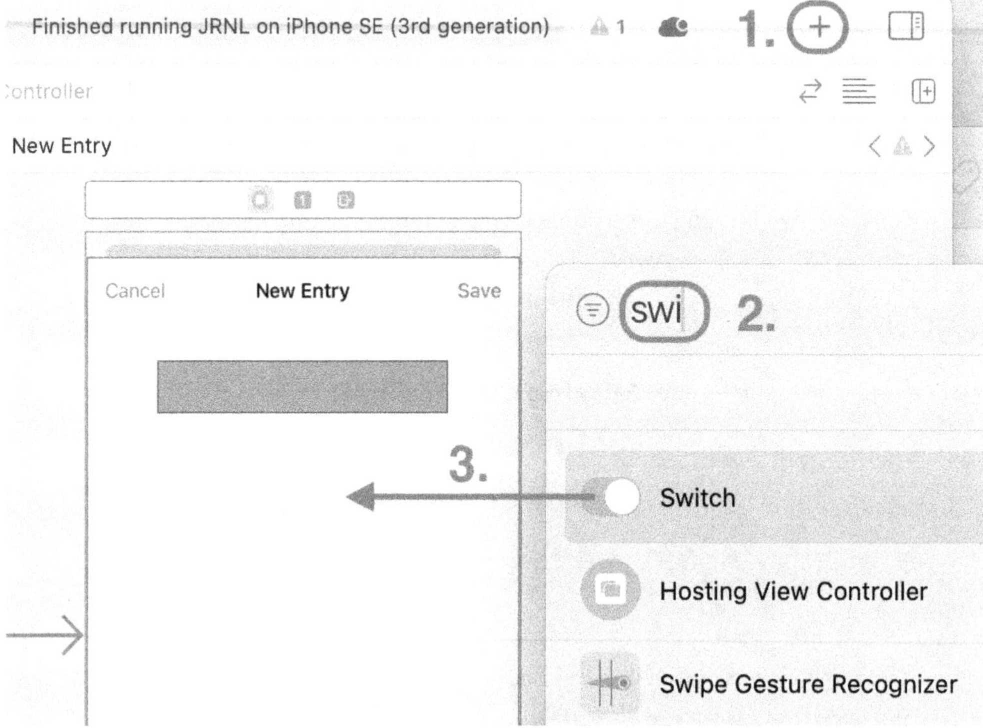

Figure 13.23: Library with the Switch object selected

Note that the switch you just added appears in the document outline and is a subview of the view for **New Entry Scene**.

2. To add a label next to the switch, drag a **Label** object from the library and position it next to the switch:

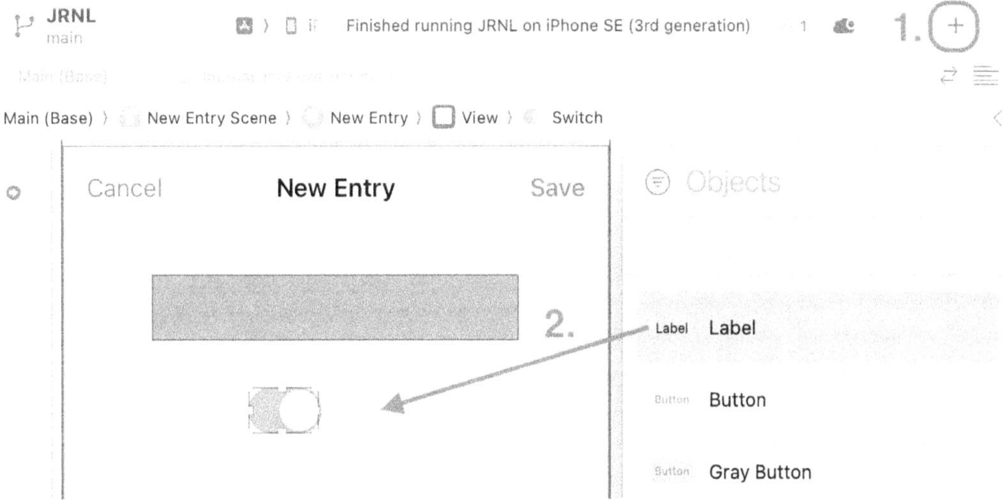

Figure 13.24: Library with Label object selected

Note that the label you just added appears in the document outline and is also a subview of the view for the **New Entry** scene.

Blue lines will appear to help you position the label the correct distance away from the switch.

3. Double-click the label and change the label text to `Get Location`:

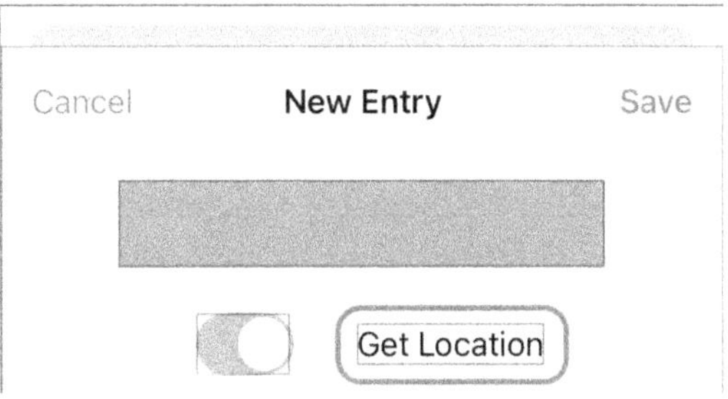

Figure 13.25: Label text changed to Get Location

4. You'll embed both the label and the switch in a horizontal stack view. In the document outline, hold down the *Shift* key, click **Switch**, and then click **Get Location** to select both the switch and the label:

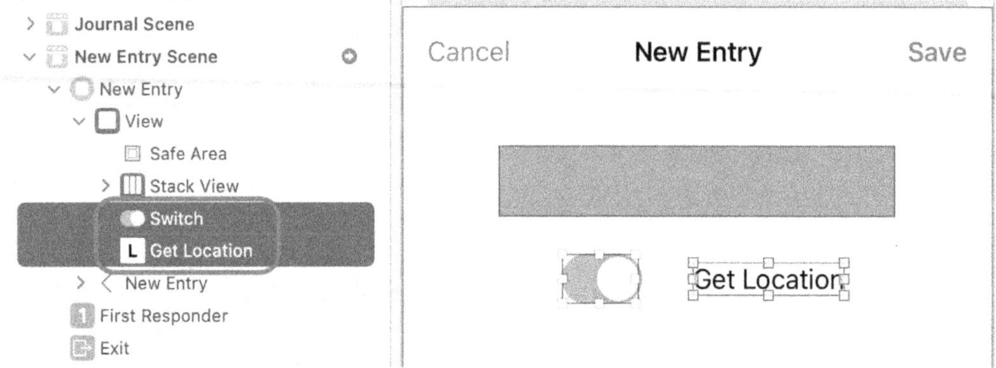

Figure 13.26: Document outline showing both label and switch selected

5. Choose **Embed In | Stack View** from the **Editor** menu.

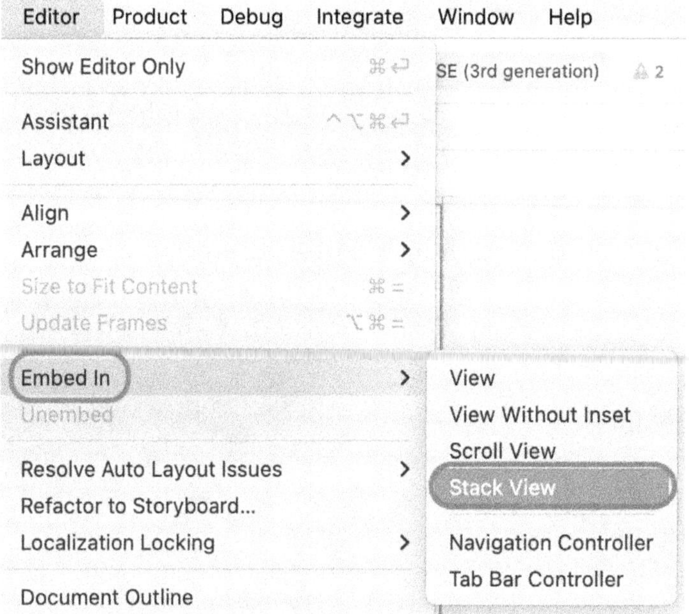

Figure 13.27: Editor menu with Embed In | Stack View selected

Both the switch and the label are now embedded in a stack view.

6. Click the Attributes inspector button, and under **Stack View**, set **Spacing** to 8:

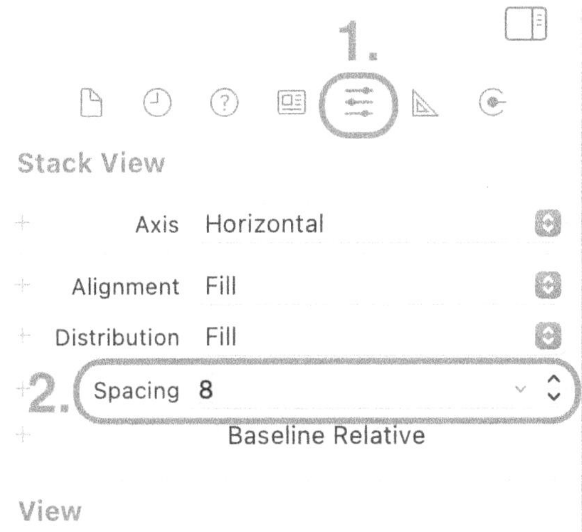

Figure 13.28: Attributes inspector with Spacing set to 8

You have successfully added a horizontal stack view containing a switch and a label to the **New Entry** scene. In the next section, you'll add UI elements to it so that a user can enter the journal title and body text.

Adding a text field and a text view to the New Entry scene

As shown in the app tour, users will enter the title and body text of a journal entry using this screen. To enter text, you can use either a text field or a text view. A text field is an instance of the `UITextField` class. It displays an editable text area and is typically limited to a single line. You'll enter the title of the journal entry using a text field. A text view is an instance of the `UITextView` class. It also displays an editable text area, but it normally displays more than one line. You'll enter the body text of the journal entry using a text view.

To add the text field and text view to the **New Entry** scene, follow these steps:

1. To add a text field to the scene, click the Library button. Type `text` into the filter field. A **Text Field** object will appear in the results. Drag it to the view of the **New Entry** scene, and position it under the horizontal stack view containing the switch and the label:

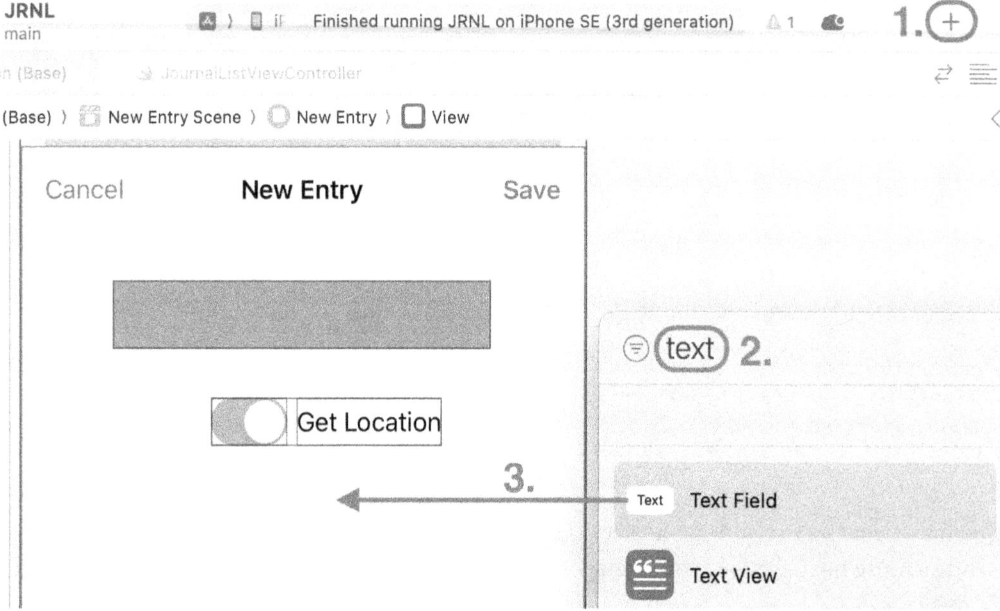

Figure 13.29: Library with the Text Field object selected

Note that the text field you just added appears in the document outline and is a subview of the view for the **New Entry** scene.

2. In the Attributes inspector, under **Text Field**, set **Placeholder** to `Journal Title`:

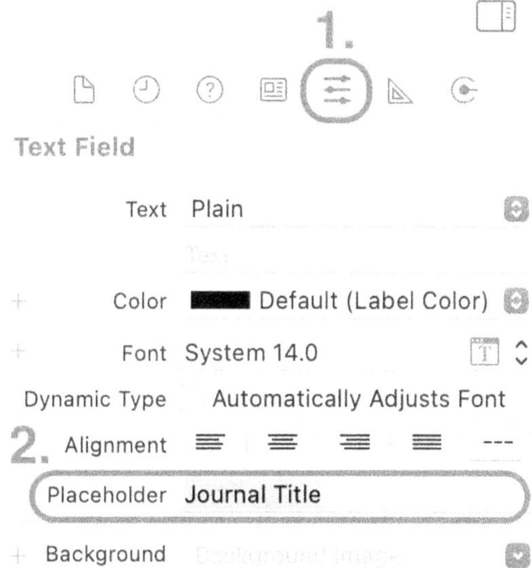

Figure 13.30: Attributes inspector with Placeholder set to Journal Title

3. To add a text view to the scene, click the Library button. Type `text` into the filter field. A **Text View** object will appear in the results. Drag it to the view of the **New Entry** scene under the text field you just added:

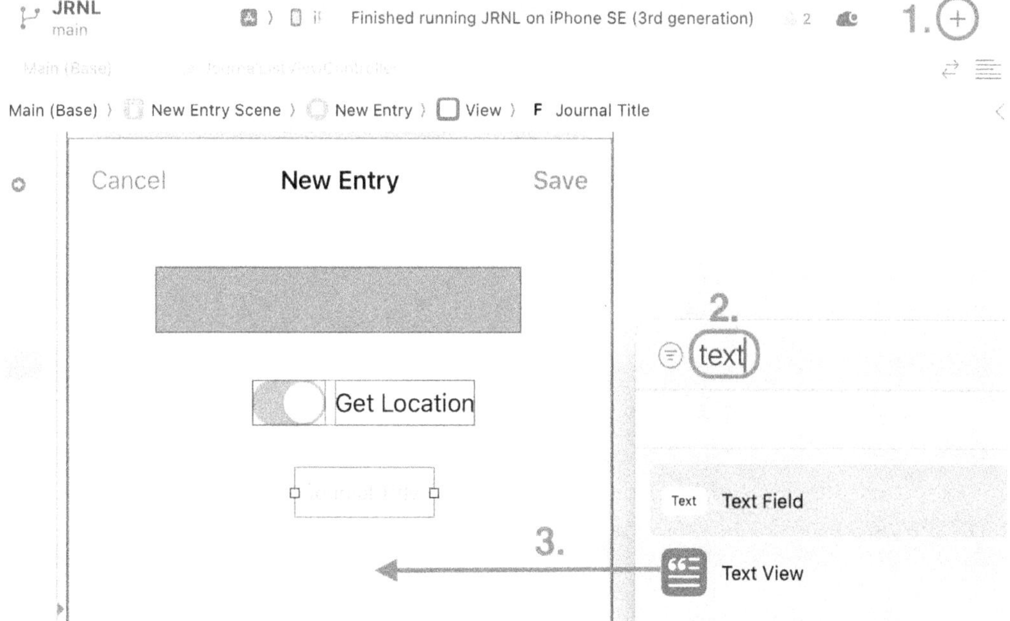

Figure 13.31: Library with Text Field object selected

Verify the text view you just added appears in the document outline and is a subview of the view for **New Entry Scene**. You may also change the default text if you wish.

4. You'll use a constraint to set the height of the text view to its default value of 128 points. With the text view selected, click the Add New Constraints button and tick the **Height** constraint. After that, click the **Add 1 Constraint** button. The text view will now look like the following screenshot:

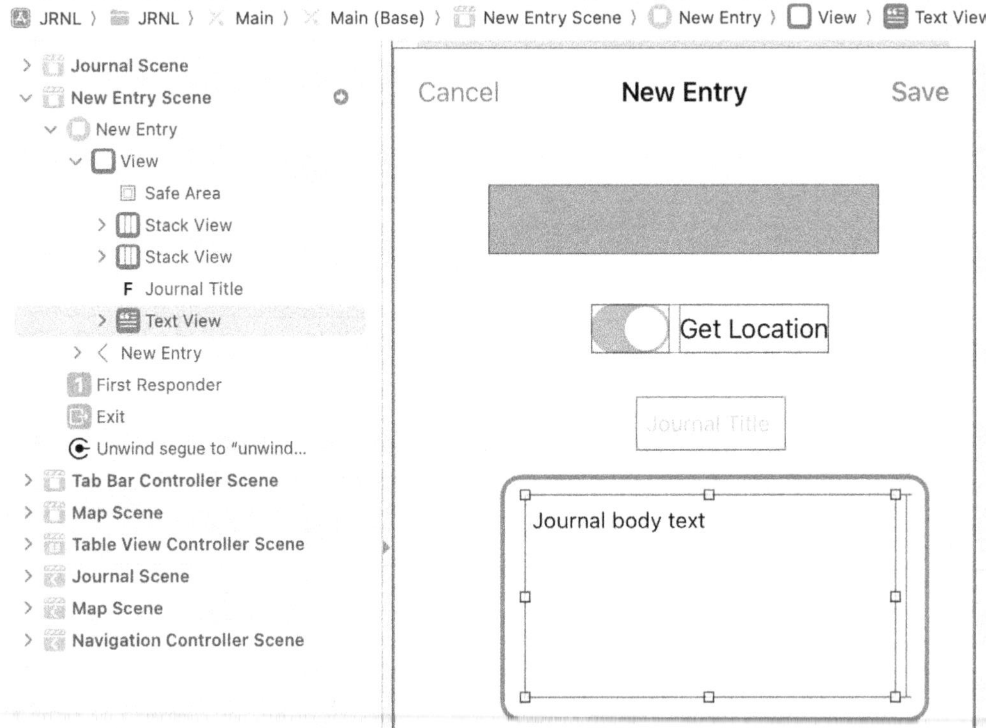

Figure 13.32: Text view with a constraint applied

Note the red outlines around the text view, as its position relative to the enclosing view is ambiguous. Don't worry about it now, as you will fix it later.

You have successfully added a text field and a text view to the New Entry scene. In the next section, you will add a UI element that lets a user take and display photos in it.

Adding an image view to the New Entry scene

As shown in the app tour, the user can use the device camera to take a photo for a journal entry. The selected photo will be shown in the Add New Journal Entry screen using an image view. In *Chapter 11, Building Your User Interface*, you added an image view to the `journalCell` table view cell. Now, you will add an image view to the New Entry scene. Follow these steps:

1. To add an image view to the scene, click the Library button. Type `imag` into the filter field. An **Image View** object will appear in the results. Drag it to the view of the **New Entry** scene under the text view:

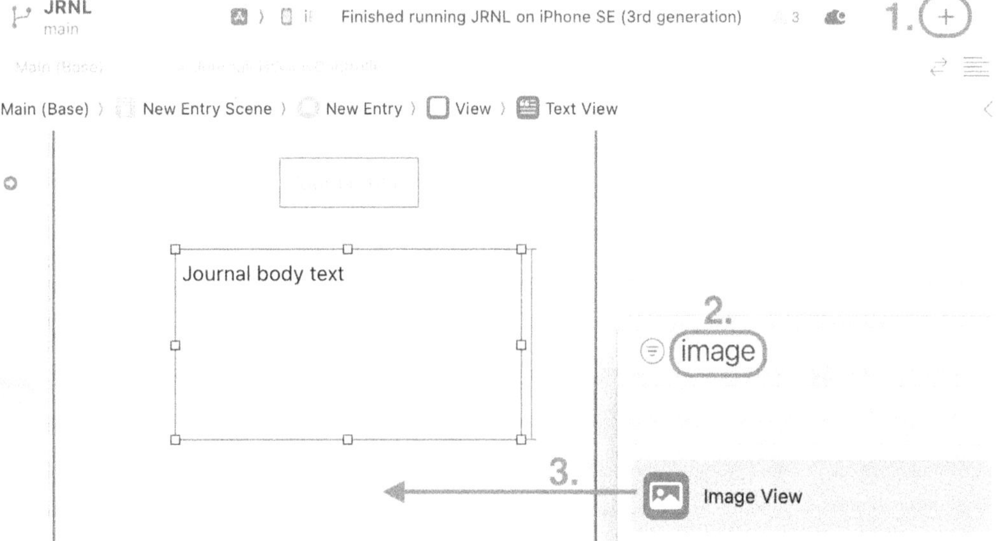

Figure 13.33: Library with the Image View object selected

   Note that the image view you just added appears in the document outline and is a subview of the view for **New Entry Scene.**

2. Click the Size inspector button. Under **View**, set both **Width** and **Height** to `200`.
3. Click the Attributes inspector button. Under **Image View**, set **Image** to `face.smiling`.

4. You'll use constraints to set the width and height for the image view. Click the **Add New Constraints** button, and tick the **Width** and **Height** constraints (their values should already be set to `200`). After that, click the **Add 2 Constraints** button. The image view will now look like the following screenshot:

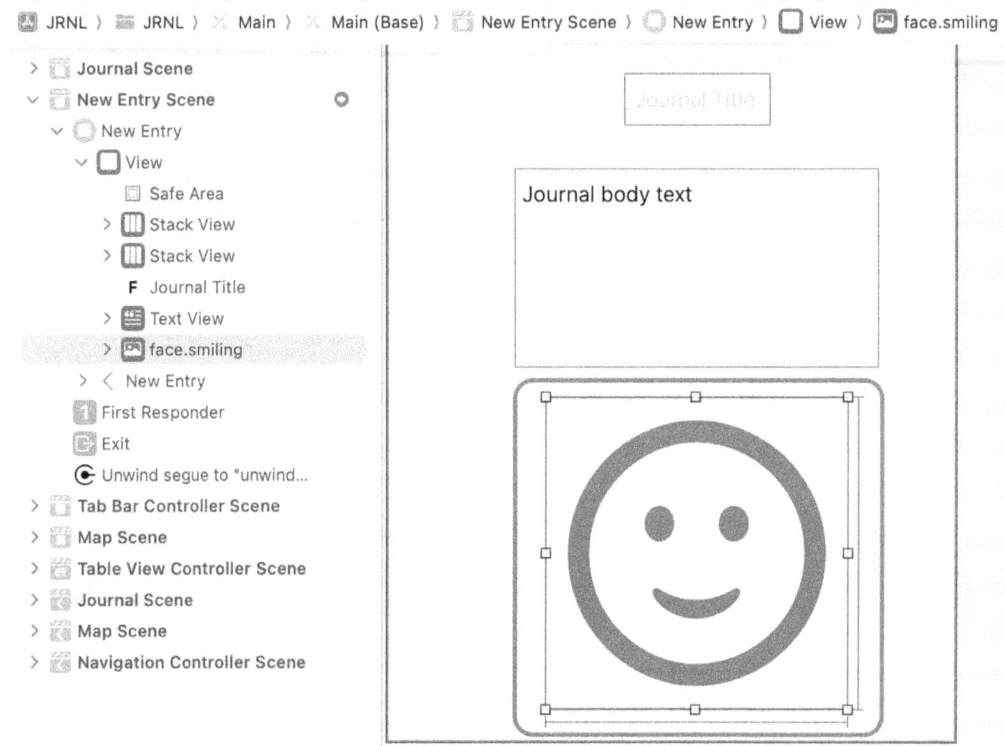

Figure 13.34: Image View with constraints applied

Note the red outlines around the image view, as its position relative to the enclosing view is ambiguous. Don't worry about it now, as you will fix it later.

All the user interface elements for the New Entry scene have been added. In the next section, you'll embed all of them in a vertical stack view to resolve the positioning issues.

Embedding user interface elements in a stack view

The **New Entry** scene now has all the required user interface elements, but the position of the elements relative to the enclosing view is ambiguous. You'll embed all the elements in a vertical stack view and use constraints to resolve the positioning issues. Follow these steps:

1. In the document outline, select all the user interface elements that you added earlier, as shown here:

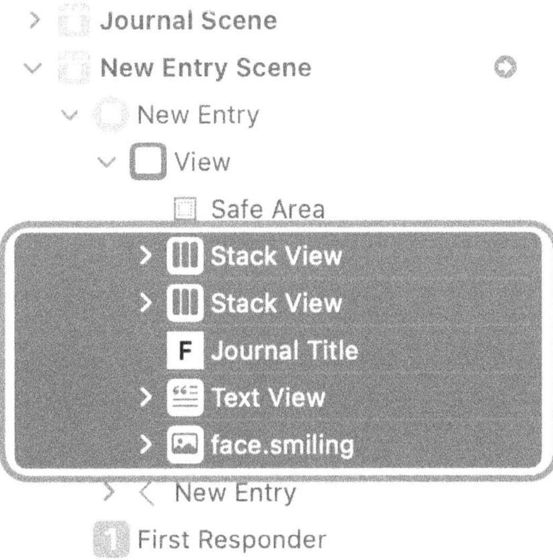

Figure 13.35: Document outline showing selected elements

2. Choose **Embed In** | **Stack View** from the **Editor** menu.
3. Select **Stack View** in the document outline. In the Attributes inspector, under **Stack View**, set **Alignment** to **Center** and **Spacing** to `8`.
4. With the stack view selected, click the Add New Constraints button. Enter the following values to set the constraints for the stack view:

    - Top: `20`
    - Left: `20`
    - Right: `20`

Constrain to Margins should already be ticked, which sets a standard margin of 8 points. When done, click the **Add 3 Constraints** button. The stack view will now look like the following screenshot:

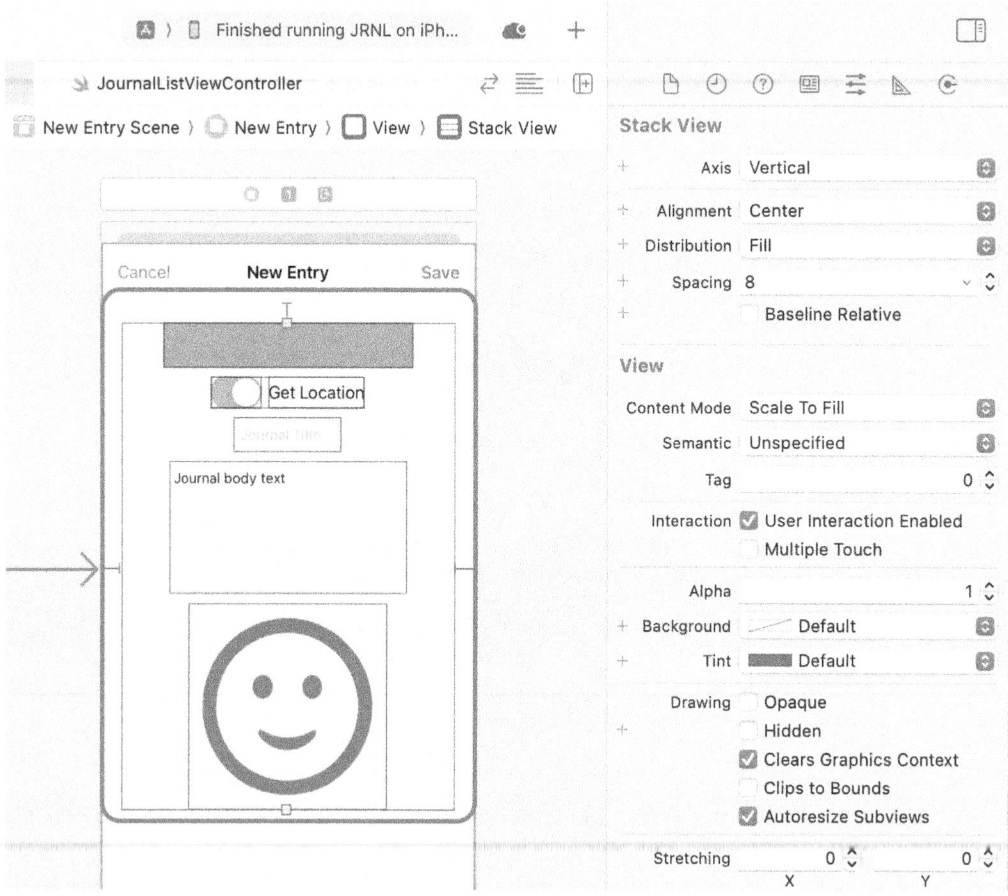

Figure 13.36: Stack view with constraints applied

Note that all the red lines are gone. The space between the top, right, and left edges of the stack view and the corresponding edges of the enclosing view have been set to `20 + 8` points. The position of the bottom edge of the stack view is automatically derived by the heights of all the elements it contains.

5. You will see that the text field does not extend to the full width of the stack view. To fix this, select **Journal Title** in the document outline, and click the Add New Constraints button. Set the right constraint to `8`, and click the **Add 1 Constraint** button. The text field will be extended to almost the full width of the stack view:

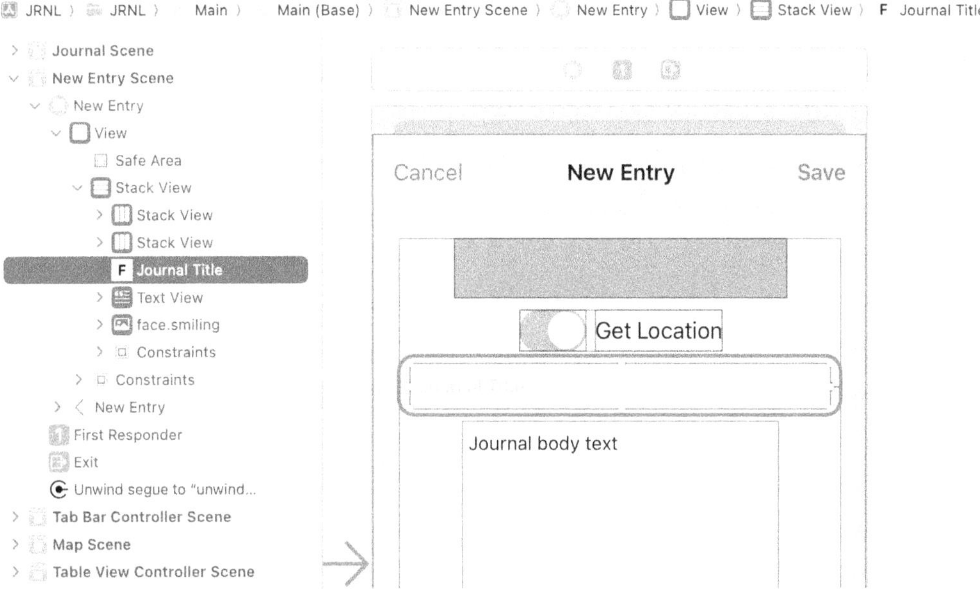

Figure 13.37: Text field in the stack view with a constraint applied

6. Note that the text view also does not extend to the full width of the stack view. Select **Text View** in the document outline, and click the Add New Constraints button. Set the right constraint to 8, and click the **Add 1 Constraint** button. The text view will be extended to almost the full width of the stack view:

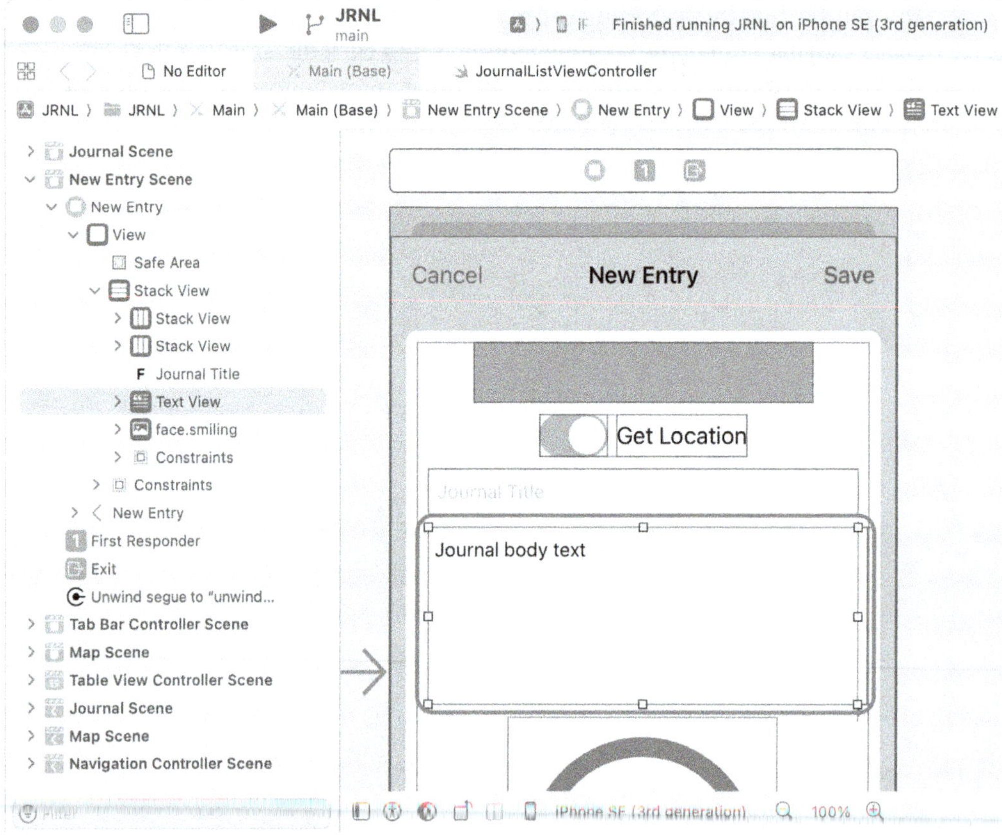

Figure 13.38: Text view in the stack view with a constraint applied

All the positioning issues have now been resolved.

7. Build and run your app, and tap the + button to display the Add New Journal Entry screen:

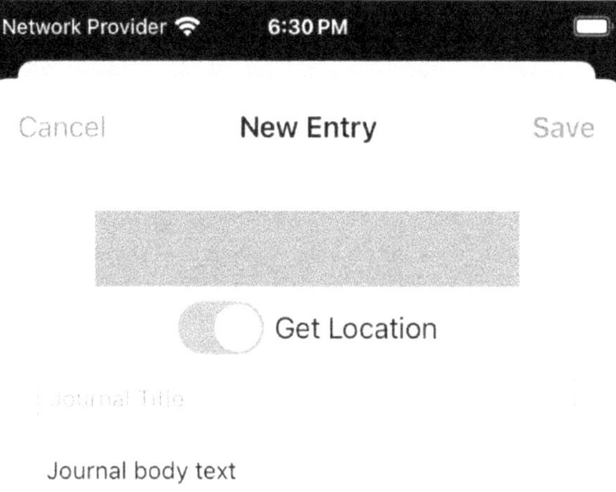

Figure 13.39: Simulator showing the Add New Journal Entry screen

You have added all the required user interface elements and constraints to the Add New Journal Entry screen. Great job! In the next section, you'll configure the static table view and add user interface elements to the Journal Entry Detail screen.

Modifying the Journal Entry Detail screen

Let's see what the Journal Entry Detail screen looks like in the app tour:

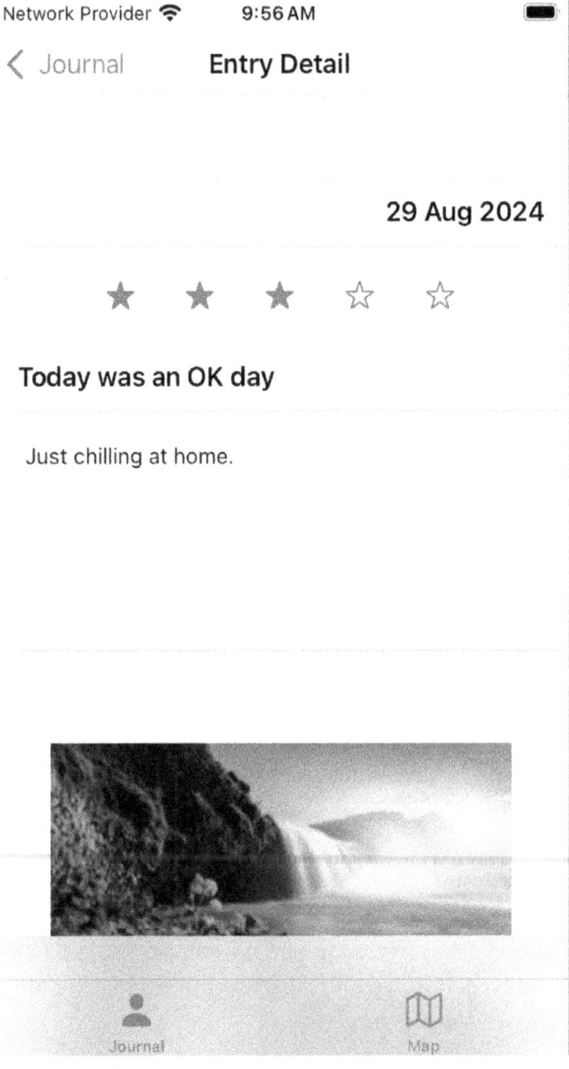

Figure 13.40: The Journal Entry Detail screen for the completed Journal app

Scrolling up reveals the remaining part of the **Journal Entry Detail** screen:

Figure 13.41: The remainder of the Journal Entry Detail screen

As you can see, the **Journal Entry Detail** screen has the following elements:

- A label showing the date
- A custom view showing star ratings
- A label for the journal entry title
- A label for the journal entry body
- An image view for a photo that you will take with your phone's camera
- An image view showing a map location

Also, you need to scroll to view the entire screen. You will now modify it to match the design shown in the app tour, beginning by setting the number and size of the table view cells in the next section.

Configuring the number and size of static table view cells

In *Chapter 12, Finishing Up Your User Interface*, you added a table view controller scene to the `Main` storyboard file and configured it to use a fixed number of cells. This will represent the Journal Entry Detail screen. Now, you will set the number and size of the cells to match the layout shown in the app tour. Follow these steps:

1. In the **Main** storyboard file, click **Table View Controller Scene** in the document outline, and click **Navigation Item**:

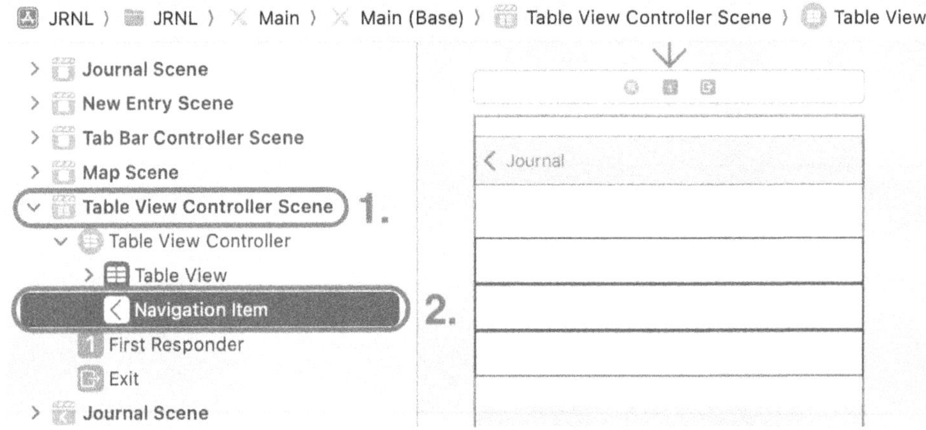

Figure 13.42: Document outline with Table View Controller Scene selected

2. Click the Attributes inspector button, and under **Navigation Item**, set **Title** to `Entry Detail`.
3. Select **Table View Section** in the document outline:

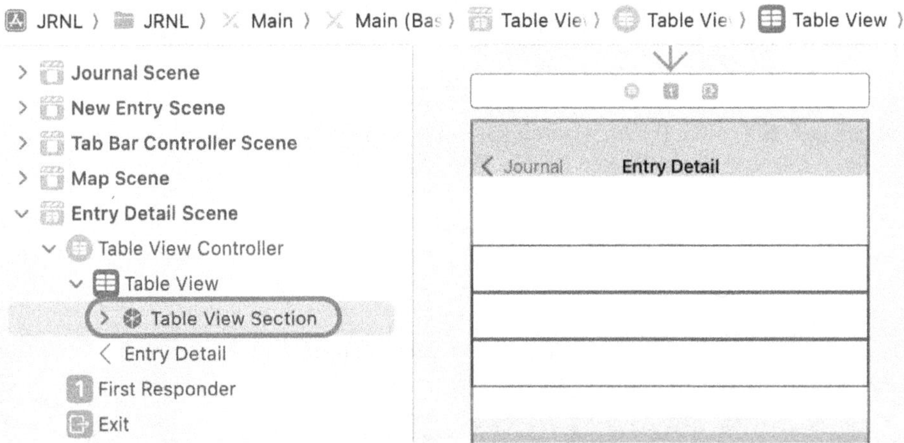

Figure 13.43: Document outline with Table View Section selected

4. In the Attributes inspector under **Table View Section**, set **Rows** to 7.

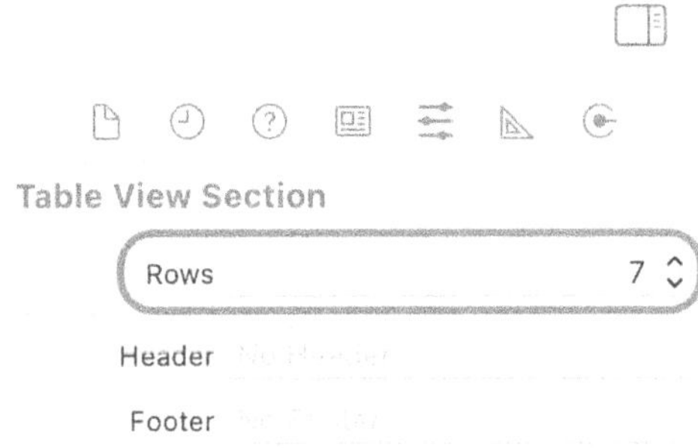

Figure 13.44: Attributes inspector showing 7 rows

5. Click the second **Table View Cell** under **Table View Section** in the document outline:

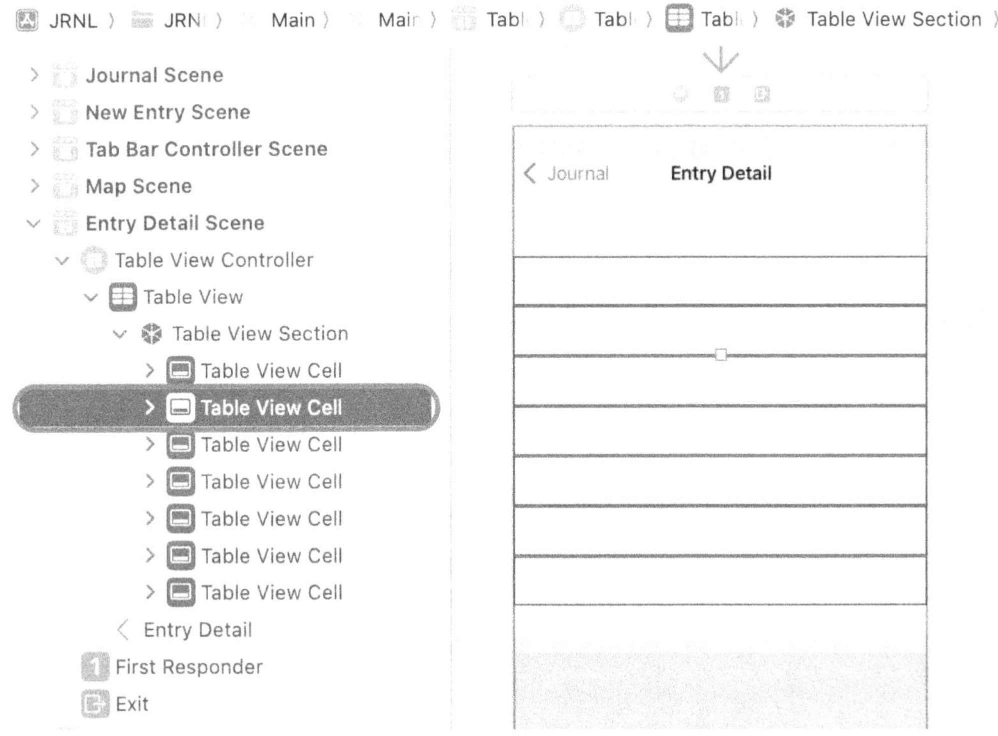

Figure 13.45: Document outline showing second Table View Cell

6. Click the Size inspector button. Under **Table View Cell**, set **Row Height** to `60`:

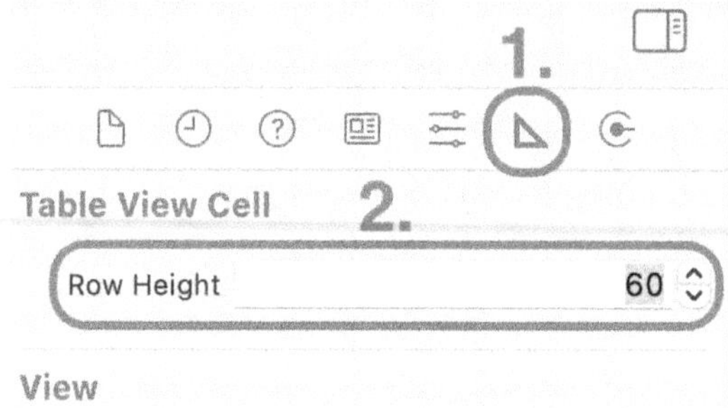

Figure 13.46: Size inspector showing Row Height set to 60

7. Click the fourth **Table View Cell** under **Table View Section** in the document outline. In the Size inspector, under **Table View Cell**, set **Row Height** to `150`.
8. Click the fifth **Table View Cell** under **Table View Section** in the document outline. In the Size inspector, under **Table View Cell**, set **Row Height** to `316`.
9. Repeat the previous step with the sixth **Table View Cell.**

You have set the number and size of the cells to match the layout shown in the app tour. In the next section, you'll add the user interface elements to each cell.

Adding user interface elements to static table view cells

In the previous section, you used a stack view to help you manage multiple user interface elements in the Add New Journal Entry screen. Here, you'll use a table view with static table view cells instead. The advantage of using a static table view is that view scrolling is built in, so it can accommodate views that are taller than the device screen. Follow these steps:

1. In the document outline, click the first **Table View Cell**, and drag a **Label** object into it from the library. In the Attributes inspector, under **Label**, use the **Font** menu to set **Style** to `Semibold` and set **Alignment** to right.
2. Click the Add New Constraints button, and set the top, left, and right constraints of the label to `0`. Make sure that **Constrain to Margins** is ticked, and click the **Add 3 Constraints** button. The label will now look like the following screenshot:

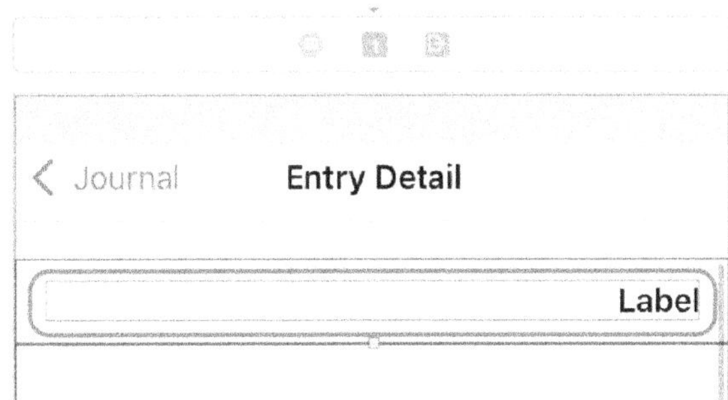

Figure 13.47: Label with the constraints applied

3. In the document outline, click the second **Table View Cell**, and drag a **Horizontal Stack View** object into it from the library. In the Attributes inspector, under **Stack View**, set **Spacing** to `8`. Under **View**, set **Background** to `System Cyan color`.
4. Click the Size inspector. Under **View**, set **Width** to `252` and **Height** to `44`.
5. Click the Add New Constraints button, and set the **Width** and **Height** constraints (their values should already be present). Click the **Add 2 Constraints** button.
6. Click the **Align** button, and tick **Horizontally in Container** and **Vertically in Container**. Click the **Add 2 Constraints** button.

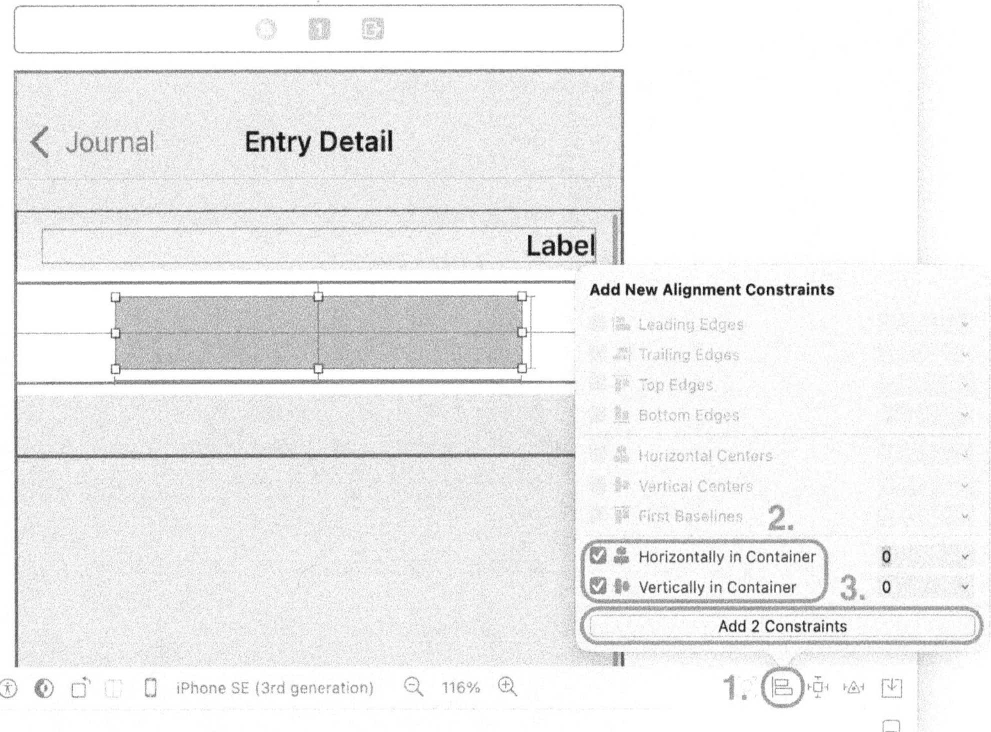

Figure 13.48: Align button and dialog box

7. The stack view will now look like the following screenshot:

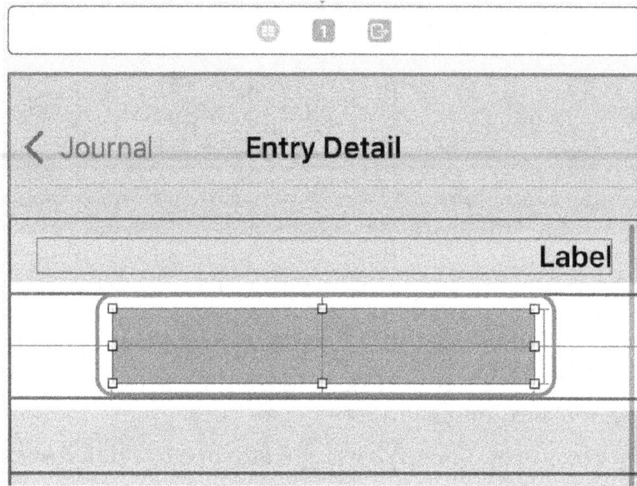

Figure 13.49: Stack view with the constraints applied

8. Click the third **Table View Cell**, and drag a **Label** object into it from the library. In the Attributes inspector, under **Label**, set **Style** to `Semibold` and **Alignment** to left.
9. Click the Add New Constraints button, and set the top, left, and right constraints of **Label** to `0`. Make sure that **Constrain to Margins** is ticked, and click the **Add 3 Constraints** button. The label will now look like the following screenshot:

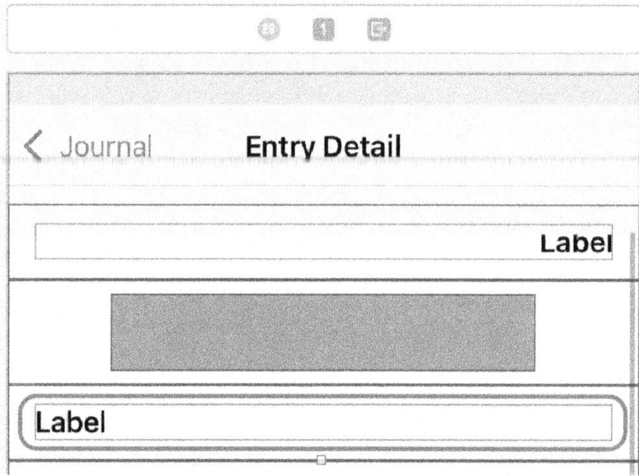

Figure 13.50: Label with the constraints applied

10. Click the fourth **Table View Cell**, and drag a **Text View** object into it from the library. In the Attributes inspector, under **Text View**, uncheck **Editable** and **Selectable.** You can also change the default text if you wish.

11. Click the Add New Constraints button, and set the top, left, right, and bottom constraints to `0`. Make sure that **Constrain to Margins** is ticked, and click the **Add 4 Constraints** button. The text view will now look like the following screenshot:

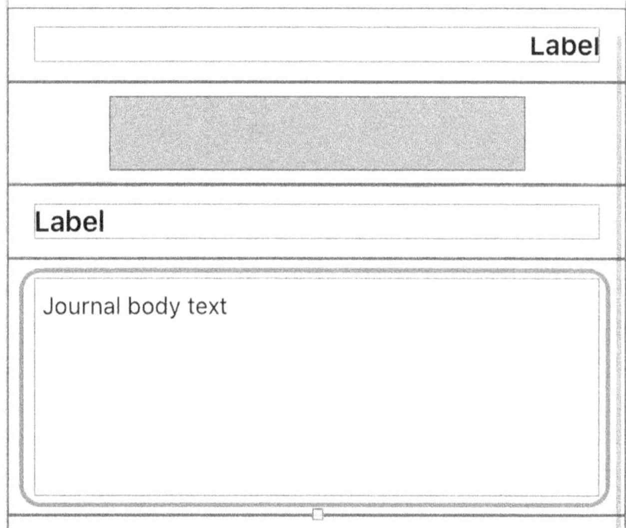

Figure 13.51: Text View with the constraints applied

12. Click the fifth **Table View Cell**, and drag an **Image View** object into it from the library. In the Attributes inspector, under **Image View**, set **Image** to `face.smiling`.
13. In the Size inspector, under **View**, set **Width** and **Height** to `300`.
14. Click the Add New Constraints button, and set the **Width** and **Height** constraints (their values should already be present). Click the **Add 2 Constraints** button.
15. Click the Align button, and tick **Horizontally in Container** and **Vertically in Container**. Click the **Add 2 Constraints** button. The image view will now look like the following screenshot:

Figure 13.52: Image View object with constraints applied

16. Click the sixth **Table View Cell**, and drag an **Image View** object into it from the library. In the Attributes inspector, under **Image View**, set **Image** to `map`.

17. Repeat *steps 13* to *15* for this image view. It will now look like the following screenshot:

Figure 13.53: Image View object with constraints applied

18. All the required user interface elements have been added to the **Entry Detail** scene. Build and run your app, and tap a row on the Journal List screen to navigate to the Journal Entry Detail screen:

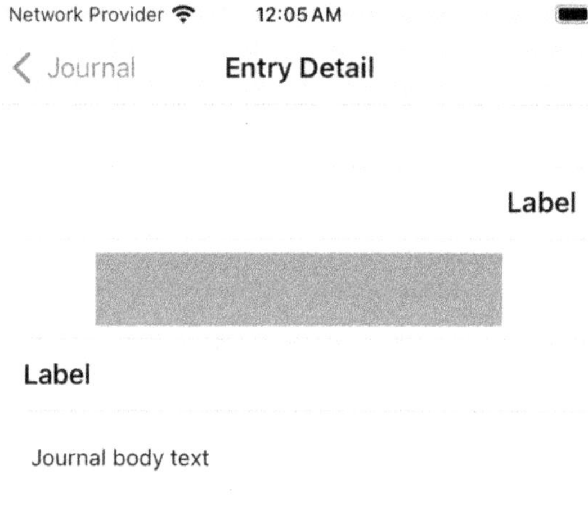

Figure 13.54: Simulator showing the Journal Entry Detail screen

19. Scroll down to see the remaining part of the Journal Entry Detail screen.

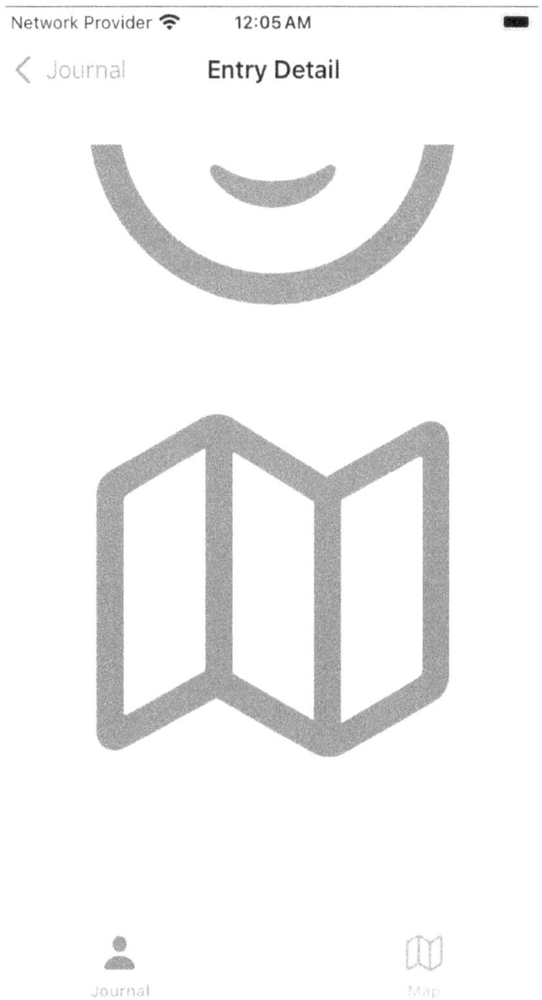

Figure 13.55: Simulator showing the remainder of the Journal Entry Detail screen

Excellent! You have modified all the screens for your app, and you're ready to add functionality to them in the next part of this book.

Summary

In this chapter, you modified the Journal List, Add New Journal Entry, and Journal Entry Detail screens to match the design shown in the app tour. For the Journal List screen, you modified the `journalCell` table view cell by adding an image view and two labels to it. You modified the Add New Journal Entry screen by adding a custom view, a switch, a text field, a text view, and an image view to it. You also configured the image view to show a default image. For the Journal Entry Detail screen, you added a text view, labels, and image views to it, configuring the image views to show default images.

You now have more experience in how to use Interface Builder to add and configure multiple user interface elements, set their sizes and positions using the Size inspector, and apply the necessary constraints using the Add New Constraints and Align buttons. This will be useful to ensure compatibility with different screen sizes and orientations. You should also be able to easily prototype the appearance and flow of your apps.

You're now finished with the storyboard and design setup. You can go through every screen that your app has and see what they look like, even though none of the screens have actual data in them. If this app was a house being built, it's as though you've built all the walls and floors, and the house is now ready to have the interior done. Great job!

This concludes *Part 2* of this book. In the next part, you'll begin to type in all the code required for your app to work. In the next chapter, you'll start by learning more about the **Model-View-Controller** design pattern. You'll also learn how table views work, which is crucial for understanding how the Journal List screen works.

Leave a review!

Thank you for purchasing this book from Packt Publishing—we hope you enjoy it! Your feedback is invaluable and helps us improve and grow. Once you've completed reading it, please take a moment to leave an Amazon review; it will only take a minute, but it makes a big difference for readers like you. Scan the QR code below or visit the link to receive a free ebook of your choice.

`https://packt.link/NzOWQ`

Part 3

Code

Welcome to *Part 3* of this book. With your user interface complete, you will now add code to implement your app's functionality. To display your data in a list, you will use a table view, and you will learn how to use an array as a data source. Next, you'll learn how to pass data from a view controller to the array used as a data source, and pass data from one view controller to another. You will also look at how to determine the device location and how to display a map containing annotations. After that, you will learn how to persist app data using JSON files. Then, you'll learn how to create custom views, use the device camera and photo library, and add search capability to your table view. Finally, you'll replace the table view with a collection view to make your app suitable for larger screens such as a Mac or iPad.

This part comprises the following chapters:

- *Chapter 14, Getting Started with MVC and Table Views*
- *Chapter 15, Getting Data into Table Views*
- *Chapter 16, Passing Data between View Controllers*
- *Chapter 17, Getting Started with Core Location and MapKit*
- *Chapter 18, Getting Started with JSON Files*
- *Chapter 19, Getting Started with Custom Views*
- *Chapter 20, Getting Started with the Camera and Photo Library*
- *Chapter 21, Getting Started with Search*
- *Chapter 22, Getting Started with Collection Views*

By the end of this part, you'll have completed the *JRNL* app. You'll have the experience of building a complete app from scratch, which will be useful as you build your own apps. Let's get started!

14

Getting Started with MVC and Table Views

In the previous chapter, you modified the Journal List screen, the Add New Journal Entry screen, and the Journal Entry Detail screen to match the app tour shown in *Chapter 10, Setting Up the User Interface*. You have completed the initial UI for the *JRNL* app, and that concluded *Part 2* of this book.

This chapter begins *Part 3* of this book, where you will focus on the code that makes your app work. In this chapter, you will learn about the **Model-View-Controller** (**MVC**) design pattern and how the different parts of an app interact with one another. Then, you'll implement a table view programmatically (which means implementing it using code instead of storyboards) using a playground, to understand how table views work. Finally, you'll revisit the table view you implemented on the Journal List screen, so you can see the differences between implementing it in a storyboard and implementing it programmatically.

By the end of this chapter, you'll understand the MVC design pattern, you'll have learned how to create a table view controller programmatically, and you'll know how to use table view delegate and data source protocols.

The following topics will be covered in this chapter:

- Understanding the MVC design pattern
- Understanding table views
- Revisiting the Journal List screen

Technical requirements

The completed Xcode playground and project for this chapter are in the `Chapter14` folder of the code bundle for this book, which can be downloaded here:

`https://github.com/PacktPublishing/iOS-18-Programming-for-Beginners-Ninth-Edition`

Check out the following video to see the code in action:

`https://youtu.be/tvnmgByShF4`

Create a new playground and call it `TableViewBasics`. You can use this playground to type in and run all the code in this chapter as you go along. Before you do, let's look at the MVC design pattern, an approach commonly used when writing iOS apps.

Understanding the MVC design pattern

The **Model-View-Controller** (**MVC**) design pattern is a common approach used when building iOS apps. MVC divides an app into three different parts:

- **Model:** This handles data storage and representation and data processing tasks.
- **View:** This includes all the things that are on the screen that the user can interact with.
- **Controller:** This manages the flow of information between the model and view.

One notable feature of MVC is that the view and model do not interact with one another; instead, all communication is managed by the controller.

For example, imagine you're at a restaurant. You look at a menu and choose something you want. Then, a waiter comes, takes your order, and sends it to the cook. The cook prepares your order, and, when it is done, the waiter takes the order and brings it out to you. In this scenario, the menu is the view, the waiter is the controller, and the cook is the model. Also, note that all interactions between you and the kitchen are only through the waiter; there is no interaction between you and the cook.

To find out more about MVC, visit `https://developer.apple.com/library/archive/documentation/General/Conceptual/DevPedia-CocoaCore/MVC.html`.

To see how MVC works in the context of an iOS app, let's learn more about view controllers. You will see what it takes to implement a view controller that is required to manage a table view, which is used on the Journal List screen.

Exploring view controllers

So far, you have implemented `JournalListViewController`, a view controller that manages the table view on the Journal List screen. However, you still haven't learned how the code you added to it works, so let's look at that now.

You may wish to re-read *Chapter 11, Building Your User Interface,* where you implemented the `JournalListViewController` class.

When an iOS app is launched, the view controller for the first screen to be displayed is loaded. The view controller has a `view` property and automatically loads the view instance assigned to its `view` property. That view may have subviews, which are also loaded.

If one of the subviews is a table view, it will have `dataSource` and `delegate` properties. The `dataSource` property is assigned to an object that provides data to the table view. The `delegate` property is assigned to an object that handles user interaction with the table view. Typically, the view controller for the table view will be assigned to the table view's `dataSource` and `delegate` properties.

The method calls that a table view will send to its view controller are declared in the `UITableViewDataSource` and `UITableViewDelegate` protocols. Remember that protocols only provide method declarations; the implementation of those method calls is in the view controller. The view controller will then get the data from the model objects and provide it for the table view. The view controller also handles user input and modifies the model objects as required.

Let's take a closer look at table views and table view protocols in the next section.

Understanding table views

The *JRNL* app uses a table view on the Journal List screen. A table view presents table view cells using rows arranged in a single column.

To learn more about table views, visit `https://developer.apple.com/documentation/uikit/uitableview`.

The data displayed by a table view is usually provided by a view controller. A view controller providing data for a table view must conform to the `UITableViewDataSource` protocol. This protocol declares a list of methods that tells the table view how many cells to display and what to display in each cell.

To learn more about the `UITableViewDataSource` protocol, visit `https://developer.apple.com/documentation/uikit/uitableviewdatasource`.

To enable user interaction, a view controller for a table view must also conform to the `UITableViewDelegate` protocol, which declares a list of methods that are triggered when a user interacts with the table view.

To learn more about the `UITableViewDelegate` protocol, visit `https://developer.apple.com/documentation/uikit/uitableviewdelegate`.

To learn how table views work, you'll implement a view controller subclass that controls a table view in your `TableViewBasics` playground. Since there is no storyboard in the playground, you can't add the UI elements using the library, as you did in the previous chapters. Instead, you will do everything programmatically.

You'll start by creating the `TableViewExampleController` class, an implementation of a view controller that manages a table view. After that, you'll create an instance of `TableViewExampleController` and make it display a table view in the playground's live view. Follow these steps:

1. Open the `TableViewBasics` playground that you have created, remove the `var` statement, and add an `import PlaygroundSupport` statement. Your playground should now contain the following:

   ```
   import UIKit
   import PlaygroundSupport
   ```

   The first `import` statement imports the API for creating iOS apps. The second statement enables the playground to display a live view, which you will use to display the table view.

2. Add the following code after the `import` statements to declare the `TableViewExampleController` class:

   ```
   class TableViewExampleController: UIViewController {
   }
   ```

   This class is a subclass of `UIViewController`, a class that Apple provides to manage views on the screen.

3. Add the following code inside the curly braces to declare a table view property and an array property in the `TableViewExampleController` class:

   ```
   class TableViewExampleController: UIViewController {
     var tableView: UITableView?
     var journalEntries: [[String]] = [
       ["sun.max","12 Sept 2024","Nice weather today"],
       ["cloud.rain","13 Sept 2024","Heavy rain today"],
       ["cloud.sun","14 Sept 2024","It's cloudy out"]
     ]
   }
   ```

   The `tableView` property is an optional property that will be assigned a `UITableView` instance. The `journalEntries` array is the model object that will be used to provide data to the table view.

You have just declared and defined the initial implementation of the `TableViewExampleController` class. Cool! In the next section, you'll learn how to set the number of cells for a table view to display, how to set the contents of each cell, and how to remove a row from a table view.

Conforming to the UITableViewDataSource protocol

A table view presents table view cells using rows arranged in a single column. However, before it can do this, it needs to know how many cells to display and what to put in each cell. To provide this information to the table view, you will make the `TableViewExampleController` class conform to the `UITableViewDataSource` protocol.

This protocol has two required methods:

- `tableview(_:numberOfRowsInSection:)` is called by the table view to determine how many table view cells to display.
- `tableView(_:cellForRowAt:)` is called by the table view to determine what to display in each table view cell.

The `UITableViewDataSource` protocol also has an optional method, `tableView(_:commit:forRowAt:)`, which is called by the table view when the user swipes left on a row. You'll use this method to handle what happens when a user wants to delete a row.

Let's add some code to make the `TableViewExampleController` class conform to the `UITableViewDataSource` protocol. Follow these steps:

1. To make the `TableViewExampleController` class adopt the `UITableViewDataSource` protocol, type a comma after the superclass name and then type `UITableViewDataSource`. Your code should look like this:

   ```
   class TableViewExampleController: UIViewController, UITableViewDataSource
   {
   ```

2. An error will appear because you haven't implemented the two required methods. Click the error icon:

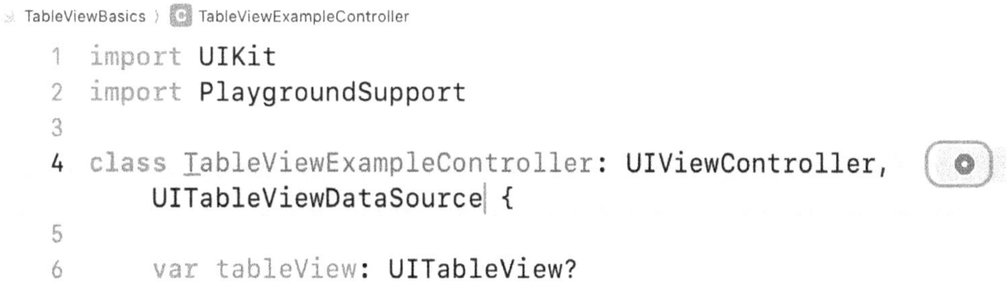

   TableViewBasics 〉 TableViewExampleController

   ```
   import UIKit
   import PlaygroundSupport

   class TableViewExampleController: UIViewController,
       UITableViewDataSource {

       var tableView: UITableView?
   ```

   Figure 14.1: Editor area showing error icon

   The error message states that the `TableViewExampleController` class does not conform to the `UITableViewDataSource` protocol.

3. Click the **Fix** button to add the stubs needed for conformance to the class:

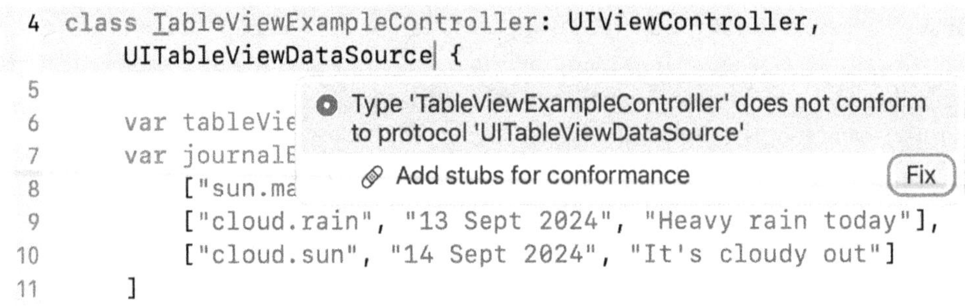

```
class TableViewExampleController: UIViewController,
    UITableViewDataSource {

    var tableVie
    var journalE
        ["sun.ma
        ["cloud.rain", "13 Sept 2024", "Heavy rain today"],
        ["cloud.sun", "14 Sept 2024", "It's cloudy out"]
    ]
```

Type 'TableViewExampleController' does not conform to protocol 'UITableViewDataSource'

Add stubs for conformance Fix

Figure 14.2: Error explanation and Fix button

4. Verify that your code looks like this:

```
class TableViewExampleController: UIViewController,
UITableViewDataSource {
  func tableView(_ tableView: UITableView,
  numberOfRowsInSection section: Int) -> Int {
    code
  }
  func tableView(_ tableView: UITableView, cellForRowAt
  indexPath: IndexPath) -> UITableViewCell  {
    code
  }
  var tableView: UITableView?
  var journalEntries: [[String]] = [
    ["sun.max","12 Sept 2024","Nice weather today"],
    ["cloud.rain","13 Sept 2024","Heavy rain today"],
    ["cloud.sun","14 Sept 2024","It's cloudy out"]
  ]
}
```

5. In a class definition, convention dictates that properties are declared at the top before any method declarations. Rearrange the code so that the property declarations are at the top, as follows:

```
class TableViewExampleController: UIViewController,
UITableViewDataSource {
  var tableView: UITableView?
  var journalEntries: [[String]] = [
    ["sun.max","12 Sept 2024","Nice weather today"],
    ["cloud.rain","13 Sept 2024","Heavy rain today"],
    ["cloud.sun","14 Sept 2024","It's cloudy out"]
  ]
```

```
func tableView(_ tableView: UITableView,
numberOfRowsInSection section: Int) -> Int {
```

6. To make the table view display a row for each element inside the `journalEntries` array, click the **code** placeholder inside the `tableView(_:numberOfRowsInSection:)` method definition and type in `journalEntries.count`. The completed method should look like this:

   ```
   func tableView(_ tableView: UITableView, numberOfRowsInSection section:
   Int) -> Int {
     journalEntries.count
   }
   ```

   `journalEntries.count` returns the number of elements inside the `journalEntries` array. Since there are three elements in it, this will make the table view display three rows.

7. To make the table view display journal entry details in each cell, click the **code** placeholder inside the `tableView(_:cellForRowAt:)` method definition and type the following:

   ```
   func tableView(_ tableView: UITableView, cellForRowAt indexPath:
   IndexPath) -> UITableViewCell {
     let cell = tableView.dequeueReusableCell(withIdentifier:
     "cell", for: indexPath)
     let journalEntry = journalEntries[indexPath.row]
     var content = cell.defaultContentConfiguration()
     content.image = UIImage(systemName: journalEntry[0])
     content.text = journalEntry[1]
     content.secondaryText = journalEntry[2]
     cell.contentConfiguration = content
     return cell
   }
   ```

   Let's break this down:

   ```
   let cell = tableView.dequeueReusableCell(withIdentifier: "cell", for:
   indexPath)
   ```

   This statement creates a new table view cell or reuses an existing table view cell and assigns it to `cell`. Imagine you have 1,000 items to display in a table view. You don't need 1,000 rows containing 1,000 table view cells—you only need just enough to fill the screen. Table view cells that scroll off the top of the screen can be reused to display items that appear at the bottom of the screen, and vice versa. As table views can display more than one type of cell, you set the cell reuse identifier to `"cell"` to identify this particular table view cell type. This identifier will be registered with the table view later.

   ```
   let journalEntry = journalEntries[indexPath.row]
   ```

The indexPath value locates the row in the table view. The first row has an indexPath containing section 0 and row 0. indexPath.row returns 0 for the first row, so this statement assigns the first element in the journalEntries array to journalEntry.

```
var content = cell.defaultContentConfiguration()
```

By default, a UITableViewCell instance can store an image, a text string, and a secondary text string. You set these by using the table view cell's content configuration property. This statement retrieves the default content configuration for the table view cell's style and assigns it to a variable, content.

```
content.image = UIImage(systemName: journalEntry[0])
content.text = journalEntry[1]
content.secondaryText = journalEntry[2]
cell.contentConfiguration = content
```

These statements update content with details from journalEntry, which is an array that has three elements. The first element is used to specify an image that is assigned to the image property. The second element is assigned to the text property. The third element is assigned to the secondaryText property. The last line assigns content to the table view cell's contentConfiguration property.

```
return cell
```

This statement returns the table view cell, which is then displayed on the screen.

The tableView(_:cellForRowAt:) method is executed for each row in the table view.

8. To handle table view cell deletion, type the following code after the implementation of tableView(_:cellForRowAt:):

```
func tableView(_ tableView: UITableView, commit editingStyle:
UITableViewCell.EditingStyle, forRowAt indexPath: IndexPath) {
  if editingStyle == .delete {
    journalEntries.remove(at: indexPath.row)
    tableView.reloadData()
  }
}
```

This removes the journalEntries element corresponding to the table view cell that the user swiped left on and reloads the table view.

The TableViewExampleController class now conforms to the UITableViewDataSource protocol. In the next section, you will make it conform to the UITableViewDelegate protocol.

Conforming to the UITableViewDelegate protocol

A user can tap on a table view cell to select it. To handle user interaction, you will make the `TableViewExampleController` class conform to the `UITableViewDelegate` protocol. You will implement one optional method from this protocol, `tableView(_:didSelectRowAt:)`, which is called by the table view when the user taps a row. Follow these steps:

1. To make the `TableViewExampleController` class adopt the `UITableViewDelegate` protocol, type a comma after `UITableViewDataSource` in the class declaration and then type `UITableViewDelegate`. Your code should look like this:

```
class TableViewExampleController: UIViewController,
UITableViewDataSource, UITableViewDelegate {
```

2. Type the following after the implementation of `tableView(_:commit:forRowAt:)`:

```
func tableView(_ tableView: UITableView, didSelectRowAt indexPath:
IndexPath) {
  let selectedJournalEntry = journalEntries[indexPath.row]
  print(selectedJournalEntry)
}
```

   This method will get the `journalEntries` array element corresponding to the tapped row and print it to the Debug area.

3. Verify that your `TableViewExampleController` class looks like this:

```
class TableViewExampleController: UIViewController,
UITableViewDataSource, UITableViewDelegate {
  var tableView: UITableView?
  var journalEntries: [[String]] = [
    ["sun.max","12 Sept 2024","Nice weather today"],
    ["cloud.rain","13 Sept 2024","Heavy rain today"],
    ["cloud.sun","14 Sept 2024","It's cloudy out"]
  ]
  func tableView(_ tableView: UITableView,
  numberOfRowsInSection section: Int) -> Int {
    journalEntries.count
  }
  func tableView(_ tableView: UITableView, cellForRowAt
  indexPath: IndexPath) -> UITableViewCell {
    let cell = tableView.dequeueReusableCell(
    withIdentifier: "cell", for: indexPath)
    let journalEntry = journalEntries[indexPath.row]
    var content = cell.defaultContentConfiguration()
```

```
      content.image = UIImage(systemName: journalEntry[0])
      content.text = journalEntry[1]
      content.secondaryText = journalEntry[2]
      cell.contentConfiguration = content
      return cell
    }
    func tableView(_ tableView: UITableView, commit
    editingStyle: UITableViewCell.EditingStyle,
    forRowAt indexPath: IndexPath) {
      if editingStyle == .delete {
        journalEntries.remove(at: indexPath.row)
        tableView.reloadData()
      }
    }
    func tableView(_ tableView: UITableView, didSelectRowAt
    indexPath: IndexPath) {
      let selectedjournalEntry =
      journalEntries[indexPath.row]
      print(selectedjournalEntry)
    }
  }
```

The `TableViewExampleController` class now conforms to the `UITableViewDelegate` protocol.

You have completed the implementation of the `TableViewExampleController` class. In the next section, you'll learn how to create an instance of this class.

Creating a TableViewExampleController instance

Now that you have declared and defined the `TableViewExampleController` class, you will write a method to create an instance of it. Follow these steps:

1. Type in the following code after the `journalEntries` variable declaration to declare a new method:

   ```
   func createTableView() {
   }
   ```

   This declares a new method, `createTableView()`, which you'll use to create an instance of a table view and assign it to the `tableView` property.

2. Type in the following code after the opening curly brace:

   ```
   tableView = UITableView(frame: CGRect(x: 0, y: 0, width: view.frame.
   width, height: view.frame.height))
   ```

   This creates a new table view instance and assigns it to `tableView`.

3. On the next line, type in the following code to set the table view's `dataSource` and `delegate` properties to an instance of `TableViewExampleController`:

    ```
    tableView?.dataSource = self
    tableView?.delegate = self
    ```

    The `dataSource` and `delegate` properties of a table view specify the object that contains the implementation of the `UITableViewDataSource` and `UITableViewDelegate` methods.

4. On the next line, type in the following code to set the table view's background color:

    ```
    tableView?.backgroundColor = .white
    ```

5. On the next line, type in the following code to set the identifier for the table view cells to `"cell"`:

    ```
    tableView?.register(UITableViewCell.self, forCellReuseIdentifier: "cell")
    ```

    This identifier will be used in the `tableView(_:cellForRowAt:)` method to identify the type of table view cells to be used.

6. On the next line, type in the following code to add the table view as a subview of the view of the `TableViewExampleController` instance:

    ```
    view.addSubview(tableView!)
    ```

7. Verify that the completed method looks like the following:

    ```
    func createTableView() {
      tableView = UITableView(frame: CGRect(x: 0, y: 0,
      width: view.frame.width,
      height: view.frame.height))
      tableView?.dataSource = self
      tableView?.delegate = self
      tableView?.backgroundColor = .white
      tableView?.register(UITableViewCell.self,
      forCellReuseIdentifier: "cell")
      view.addSubview(tableView!)
    }
    ```

Now you must determine when to call this method. View controllers have a `view` property. The view assigned to the `view` property will be automatically loaded when the view controller is loaded. After a view has been loaded successfully, the view controller's `viewDidLoad()` method will be called. You will override the `viewDidLoad()` method in your `TableViewControllerExample` class to call `createTableView()`. Type in the following code just before the `createTableView()` method:

```
override func viewDidLoad() {
   super.viewDidLoad()
   view.bounds = CGRect(x: 0, y: 0, width: 375,
```

```
    height: 667)
    createTableView()
  }
```

This sets the size of the live view, creates a table view instance, assigns it to `tableView`, and adds it as a subview to the view of the `TableViewExampleController` instance. The table view then calls the data source methods to determine how many table view cells to display and what to display in each cell. `tableView(_:numberOfRowsInSection:)` returns the number of elements in `journalEntries`, so three table view cells will be displayed. `tableView(_:cellForRowAt:)` creates the cell, creates a new cell configuration, sets the properties of the cell configuration, and assigns the updated configuration to the cell.

Verify that your completed playground looks like this:

```
import UIKit
import PlaygroundSupport
class TableViewExampleController: UIViewController, UITableViewDataSource,
UITableViewDelegate {
  var tableView: UITableView?
  var journalEntries: [[String]] = [
    ["sun.max","12 Sept 2024","Nice weather today"],
    ["cloud.rain","13 Sept 2024","Heavy rain today"],
    ["cloud.sun","14 Sept 2024","It's cloudy out"]
  ]
  override func viewDidLoad() {
    super.viewDidLoad()
    view.bounds = CGRect(x: 0, y: 0, width: 375, height: 667)
    createTableView()
  }
  func createTableView() {
    tableView = UITableView(frame: CGRect(x: 0, y: 0,
    width: view.frame.width,
    height: view.frame.height))
    tableView?.dataSource = self
    tableView?.delegate = self
    tableView?.backgroundColor = .white
    tableView?.register(UITableViewCell.self,
    forCellReuseIdentifier: "cell")
    view.addSubview(tableView!)
  }
```

```
    func tableView(_ tableView: UITableView,
    numberOfRowsInSection section: Int) -> Int {
      journalEntries.count
    }
    func tableView(_ tableView: UITableView, cellForRowAt
    indexPath: IndexPath) -> UITableViewCell {
      let cell = tableView.dequeueReusableCell(withIdentifier:
      "cell", for: indexPath)
      let journalEntry = journalEntries[indexPath.row]
      var content = cell.defaultContentConfiguration()
      content.image = UIImage(systemName: journalEntry[0])
      content.text = journalEntry[1]
      content.secondaryText = journalEntry[2]
      cell.contentConfiguration = content
      return cell
    }
    func tableView(_ tableView: UITableView, commit editingStyle:
    UITableViewCell.EditingStyle, forRowAt indexPath: IndexPath) {
      if editingStyle == .delete {
        journalEntries.remove(at: indexPath.row)
        tableView.reloadData()
      }
    }
    func tableView(_ tableView: UITableView, didSelectRowAt
    indexPath: IndexPath) {
      let selectedjournalEntry = journalEntries[indexPath.row]
      print(selectedjournalEntry)
    }
  }
```

Now it's time to see it in action. Follow these steps:

1. Type the following after all the other code in the playground:

   ```
   PlaygroundPage.current.liveView = TableViewExampleController()
   ```

   This command creates the instance of `TableViewExampleController` and displays its view in the playground's live view. The `createTableView()` method will create a table view and add it as a subview to the `TableViewExampleController` instance's view, and it will appear on the screen.

2. To see a representation of the table view on your screen, the playground's live view must be enabled. Click the Adjust Editor Options button and verify that **Live View** is selected:

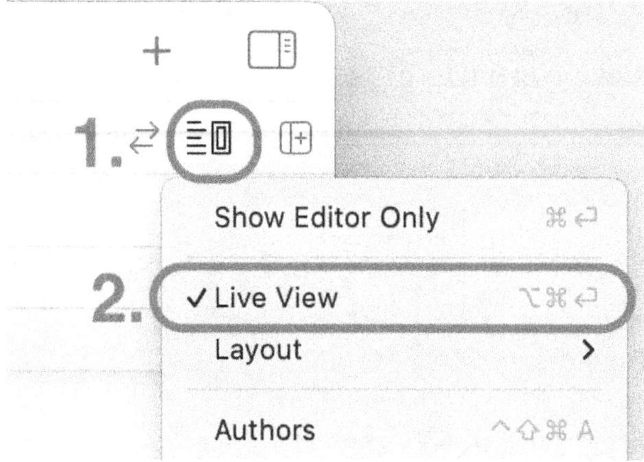

Figure 14.3: Adjust Editor Options menu with Live View selected

3. Run your code and verify that the table view is displaying three table view cells in the live view:

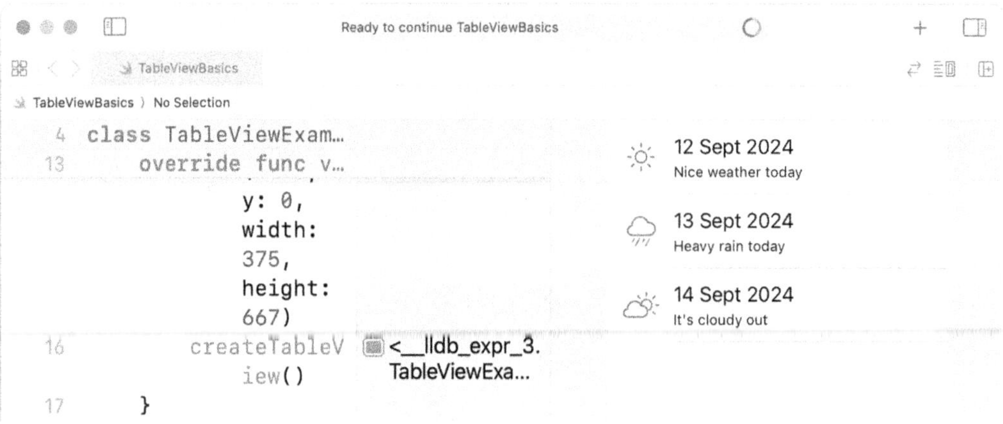

Figure 14.4: Playground live view showing table view with three table view cells

4. Tap a row. The journal entry details for that row will be printed in the Debug area:

Figure 14.5: Debug area showing journal entry details

5. Swipe left on a row. A **Delete** button will appear:

Figure 14.6: Table view row showing a Delete button

6. Tap the **Delete** button to remove the row from the table view:

Figure 14.7: Table view with one row removed

You've just created a view controller for a table view, created an instance of it, and displayed a table view in the playground's live view. Good job!

In the next section, you'll revisit how view controllers are used on the Journal List screen that you implemented in *Chapter 11, Building Your User Interface*. Using what you have learned in this section as a reference, you should have a better understanding of how it works.

Revisiting the Journal List screen

Remember the `JournalListViewController` class in *Chapter 11, Building Your User Interface*? This is an example of a view controller that manages a table view. Note that the code for this class is very similar to that in your playground. The differences are as follows:

- You created and assigned the table view to the `tableView` in `TableViewExampleController` programmatically, instead of using the assistant.
- You set the dimensions of the table view programmatically in `UITableView(frame:)`, instead of using the Size inspector and constraints.
- You connected the data source and delegate outlets to the view controller programmatically, instead of using the Connections inspector.
- You set the reuse identifier and UI element color programmatically, instead of using the Attributes inspector.
- You added the table view as a subview of the view for `TableViewExampleController` programmatically, instead of dragging in a **Table View** object from the library.

You may wish to open the *JRNL* project and review *Chapter 11, Building Your User Interface*, once more, to compare the table view implementation using the storyboard and the table view implementation done programmatically as you have done in this chapter.

Summary

In this chapter, you learned about the MVC design pattern and table view controllers in detail. You then revisited the table view used on the Journal List screen and learned how table view controllers work.

You should now understand the MVC design pattern, how to create a table view controller, and how to use the table view data source and delegate protocols. This will enable you to implement table view controllers for your own apps.

Up to this point, you have set up the views and view controllers for the Journal List screen, but it just displays a column of cells. In the next chapter, you're going to implement the model objects for the Journal List screen so it can display a list of journal entries. To do this, you will create structures to store data and provide it to the `JournalListViewController` instance so that it may be displayed by the table view on the Journal List screen.

Join us on Discord!

Read this book alongside other users, experts, and the author himself. Ask questions, provide solutions to other readers, chat with the author via Ask Me Anything sessions, and much more. Scan the QR code or visit the link to join the community.

`https://packt.link/ios-Swift`

15

Getting Data into Table Views

In the previous chapter, you learned about the **Model-View-Controller** (**MVC**) design pattern and table views. You also reviewed the table view in the Journal List screen. At this point, the Journal List screen displays cells that do not contain any data. As shown in the app tour in *Chapter 10, Setting Up the User Interface*, it should display a list of journal entries.

In this chapter, you're going to implement the model objects for the Journal List screen to make it display a list of journal entries. You'll start by learning about the model objects that you will use. Then, you'll create a Swift class that can store journal entry instances. After that, you'll create a static method that can return sample journal entry instances. This array will then be used as the data source for the table view on the Journal List screen.

By the end of this chapter, you'll have learned how to create model objects, how to create sample data, and how to configure view controllers to populate table views.

The following topics will be covered in this chapter:

- Understanding model objects
- Creating a class to represent a journal entry
- Creating sample data
- Displaying data in a collection view

Technical requirements

You will continue working on the `JRNL` project that you modified in *Chapter 13, Modifying App Screens*. The resource files and completed Xcode project for this chapter are in the `Chapter15` folder of the code bundle for this book, which can be downloaded here:

`https://github.com/PacktPublishing/iOS-18-Programming-for-Beginners-Ninth-Edition`

Check out the following video to see the code in action:

`https://youtu.be/TmQOr3Qy954`

Let's start by examining the model objects required to store journal entry data in the next section.

Understanding model objects

As you learned in *Chapter 14, Getting Started with MVC and Table Views*, a common design pattern for iOS apps is Model-View-Controller, or MVC. To recap, MVC divides an app into three different parts:

- **Model**: This handles data storage, representation, and data processing tasks.
- **View**: This is anything that is on the screen that the user can interact with.
- **Controller**: This manages the flow of information between the model and the view.

Let's revisit the design of the Journal List screen that you saw during the app tour, which looks like this:

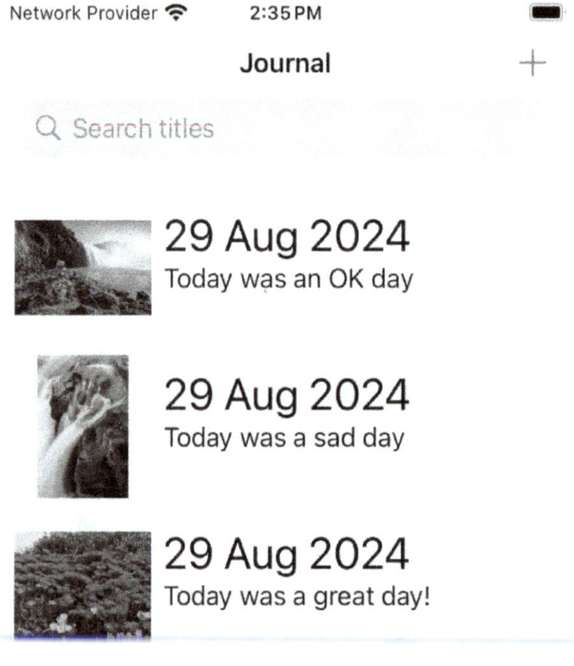

Figure 15.1: Simulator showing the Journal List screen from the app tour

Build and run your app, and the Journal List screen will look like this:

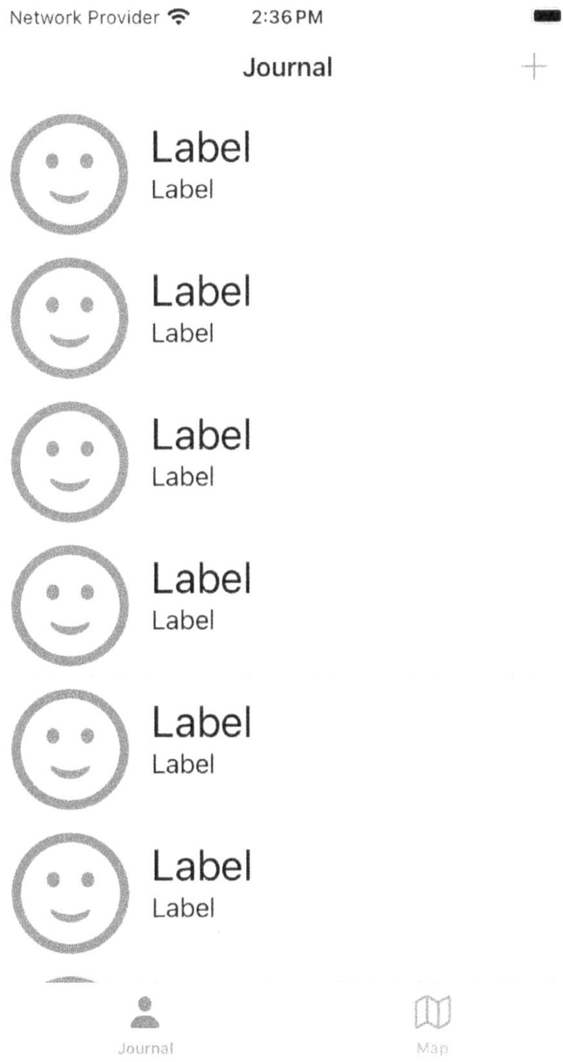

Figure 15.2: Simulator showing the Journal List screen from your app

As you can see, all the cells currently display placeholders. Based on the MVC design pattern, you have partly completed the implementation of the views required (the table view) and the controller (the `JournalListViewController` class). You will need to create a custom `UITableViewCell` instance to manage what your table view cell will display, and you need to add model objects that will provide the journal entry data.

Each journal entry will store the following:

- The date the entry was made
- A rating value

- Title text
- Body text
- An optional photo
- An optional geographical location

In *Chapter 14, Getting Started with MVC and Table Views*, you used an array of `String` arrays to represent journal entries. However, a `String` array can only store strings, and you must be able to store data types other than `String`. To resolve this, you will create a class named `JournalEntry` to store all the data required by a journal entry. Next, you will create a static method that returns sample data stored in `JournalEntry` instances. After that, you will create a custom `UITableView` class to manage the data displayed by the table view cells. Finally, you will modify the `JournalListViewController` class so that it can provide data for the table view to display.

Creating a class to represent a journal entry

To create a model object that can represent a journal entry in your app, you will add a new file to your project, `JournalEntry.swift`, and declare a `JournalEntry` class that has the required properties for a journal entry. Before you do so, you'll configure your project to use Swift 6 and change the main project folder to a group. Follow these steps:

1. In the Project navigator, click the **JRNL** icon. Click the **JRNL** target and click **Build Settings**. Scroll down to **Swift Compiler – Language** and set **Swift Language Version** to **Swift 6**:

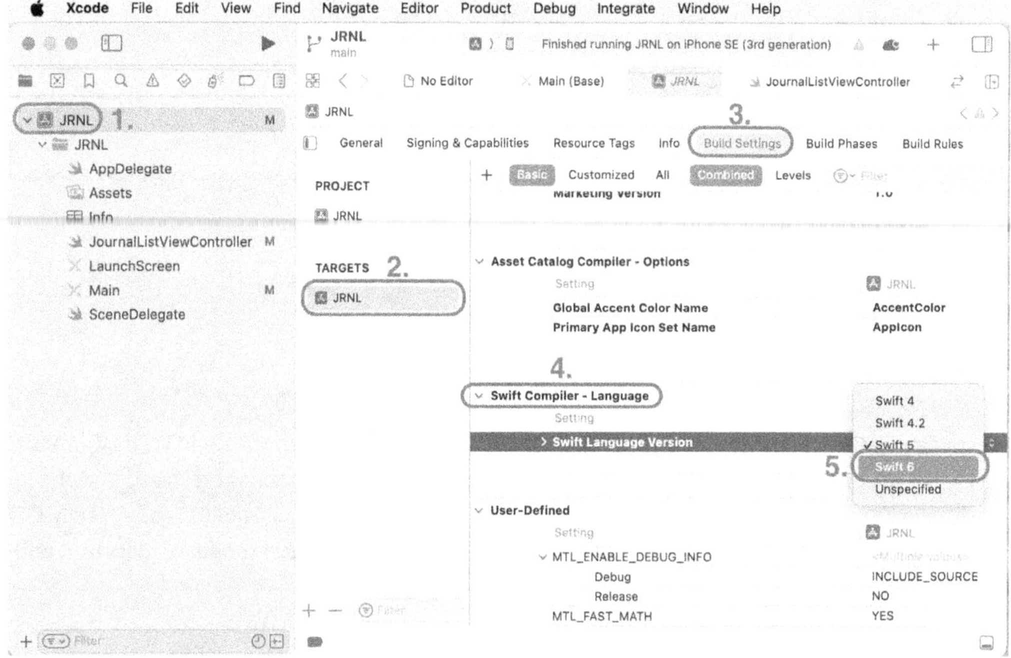

Figure 15.3: Editor area showing Build Settings with Swift 6 set

2. Right-click on the blue **JRNL** project folder under the **JRNL** icon and choose **Convert to Group**:

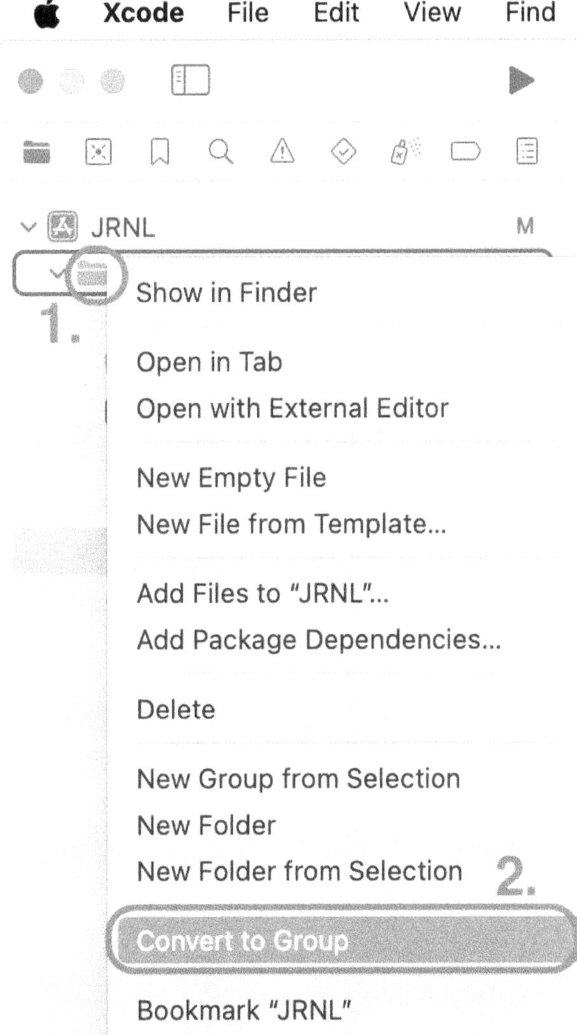

Figure 15.4: Pop-up menu with Convert to Group selected

The folder color will change from blue to dark gray. Making the folder a group will allow you to rearrange the order of the files in it.

To learn more about the differences between folders and groups in Xcode, visit: `https://developer.apple.com/documentation/xcode/managing-files-and-folders-in-your-xcode-project`.

3. Reorder the files in the group until they look like the screenshot below:

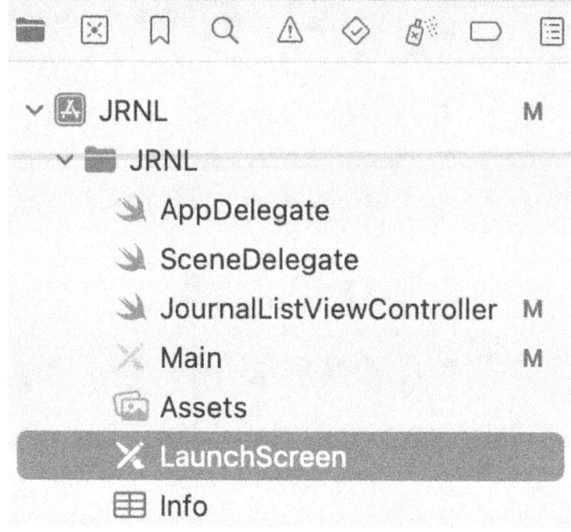

Figure 15.5: Project navigator showing reordered files

4. Right-click on the **JournalListViewController** file in the Project navigator and select **New Group from Selection**:

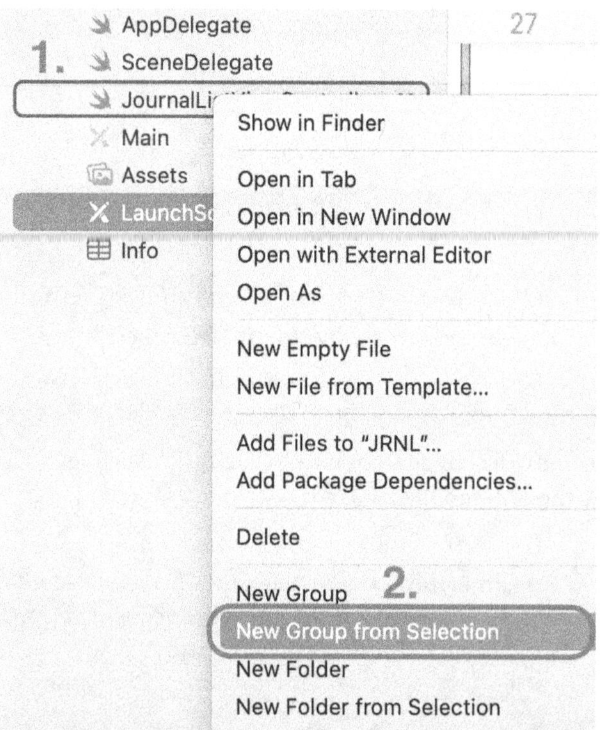

Figure 15.6: Pop-up menu with New Group from Selection selected

This will create a new group containing the `JournalListViewController` file.

5. Replace the placeholder text for the group name with `Journal List Screen` and press *Return*:

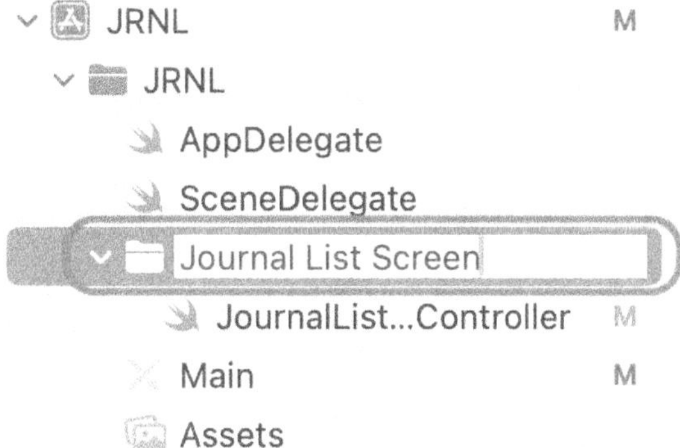

Figure 15.7: Project navigator showing the Journal List Screen group

6. You will now create groups for the model and view objects used in the Journal List screen. Right-click the **Journal List Screen** group and choose **New Group**:

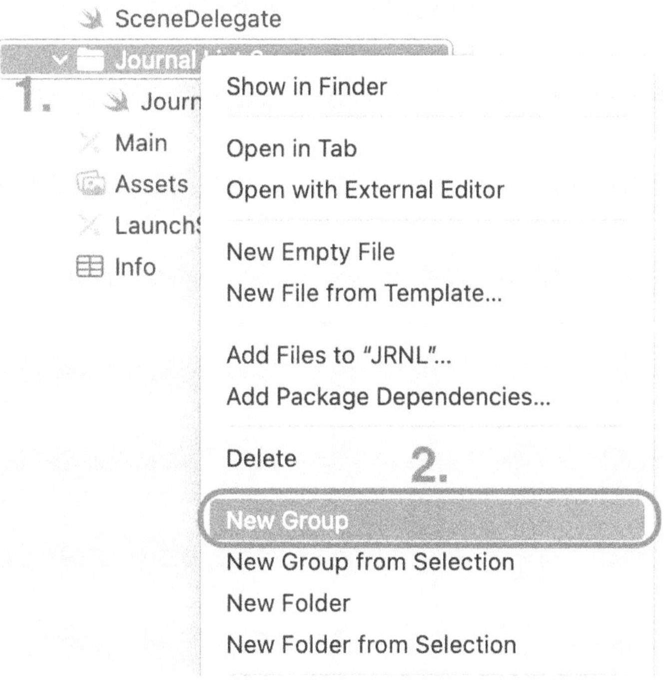

Figure 15.8: Pop-up menu with New Group selected

7. Replace the placeholder text with `Model` and press *Return*:

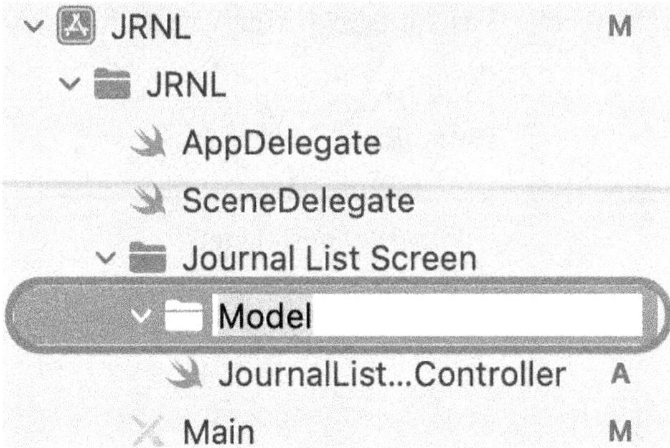

Figure 15.9: Project navigator showing the Model group

8. Create another folder by repeating *Step 3* and replacing the placeholder text with `View`. The Project navigator should look like the following screenshot:

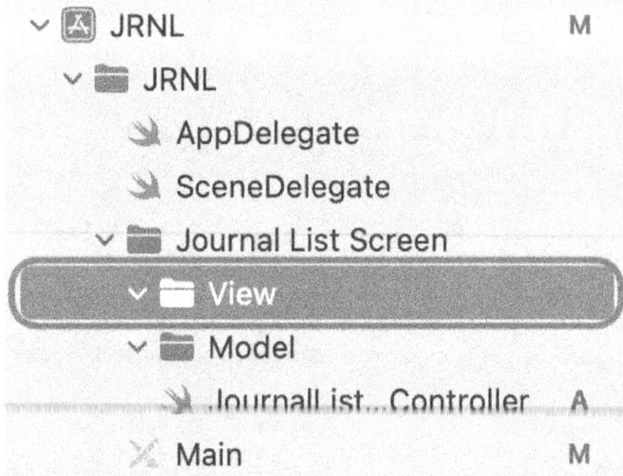

Figure 15.10: Project navigator showing the View and Model groups

9. You'll now create a file that will contain the implementation of the `JournalEntry` class. Right-click the **Model** folder and choose **New File from Template...**:

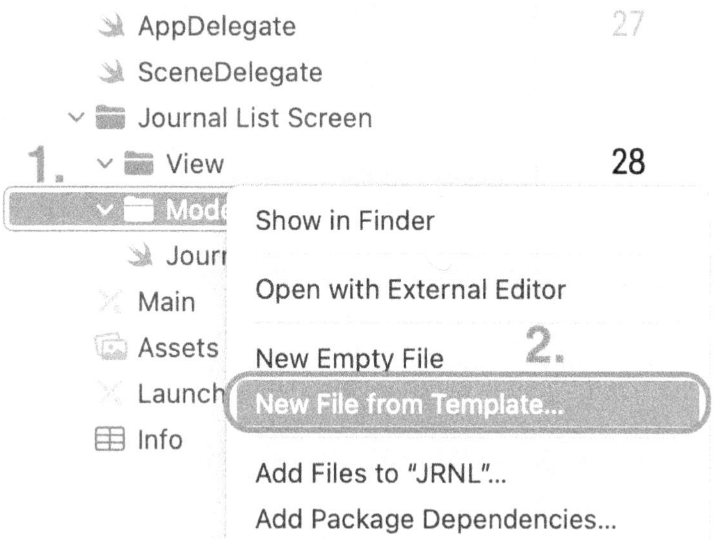

Figure 15.11: Pop-up menu with New File from Template… selected

10. **iOS** should already be selected. Choose **Swift File** and click **Next**:

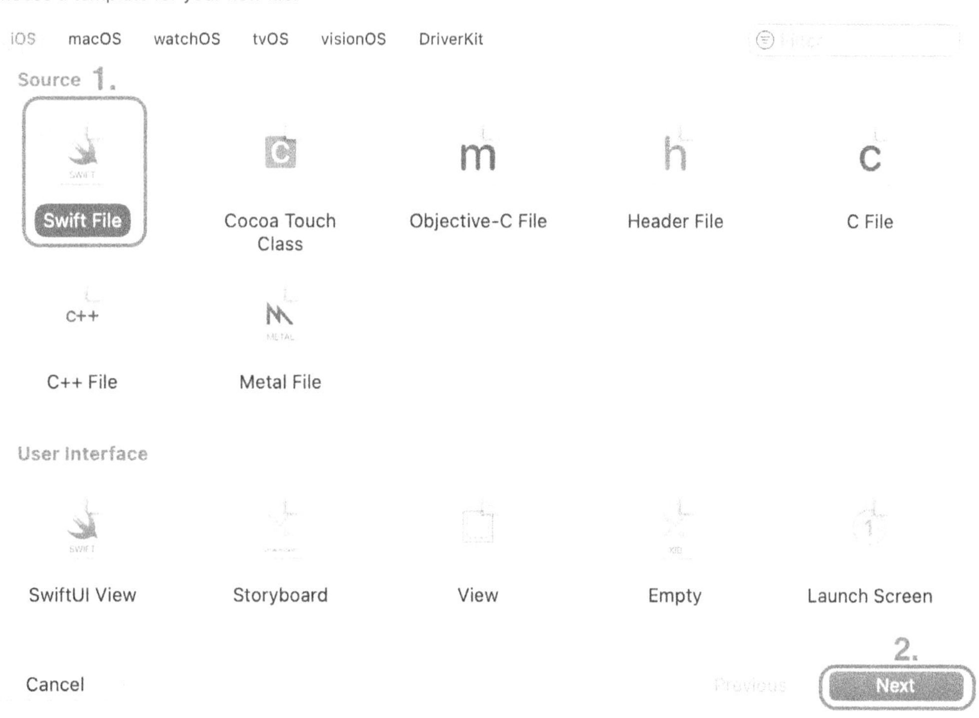

Figure 15.12: Choose a template for your new file: screen with Swift File selected

11. Name the file `JournalEntry` and then click **Create**. The file will appear in the Project navigator and its contents will appear in the **Editor** area:

Figure 15.13: Editor area showing the contents of the JournalEntry file

The only line in this file is an `import` statement.

The `import` statement allows you to import other code libraries into your project, giving you the ability to use classes, properties, and methods from them. Foundation is one of Apple's core frameworks, and you can learn more about it here: `https://developer.apple.com/documentation/foundation`.

12. Modify the `import` statement to import `UIKit`:

```
import UIKit
```

`UIKit` provides the infrastructure required for iOS apps. You can read more about it here: `https://developer.apple.com/documentation/uikit`.

13. Add the following code after the `import` statement to declare a class named `JournalEntry`:

```
class JournalEntry {

}
```

14. Add the following code after the opening curly brace of the `JournalEntry` class to add the desired properties for this class:

```
class JournalEntry {
  // MARK: - Properties
  let date: Date
  let rating: Int
```

```
    let title: String
    let body: String
    let photo: UIImage?
    let latitude: Double?
    let longitude: Double?
  }
```

Let's break this down:

- The `date` property is of type `Date` and will store the date the journal entry was made.
- The `rating` property is of type `Int` and will store the number of stars for the journal entry.
- The `title` property is of type `String` and will store the title text of the journal entry.
- The `body` property is of type `String` and will store the body text of the journal entry.
- The `photo` property is of type `UIImage?` and will store a photo. This is an optional property because not all journal entries require a photo.
- The `lat` and `long` properties are of type `Double?` and will store the location where the journal entry was made. These are optional properties because not all journal entries require a location.

An error will appear because your class does not have an initializer.

15. Add the following code to implement an initializer after the `longitude` property:

```
// MARK: - Initialization
init?(rating: Int, title: String, body: String, photo: UIImage? = nil,
latitude: Double? = nil, longitude: Double? = nil) {
  if title.isEmpty || body.isEmpty || rating < 0 ||
  rating > 5 {
    return nil
  }
  self.date = Date()
  self.rating = rating
  self.title = title
  self.body = body
  self.photo = photo
  self.latitude = latitude
  self.longitude = longitude
}
```

Classes are covered in *Chapter 7, Classes, Structures, and Enumerations*.

Let's break this down:

```
init?(rating: Int, title: String, body: String, photo: UIImage? = nil,
latitude: Double? = nil, longitude: Double? = nil) {
```

The initializer for the `JournalEntry` class has arguments for an `Int` value, two `String` values, an optional `UIImage` value, and two optional `Double` values. The default value for all the optional values is `nil`. The question mark after the `init` keyword means that this is a **failable initializer**; it will not create a `JournalEntry` instance if certain conditions are not met.

```
if title.isEmpty || body.isEmpty || rating < 0 || rating > 5 {
  return nil
}
```

The initializer will fail to create a `JournalEntry` instance if any or all of the following conditions return `true`; `title` is empty, body is empty, `rating` is less than `0`, and `rating` is greater than `5`.

```
self.date = Date()
```

The current date is assigned to the `date` property when the `JournalEntry` instance is created.

```
self.rating = rating
self.Title = entryTitle
self.Body = entryBody
self.photo = photo
self.latitude = latitude
self.longitude = longitude
```

This assigns the argument values to the corresponding properties of the `JournalEntry` instance. Note the use of `self` to differentiate properties from arguments having the same name.

16. The `MARK: -` statements make it easy to navigate through your code. Click the last part of the path that is visible under the toolbar, and you will see both **Properties** and **Initialization** sections displayed in a menu. This enables you to easily go to these sections:

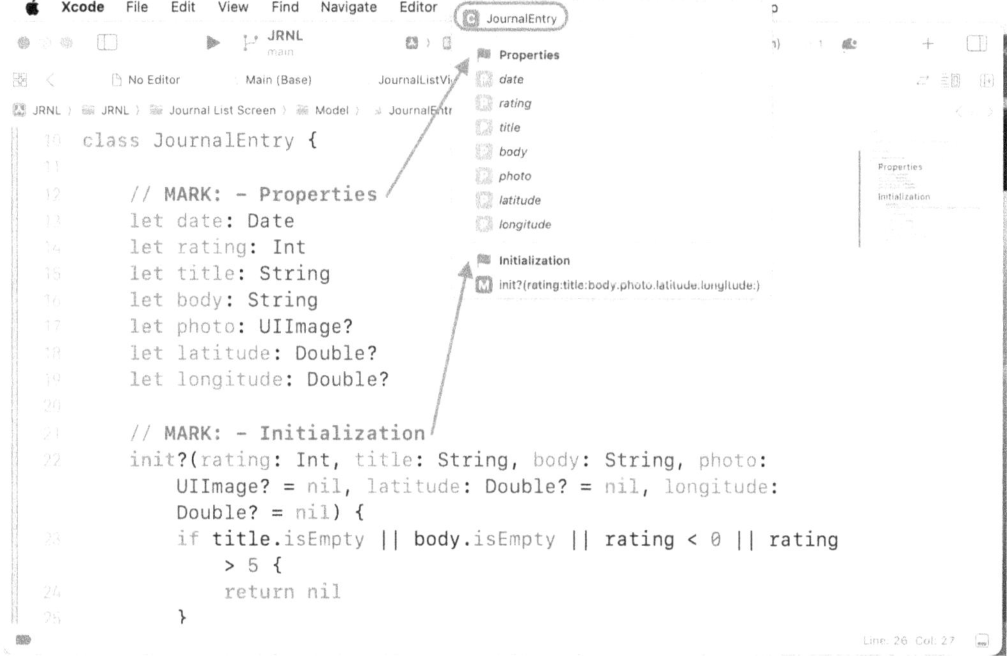

Figure 15.14: Menu showing the Properties and Initialization sections

At this point, you have a class, `JournalEntry`, that can store all the details of a single journal entry. In the next section, you'll create a static method that returns sample `JournalEntry` instances.

Creating sample data

As you saw in *Chapter 14, Getting Started with MVC and Table Views*, you can use an array as the data source for a table view. You will now create an extension containing a static method that will return an array containing three sample journal entries.

Click the `JournalEntry` file in the Project navigator, and type in the following after all other code in the file:

```
// MARK: - Sample data
extension JournalEntry {
  static func createSampleJournalEntryData() -> [JournalEntry] {
    let photo1 = UIImage(systemName: "sun.max")
    let photo2 = UIImage(systemName: "cloud")
    let photo3 = UIImage(systemName: "cloud.sun")
    guard let journalEntry1 =  JournalEntry(rating: 5,
    title: "Good", body: "Today is a good day",
    photo: photo1) else {
      fatalError("Unable to instantiate journalEntry1")
    }
```

```
        guard let journalEntry2 = JournalEntry(rating: 0,
        title: "Bad", body: "Today is a bad day",
        photo: photo2) else {
          fatalError("Unable to instantiate journalEntry2")
        }
        guard let journalEntry3 = JournalEntry(rating: 3,
        title: "Ok", body: "Today is an Ok day",
        photo: photo3) else {
          fatalError("Unable to instantiate journalEntry3")
        }
        return [journalEntry1, journalEntry2, journalEntry3]
    }
}
```

This extension contains a `createSampleJournalEntryData()` method that creates three `UIImage` instances using the symbols from Apple's `SFSymbols` library, creates three `JournalEntry` instances, adds them to an array, and returns the array. The `static` keyword means that it is a method on the `JournalEntry` type instead of a `JournalEntry` instance method.

To learn more about type and instance methods, see this link: `https://docs.swift.org/swift-book/documentation/the-swift-programming-language/methods/`

You have now completed the implementation of the `JournalEntry` class. You have also added a static method that will generate three sample journal entries. In the next section, you'll modify the `JournalListViewController` class to use the array returned by this method to populate the table view.

Displaying data in a table view

In *Chapter 14, Getting Started with MVC and Table Views*, you used a table view cell configuration to set the data to be displayed by the table view cells. You will not be able to do the same here because you are using a custom table view cell that you implemented in *Chapter 13, Modifying App Screens*.

So far in this chapter, you've implemented a static method that returns an array containing three `JournalEntry` instances. You will now modify the `JournalListViewController` class to use that array as the data source for the table view on the Journal List screen. To do so, you will do the following:

- Create a custom `UITableViewCell` instance and assign it as the identity for the `journalCell` table view cells.
- Modify the `JournalListViewController` class to get sample data from the `createSampleJourneyEntryData` static method and assign it to a `journalEntries` array.
- Modify the data source methods in the `JournalListViewController` class to populate the table view cells using data from the `journalEntries` array.

You'll begin by creating a custom `UITableViewCell` instance in the next section.

Creating a custom UITableViewCell subclass

At present, the table view on the Journal List screen displays 10 table view cells that do not contain any data. You need a way to set the values for the image view and the labels in the table view cells, so you will create a new `UITableViewCell` subclass, `JournalEntryTableViewCell`, for this purpose. You will assign this class as the identity of the table view cells in the Journal List screen. Follow these steps:

1. In the Project navigator, right-click on the **View** folder and select **New File from Template...**.
2. **iOS** should already be selected. Choose **Cocoa Touch Class**, then click **Next**:

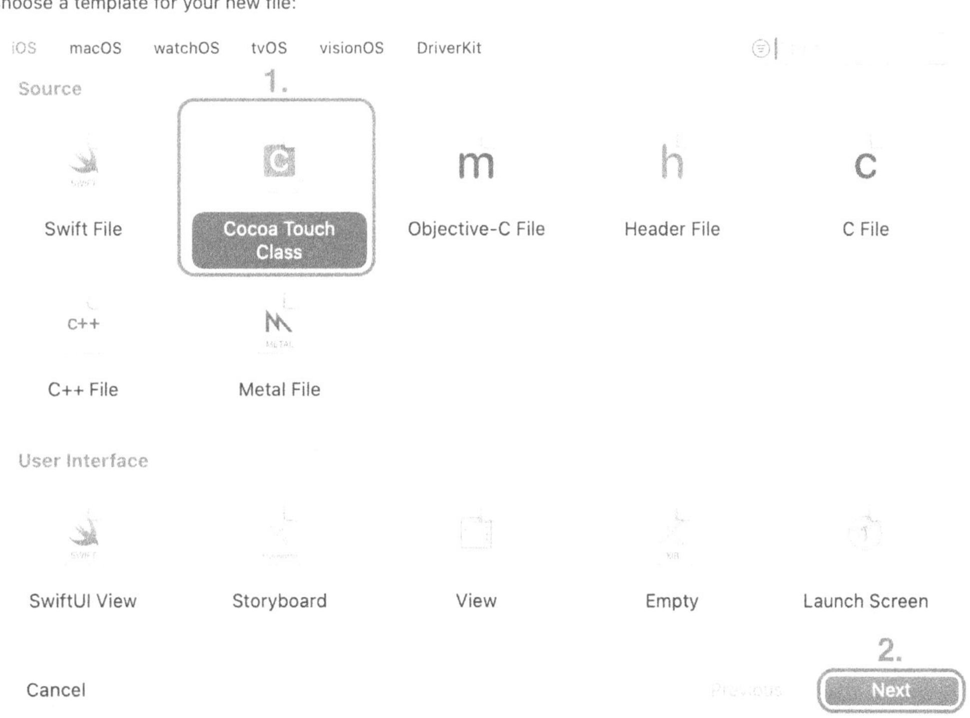

Figure 15.15: Choose a template for your new file screen

Using the **Cocoa Touch Class** template will allow you to set the superclass and automatically insert boilerplate code for the class that you will create.

3. The **Choose options for your new file** screen will appear:

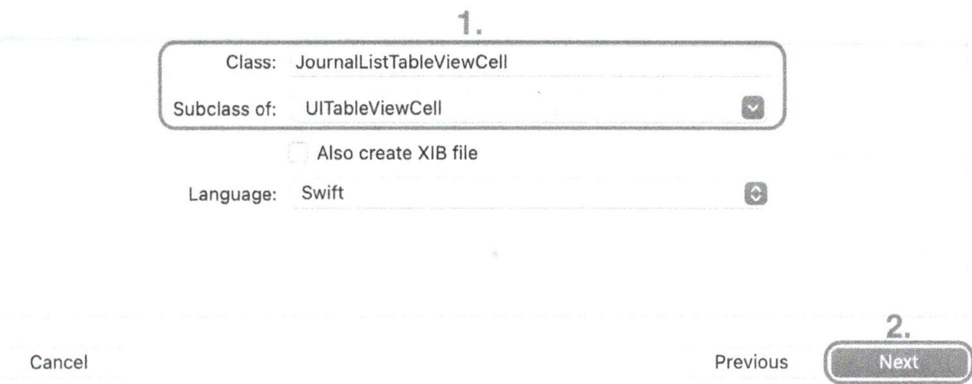

Figure 15.16: Choose options for your new file screen

Configure the class as follows:

- **Class:** `JournalListTableViewCell`
- **Subclass:** `UITableViewCell`
- **Also create XIB:** Unchecked
- **Language:** `Swift`

Click **Next** when you're done.

4. Click **Create**, and a new file, `JournalListTableViewCell`, will be added to the **View** group in your project. Inside it you will see the following code:

```
import UIKit
class JournalListTableViewCell: UITableViewCell {
  override func awakeFromNib() {
    super.awakeFromNib()
    // Initialization code
  }
  override func setSelected(_ selected: Bool,
  animated: Bool) {
    super.setSelected(selected, animated: animated)
    // Configure the view for the selected state
  }
}
```

5. Remove all the code from the `JournalListTableViewCell` class declaration as shown below:

```
class JournalListTableViewCell: UITableViewCell {

}
```

6. To create three properties corresponding to the subviews of the `journalCell` table view cell, type in the following code between the curly braces of the `JournalEntry` class declaration:

```
// MARK: - Properties
@IBOutlet var photoImageView: UIImageView!
@IBOutlet var dateLabel: UILabel!
@IBOutlet var titleLabel: UILabel!
```

7. The implementation of the `JournalListTableViewCell` class is complete. You'll now assign this class as the identity of the `journalCell` table view cell. Click the **Main** storyboard file in the Project navigator and click **journalCell** under the **Journal Scene** in the document outline:

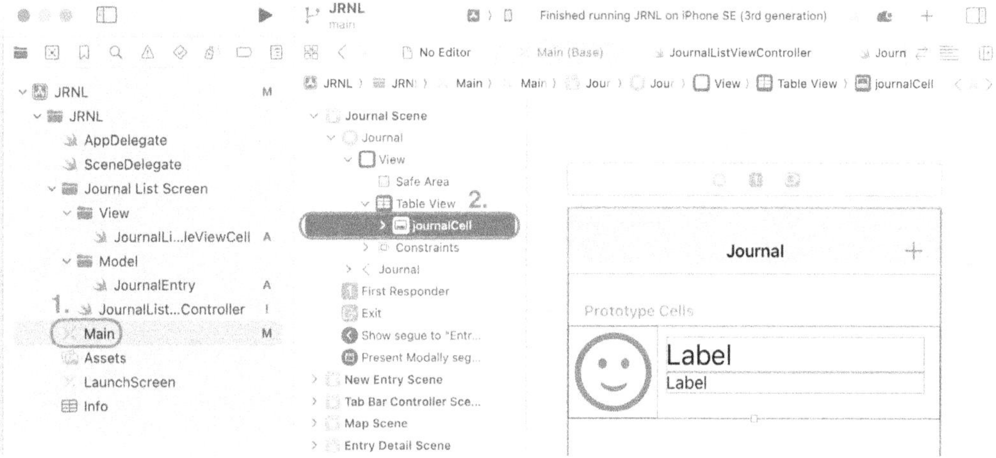

Figure 15.17: Document outline showing journalCell selected

8. Click the Identity inspector button. In the **Custom Class** section, set **Class** to `JournalListTableViewCell`. This sets a `JournalListTableViewCell` instance as the custom table view subclass for `journalCell`. Press *Return* when this is done:

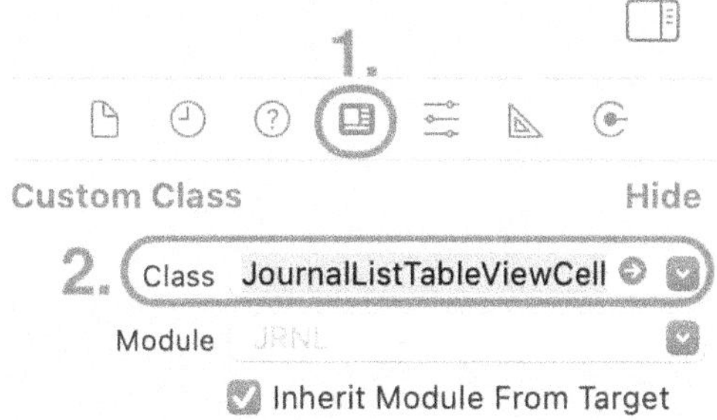

Figure 15.18: Identity inspector showing Class settings for journalCell

You've just declared and defined the `JournalListTableViewCell` class and assigned it as the custom table view cell subclass for the `journalCell` table view cell. In the next section, you'll connect this class to the image view and the labels in the `journalCell` table view cell, so you can control what they display.

Connecting the outlets in journalCell

To manage what is being displayed by the table view cells in the Journal List screen, you'll use the Connections inspector to connect the image view and labels in the `journalCell` table view cell to outlets in the `JournalListTableViewCell` class. Follow these steps:

1. With **journalCell** selected in the document outline, click the Connections inspector to display its outlets.

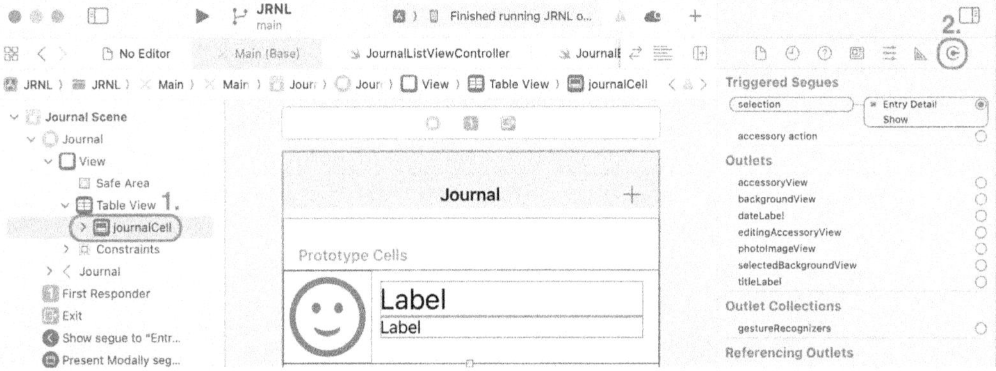

Figure 15.19: Connections inspector showing outlets for journalCell

2. Drag from the **photoImageView** outlet to the image view in the table view cell:

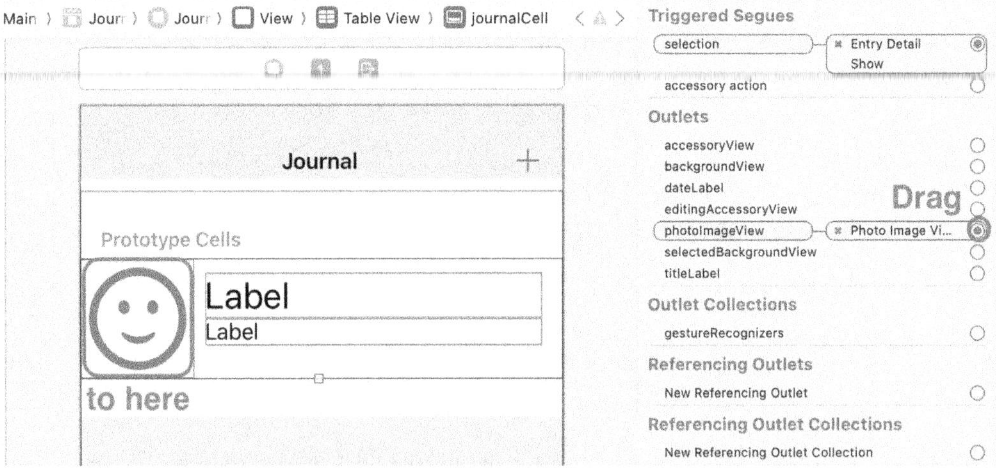

Figure 15.20: Connections inspector showing photoImageView outlet

3. Drag from the **dateLabel** outlet to the top label in the table view cell.

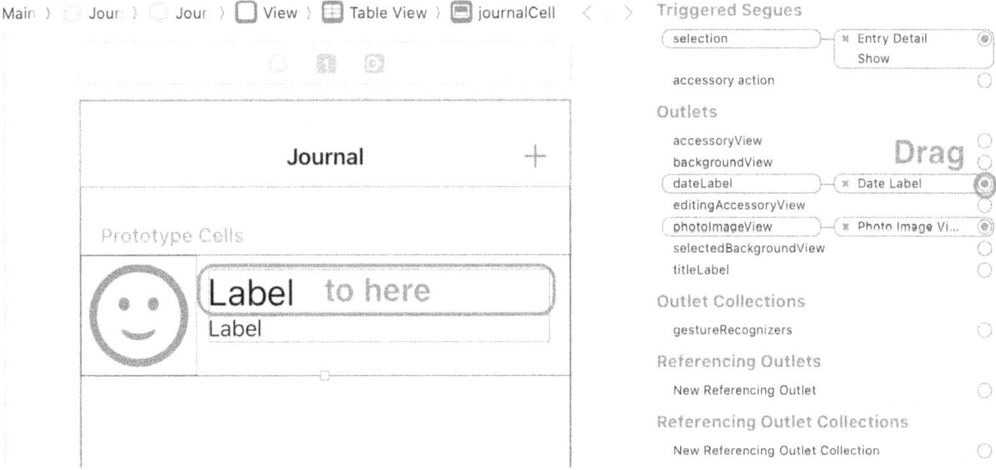

Figure 15.21: Connections inspector showing dateLabel outlet

4. Drag from the **titleLabel** outlet to the bottom label in the table view cell:

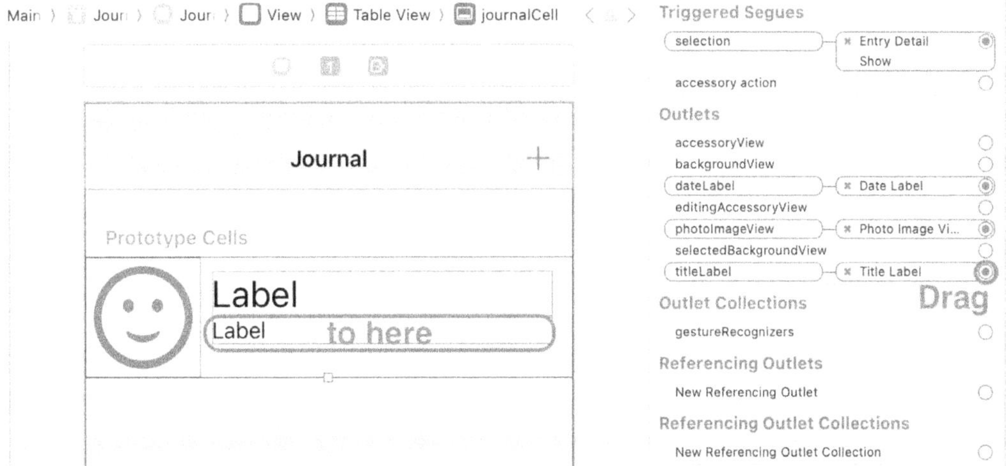

Figure 15.22: Connections inspector showing titleLabel outlet

Remember that if you make a mistake, you can click the **x** to break the connection and drag from the outlet to the UI element once more.

The `journalCell` table view cell in the `Main` storyboard file has now been set up with a custom table view subclass, `JournalListTableViewCell`. The outlets for the table view cell's image view and labels have also been created and assigned. Now you will be able to set the `photoImageView`, `dateLabel`, and `titleLabel` outlets to display a photo, a date, and a title in each cell when the app is run.

In the next section, you'll update the table view data source methods in the `JournalListViewController` class to provide the number of table view cells to be displayed in the table view, as well as to provide the journal entry photo, date, and title for each cell.

Updating the data source methods in JournalListViewController

The data source methods in the `JournalListViewController` class are currently set to display 10 table view cells, with each cell containing an image view displaying a smiley face and two labels. You'll update them to get the number of cells to display, and the data to put in each cell, from the `SampleJournalEntryData` instance. Follow these steps:

1. Click the **JournalListViewController** file in the Project navigator.
2. Rearrange the code in the `JournalListViewController` class so that the `tableView` outlet and the `viewDidLoad()` method are located before the table view delegate methods:

```
class JournalListViewController: UIViewController, UITableViewDataSource,
UITableViewDelegate {
  @IBOutlet var tableView: UITableView!
  override func viewDidLoad() {
    super.viewDidLoad()
  }
  func tableView(_ tableView: UITableView,
  numberOfRowsInSection section: Int) -> Int {
```

3. Add a `MARK` statement before the property declarations as follows:

```
// MARK: - Properties
@IBOutlet var tableView: UITableView!
```

4. Add a `MARK` statement before the `viewDidLoad()` method as follows:

```
// MARK: - View controller lifecycle
@override func viewDidLoad() {
```

5. Add a `MARK` statement before the table view data source methods as follows:

```
// MARK: - UITableViewDataSource
func tableView(_ tableView: UITableView, numberOfRowsInSection section:
```

6. Add a `MARK` statement before `UnwindNewEntryCancel(segue:)` as follows:

```
// MARK: - Methods
@IBAction func unwindNewEntryCancel(segue: UIStoryboardSegue) {
```

7. Verify that the code in `JournalListViewController` appears as follows:

```
class JournalListViewController: UIViewController, UITableViewDataSource,
UITableViewDelegate {
  // MARK: - Properties
```

```
    @IBOutlet var tableView: UITableView!
    // MARK: - View controller lifecycle
    override func viewDidLoad() {
      super.viewDidLoad()
    }
    // MARK: - UITableViewDataSource
    func tableView(_ tableView: UITableView,
    numberOfRowsInSection section: Int) -> Int {
      10
    }
    func tableView(_ tableView: UITableView, cellForRowAt
    indexPath: IndexPath) -> UITableViewCell {
      tableView.dequeueReusableCell(withIdentifier:
      "journalCell", for: indexPath)
    }
    // MARK: - Methods
    @IBAction func unwindNewEntryCancel(segue:
    UIStoryboardSegue) {
    }
}
```

8. Type the following code after the `tableView` outlet declaration to create a `journalEntries` property, which will contain an array of `JournalEntry` instances:

```
@IBOutlet var tableView: UITableView!
private var journalEntries: [JournalEntry] = []
```

The `private` keyword restricts the use of the `journalEntries` array to the `JournalListViewController` class.

You can learn more about Access Control in Swift at this link: https://docs.swift.org/swift-book/documentation/the-swift-programming-language/accesscontrol/.

9. Modify the `viewDidLoad()` method as shown to populate the `journalEntries` array when the app is launched:

```
override func viewDidLoad() {
  super.viewDidLoad()
  journalEntries =
  JournalEntry.createSampleJournalEntryData()
}
```

The `createSampleJournalEntryData()` method will create three `JournalEntry` instances and assign them to the `journalEntries` array.

10. Update `tableView(_:numberOfRowsInSection:)` as shown here. This will make the table view display a `journalCell` for each element in the `journalEntries` array:

```
func tableView(_ tableView: UITableView, numberOfRowsInSection section:
Int) -> Int {
  journalEntries.count
}
```

11. Update `tableView(_:cellForRowAt:)` as shown to set the image view and labels for each cell using data from the corresponding element in the `journalEntries` array:

```
func tableView(_ tableView: UITableView, cellForRowAt indexPath:
IndexPath) -> UITableViewCell {
  let journalCell =
  tableView.dequeueReusableCell(withIdentifier:
  "journalCell", for: indexPath) as!
  JournalListTableViewCell
  let journalEntry = journalEntries[indexPath.row]
  journalCell.photoImageView.image = journalEntry.photo
  journalCell.dateLabel.text = journalEntry.date.formatted(
    .dateTime.month().day().year()
  )
  journalCell.titleLabel.text = journalEntry.title
  return journalCell
}
```

Let's break this down:

```
let journalCell = tableView.dequeueReusableCell(withIdentifier:
"journalCell", for: indexPath) as! JournalListTableViewCell
```

This statement specifies the cell that is dequeued is cast as an instance of `JournalListTableViewCell`.

You can learn more about the `as!` operator at this link: `https://developer.apple.com/swift/blog/?id=23`.

```
let journalEntry = journalEntries[indexPath.row]
```

This statement gets the `JournalEntry` instance that corresponds to the current cell in the table view. In other words, the first table view cell in the table view corresponds to the first `JournalEntry` instance in the `journalEntries` array, the second table view cell corresponds to the second `JournalEntry` instance, and so on.

```
journalCell.photoImageView.image = journalEntry.photo
```

This statement gets the photo from the `JournalEntry` instance and assigns it to the image of the `journalCell` instance's `photoImageView` property.

```
journalCell.dateLabel.text = journalEntry.date.formatted(
  .dateTime.month().day().year()
)
```

This statement gets the date from the `JournalEntry` instance, formats it into a string, and assigns it to the text for the `journalCell` instance's `dateLabel` property.

```
journalCell.titleLabel.text = journalEntry.title
```

This statement gets the string stored in `title` from the `JournalEntry` instance and assigns it to the text for the `journalCell` instance's `titleLabel` property.

```
return journalCell
```

This statement returns the populated `journalCell` instance for display in the table view.

12. Build and run the app. You'll see the table view in the Journal List screen display text and images for each `JournalEntry` instance in the `journalEntries` array:

Figure 15.23: Simulator showing the Journal List screen

Tapping on a row displays the Journal Entry Detail screen, but this screen does not display any data from the selected journal entry yet. You will address this in the next chapter.

Congratulations! At this point, the Journal List screen displays text and images from the `journalEntries` array. But you can't add or remove journal entries from the `journalEntries` array yet. You will learn how to do this in the next chapter.

Summary

In this chapter, you implemented the model objects for the Journal List screen to make it display a list of journal entries. You learned about the model objects that you will use, created a Swift class that can be used to store journal entry instances, and created a static method returning sample journal entries. You then created a custom instance of `UITableViewCell` for your table view and used the method returning sample journal entries to populate an array. This array is then used as the data source for the table view in the Journal List screen.

You now know how to create model objects, how to create sample data, and how to configure view controllers to populate table views using that sample data. This will be useful should you wish to create your own apps that use table views.

In the next chapter, you'll learn how to add and remove journal entries from the Journal List screen. You'll also learn how to pass data between view controllers.

Join us on Discord!

Read this book alongside other users, experts, and the author himself. Ask questions, provide solutions to other readers, chat with the author via Ask Me Anything sessions, and much more. Scan the QR code or visit the link to join the community.

`https://packt.link/ios-Swift`

16

Passing Data between View Controllers

In the previous chapter, you configured the `JournalListViewController` class, the view controller for the Journal List screen, to display journal entries from a structure containing sample data in a table view.

In this chapter, you'll learn how to pass data from one view controller to another. You'll start by implementing a view controller for the Add New Journal Entry screen, then add code to pass data from the Add New Journal Entry screen to the Journal List screen. Next, you'll learn how to remove journal entries while you're on the Journal List screen. After that, you'll learn about **text field** and **text view delegate** methods, and finally, you'll pass data from the Journal List screen to the Journal Entry Detail screen.

By the end of this chapter, you'll have learned how to pass data between view controllers and how to use text field and text view delegate methods. This will enable you to easily pass data between view controllers in your own apps.

The following topics will be covered in this chapter:

- Passing data from the Add New Journal Entry screen to the Journal List screen
- Removing rows from a table view
- Exploring text field and text view delegate methods
- Passing data from the Journal List screen to the Journal Entry Detail screen

Technical requirements

You will continue working on the `JRNL` project that you modified in the previous chapter.

The playground and completed Xcode project for this chapter are in the `Chapter16` folder of the code bundle for this book, which can be downloaded here:

`https://github.com/PacktPublishing/iOS-18-Programming-for-Beginners-Ninth-Edition`

Check out the following video to see the code in action:

`https://youtu.be/9207NgVoT2Q`

Let's begin by learning how data is passed between the Add New Journal Entry screen and the Journal List screen in the next section.

Passing data from the Add New Journal Entry screen to the Journal List screen

As shown in the app tour in *Chapter 10, Setting Up the User Interface*, the Add New Journal Entry screen allows the user to enter data to create a new journal entry. To do so, the user will click the + button in the top-right corner of the Journal List screen to display the Add New Journal Entry screen. The user will then enter the details for the new journal entry. Clicking the **Save** button will dismiss the Add New Journal Entry screen and a new row containing a table view cell is added to the table view on the Journal List screen. The table view cell will display the photo, date, and title of the newly added journal entry.

For this to work, you'll implement the `prepare(for:sender:)` method for the view controller managing the Add New Journal Entry screen. This method is triggered when you go from one view controller to another. With this method, you'll create a new journal entry using the information the user entered and assign it to a variable. You'll implement an unwind method in the `JournalListViewController` class so you will be able to access this variable when you're on the Journal List screen. Then, you will add the new journal entry obtained from this variable to the `journalEntries` array and then redraw the table view.

To learn more about the `prepare(for:sender:)` method, see: `https://developer.apple.com/documentation/uikit/uiviewcontroller/1621490-prepare`.

In the next section, you'll create a new view controller instance to manage the Add New Journal Entry screen.

Creating the AddJournalEntryViewController class

At present, the Add New Journal Entry screen does not have a view controller. You'll add a new file to your project and implement the `AddJournalEntryViewController` class, assign it to the **New Entry Scene**, and connect the outlets. Follow these steps:

1. Open your `JRNL` project from the previous chapter. In the Project navigator, create a new group by right-clicking the **JRNL** group and choosing **New Group**.
2. Name this group `Add New Journal Entry Screen` and move it so it is below the **Journal List Screen** group.
3. Right-click on the **Add New Journal Entry Screen** group and select **New File from Template...**.
4. **iOS** should already be selected. Choose **Cocoa Touch Class** and click **Next**.

5. Configure the class with the following details:
    - **Class:** `AddJournalEntryViewController`
    - **Subclass of:** `UIViewController`
    - **Also create XIB:** Unchecked
    - **Language: Swift**

    Click **Next** when you're done.
6. Click **Create** and the `AddJournalEntryViewController` file will appear in the Project navigator.

The `AddJournalEntryViewController` file has now been created, with the `AddJournalEntryViewController` class declaration inside it. You'll set this class as the custom class of the view controller scene that's presented when you tap the + button in the Journal List screen. Follow these steps:

1. Click the **Main** storyboard file in the Project navigator and click **New Entry Scene** in the document outline:

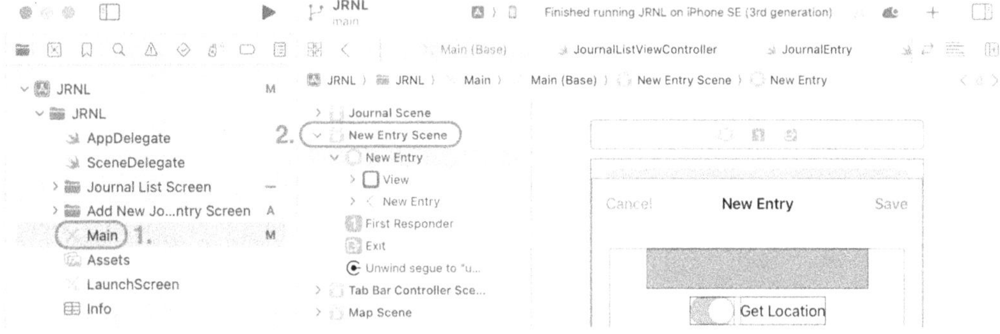

Figure 16.1: Editor area showing New Entry Scene

2. Click the Identity inspector button and, under **Custom Class**, set **Class** to `AddJournalEntryViewController`:

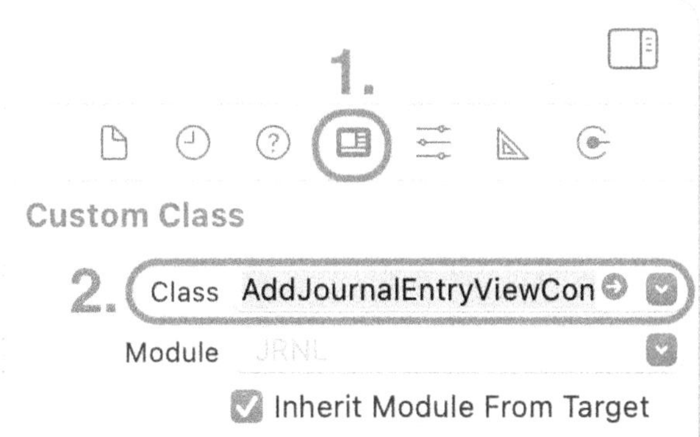

Figure 16.2: Identity inspector settings for New Entry Scene

Great! In the next section, let's connect the user interface elements in the **New Entry Scene** to outlets in the `AddJournalEntryViewController` class. By doing this, the `AddJournalEntryViewController` instance will be able to access the data the user enters on the Add New Journal Entry screen.

Connecting the UI elements to the AddJournalEntryViewController class

Currently, the `AddJournalEntryViewController` instance for the Add New Journal Entry screen has no way of communicating with the UI elements in it. You'll add outlets in the `AddJournalEntryViewController` class and assign the corresponding UI elements in the **New Entry Scene** to each outlet. Follow these steps:

1. In the Project navigator, click the **AddJournalEntryViewController** file and add the following properties to the `AddJournalEntryViewController` class after the opening curly brace:

```
// MARK: - Properties
@IBOutlet var titleTextField: UITextField!
@IBOutlet var bodyTextView: UITextView!
@IBOutlet var photoImageView: UIImageView!
```

2. Click the **Main** storyboard file and click **New Entry Scene** in the document outline.
3. Click the Connections inspector to display all the outlets for **New Entry Scene.** Drag from the **titleTextField** outlet to the text field in **New Entry Scene:**

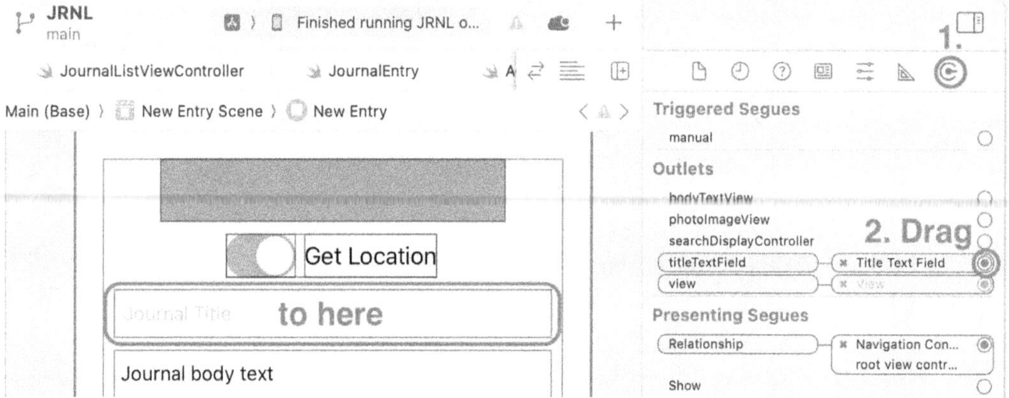

Figure 16.3: Connections inspector showing the titleTextField outlet

4. Drag from the **bodyTextView** outlet to the text view in **New Entry Scene**:

Figure 16.4: Connections inspector showing the bodyTextView outlet

5. Drag from the **photoImageView** outlet to the image view in **New Entry Scene**:

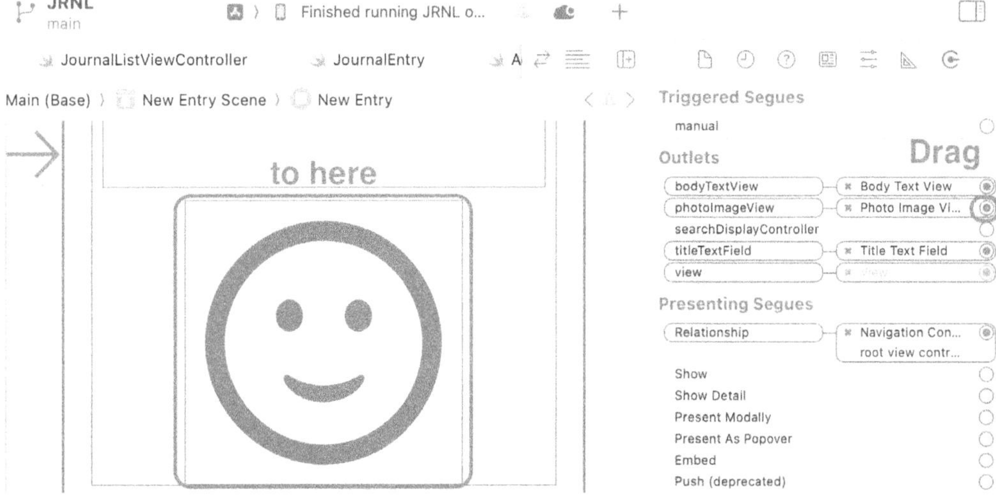

Figure 16.5: Connections inspector showing the photoImageView outlet

Remember that if you make a mistake, you can click the **x** to break the connection and drag from the outlet to the UI element once more.

With that, you've connected the UI elements in the **New Entry Scene** to the outlets in the `AddJournalEntryViewController` class. In the next section, you'll implement the code to create a `JournalEntry` instance when the user clicks the **Save** button.

Creating a JournalEntry instance from user input

You have implemented the `AddJournalEntryViewController` class and connected the outlets in this class to the text field, text view, and image view in the **New Entry Scene**. When the user enters data into the text field and text view, you can use this information to create a new journal entry.

When a view controller is about to transition to another view controller, the view controller's `prepare(for:sender:)` method is called. You'll implement this method to create a new `JournalEntry` instance, which can then be passed to the view controller for the Journal List screen. Follow these steps:

1. In the Project navigator, click the **AddJournalEntryViewController** file and add a `newJournalEntry` property to the `AddJournalEntryViewController` class after the outlet declarations:

    ```
    //MARK: - Properties
    @IBOutlet var titleTextField: UITextField!
    @IBOutlet var bodyTextView: UITextView!
    @IBOutlet var photoImageView: UIImageView!
    var newJournalEntry: JournalEntry?
    ```

    The `JournalEntry` instance created using the data entered by the user will be assigned to this property.

2. Uncomment the `prepare(for:sender:)` method in this class. It should look like the following:

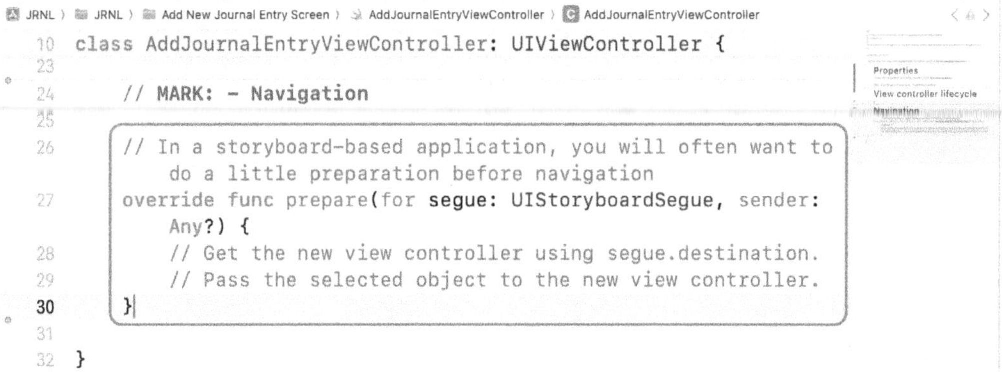

    ```
    JRNL 〉 JRNL 〉 Add New Journal Entry Screen 〉 AddJournalEntryViewController 〉 AddJournalEntryViewController
    class AddJournalEntryViewController: UIViewController {

        // MARK: - Navigation

        // In a storyboard-based application, you will often want to
            do a little preparation before navigation
        override func prepare(for segue: UIStoryboardSegue, sender:
            Any?) {
            // Get the new view controller using segue.destination.
            // Pass the selected object to the new view controller.
        }

    }
    ```

Figure 16.6: Editor area showing the prepare(for:sender:) method

3. Add the following code between the curly braces of this method:

```
let title = titleTextField.text ?? ""
let body = bodyTextView.text ?? ""
let photo = photoImageView.image
let rating = 3
newJournalEntry = JournalEntry(rating: rating, title: title, body: body,
photo: photo)
```

   This will assign the strings from the text field and text view and the image from the image view to `title`, `body`, and `photo`, respectively. Since the custom rating control shown in the app tour has not been implemented yet, a placeholder value is assigned to `rating`. A new `JournalEntry` instance is then created using these constants and assigned to the `newJournalEntry` property.

You've now added code that will create a `JournalEntry` instance before the Add New Journal Entry screen transitions to the Journal List screen. In the next section, you'll modify the `JournalListViewController` class to get the new `JournalEntry` instance and add it to the `journalEntries` array.

Updating the table view with a new journal entry

On the Journal List screen, journal entries are displayed in a table view. The table view gets its data from the `journalEntries` array contained in the `sampleJournalEntryData` structure. You will add code to the `JournalListViewController` class to get the `JournalEntry` instance assigned to the `newJournalEntry` property. After that, you will insert this instance to the `journalEntries` array. Follow these steps:

1. In the Project navigator, click the **JournalListViewController** file and add the following code before the closing curly brace:

```
@IBAction func unwindNewEntrySave(segue: UIStoryboardSegue) {
  if let sourceViewController = segue.source as?
  AddJournalEntryViewController, let newJournalEntry =
  sourceViewController.newJournalEntry {
    journalEntries.insert(newJournalEntry, at: 0)
    tableView.reloadData()
  }
}
```

   This method checks to see if the source view controller is an instance of the `AddJournalEntryViewController` class, and if it is, gets the `JournalEntry` instance from the `newJournalEntry` property. This instance is then inserted as the first item in the `journalEntries` array. After that, the `tableView.reloadData()` statement will redraw the table view.

2. Click the **Main** storyboard file and expand the **New Entry Scene** in the document outline. *Ctrl + Drag* from the **Save** button in the document outline to the scene exit and choose **unwind-NewEntrySaveWithSegue:** from the pop-up menu.

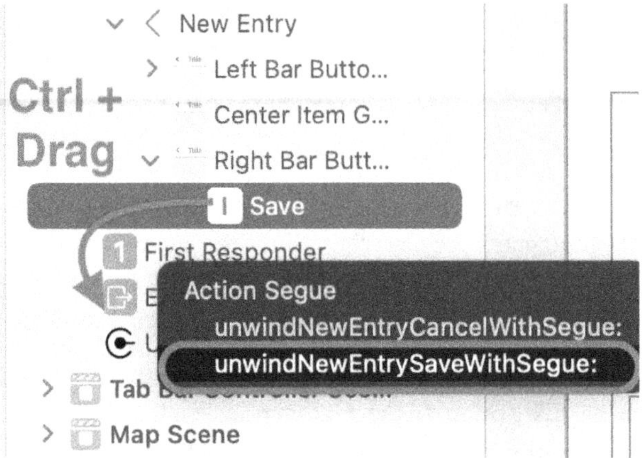

Figure 16.7: Pop-up menu showing unwindNewEntrySaveWithSegue: selected

When you run your project, tapping the **Save** button will transition from the Add New Entry screen to the Journal List screen, and execute the `unwindNewEntrySave(segue:)` method.

3. Build and run your project and click the + button. Enter some sample text in the text field and text view. Click **Save.**

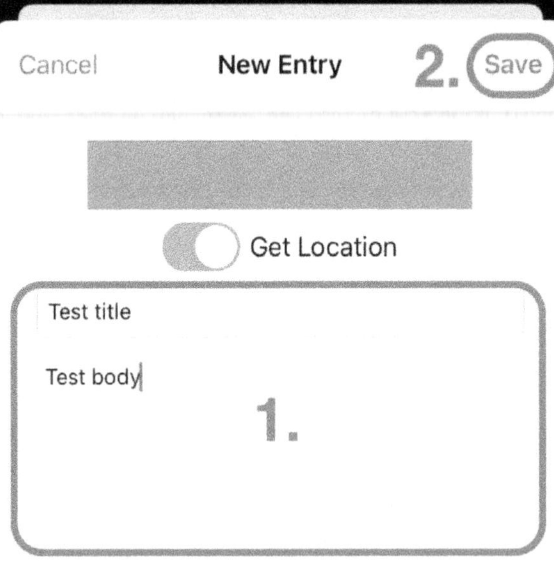

Figure 16.8: Simulator with the Save button highlighted

4. The new journal entry will appear in the table view when the Journal List screen reappears.

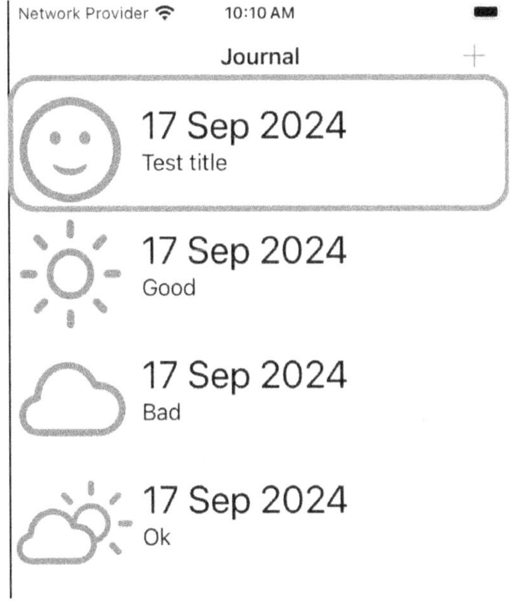

Figure 16.9: Simulator with new table view cell highlighted

Excellent! You have successfully implemented a view controller for the Add New Journal Entry screen and can now add new journal entries, which will appear on the Journal List screen. In the next section, you'll implement the code that will let you remove journal entries from the table view on the Journal List screen.

Removing rows from a table view

As you have learned in *Chapter 14, Getting Started with MVC and Table Views,* table view row deletion is handled by the `tableView(_:commit:forRowAt:)` method, which is one of the methods declared in the `UITableViewDataSource` protocol.

You'll implement this method in the `JournalListViewController` class. Follow these steps:

1. In the Project navigator, click the **JournalListViewController** file and add the following code to the `JournalListViewController` class after the existing table view data source methods:

```
func tableView(_ tableView: UITableView, commit editingStyle:
UITableViewCell.EditingStyle, forRowAt indexPath: IndexPath) {
  if editingStyle == .delete {
    journalEntries.remove(at: indexPath.row)
    tableView.reloadData()
  }
}
```

This will allow you to swipe left to display a **Delete** button, and when you tap the **Delete** button, the corresponding `JournalEntry` instance will be removed from the `journalEntries` array, and the table view will be redrawn.

2. Build and run your project. Swipe left on any row to reveal a **Delete** button:

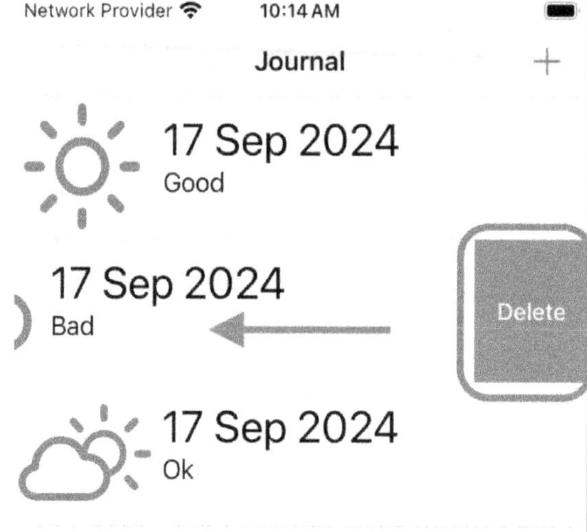

Figure 16.10: Simulator showing the Delete button on the Journal List screen

3. Tap the **Delete** button and the row will be removed from the table view:

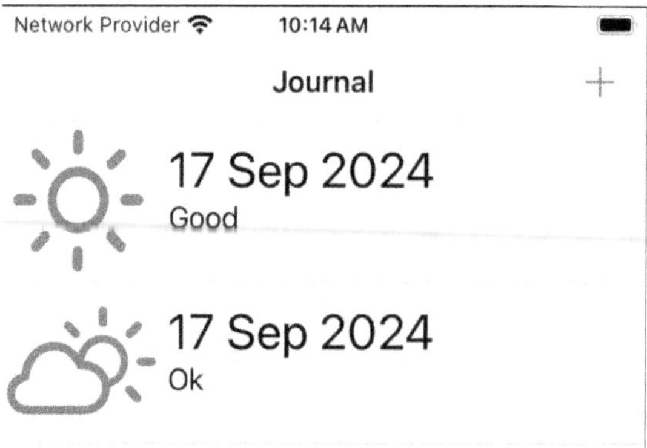

Figure 16.11: Simulator showing the redrawn table view

With that, you have successfully implemented a way to remove rows from a table view! Awesome! In the next section, you'll learn more about text field and text view delegate methods, which will be useful when you're entering data in the Add New Journal Entry screen.

Exploring text field and text view delegate methods

Currently, there are a couple of issues with the Add New Journal Entry screen. The first issue is that it's not possible to dismiss the software keyboard once it has appeared on the screen. The second issue is that you can click the **Save** button even when the text field and text view are empty.

To make it easier to work with text fields, Apple has implemented `UITextFieldDelegate`, a protocol declaring a set of optional methods to manage the editing and validation of text in a text field object. Apple has also implemented `UITextViewDelegate`, a protocol declaring methods for receiving editing-related messages for text view objects.

You can learn more about the `UITextFieldDelegate` protocol at this link: `https://developer.apple.com/documentation/uikit/uitextfielddelegate`.

You can learn more about the `UITextViewDelegate` protocol at this link: `https://developer.apple.com/documentation/uikit/uitextviewdelegate`.

You'll implement methods from the `UITextFieldDelegate` and `UITextViewDelegate` protocols to the `AddJournalEntryViewController` class so that the user can dismiss the software keyboard once data entry is complete. Follow these steps:

1. In the Project navigator, click the **AddJournalEntryViewController** file. Add an extension after the closing curly brace of the `AddJournalEntryViewController` class declaration to make it conform to the `UITextFieldDelegate` and `UITextViewDelegate` protocols:

   ```
   extension AddJournalEntryViewController: UITextFieldDelegate,
   UITextViewDelegate {
   }
   ```

2. Modify the `viewDidLoad()` method as follows to set the `AddJournalEntryViewController` instance as the delegate for the text field and text view:

   ```
   override func viewDidLoad() {
     super.viewDidLoad()
     titleTextField.delegate = self
     bodyTextView.delegate = self
   }
   ```

   This means that the implementation of the text field and text view delegate methods are in the `AddJournalEntryViewController` class.

3. Add the following code after the opening curly brace of the extension to dismiss the software keyboard when the *return* key is tapped after you have finished entering text in the text field:

   ```
   extension AddJournalEntryViewController: UITextFieldDelegate,
   UITextViewDelegate {
     // MARK: - UITextFieldDelegate
   ```

```
    func textFieldShouldReturn(_ textField: UITextField) ->  Bool {
      textField.resignFirstResponder()
      return true
    }
}
```

4. Add the following code after the `textFieldShouldReturn(_:)` method to dismiss the software keyboard when the *return* key is tapped after you have finished entering text in the text view:

```
//MARK: - UITextViewDelegate
func textView(_ textView: UITextView, shouldChangeTextIn range: NSRange,
replacementText text: String) -> Bool {
  if (text == "\n") {
    textView.resignFirstResponder()
  }
  return true
}
```

When you tap a text field or text view on the screen, it gains first responder status, and a software keyboard pops up from the bottom of the screen. Anything you type on the keyboard will go to whichever object has first responder status. After implementing the preceding methods, tapping *return* on the software keyboard while in a text field or text view will tell it to resign the first responder status, which will automatically make the keyboard disappear.

5. Build and run your app and tap the text field. If the software keyboard does not appear, choose **Keyboard** | **Toggle Software Keyboard** from Simulator's **I/O** menu:

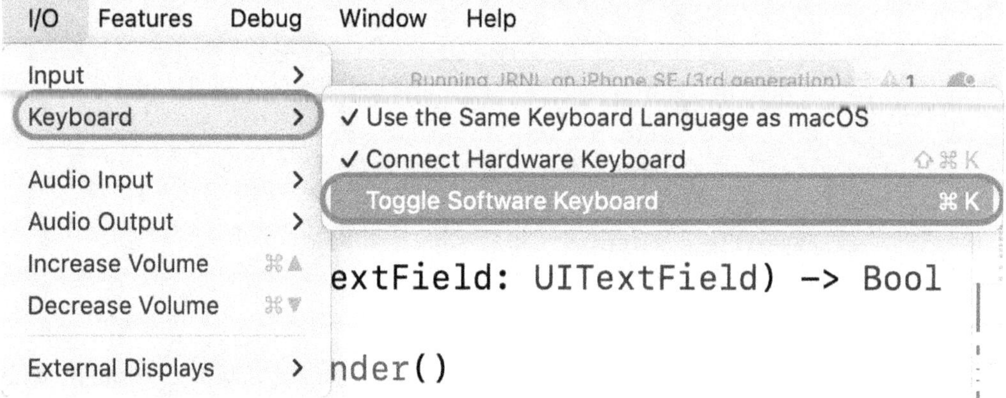

Figure 16.12: Simulator I/O menu with Keyboard | Toggle Software Keyboard selected

6. Use the software keyboard to type some text into the text field or text view:

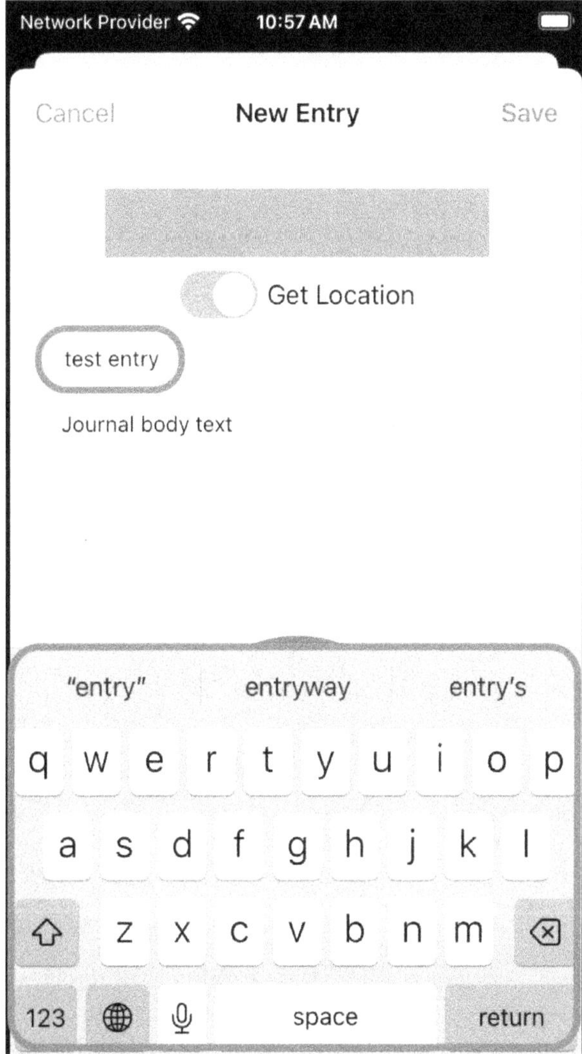

Figure 16.13: Simulator showing the software keyboard

7. After typing some text in either the text field or the text view, tap *return* on the software keyboard, and it should automatically disappear.

The first issue has been resolved, and the user is now able to dismiss the software keyboard. Great! Now you will modify your app so that the user can only tap **Save** if there is text in the text field and text view. Follow these steps:

1. To enable or disable the **Save** button, you will need to be able to set its state. Type in the following after the outlet declarations to create an outlet for the **Save** button:

```
@IBOutlet var titleTextField: UITextField!
@IBOutlet var bodyTextView: UITextView!
@IBOutlet var photoImageView: UIImageView!
@IBOutlet var saveButton: UIBarButtonItem!
var newJournalEntry: JournalEntry?
```

2. In the Project navigator, click the **Main** storyboard file and click **New Entry Scene** in the document outline.
3. Click the Connections inspector button and drag from the **saveButton** outlet to the **Save** button in the **New Entry** scene:

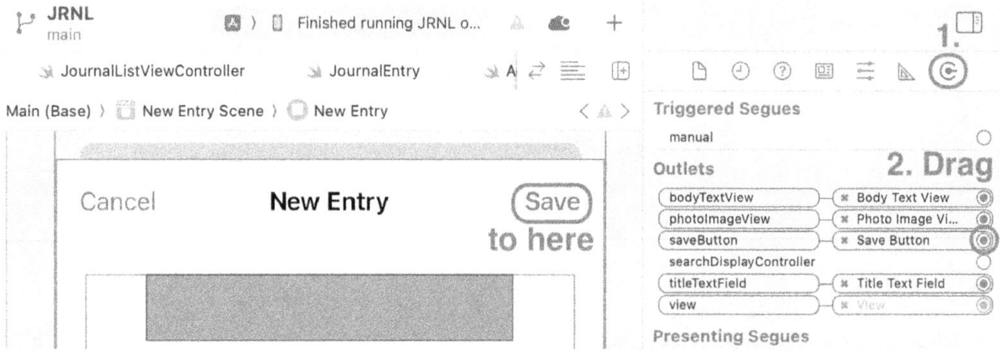

Figure 16.11: Connections inspector showing the saveButton outlet

4. In the Project navigator, click the **AddJournalEntryViewController** file. Add the `UITextFieldDelegate` method shown after the `textFieldShouldReturn(_:)` method:

```
func textFieldShouldReturn(_ textField: UITextField) -> Bool {
  textField.resignFirstResponder()
  return true
}
func textFieldDidBeginEditing(_ textField: UITextField) {
  saveButton.isEnabled = false
}
```

This method disables the **Save** button when the user starts editing text in the text field.

5. Add the `UITextViewDelegate` method shown after the `textView(_:shouldChangeTextIn range:replacementText:)` method:

```
func textView(_ textView: UITextView, shouldChangeTextIn range: NSRange,
replacementText text: String) -> Bool {
  if(text == "\n") {
    textView.resignFirstResponder()
  }
  return true
}
func textViewDidBeginEditing(_ textView: UITextView) {
  saveButton.isEnabled = false
}
```

This method disables the **Save** button when the user starts editing text in the text view.

6. Add a method to enable the **Save** button if there is text in the text field or text view before the closing curly brace:

```
// MARK: - Private methods
private func updateSaveButtonState() {
  let textFieldText = titleTextField.text ?? ""
  let textViewText = bodyTextView.text ?? ""
  saveButton.isEnabled = !textFieldText.isEmpty &&
  !textViewText.isEmpty
}
```

The `private` keyword means that this method is only accessible within this class.

7. Add the `UITextFieldDelegate` method shown after the `textFieldDidBeginEditing(_:)` method:

```
func textFieldDidBeginEditing(_ textField: UITextField) {
  saveButton.isEnabled = false
}
func textFieldDidEndEditing(_ textField: UITextField) {
  updateSaveButtonState()
}
```

This method calls `updateSaveButtonState()` after the text field resigns the first responder status.

8. Add the `UITextViewDelegate` methods shown after the `textViewDidBeginEditing(_:)` method:

```
func textViewDidBeginEditing(_ textView: UITextView) {
  saveButton.isEnabled = false
}
func textViewDidEndEditing(_ textView: UITextView) {
```

```
    updateSaveButtonState()
  }
  func textViewDidChange(_ textView: UITextView) {
    updateSaveButtonState()
  }
```

These methods call `updateSaveButtonState()` after the text view resigns the first responder status and when the contents of the text view change.

9. In the `viewDidLoad()` method, call `updateSaveButtonState()` to disable the **Save** button when the Add New Journal Entry screen first appears:

```
override func viewDidLoad() {
  super.viewDidLoad()
  titleTextField.delegate = self
  bodyTextView.delegate = self
  updateSaveButtonState()
}
```

10. Build and run your project and tap the + button to go to the Add Entry screen. The **Save** button will be disabled:

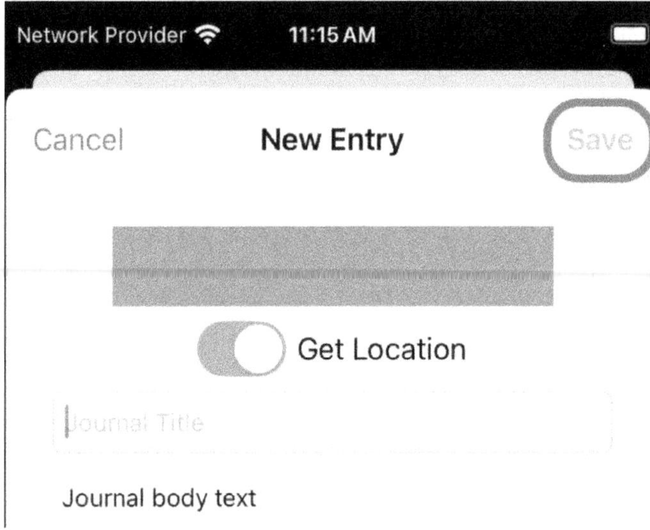

Figure 16.15: Simulator showing the Save button disabled

11. Enter some text in the text field and press the *return* key. As there is already placeholder text in the text view, the **Save** button will be enabled.

Both issues with the Add New Journal Entry screen have been resolved. Fantastic! In the next section, you'll learn how to pass data from the Journal List screen to the Journal Entry Detail screen when you tap a table view row.

Passing data from the Journal List screen to the Journal Entry Detail screen

As shown in the app tour in *Chapter 10, Setting Up the User Interface*, the Journal Entry Detail screen allows the user to view the details of a journal entry when a table view cell on the Journal List screen is tapped. For this to work, you'll create a view controller subclass to manage the Journal Entry Detail screen. Next, you'll implement the `prepare(for:sender:)` method for the `JournalListViewController` class to get the `JournalEntry` instance corresponding to the row that was tapped. You will then pass this instance to the view controller instance managing the Journal Entry Detail screen.

You'll start by creating a new view controller instance to manage the Journal Entry Detail screen in the next section.

Creating the JournalEntryDetailViewController class

At present, the Journal Entry Detail screen does not have a view controller. You'll add a new file to your project and implement the `JournalEntryDetailViewController` class, assign it as the identity for the **Entry Detail Scene**, and connect the outlets. Follow these steps:

1. In the Project navigator, create a new group by right-clicking the **JRNL** group and choosing **New Group**. Name this group **Journal Entry Detail Screen** and move it below the **Add New Journal Entry Screen** group.
2. Right-click on the **Journal Entry Detail Screen** group and select **New File from Template...**.
3. **iOS** should already be selected. Choose **Cocoa Touch Class** and click **Next**.
4. Configure the class with the following details:
    - **Class:** `JournalEntryDetailViewController`
    - **Subclass:** `UITableViewController`
    - **Also create XIB:** Unchecked
    - **Language:** `Swift`

    Click **Next**.
5. Click **Create** and the `JournalEntryDetailViewController` file will appear in the Project navigator.

With that, the `JournalEntryDetailViewController` file has been created, with the `JournalEntryDetailViewController` class declaration inside it. Now you'll set the identity of the view controller scene that's presented when you tap a table view cell on the Journal List screen. Follow these steps:

1. Click the **Main** storyboard file in the Project navigator and choose **Entry Detail Scene** in the document outline.

2. Click the Identity inspector button and, under **Custom Class**, set **Class** to `JournalEntryDetailViewController`:

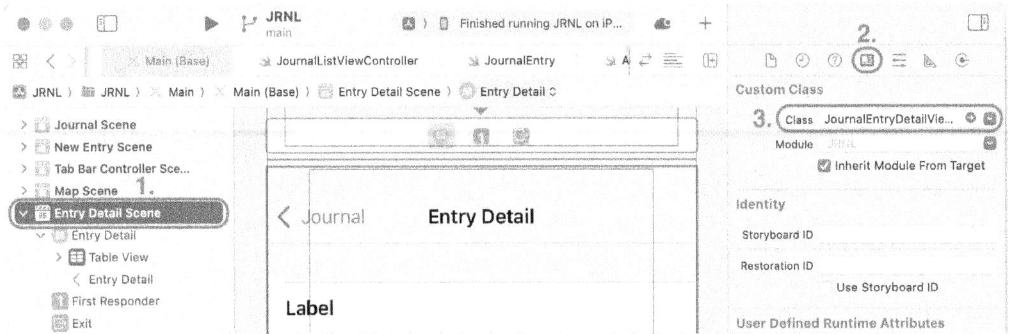

Figure 16.16: Identity inspector settings for Entry Detail Scene

Cool! In the next section, you'll connect the user interface elements in the **Entry Detail Scene** to outlets in the `JournalEntryDetailViewController` class. By doing this, the `JournalEntryDetailViewController` instance will be able to display the details for the journal entry.

Connecting the UI elements to the JournalEntryDetailViewController class

Currently, the `JournalEntryDetailViewController` instance for the Journal Entry Detail screen has no way of communicating with the UI elements in it. You'll create outlets in the `JournalEntryDetailViewController` class and assign the corresponding UI elements in the **Entry Detail Scene** to each outlet. Follow these steps:

1. In the Project navigator, click the **JournalEntryDetailViewController** file and remove all the code between the curly braces except for the `viewDidLoad()` method.
2. Add the following properties to the `JournalEntryDetailViewController` class after the opening curly brace:

```
// MARK: - Properties
@IBOutlet var dateLabel: UILabel!
@IBOutlet var titleLabel: UILabel!
@IBOutlet var bodyTextView: UITextView!
@IBOutlet var photoImageView: UIImageView!
```

3. Click the **Main** storyboard file and select **Entry Detail Scene** in the document outline.
4. Click the Connections inspector button and drag from the **dateLabel** outlet to the first label in **Entry Detail Scene**:

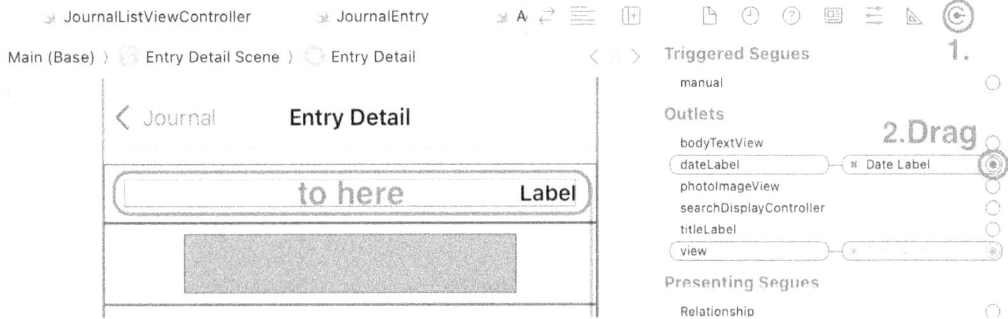

Figure 16.17: Connections inspector showing the dateLabel outlet

5. Drag from the **titleLabel** outlet to the second label in **Entry Detail Scene**:

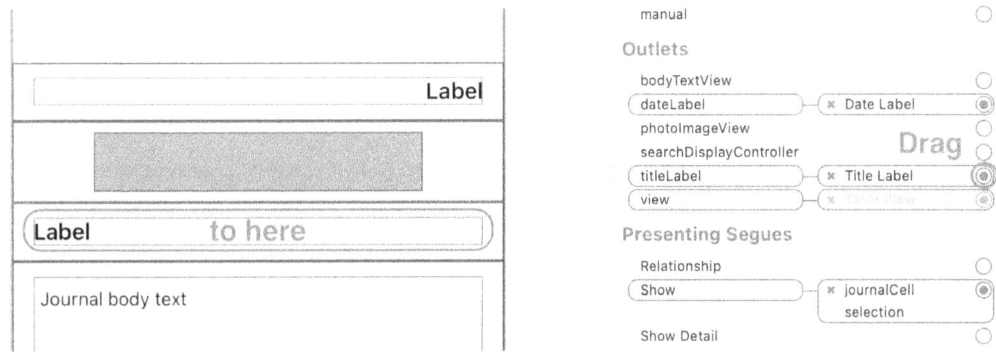

Figure 16.18: Connections inspector showing the titleLabel outlet

6. Drag from the **bodyTextView** outlet to the text view in **Entry Detail Scene**:

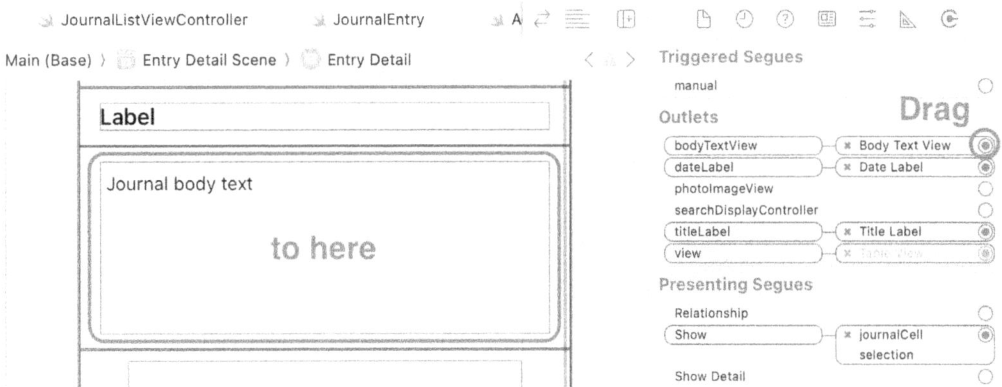

Figure 16.19: Connections inspector showing the bodyTextView outlet

7. Drag from the **photoImageView** outlet to the image view in **Entry Detail Scene**:

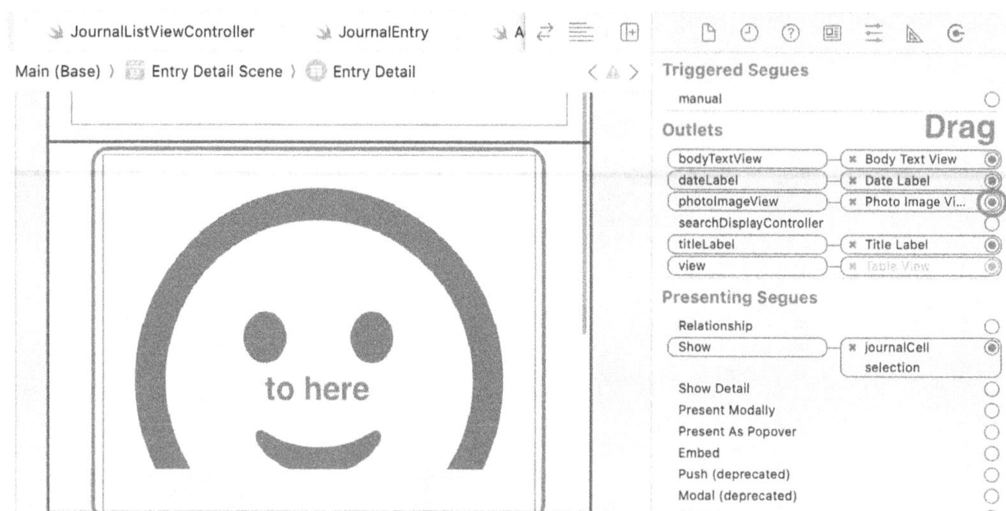

Figure 16.20: Connections inspector showing the photoImageView outlet

Remember that if you make a mistake, you can click the **x** to break the connection and drag from the outlet to the UI element once more.

You've now successfully connected the UI elements in the **Entry Detail Scene** to the outlets in the `JournalEntryDetailViewController` class. In the next section, you will implement the code to display the details of a `JournalEntry` instance when the user taps a table view cell on the Journal List screen.

Displaying the details of a journal entry

Until this point, you have implemented the `JournalEntryDetailViewController` class and connected the outlets in this class to the labels, text view, and image view in the **Entry Detail Scene**. When the user taps a table view cell on the Journal List screen, you'll get the corresponding `JournalEntry` instance from the data source and pass it to the `JournalEntryDetailViewController` instance to display on the Journal Entry Detail screen. To do this, you will implement the `prepare(for:sender:)` method in the `JournalListViewController` class. Follow these steps:

1. In the Project navigator, click the **Main** storyboard file and click the segue connecting **Journal Scene** and **Entry Detail Scene**. Click the Attributes inspector button and, under **Storyboard Segue**, set **Identifier** to `entryDetail`:

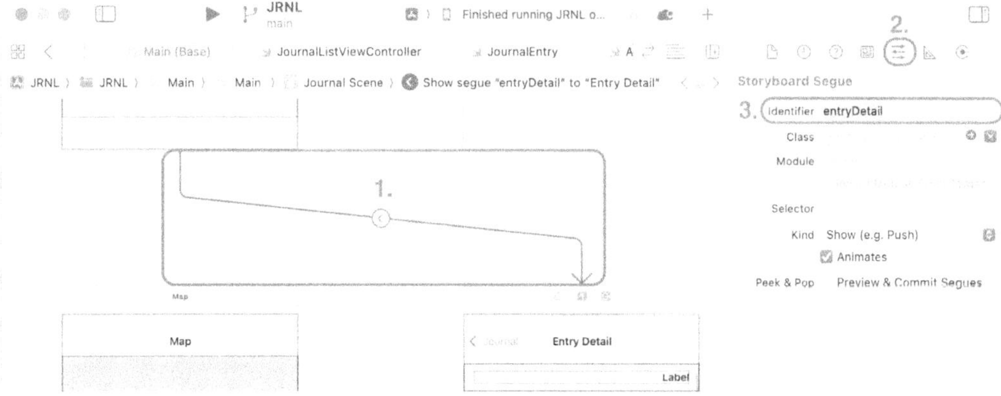

Figure 16.21: Attributes inspector showing Identifier set to entryDetail

You will use this identifier later to identify the segue used to go from the **Journal Scene** to the **Entry Detail Scene**.

2. Click the **JournalEntryDetailViewController** file in the Project navigator and add the following property to the `JournalEntryDetailViewController` class after the outlet declarations:

```
//MARK: - Properties
@IBOutlet var dateLabel: UILabel!
@IBOutlet var titleLabel: UILabel!
@IBOutlet var bodyTextView: UITextView!
@IBOutlet var photoImageView: UIImageView!
var selectedJournalEntry: JournalEntry?
```

The `JournalEntry` instance that you pass to the `JournalEntryDetailViewController` instance will be assigned to the `selectedJournalEntry` property.

3. Click the **JournalListViewController** file in the Project navigator and implement the `prepare(for:sender:)` method in the `JournalListViewController` class after the table view delegate methods, as shown:

```
// MARK: - Navigation
override func prepare(for segue: UIStoryboardSegue, sender: Any?) {
  super.prepare(for: segue, sender: sender)
  guard segue.identifier == "entryDetail" else {
    return
  }
  guard let journalEntryDetailViewController =
  segue.destination as?
  JournalEntryDetailViewController,
  let selectedJournalEntryCell = sender as?
  JournalListTableViewCell,
```

```
    let indexPath = tableView.indexPath(for:
    selectedJournalEntryCell) else {
      fatalError("Could not get indexPath")
    }
    let selectedJournalEntry = journalEntries[indexPath.row]
    journalEntryDetailViewController.selectedJournalEntry =
    selectedJournalEntry
  }
```

Let's break this down:

```
guard segue.identifier == "entryDetail" else {
  return
}
```

This code checks to see if the correct segue is being used, if not, the method exits.

```
guard let journalEntryDetailViewController = segue.destination as?
JournalEntryDetailViewController,
let selectedJournalEntryCell = sender as? JournalListTableViewCell,
let indexPath = tableView.indexPath(for: selectedJournalEntryCell) else {
  fatalError("Could not get indexpath")
}
```

This code checks that the destination view controller is an instance of `JournalEntryDetailViewController`, gets the table view cell the user tapped, and gets the index path of that cell.

```
let selectedJournalEntry = journalEntries[indexPath.row]
```

This statement gets the corresponding `JournalEntry` instance from the `journalEntries` array.

```
journalEntryDetailViewController.selectedJournalEntry =
selectedJournalEntry
```

This statement assigns the `JournalEntry` instance to the destination view controller's `selectedJournalEntry` property.

You've now added code that will pass the journal entry corresponding to the table view cell tapped by the user to the `JournalEntryDetailViewController` instance when transitioning from the Journal List screen to the Journal Entry Detail screen. In the next section, you'll modify the `JournalEntryDetailViewController` class to display the details of a journal entry.

Displaying the details of a selected journal entry

When transitioning to the Journal Entry Detail screen, the `JournalEntry` instance corresponding to the table view cell that the user tapped will be assigned to the `JournalEntryDetailViewController` instance's `selectedJournalEntry` property. You will add code to the `JournalEntryDetailViewController` class to access this property and display the details for a journal entry. Follow these steps:

1. In the Project navigator, click the **JournalEntryDetailViewController** file and modify the `JournalEntryDetailViewController` class's `viewDidLoad()` method as shown to display the details of the journal entry:

```
override func viewDidLoad() {
  super.viewDidLoad()
  dateLabel.text = selectedJournalEntry?.date.formatted(
    .dateTime.day().month(.wide).year()
  )
  titleLabel.text = selectedJournalEntry?.title
  bodyTextView.text = selectedJournalEntry?.body
  photoImageView.image = selectedJournalEntry?.photo
}
```

As you can see, the properties of the `JournalEntry` instance that was passed to this view controller earlier are used to populate the user interface elements. Note that the `date` property needs to be formatted into a string before it can be assigned to the `dateLabel text` property.

2. Build and run your project and tap a table view cell. The details of the journal entry corresponding to that table view cell will be displayed on the Journal Entry Detail screen:

Figure 16.22: Simulator displaying the Journal Entry Detail screen

Congratulations! You have successfully implemented a view controller for the Journal Entry Detail screen. You will now be able to display journal entry details in it when the user taps a table view cell on the Journal List screen.

Summary

In this chapter, you learned how to pass data from one view controller to another. You implemented a view controller for the Add New Journal Entry screen, then you added code to pass data from the Add New Journal Entry screen to the Journal List screen. Next, you learned how to remove journal entries while you're on the Journal List screen. After that, you learned about text field and text view delegate methods, and finally, you learned how to pass data from the Journal List screen to the Journal Entry Detail screen.

You now know how to pass data between view controllers and how to use text field and text view delegate methods. This will enable you to easily pass data between view controllers in your own apps. Cool!

In the next chapter, you will add a view controller to the Map screen and configure it to display journal entry locations using map annotations. You'll also configure the map annotations to display the Journal Entry Detail screen when a button in the annotation callout is tapped.

Join us on Discord!

Read this book alongside other users, experts, and the author himself. Ask questions, provide solutions to other readers, chat with the author via Ask Me Anything sessions, and much more. Scan the QR code or visit the link to join the community.

`https://packt.link/ios-Swift`

17

Getting Started with Core Location and MapKit

In the previous chapter, you learned how to pass data from the Add New Journal Entry screen to the Journal List screen, and from the Journal List screen to the Journal Entry Detail screen. You also learned about the `UITextFieldDelegate` and `UITextViewDelegate` methods.

In this chapter, you'll learn how to get your device location using Apple's **Core Location** framework, and how to set map regions, display map annotations, and create map snapshots using Apple's **MapKit** framework. This will come in handy if you're planning to build apps that use maps, such as *Apple Maps* or *Waze*.

First, you'll modify the Add New Journal Entry screen so that the user can add their current location to a new journal entry. Next, you'll create a `MapViewController` class (a view controller for the Map screen) and configure it to display a map region centered on your location. Then, you'll update the `JournalEntry` class to conform to the **MKAnnotation** protocol, which lets you add journal entries as map annotations to a map. After that, you'll modify the `MapViewController` class to display a pin for each journal entry within the map region that you set earlier. You'll configure the pins to display callouts and configure buttons in the callouts to display the Journal Entry Detail screen when tapped. Finally, you'll modify the `JournalEntryViewController` class to display a map snapshot showing the location where the journal entry was made in the Journal Entry Detail screen.

By the end of this chapter, you'll have learned how to use Core Location to get your device location and how to use MapKit to specify a map region, add map annotation views to a map, and create map snapshots.

The following topics will be covered in this chapter:

- Getting your device location using the Core Location framework
- Updating the `JournalEntry` class to conform to the `MKAnnotation` protocol
- Displaying annotation views on the Map screen
- Displaying a map snapshot on the Journal Entry Detail screen

Technical requirements

You will continue working on the JRNL project that you modified in the previous chapter.

The resource files and completed Xcode project for this chapter are in the Chapter17 folder of the code bundle for this book, which can be downloaded here:

https://github.com/PacktPublishing/iOS-18-Programming-for-Beginners-Ninth-Edition

Check out the following video to see the code in action:

https://youtu.be/6WE5Ed6jIWk

In the next section, you'll learn how the Core Location framework is used to get your device location.

Getting your device location using the Core Location framework

Every iPhone has multiple means of determining its location, including Wi-Fi, GPS, Bluetooth, magnetometer, barometer, and cellular hardware. Apple created the Core Location framework to gather location data using all available components on an iOS device.

To learn more about Core Location, see https://developer.apple.com/documentation/corelocation.

To configure your app to use Core Location, you will need to create an instance of CLLocationManager, which is used to configure, start, and stop location services. Next, you will create an instance of CLLocationUpdate, a structure that contains location information delivered by the Core Location framework. Calling the CLLocationUpdate type's liveUpdates method tells Core Location to start delivering location updates containing the user's location, authorization status, and location availability.

To learn more about the CLLocationManager class, see https://developer.apple.com/documentation/corelocation/configuring_your_app_to_use_location_services.

You can watch Apple's WWDC 2023 video on streamlined location updates here: https://developer.apple.com/videos/play/wwdc2023/10180/.

Since location information is considered sensitive user data, you'll also need to obtain authorization to use location services. You can also check the authorization status of location updates delivered by the CLLocationUpdate type's liveUpdates method.

To learn more about requesting authorization to use location services, see `https://developer.apple.com/documentation/corelocation/requesting_authorization_to_use_location_services`.

You can watch Apple's WWDC 2024 video on *What's new in location authorization* here: `https://developer.apple.com/videos/play/wwdc2024/10212/`.

In the next section, you'll modify the `AddJournalEntryViewController` class so that the user can assign a location when they create a new journal entry.

Modifying the AddJournalEntryViewController class

At present, the Add New Journal Entry screen has a **Get Location** switch, but it doesn't do anything yet. You'll add an outlet to the `AddJournalEntryViewController` class for this switch and modify it to add your location to the `JournalEntry` instance when the switch is on. Follow these steps:

1. In the Project navigator, click the `AddJournalEntryViewController` file. In this file, add an `import` statement after the `import UIKit` statement to import the Core Location framework:

   ```
   import UIKit
   import CoreLocation
   ```

2. Add an extension for the `AddJournalEntryViewController` class after all other code in the file:

   ```
   extension AddJournalEntryViewController {
     // MARK: - CoreLocation
   }
   ```

   You'll implement code to ask for permission to use the user's private data and to determine the user's location in this extension later.

3. Add outlets for the Get Location switch and the label next to it after all the other outlets in the `AddJournalEntryViewController` class:

   ```
   @IBOutlet var photoImageView: UIImageView!
   @IBOutlet var saveButton: UIBarButtonItem!
   @IBOutlet var getLocationSwitch: UISwitch!
   @IBOutlet var getLocationSwitchLabel: UILabel!
   var newJournalEntry: JournalEntry?
   ```

4. Add properties to store an instance of the `CLLocationManager` class, an asynchronous task that will manage location updates, and the current device location after all other property declarations:

   ```
   var newJournalEntry: JournalEntry?
   private let locationManager = CLLocationManager()
   ```

```
private var locationTask = Task<Void, Error>?
private var currentLocation: CLLocation?
```

All these properties are private as they will only be used within this class.

5. Implement a method to determine the user's location in the AddJournalEntryViewController extension before the closing curly brace:

```
private func fetchUserLocation() {
  locationManager.requestWhenInUseAuthorization()
  self.locationTask = Task {
    for try await update in
    CLLocationUpdate.liveUpdates() {
      if let location = update.location  {
        updateCurrentLocation(location)
      } else if update.authorizationDenied {
        failedToGetLocation(message: "Check Location
        Services settings for JRNL in Settings > Privacy
        & Security.")
      } else if update.locationUnavailable {
        failedToGetLocation(message: "Location
        Unavailable")
      }
    }
  }
}
```

Let's break this down:

```
locationManager.requestWhenInUseAuthorization()
```

This statement asks the user for permission to use their location information.

```
self.locationTask = Task {
```

This statement assigns an asynchronous task that will continuously obtain location updates to the locationTask property.

```
for try await update in
CLLocationUpdate.liveUpdates() {
```

This statement gets the updates provided by CLLocation.liveUpdates() and assigns each to an update instance.

```
if let location = update.location  {
  updateCurrentLocation(location)
}
```

This statement gets the user location from the update instance and calls the updateCurrentLocation method.

```
else if update.authorizationDenied {
failedToGetLocation(message: "Check Location
Services settings for JRNL in Settings > Privacy & Security."))}
```

This statement will call the `failedToGetLocation(message:)` method if the user did not give authorization to use their private data.

```
else if update.locationUnavailable {
failedToGetLocation(message: "Location
Unavailable")
}
```

This statement will call the `failedToGetLocation(message:)` method if the user's location is not available.

You will see error messages because the methods called by `fetchUserLocation()` have not been implemented yet.

6. Before the closing curly brace of the `AddJournalEntryViewController` extension, implement the `updateCurrentLocation(_:)` method called by the `fetchUserLocation()` class:

```
private func updateCurrentLocation(_ location: CLLocation) {
  let interval = location.timestamp.timeIntervalSinceNow
  if abs(interval) < 30 {
    self.locationTask?.cancel()
    getLocationSwitchLabel.text = "Done"
    let lat = location.coordinate.latitude
    let long = location.coordinate.longitude
    currentLocation = CLLocation(latitude: lat,
    longitude: long)
  }
}
```

This method first gets the timestamp of `location`. This location may be an old, cached location, and not actually the current location, so the timestamp is compared with the current date and time. If the duration is less than 30 seconds, this indicates that the user's location is current. In this case, the asynchronous task assigned to `locationTask` is canceled, the Get Location switch label's text is set to `Done`, and the `currentLocation` property is set to this location.

7. Before the closing curly brace of the `AddJournalEntryViewController` extension, implement the `failedToGetLocation(message:)` method called by the `fetchUserLocation()` class:

```
private func failedToGetLocation(message: String) {
```

```
    self.locationTask?.cancel()
    getLocationSwitch.setOn(false, animated: true)
    getLocationSwitchLabel.text = "Get location"
    let alertController = UIAlertController(title:
    "Failed to get location", message: message,
    preferredStyle: .alert)
    let okAction = UIAlertAction(title: "OK", style:
    .default)
    alertController.addAction(okAction)
    present(alertController, animated: true)
  }
```

This method will cancel the asynchronous task assigned to `locationTask`, reset the values of the Get Location switch and label to their initial values, and display an alert configured with an appropriate error message.

8. In the `AddJournalEntryViewController` class, implement an action to be performed when the Get Location switch's value is changed before the closing curly brace:

```
  // MARK: - Actions
  @IBAction func locationSwitchValueChanged(_ sender: UISwitch) {
    if getLocationSwitch.isOn {
      getLocationSwitchLabel.text = "Getting location..."
      fetchUserLocation()
    } else {
      currentLocation = nil
      getLocationSwitchLabel.text = "Get location"
      self.locationTask?.cancel()
    }
  }
```

If the Get Location switch is turned on, the switch label text is set to `Getting location...` and the `fetchUserLocation()` method is called. If the switch is turned off, `currentLocation` will be set to `nil`, the switch label text is reset to `Get location`, and the asynchronous task assigned to `locationTask` is canceled.

9. Modify the `prepare(for:sender:)` method to add location information to the `JournalEntry` instance:

```
override func prepare(for segue: UIStoryboardSegue, sender: Any?) {
  let title = titleTextField.text ?? ""
  let body = bodyTextView.text ?? ""
  let photo = photoImageView.image
  let rating = 3
  let lat = currentLocation?.coordinate.latitude
  let long = currentLocation?.coordinate.longitude
  newJournalEntry = JournalEntry(rating: rating, title:
  title, body: body, photo: photo, latitude: lat,
  longitude: long)
}
```

10. Modify the `updateSaveButtonState()` method to enable the **Save** button only after the location has been found if the Get Location switch is on:

```
private func updateSaveButtonState() {
  let textFieldText = titleTextField.text ?? ""
  let textViewText = bodyTextView.text ?? ""
  let textIsValid = !textFieldText.isEmpty &&
  !textViewText.isEmpty
  if getLocationSwitch.isOn {
    saveButton.isEnabled = textIsValid
    && currentLocation != nil
  } else {
    saveButton.isEnabled = textIsValid
  }
}
```

11. Call `updateSaveButtonState()` in `updateCurrentLocation(_:)` so that the **Save** button state will be updated once the location has been found:

```
      currentLocation = CLLocation(latitude: lat,
      Longitude: long
      updateSaveButtonState()
  }
}
```

12. In the Project navigator, click the `Main` storyboard file. Click **New Entry Scene** in the document outline.

13. Most journal entries will probably not require a location, so you will set the default value for the Get Location switch to **off**. Click the Get Location switch and click the Attributes inspector button. Under **Switch**, set **State** to **Off**:

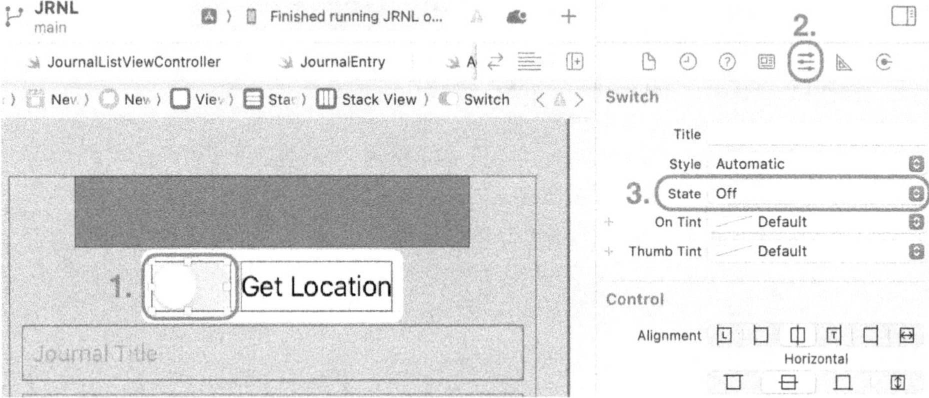

Figure 17.1: Attributes inspector showing the Get Location switch state set to Off

14. Click **New Entry Scene** in the document outline and click the Connections inspector button. Connect the **getLocationSwitch** outlet to the Get Location switch in **New Entry Scene**:

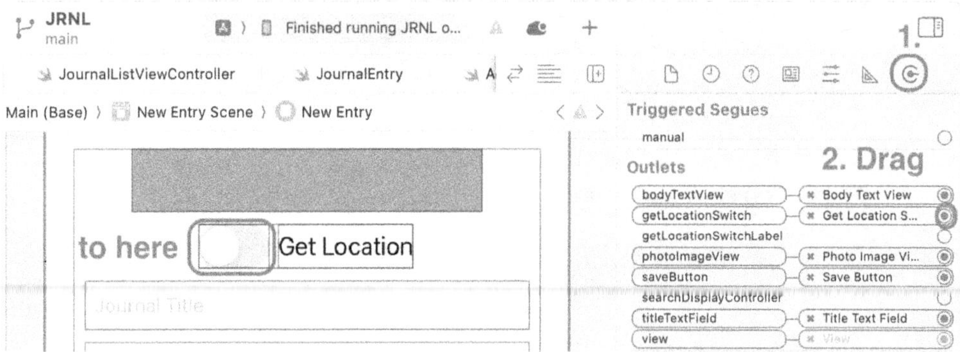

Figure 17.2: Connections inspector showing the getLocationSwitch outlet

15. Connect the **getLocationSwitchLabel** outlet to the label next to the Get Location switch in **New Entry Scene**:

Figure 17.3: Connections inspector showing the getLocationSwitchLabel outlet

16. Connect the **locationSwitchValueChanged** action to the Get Location switch, and choose **Value Changed** from the pop-up menu:

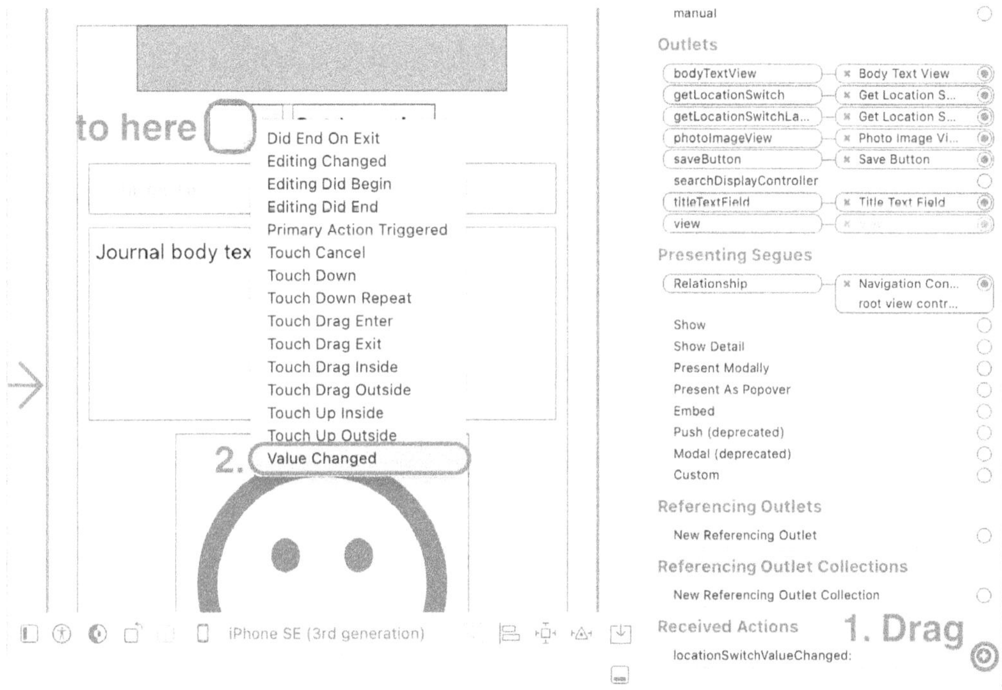

Figure 17.4: Attributes inspector showing the getLocationSwitchValueChanged outlet

You have completed modifying the `AddJournalEntryViewController` class. In the next section, you'll learn how to configure your app to access user data.

Modifying the Info.plist file

Since your app uses user data, you will need to ask the user for permission to use it. To do so, you will add a new setting to your app's `Info.plist` file. Follow these steps:

1. In the Project navigator, click the `Info.plist` file. If you move your pointer over the **Information Property List** row, you'll see a small + button. Click it to create a new row:

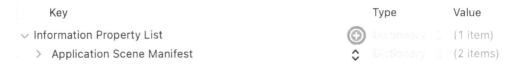

Figure 17.5: Editor area showing contents of Info.plist

2. In the new row, set the **Key** to **Privacy - Location When In Use Usage Description** and set the **Value** to `This app uses your location for journal entries.` Your `Info.plist` file should look like the following when you're done:

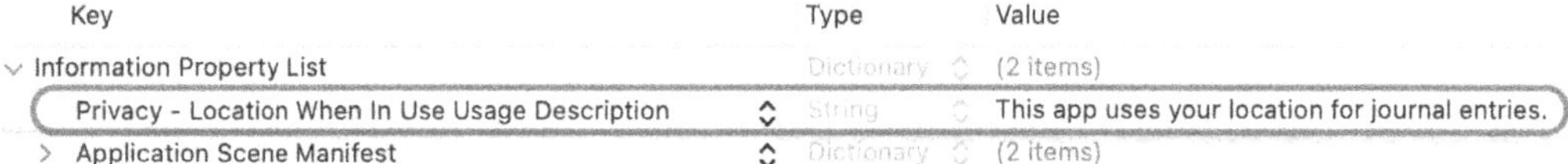

Key
Type
Value
Information Property List
Dictionary
(2 items)
Privacy - Location When In Use Usage Description
String
This app uses your location for journal entries.
Application Scene Manifest
Dictionary
(2 items)

Figure 17.6: Info.plist with a new row added

3. Launch Simulator and choose **Location** | **Apple** from Simulator's **Features** menu to simulate a location:

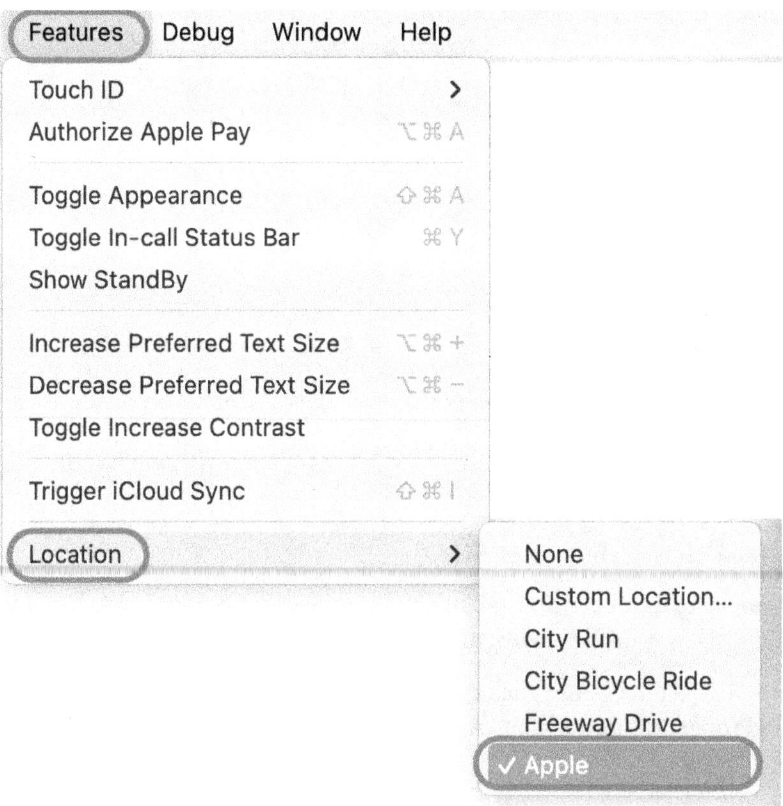

Figure 17.7: Location | Apple selected from Simulator's Features menu

4. Build and run your app and click the + button to display the Add New Journal Entry screen. When prompted, tap the **Allow While Using App** button:

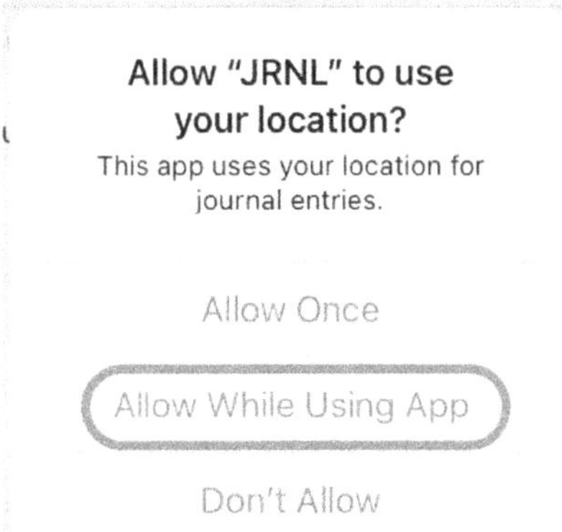

Figure 17.8: Alert showing Allow While Using App highlighted

Note that this alert will appear the first time you launch your app during this chapter, and the setting you picked will be used by default during subsequent launches.

5. Enter the journal entry title and body, and set the Get Location switch to on:

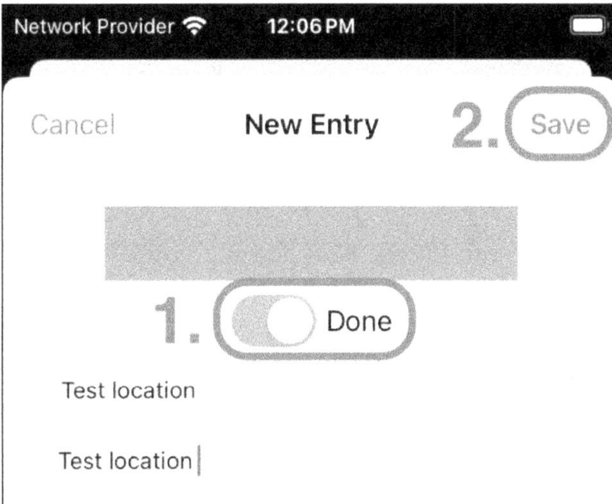

Figure 17.9: Add New Journal Entry screen showing the Get Location switch

Once the location has been determined, the label next to the Get Location switch will display **Done**, and the **Save** button will be active.

6. Click **Save**. You will be returned to the Journal List screen.

At this point, the Get Location switch and the label next to it have been connected to outlets in the `AddJournalEntryViewController` class, the method for getting the device location has been assigned to the Get Location switch, and all the code required to add your location to the new `JournalEntry` instance has been added. Great!

In the next section, you'll create a view controller for the Map screen and configure it to display your current device location.

Creating the MapViewController class

In *Chapter 12, Finishing Up Your User Interface*, you added a map view to the Map screen. A map view is an instance of the `MKMapView` class. You can see what it looks like in the Apple Maps app.

To learn more about `MKMapView`, see `https://developer.apple.com/documentation/mapkit/mkmapview`.

When you build and run your app, you will see a map on the screen. The part of the map that is visible onscreen can be specified by setting the map view's `region` property.

To learn more about regions and how to make them, see `https://developer.apple.com/documentation/mapkit/mkmapview/1452709-region`.

You'll create a new class, `MapViewController`, to be the view controller for the Map screen, and you'll use Core Location to determine the center point of the map region that will be displayed. Follow these steps:

1. Create a new group inside your project by right-clicking the **JRNL** group and choosing **New Group**. Name this group `Map Screen` and move it so it is below the **Journal Entry Detail Screen** group.
2. Right-click on the **Map Screen** group and select **New File From Template...**.
3. **iOS** should already be selected. Choose **Cocoa Touch Class** and click **Next**.
4. Configure the class with the following details:
    - **Class:** `MapViewController`
    - **Subclass:** `UIViewController`
    - **Also create XIB:** Unchecked
    - **Language: Swift**

    Click **Next** when you're done.
5. Click **Create**. The `MapViewController` file appears in the Project navigator, and its contents appear in the Editor area.

6. Add code to import the `Core Location` and `MapKit` frameworks after the existing `import` statement:

```
import UIKit
import CoreLocation
import MapKit
```

7. Add a new extension for the `MapViewController` class after all other code in the file:

```
extension MapViewController {
// MARK: - CoreLocation
}
```

You'll implement code to ask for permission to use the user's private data and to determine the user's location in this extension later.

8. In the `MapViewController` class, add an outlet for the map view after the opening curly brace:

```
// MARK: - Properties
@IBOutlet var mapView: MKMapView!
```

9. Add properties to hold a `CLLocationManager` instance and an asynchronous task after the `mapView` property:

```
@IBOutlet var mapView: MKMapView!
private let locationManager = CLLocationManager()
private var locationTask = Task<Void, Error>?
```

10. Implement a method to determine the user's location in the `MapViewController` extension before the closing curly brace:

```
private func fetchUserLocation() {
  locationManager.requestWhenInUseAuthorization()
  self.navigationItem.title = "Getting location..."
  self.locationTask = Task {
    for try await update in CLLocationUpdate.liveUpdates() {
      if let location = update.location  {
        updateMapWithLocation(location)
      } else if update.authorizationDenied {
        failedToGetLocation(message: "Check Location Services
        settings for JRNL in Settings > Privacy & Security.")
      } else if update.locationUnavailable {
        failedToGetLocation(message: "Location Unavailable")
      }
    }
  }
}
```

This method is similar to the `fetchUserLocation()` method you implemented in the `AddJournalEntryViewController` extension. First, it will ask for permission to use private user data and set the title of the Map screen to `Getting location...`. Next, an asynchronous task to continuously determine the user's location will be assigned to `locationTask`, and if the user's location is found, the map will be updated to show the user's location. Otherwise, an alert with an appropriate error message will be displayed.

Note that since the methods called by `fetchUserLocation()` have not been implemented yet, you will see error messages.

11. Implement the missing methods in the `MapViewController` extension before the closing curly brace:

```
private func updateMapWithLocation(_ location: CLLocation) {
  let interval = location.timestamp.timeIntervalSinceNow
  if abs(interval) < 30 {
    self.locationTask?.cancel()
    let lat = location.coordinate.latitude
    let long = location.coordinate.longitude
    navigationItem.title = "Map"
    mapView.region = MKCoordinateRegion(center:
    CLLocationCoordinate2D(latitude: lat,
    longitude: long), span: MKCoordinateSpan(
    latitudeDelta: 0.01, longitudeDelta: 0.01))
  }
}
private func failedToGetLocation(message: String) {
  self.locationTask?.cancel()
  navigationItem.title = "Location not found"
  let alertController = UIAlertController(title:
  "Failed to get location", message: message,
  preferredStyle: .alert)
  let okAction = UIAlertAction(title: "OK",
  style: .default)
  alertController.addAction(okAction)
  present(alertController, animated: true)
}
```

These methods are similar to the `updateCurrentLocation(location:)` and `failedToGetLocation(message:)` methods you implemented earlier in the `AddJournalEntryViewController` extension.

The `updateMapWithLocation(location:)` method will set the title of the Map screen to `Map`, create a map region centered on the user's location, and assign it as the region for the `mapView` property.

The `failedToGetLocation(message:)` method will set the title of the Map screen to `Location not found` and display an appropriate error message if permission to use user data was denied or if the location is unavailable.

12. Modify the `viewDidLoad()` method for the `MapViewController` class to call `fetchUserLocation()` as shown:

```
override func viewDidLoad() {
  super.viewDidLoad()
  fetchUserLocation()
}
```

13. Click the `Main` storyboard file in the Project navigator and click the first **Map Scene** in the document outline. Click the Identity inspector button, and under **Custom Class**, set **Class** to `MapViewController`:

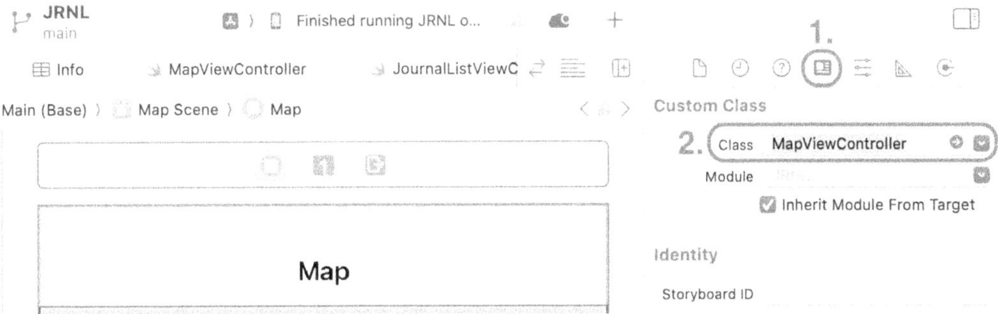

Figure 17.10: Identity inspector settings for Map scene

14. Click the Connections inspector to display all the outlets for **Map Scene**. Drag from the **mapView** outlet to the map view in **Map Scene**:

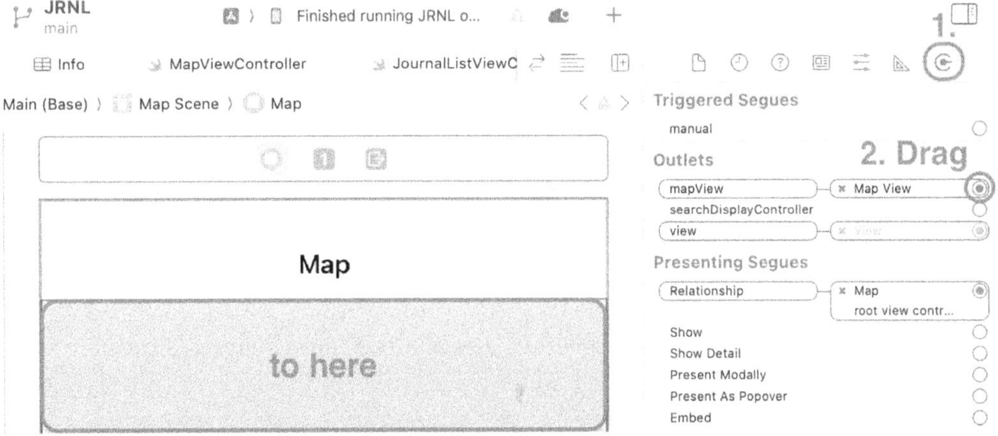

Figure 17.11: Connections inspector showing the mapView outlet

Remember that if you make a mistake, you can click the **x** to break the connection and drag from the outlet to the UI element once more.

15. Build and run your app, and verify that **Apple** is selected from Simulator's **Features** | **Location** menu. Tap the **Map** tab button to display a map region centered on the location you selected, which, in this case, is the Apple campus:

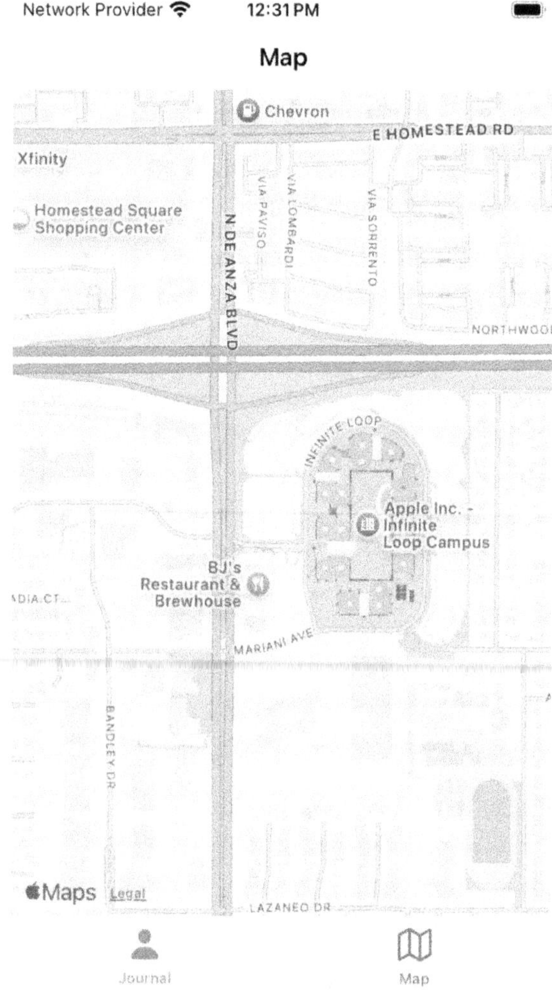

Figure 17.12: Simulator showing a map centered on your location

You can simulate any location in Simulator by choosing **Features** | **Location** | **Custom Location** and entering the longitude and latitude of the desired location.

Since `viewDidLoad()` is only called once when the `MapViewController` instance loads its view, the map will not be updated if the user's location changes after it was initially set. Also, you'll notice that it takes a long time for the location to be determined if you're running the app on an actual iOS device. You'll address both of these issues in the next chapter.

You have successfully created a new view controller for the Map screen and configured it to display a map region centered on your device location. Excellent! In the next section, you'll learn about the `MKAnnotation` protocol, and how to make a class conform to it.

Updating the JournalEntry class to conform to the MKAnnotation protocol

When you use the *Maps* app on iPhone, you can tap and hold on the map to drop a pin:

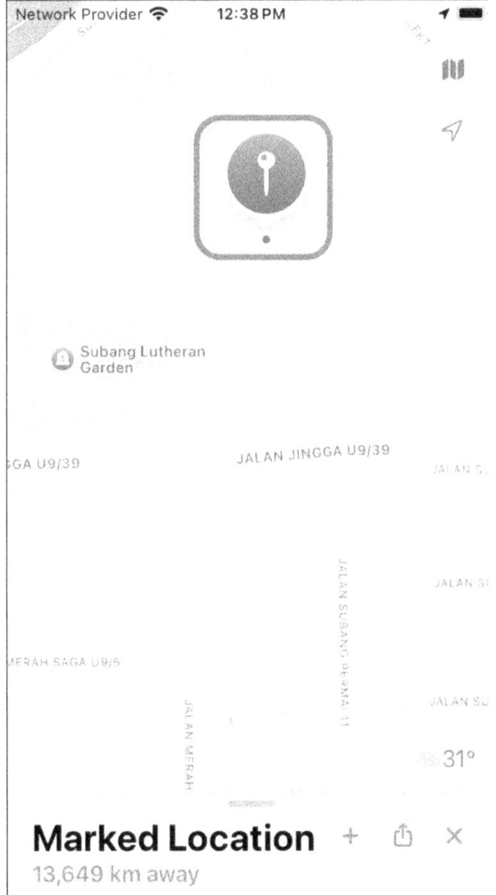

Figure 17.13: Maps app showing a dropped pin

To add a pin to a map view for your own apps, you need a class that conforms to the `MKAnnotation` protocol. This protocol allows you to associate an instance of that class with a specific location.

To learn more about the MKAnnotation protocol, see https://developer.apple.com/documentation/mapkit/mkannotation.

Any class can adopt the MKAnnotation protocol by implementing a coordinate property, which contains a location. Optional MKAnnotation protocol properties are title, a string containing the annotation's title, and subtitle, a string containing the annotation's subtitle.

When an instance of a class conforming to the MKAnnotation protocol is in the region of the map that is visible onscreen, the map view asks its delegate (usually a view controller) to provide a corresponding instance of the MKAnnotationView class. This instance appears as a pin on the map.

To learn more about MKAnnotationView, see https://developer.apple.com/documentation/mapkit/mkannotationview.

If the user scrolls the map and the MKAnnotationView instance goes off screen, it will be put into a reuse queue and recycled later, like the way table view cells and collection view cells are recycled.

To represent journal entry locations on the Map screen, you will modify the JournalEntry class to make it conform to the MKAnnotation protocol. This class will have a coordinate property to store the journal entry's location, a title property to store the journal entry date, and a subtitle property to store the journal entry title. You will use the JournalEntry instance's latitude and longitude properties to compute the value that will be assigned to the coordinate property. Follow these steps:

1. In the Project navigator, click the JournalEntry file (inside the **Journal List Screen** | **Model** group). Type the following after the import UIKit statement to import the MapKit framework:

   ```
   import UIKit
   import MapKit
   ```

   This lets you use the MKAnnotation protocol in your code.

2. The MKAnnotation protocol has an optional title property, which you will use later. You will change the name of the title property in the JournalEntry class to entryTitle so you don't have two properties with the same name. Right-click on the title property and choose **Refactor** | **Rename** from the pop-up menu.
3. Set the new name to entryTitle, as shown, and click **Rename**:

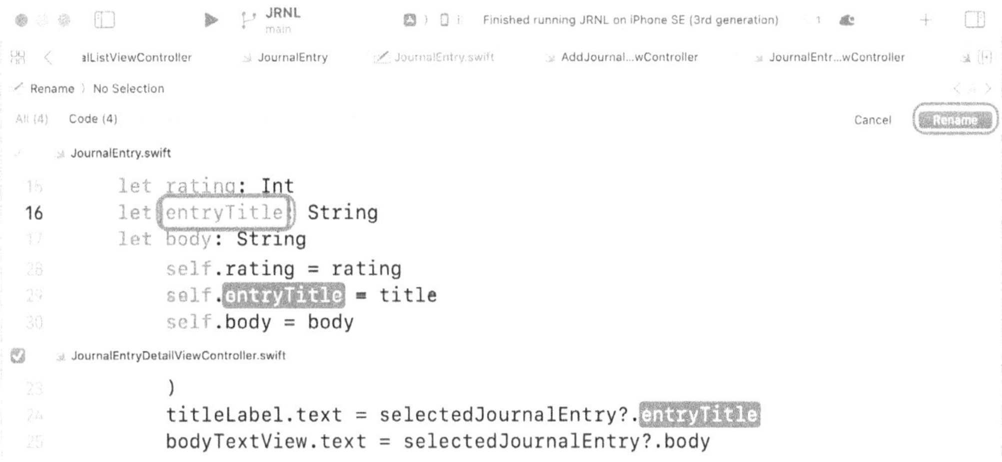

Figure 17.14: Refactoring the title property to entryTitle

4. Modify the `JournalEntry` class declaration as follows to make it a subclass of the `NSObject` class and to adopt the `MKAnnotation` protocol:

```
class JournalEntry: NSObject, MKAnnotation {
```

5. You'll see an error because you have not yet implemented the `coordinate` property, which is required to conform to the `MKAnnotation` protocol. Type the following after the initializer:

```
// MARK: - MKAnnotation
var coordinate: CLLocationCoordinate2D {
  guard let lat = latitude, let long = longitude else {
    return CLLocationCoordinate2D()
  }
  return CLLocationCoordinate2D(latitude: lat,
  longitude: long)
}
```

The `coordinate` property is of the `CLLocationCoordinate2D` type, and it holds a geographical location. The value of the `coordinate` property is not assigned directly; the `guard` statement gets the latitude and longitude values from the `latitude` and `longitude` properties, which are then used to create the value for the `coordinate` property. Such properties are called **computed properties.**

6. To implement the optional `title` property, type the following after the `coordinate` property:

```
var title: String? {
  date.formatted(
    .dateTime.day().month().year()
  )
}
```

This is a computed property that returns the journal entry date formatted as a string.

7. To implement the optional `subtitle` property, type the following after the `title` property:

```
var subtitle: String? {
  entryTitle
}
```

This is a computed property that returns the journal entry title.

At this point, you've modified the `JournalEntry` class to conform to the `MKAnnotation` protocol. In the next section, you'll modify the `MapViewController` class to add an array of `JournalEntry` instances to a map view, and any instance within the region displayed by the map view will appear as a pin on the Map screen.

Displaying annotation views on the Map screen

The Map screen at present displays a map region centered on your device location. Now that the map region has been set, you can determine which `JournalEntry` instances are in this region based on their `coordinate` property. Remember that the `JournalEntry` class conforms to `MKAnnotation`. As the view controller for the map view, the `MapViewController` class is responsible for providing an `MKAnnotationView` instance for any `MKAnnotation` instance within this region. You will now modify the `MapViewController` class to get an array of `JournalEntry` instances from the `SampleJournalEntryData` structure and add it to the map view. Follow these steps:

1. In the Project navigator, click the **JournalEntry** file. In the `createSampleJournalEntryData()` method, modify the statement that creates the `journalEntry2` instance as shown:

```
guard let journalEntry2 = JournalEntry(rating: 0, title: "Bad", body:
"Today is a bad day", photo: photo2, latitude: 37.3318, longitude:
-122.0312) else {
  fatalError("Unable to instantiate journalEntry2")
}
```

This instance now has values for its `latitude` and `longitude` properties, which will be used to set its `coordinate` property. The values used are for a location close to the Apple campus, which you will set in Simulator when you run the app.

You can use whatever location you wish, but you will need to make sure that the location is close to the center point of the map, otherwise, the pin will not be displayed.

2. In the Project navigator, click the **MapViewController** file. Add an extension after all other code in the file to make the `MapViewController` class conform to the `MKMapViewDelegate` protocol:

```
extension MapViewController: MKMapViewDelegate {


}
```

3. Just after the `locationTask` property declaration, add the following code to create a private property, `annotations`, that will hold an array of `JournalItem` instances:

```
private var locationTask: Task<Void, Error>?
private var annotations: [JournalEntry] = []
```

There is currently no connection between the Journal List and Map screens. This means that any journal entries that you add using the Add New Journal Entry screen will not appear on the Map screen. You will create a shared instance that will be used by both view controllers in the next chapter.

4. In the `viewDidLoad()` method, set the map view's `delegate` property to an instance of `MapViewController` before the closing curly brace:

```
    fetchUserLocation()
    mapView.delegate = self
}
```

5. On the next line, populate the `journalEntries` array by calling the `sampleJournalEntryData` structure's `createSampleJournalEntryData()` method:

```
    mapView.delegate = self
    annotations = JournalEntry.createSampleJournalEntryData()
}
```

6. On the next line, add the following statement to add all the sample journal entries (which conform to the `MKAnnotation` protocol) to the map view:

```
    annotations = JournalEntry.createSampleJournalEntryData()
    mapView.addAnnotations(annotations)
}
```

The map view's delegate (the `MapViewController` class in this case) will now automatically provide an `MKAnnotationView` instance for every `JournalItem` instance within the map region displayed on the Map screen.

7. Build and run your app, and verify the location has been set to **Apple** using Simulator's **Features | Location** menu. You should see a single pin (`MKAnnotationView` instance) on the Map screen:

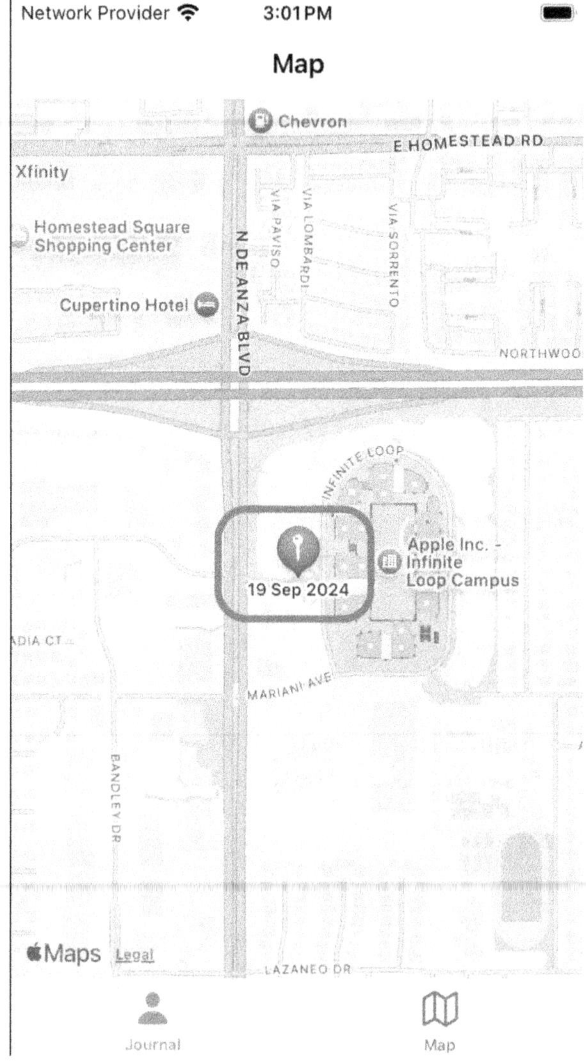

Figure 17.15: iOS Simulator showing a standard MKAnnotationView instance

The Map screen can now display pins, but tapping a pin just makes it bigger. You will add code to make pins display a callout with a button in the next section.

Since `viewDidLoad()` is only called once when the `MapViewController` instance loads its view, any journal entries with locations added to the `journalEntries` array after that will not be added as annotations to the map. You'll address this issue in the next chapter.

Configuring a pin to display a callout

Currently, the Map screen displays standard MKAnnotationView instances, which look like pins. Tapping a pin just makes it bigger. An MKAnnotationView instance can be configured to display callout bubbles when tapped. To make it do so, you will implement the mapView(_:viewFor:) method, an optional MKMapViewDelegate protocol method. Follow these steps:

1. Click the **MapViewController** file in the Project navigator. In the MKMapViewDelegate extension, add the following method after the opening curly brace:

```
// MARK: - MKMapViewDelegate
func mapView(_ mapView: MKMapView, viewFor annotation: any MKAnnotation)
-> MKAnnotationView? {
  let identifier = "mapAnnotation"
  guard annotation is JournalEntry else {
    return nil
  }
  if let annotationView =
  mapView.dequeueReusableAnnotationView(withIdentifier:
  identifier) {
    annotationView.annotation = annotation
    return annotationView
  } else {
    let annotationView =
    MKMarkerAnnotationView(annotation:annotation,
    reuseIdentifier:identifier)
    annotationView.canShowCallout = true
    let calloutButton = UIButton(type: .detailDisclosure)
    annotationView.rightCalloutAccessoryView =
    calloutButton
    return annotationView
  }
}
```

Let's break this down:

```
func mapView(_ mapView: MKMapView, viewFor annotation: any MKAnnotation)
-> MKAnnotationView?
```

This is one of the methods specified in the MKMapViewDelegate protocol. It's triggered when an MKAnnotation instance is within the map region, and it returns an MKAnnotationView instance, which the user will see on the screen.

```
let identifier = "mapAnnotation"
```

A constant, identifier, is assigned the "MapAnnotation" string. This will be the reuse identifier for the MKAnnotationView instance.

```
guard annotation is JournalEntry else {
  return nil
}
```

This guard statement checks to see if the annotation is a JournalEntry instance, and returns nil if it is not.

```
if let annotationView =
mapView.dequeueReusableAnnotationView(withIdentifier:
identifier) {
  annotationView.annotation = annotation
  return annotationView
```

This if statement checks to see whether there is an existing MKAnnotationView instance that was initially visible but is no longer on the screen. If there is, it can be reused and is assigned to the annotationView constant. The JournalItem instance is then assigned to the annotation property of annotationView and the annotationView is returned.

```
} else {
    let annotationView =
    MKMarkerAnnotationView(annotation:annotation,
    reuseIdentifier:identifier)
```

The else clause is executed if there are no existing MKAnnotationView instances that can be reused. A new MKAnnotationView instance is created with the reuse identifier specified earlier (MapAnnotation).

```
annotationView.canShowCallout = true
let calloutButton = UIButton(type: .detailDisclosure)
annotationView.rightCalloutAccessoryView =
calloutButton
```

The MKAnnotationView instance is configured with a callout. When you tap a pin on the map, a callout bubble will appear showing the title (journal entry date), subtitle (journal entry title), and a button. You'll program the button later to present the Journal Entry Detail screen.

```
return annotationView
```

The custom MKAnnotationView instance is returned.

To learn more about the mapView(_:viewFor:) method, see https://developer.apple.com/documentation/mapkit/mkmapviewdelegate/1452045-mapview.

2. Build and run your app, and set the location to **Apple** using Simulator's **Features** | **Location** menu. You should see a single pin on the Map screen. Tap the pin to display a callout:

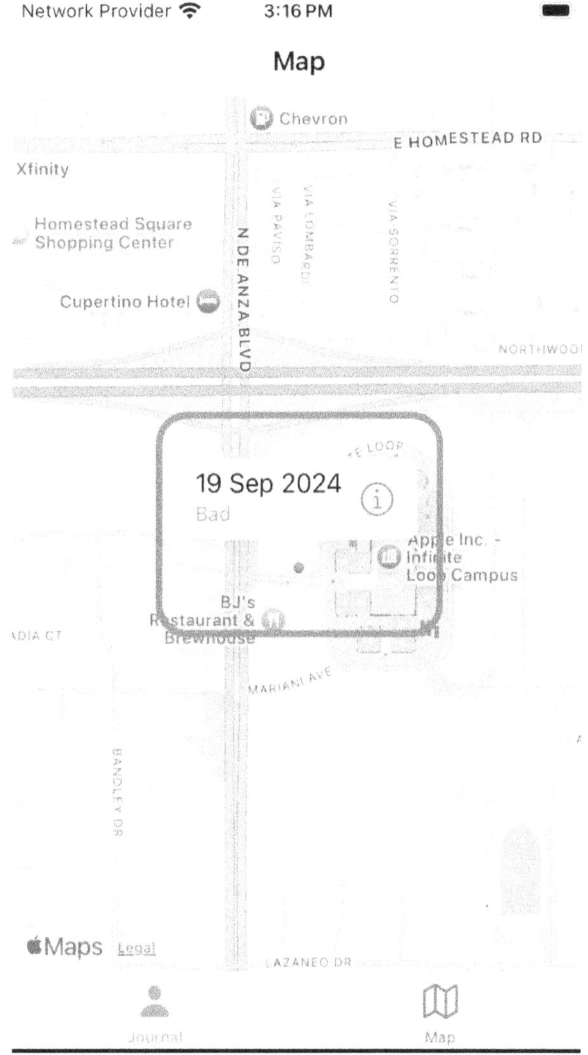

Figure 17.16: iOS Simulator showing a callout when a pin is tapped

You have successfully created a custom `MKAnnotationView` that displays a callout when tapped, but tapping the button in the callout bubble doesn't do anything yet. You'll configure the button to present the Journal Entry Detail screen in the next section.

Going from the Map screen to the Journal Entry Detail screen

At this point, the Map screen now displays an `MKAnnotationView` instance, and tapping it displays a callout bubble showing journal entry details. The button in the callout bubble doesn't work yet, though.

To present the Journal Entry Detail screen from the callout button, you'll add a segue between the Map screen and the Journal Entry Detail screen and you will implement the `mapView(_:annotationView:calloutAccessoryControlTapped:)` method, an optional `MKMapViewDelegate` protocol method, to perform that segue when the callout button is tapped. Follow these steps:

1. In the Project navigator, click the **Main** storyboard file. Find the **Map** icon under **Map Scene** in the document outline. *Ctrl + Drag* from the **Map** icon to the **Entry Detail Scene** and choose **Show** from the pop-up menu to add a segue between the **Map Scene** and the **Entry Detail Scene**:

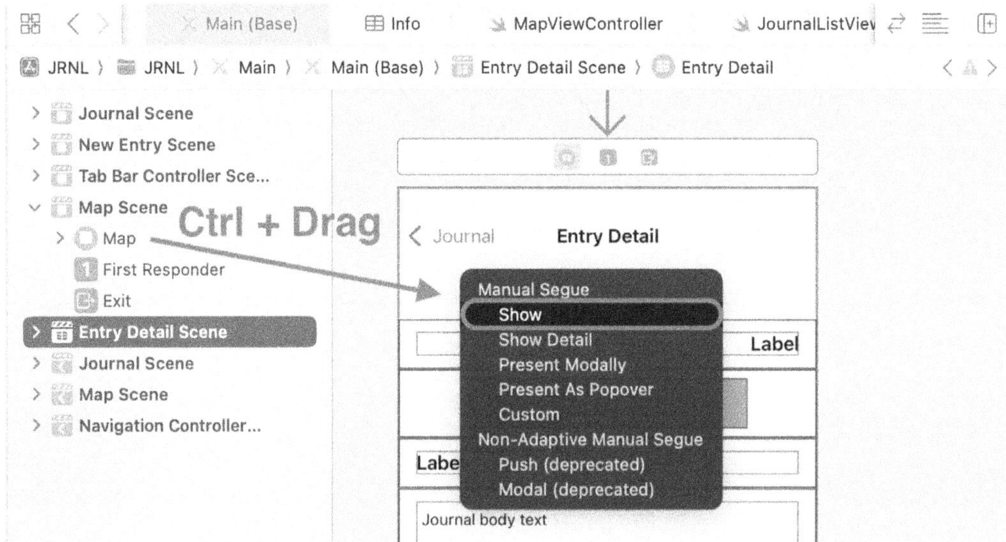

Figure 17.17: Segue pop-up menu

2. You will set an identifier for this segue so that the `mapView(_:annotationView:calloutAccessoryControlTapped:)` method knows which segue to perform. Select the segue connecting the **Map Scene** to the **Entry Detail Scene**:

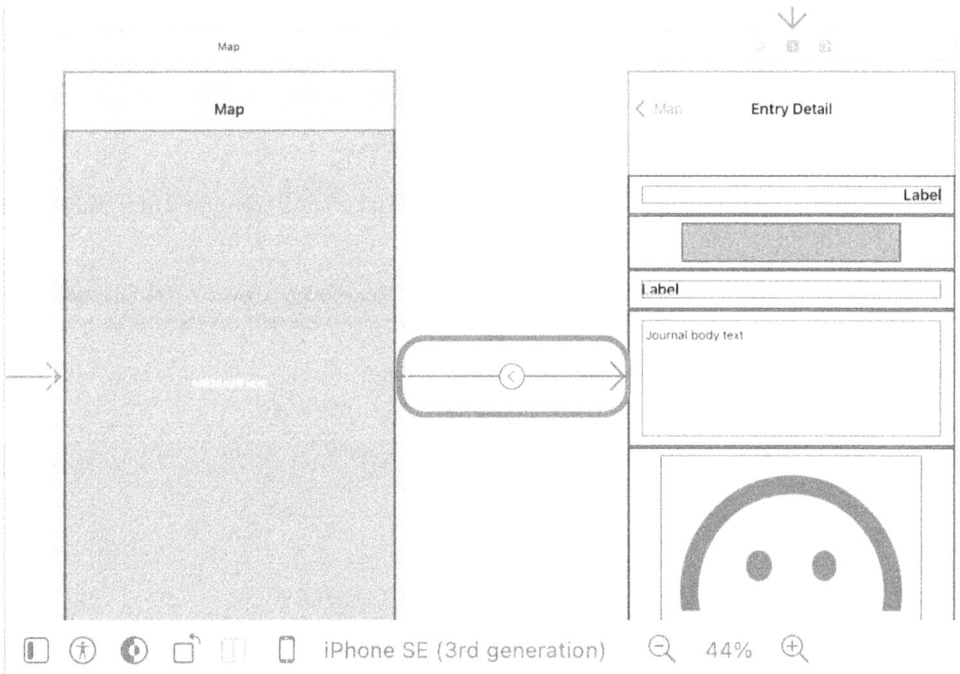

Figure 17.18: Segue between Map scene and Entry Detail scene

3. In the Attributes inspector, under **Storyboard Segue**, set **Identifier** to `showMapDetail`:

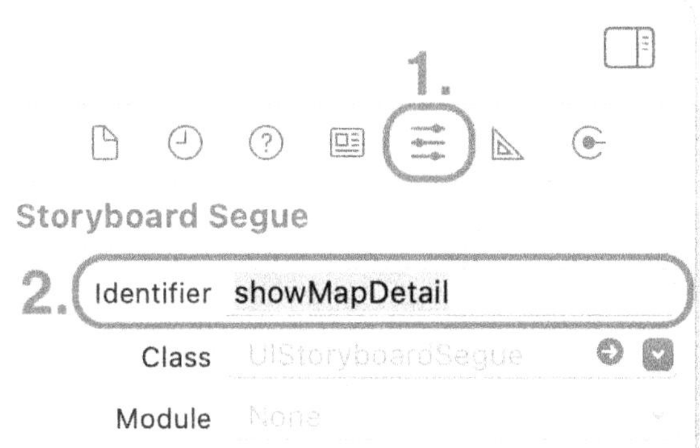

Figure 17.19: Attributes inspector settings for the showDetail segue

4. Click the **MapViewController** file in the Project navigator. In the `MapViewController` class, add a property to store a journal entry after all other property declarations:

```
private var annotations: [JournalEntry] = []
private var selectedAnnotation: JournalEntry?
```

   This property will store the `JournalEntry` instance for the `MKAnnotationView` instance that was tapped.

5. Add `mapView(_:annotationView:calloutAccessoryControlTapped:)` after the `mapView(_:viewFor:)` method:

```
func mapView(_ mapView: MKMapView, annotationView view: MKAnnotationView,
calloutAccessoryControlTapped control: UIControl) {
  guard let annotation = mapView.selectedAnnotations.first
  else {
    return
  }
  selectedAnnotation = annotation as? JournalEntry
  self.performSegue(withIdentifier: "showMapDetail",
  sender: self)
}
```

   This method is triggered when the user taps the callout bubble button. The annotation for the annotation view will be assigned to the `selectedAnnotation` property, and the segue with the `showMapDetail` identifier will be performed, which presents the Journal Entry Detail screen.

To learn more about the `mapView(_:annotationView:calloutAccessoryControlTapped:)` method, see `https://developer.apple.com/documentation/mapkit/mkmapviewdelegate/1616211-mapview`.

6. Build and run your app and set the location to **Apple** using Simulator's **Features | Location** menu. You should see a single pin on the Map screen. Tap the pin to display a callout and tap the button inside the callout.
7. The Journal Entry Detail screen appears, but it does not contain any details about the journal entry:

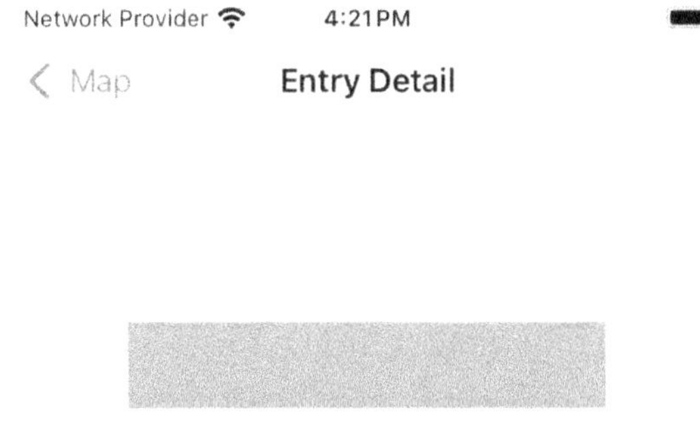

Figure 17.20: iOS Simulator showing a blank Journal Entry Detail screen

8. To make the Journal Entry Detail screen display the details of a journal entry, you will use the `prepare(for:sender:)` method to pass the selected journal entry to the Journal Entry Detail screen's view controller. In the `MapViewController` class, uncomment and modify the `prepare(for:sender:)` method as shown:

```
// MARK: - Navigation
override func prepare(for segue: UIStoryboardSegue, sender: Any?) {
  super.prepare(for: segue, sender: sender)
  guard segue.identifier == "showMapDetail" else {
    fatalError("Unexpected segue identifier")
  }
  guard let entryDetailViewController = segue.destination
  as? JournalEntryDetailViewController else {
    fatalError("Unexpected view controller")
  }
  entryDetailViewController.selectedJournalEntry =
  selectedAnnotation
}
```

As you have learned before, the `prepare(for:sender:)` method is executed by a view controller before transitioning to another view controller. In this case, this method is called before the Map screen transitions to the Journal Entry Detail screen. If the segue's identifier is `showMapDetail` and the segue destination is a `JournalEntryDetailViewController` instance, `selectedAnnotation` will be assigned to the `selectedJournalEntry` property for the `JournalEntryDetailViewController` instance.

9. Build and run your app and verify the location has been set to **Apple** using Simulator's **Features | Location** menu. You should see a single pin on the Map screen. Tap the pin to display a callout and tap the button inside the callout. The Journal Entry Detail screen appears, and it displays the details of the journal entry that you tapped on the Map screen.

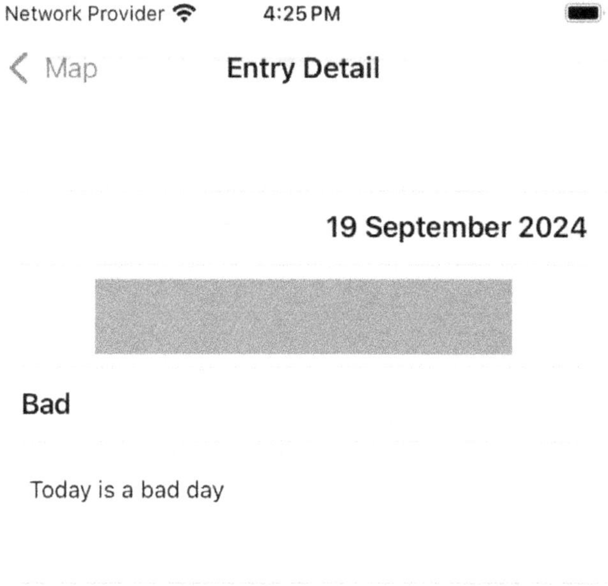

Figure 17.21: Simulator showing the Journal Entry Detail screen

You have connected the Journal Entry Detail screen to the Map screen, and have successfully passed data from a selected journal entry on the Map screen to the Journal Entry Detail screen. Fantastic! In the next section, you'll configure the Journal Entry Detail screen to display a map snapshot of the journal entry's location.

Displaying a map snapshot on the Journal Entry Detail screen

The Map screen currently displays a single pin representing a journal entry. When you tap the pin on the Map screen and tap the callout button, the details of the journal entry are displayed on the Journal Entry Detail screen, but the second image view on the Journal Entry Detail screen currently displays a placeholder map image. You can capture a map region and convert it into an image using the `MKMapSnapshotter` class.

For more information on the `MKMapSnapshotter` class, see `https://developer.apple.com/documentation/mapkit/mkmapsnapshotter`.

To configure the region and appearance of the map that is captured in the snapshot, an `MKMapSnapshotter.Options` object is used.

For more information on the `MKMapSnapshotter.Options` object, see `https://developer.apple.com/documentation/mapkit/mkmapsnapshotter/options`.

You will connect the second image view in the **Entry Detail Scene** to an outlet in the `JournalEntryDetailViewController` class and replace the placeholder image with a map snapshot showing the location of the journal entry. Follow these steps:

1. In the Project navigator, click the `JournalEntryDetailViewController` file. Type the following after the `import UIKit` statement to import the `MapKit` framework:

```
import UIKit
import MapKit
```

2. Add the following outlet after all other outlets in the `JournalEntryDetailViewController` class:

```
@IBOutlet var bodyTextView: UITextView!
@IBOutlet var photoImageView: UIImageView!
@IBOutlet var mapImageView: UIImageView!
var selectedJournalEntry: JournalEntry?
```

3. Add a method to generate the map snapshot before the closing curly brace:

```
// MARK: - Private methods
private func getMapSnapshot() {
  guard let lat = selectedJournalEntry?.latitude, let
  long = selectedJournalEntry?.longitude else {
    self.mapImageView.image = nil
    return
  }
  let options = MKMapSnapshotter.Options()
  options.region = MKCoordinateRegion(center:
  CLLocationCoordinate2D(latitude: lat, longitude: long),
  span: MKCoordinateSpan(latitudeDelta: 0.01,
  longitudeDelta: 0.01))
  options.size = CGSize(width: 300, height: 300)
  options.preferredConfiguration =
  MKStandardMapConfiguration()
  let snapshotter = MKMapSnapshotter(options: options)
  snapshotter.start { snapshot, error in
    if let snapshot {
```

```
            self.mapImageView.image = snapshot.image
        } else if let error {
            print("snapshot error:
            \(error.localizedDescription)")
        }
    }
}
```

This method checks to see if `journalEntry` has values in its `latitude` and `longitude` properties. If it does, then an `MKMapSnapShotter.Options` object is created, configured, and assigned to an `MKMapSnapshotter` object. The `MKMapSnapshotter` object is then used to generate the map snapshot, which will be assigned to the `image` property of the `mapImageView` property.

4. Call the `getMapSnapshot()` method in the `viewDidLoad()` method before the closing curly brace:

```
    photoImageView.image = journalEntry?.photo
    getMapSnapshot()
}
```

5. In the Project navigator, click the **Main** storyboard file and click **Entry Detail Scene** in the document outline. Click the Connections inspector button and connect the **mapImageView** outlet to the second image view in the **Entry Detail Scene**:

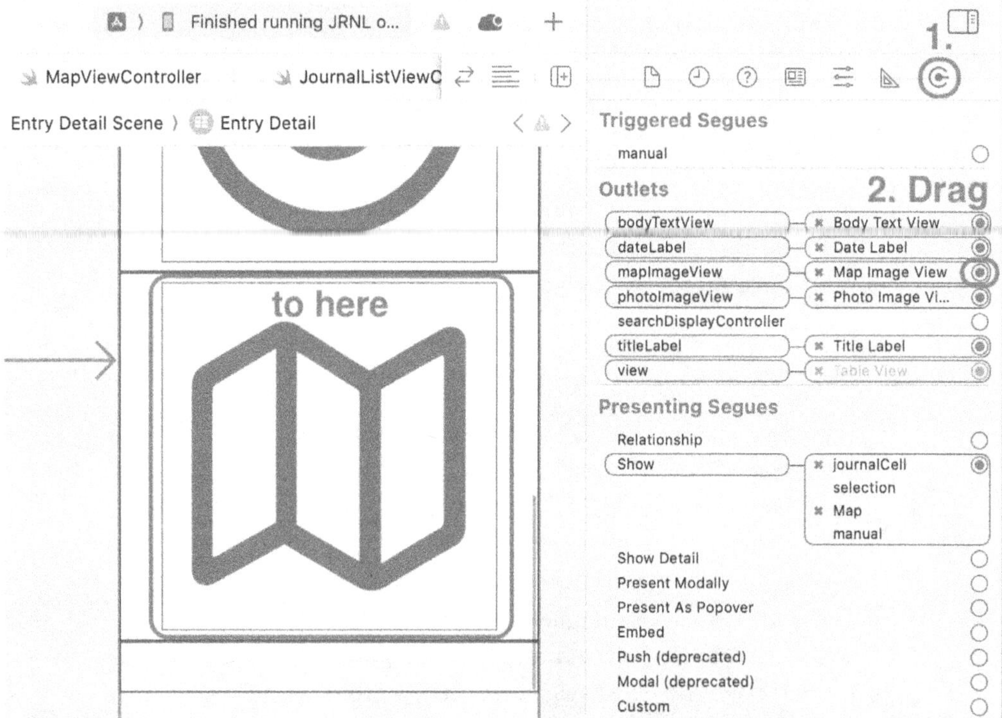

Figure 17.22: Connections inspector showing the mapImageView outlet

It may be easier to drag to the image view in the document outline.

6. Build and run your app and verify the location has been set to **Apple** using Simulator's **Features | Location** menu. You should see a single pin on the Map screen. Tap the pin to display a callout and tap the button inside the callout. The Journal Entry Detail screen appears, and it displays the details of the journal entry that you tapped on the Map screen. Scroll down and you will see the map snapshot in the second image view:

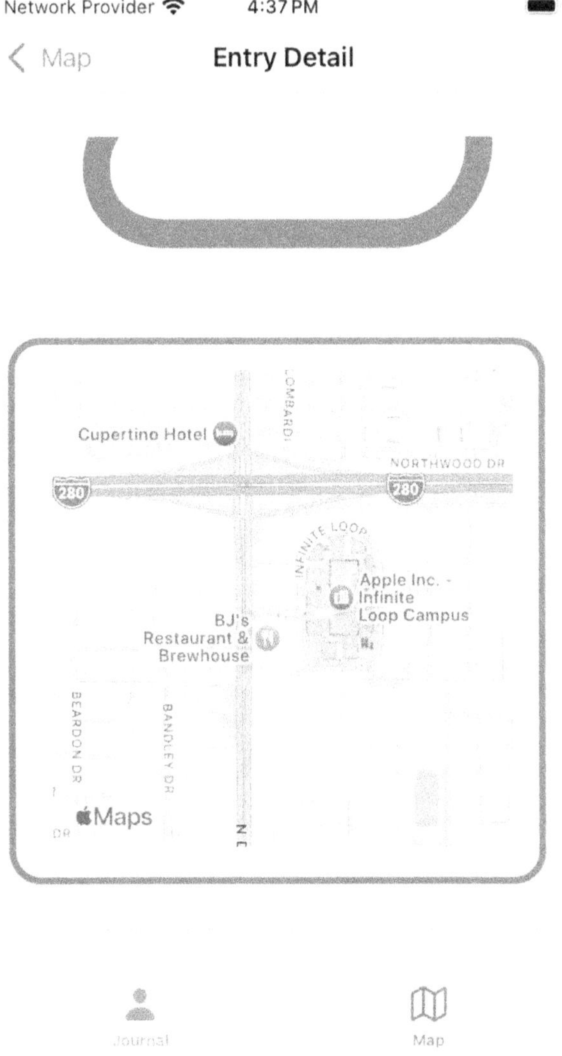

Figure 17.23: Simulator showing the Journal Entry Detail screen with a map snapshot

The Journal Entry Detail screen can now display the map snapshot showing the location of a journal entry. Cool!

Summary

In this chapter, you modified the Add New Journal Entry screen so that the user can add their current location to a new journal entry. Next, you created a `MapViewController` class and configured it to display a custom map region centered on your location. Then, you updated the `JournalEntry` class to conform to the `MKAnnotation` protocol. After that, you modified the `MapViewController` class to display a pin for each journal entry within the map region. You configured the pins to display callouts and configured buttons in the callouts to display the Journal Entry Detail screen when tapped. Finally, you modified the `JournalEntryViewController` class to display a map snapshot for the journal entry on the Journal Entry Detail screen.

You now know how to get your device location using Apple's Core Location framework, how to create custom map regions and display map annotations using Apple's MapKit framework, and how to create map snapshots, which will come in handy if you're planning to build apps that use maps, such as *Apple Maps* or *Waze*.

In the next chapter, you'll learn how to create a shared data instance, and how to load and save data from JSON files.

Join us on Discord!

Read this book alongside other users, experts, and the author himself. Ask questions, provide solutions to other readers, chat with the author via Ask Me Anything sessions, and much more. Scan the QR code or visit the link to join the community.

`https://packt.link/ios-Swift`

18

Getting Started with JSON Files

In the previous chapter, you modified the Add Journal Entry screen so that the user can add their current location to a new journal entry, and configured the Map screen to display a region centered on your current location as well as pins representing the locations where the journal entries are made. However, since the `MapViewController` instance does not have access to the `journalEntries` array in the `JournalListViewcontroller` instance, newly added journal entries do not appear on the Map screen as pins. Also, all newly added journal entries are lost when you quit the app.

In this chapter, you will create a **singleton**, `SharedData`, that will provide journal entry data to both the Journal List and Map screens. This class will also be used to load journal entry data from a file on your device when the app starts up and save journal entry data to a file on your device when you add or delete journal entries.

You'll start by creating the `SharedData` class and configuring your app to use it. Next, you'll modify the `JournalEntry` class to be compatible with the **JSON** format, so you can save journal entries to a JSON file and load journal entries from a JSON file. After that, you'll add methods to save journal entry data when you add or delete journal entries, and to load journal entry data when your app is starting up.

By the end of this chapter, you'll know how to create a class to store, load, and save data from JSON files for use in your own apps.

The following topics will be covered in this chapter:

- Creating a singleton
- Modifying the `JournalEntry` class to be JSON-compatible
- Loading and saving JSON data

Technical requirements

You will continue working on the `JRNL` project that you modified in the previous chapter.

The resource files and completed Xcode project for this chapter are in the `Chapter18` folder of the code bundle for this book, which can be downloaded here:

`https://github.com/PacktPublishing/iOS-18-Programming-for-Beginners-Ninth-Edition`

Check out the following video to see the code in action:

https://youtu.be/1J4zuzzyjYE

Let's start by creating a new singleton to store the data used by your app.

Creating a singleton

At present, when you add new journal entries to your app, they will appear on the Journal List screen, but when you switch to the Map screen, the newly added journal entries are not present. This is because the `MapViewController` instance does not have access to the `journalEntries` array in the `JournalListViewcontroller` instance. To solve this issue, you'll create a new singleton to store your app data. A singleton is created once and then referenced throughout your app. This means that the `JournalListViewController` class and the `MapViewController` class will be getting their data from a single source.

For more information on singletons, see `https://developer.apple.com/documentation/swift/managing-a-shared-resource-using-a-singleton`.

You will create a singleton named `SharedData` and configure the `JournalListViewController` and `MapViewController` classes to use it. Follow these steps:

1. In the Project navigator, move the **Model** group that is inside the **Journal List Screen** group to a new location just below the **SceneDelegate** file:

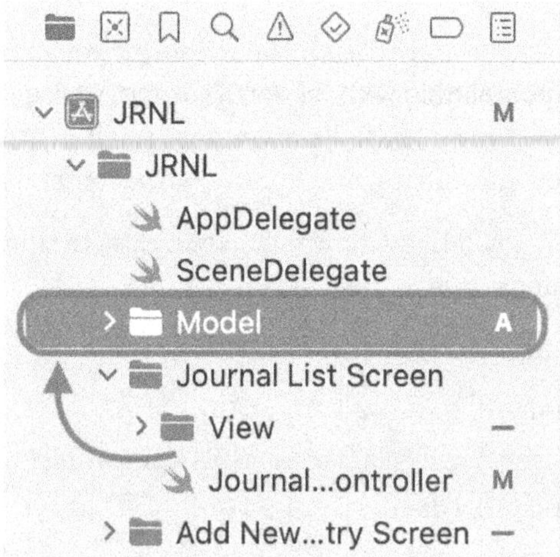

Figure 18.1: Model group moved to a new location

This reflects the fact that the model objects are no longer solely used by the Journal List screen, but are used by the entire app.

2. Right-click the **Model** group and choose **New File from Template...**.

3. **iOS** should already be selected. Choose **Swift File** and click **Next**.
4. Name the file `SharedData` and then click **Create**. It will appear in the Project navigator and its contents will appear in the Editor area.
5. Replace the contents of this file with the following code to declare and define the `SharedData` class:

```
import UIKit
class SharedData {
  // MARK: - Properties
  @MainActor static let shared = SharedData()
  private var journalEntries: [JournalEntry] = []
  // MARK: - Initializers
  private init() {
  }
  // MARK: - Access methods
  func numberOfJournalEntries() -> Int {
    journalEntries.count
  }
  func journalEntry(at index: Int) -> JournalEntry {
    journalEntries[index]
  }
  func allJournalEntries() -> [JournalEntry] {
    journalEntries
  }
  func addJournalEntry(_ newJournalEntry: JournalEntry) {
    journalEntries.insert(newJournalEntry, at: 0)
  }
  func removeJournalEntry(at index: Int) {
    journalEntries.remove(at: index)
  }
}
```

Let's break this down:

```
@MainActor static let shared = SharedData()
```

This statement creates a single instance of this class, which means that the only instance of `SharedData` in your app is stored in the `shared` property. This property is marked with `@MainActor` to ensure that it should only be accessed from the main queue.

For more information, watch Apple's WWDC 2022 video titled *Eliminate data races using Swift Concurrency* here: `https://developer.apple.com/videos/play/wwdc2022/110351/`.

```
private var journalEntries: [JournalEntry] = []
```

This statement creates an empty array named `journalEntries` that will be used to store `JournalEntry` instances. The private keyword means that the `journalEntries` array may only be modified by methods in the `SharedData` class. This is to ensure that no other part of your app can make changes to the `journalEntries` array.

```
private init() {
}
```

The `init()` method has an empty body. This prevents the accidental creation of a `SharedData()` instance.

```
func numberOfJournalEntries() -> Int {
  journalEntries.count
}
```

This method returns the number of items in the `journalEntries` array.

```
func journalEntry(at index: Int) -> JournalEntry {
  journalEntries[index]
}
```

This method returns the `JournalEntry` instance located at the specified index in the `journalEntries` array.

```
func allJournalEntries() -> [JournalEntry] {
  journalEntries
}
```

This method returns a copy of the `JournalEntries` array.

```
func addJournalEntry(_ newJournalEntry: JournalEntry) {
  journalEntries.insert(newJournalEntry, at: 0)
}
```

This method inserts the `JournalEntry` instance that was passed into the `JournalEntries` array at index `0`.

```
func removeJournalEntry(at index: Int) {
  journalEntries.remove(at: index)
}
```

This method removes the `JournalEntry` instance at the specified index from the `JournalEntries` array.

Now that you have created the `SharedData` class, you'll modify your app to use it. Follow these steps:

1. In the Project navigator, click the **JournalListViewController** file. Remove the `journalEntries` property from the `JournalListViewController` class:

```
//MARK: - Properties
@IBOutlet var tableView: UITableView!
private var journalEntries: [JournalEntry] = [] // remove
```

2. In `viewDidLoad()`, remove the statement that creates the sample data and appends it to the `journalEntries` array:

```
override func viewDidLoad() {
  super.viewDidLoad()
  journalEntries =
  JournalEntry.createSampleJournalEntryData() // remove
}
```

3. Modify the `tableView(_:numberOfRowsInSection:)` method to get the number of rows for the table view from `SharedData`:

```
func tableView(_ tableView: UITableView, numberOfRowsInSection section:
Int) -> Int {
  SharedData.shared.numberOfJournalEntries()
}
```

4. Modify the `tableView(_:cellForRowAt:)` method to get the required `JournalEntry` instance from `SharedData`:

```
let journalCell = tableView.dequeueReusableCell(withIdentifier:
"journalCell", for: indexPath) as! JournalListTableViewCell
let journalEntry = SharedData.shared.journalEntry(at: indexPath.row)
journalCell.photoImageView.image = journalEntry.photo
```

5. Modify the `tableView(_:commit:forRowAt:)` method to remove the selected `JournalEntry` instance from `SharedData`:

```
if editingStyle == .delete {
  SharedData.shared.journalEntry(at: indexPath.row)
  tableView.reloadData()
}
```

6. Modify the `prepare(for:sender:)` method to use `SharedData` to get the selected `JournalEntry` instance:

```
let selectedJournalEntry = SharedData.shared.journalEntry(at: indexPath.
row)
journalEntryDetailViewController.selectedJournalEntry =
selectedJournalEntry
```

7. Modify the `unwindNewEntrySave(segue:)` method to add a new `JournalEntry` instance to `SharedData`:

```
if let sourceViewController = segue.source as?
AddJournalEntryViewController, let newJournalEntry =
sourceViewController.newJournalEntry {
  SharedData.shared.addjournalEntry(newJournalEntry)
  tableView.reloadData()
}
```

You have made all the required changes to the `JournalListViewController` class. Now you will modify the `MapViewController` class to use `SharedData`. As noted in the previous chapter, when running your app on an actual device, it takes a long time to determine the device's location, and the map on the Map screen will not be updated if the user's location changes. You'll address both of these issues as well. Follow these steps:

1. In the Project navigator, click the **MapViewController** file. Remove the `annotations` property from the `MapViewController` class:

```
//MARK: - Properties
@IBOutlet var mapView: MKMapView!
let locationManager = CLLocationManager()
private var locationTask: Task<Void, Error>?
private var annotations: [JournalEntry] = [] // remove
var selectedJournalEntry: JournalEntry?
```

2. Modify the `viewDidLoad()` method by removing the highlighted statements:

```
override func viewDidLoad() {
  super.viewDidLoad()
  fetchUserLocation() // remove
  mapView.delegate = self
  annotations = JournalEntry.
  createSampleJournalEntryData() // remove
  mapView.addAnnotations(annotations) // remove
}
```

3. To reduce the time taken to determine the user's location, add a statement to `fetchUserLocation()` as shown to set the location manager instance's accuracy to `kCLLocationAcccuracyKilometer`:

   ```
   locationManager.requestWhenInUseAuthorization()
   locationManager.desiredAccuracy = kCLLocationAccuracyKilometer
   self.navigationItem.title = "Getting location..."
   ```

   The default value of this property is `kCLLocationAccuracyBest`, which takes a relatively long time to determine. This trade-off is acceptable since the *JRNL* app does not require the highest level of accuracy when displaying annotations on the map.

4. To update the user's location whenever the Map screen appears, first implement the following method after the `viewDidLoad()` method:

   ```
   override func viewIsAppearing(_ animated: Bool) {
     super.viewIsAppearing(animated)
     fetchUserLocation()
   }
   ```

The `viewIsAppearing()` view controller lifecycle method was introduced during WWDC 2023. You can learn more about this method at this link: `https://developer.apple.com/documentation/uikit/uiviewcontroller/4195485-viewisappearing`.

5. Add the following statement to the `updateMapWithLocation(_:)` method as shown so that the map view gets all the annotations from `SharedData` after the user's location has been determined and the map region has been set:

   ```
       mapView.region = MKCoordinateRegion(center:
       CLLocationCoordinate2D(latitude: lat, longitude: long),
       span: MKCoordinateSpan(latitudeDelta: 0.01,
       longitudeDelta: 0.01))
       mapView.addAnnotations(SharedData.shared
       .allJournalEntries())
     }
   }
   ```

With this change, if you are on the Journal List screen, tapping the **Map** tab bar button will update the user's location, redraw the map on the Map screen, and reload the map annotations.

You have made all the required changes to the `MapViewController` class. Now let's test your app. Follow these steps:

1. Launch Simulator, and choose **Location** | **Apple** from Simulator's **Features** menu to simulate a location. Build and run your app.

Click the + button and add a new journal entry. Make sure the **Get Location** switch is on:

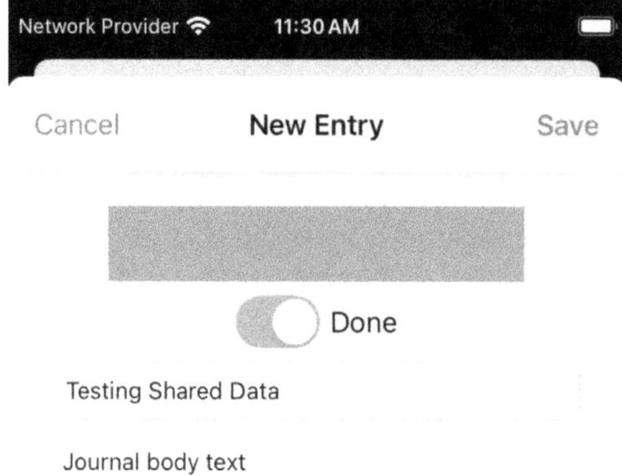

Figure 18.2: Simulator showing Add New Journal Entry screen

2. Tap the **Map** tab button to go to the Map screen:

Figure 18.3: Simulator showing the Map tab button

3. Note the journal entry you added earlier is visible as a pin on the Map screen. Tap the pin and then tap the callout button:

Figure 18.4: Simulator showing pin callout button

The journal entry details are displayed on the Journal Entry Detail screen:

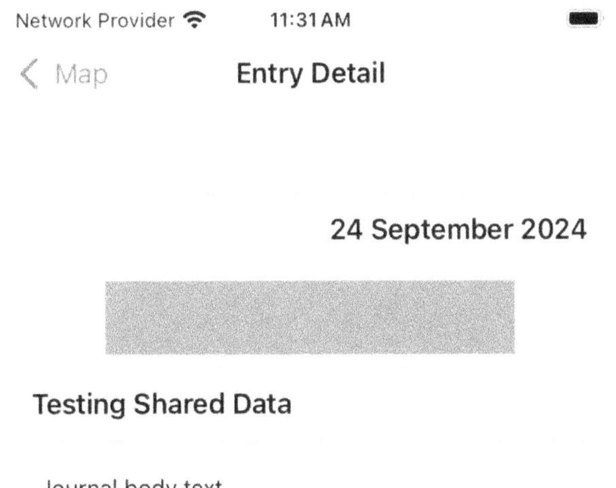

Figure 18.5: Simulator showing Journal Entry Detail screen

You have successfully created a singleton and configured your app to use it, but the data is lost once the app quits. Later, you will write code to save journal entries to your device storage. But before you can do that, you'll modify the `JournalEntry` class so that the data in it can be stored in JSON format. You'll do this in the next section.

Modifying the JournalEntry class to be JSON-compatible

At present, all app data is lost when you quit the app. You will need to implement a way to save your app data. iOS provides many ways to store your app data. One of them is converting the data to **JavaScript Object Notation (JSON)** format, and then writing it as a file to your device storage. JSON is a way to structure data in a file that can be easily read by both people and computers.

To help you to understand the JSON format, look at the sample shown below:

```
[
  {
    "dateString": "May 17, 2023"
    "rating": 5
    "entryTitle": "Good"
    "entryBody": "Today is a good day"
    "photoData": "<photo data for the sun.max image>"
    "latitude":
    "longitude":
  },
  {
    "dateString": "May 17, 2023"
    "rating": 0
    "entryTitle": "Bad"
    "entryBody": "Today is a bad day"
    "photoData": "<photo data for the cloud image>"
    "latitude": 37.331354
    "longitude": -122.031791
  },
  {
    "dateString": "May 17, 2023"
    "rating": 3
    "entryTitle": "Good"
    "entryBody": "Today is a good day"
    "photoData": "<photo data for the cloud.sun>"
    "latitude":
    "longitude":
  }
]
```

This sample is a representation of the `journalEntries` array in JSON format. As you can see, it starts with an opening square bracket, and each item inside consists of key-value pairs containing journal entry information, enclosed by curly braces and separated by commas.

At the very end of the file, you can see a closing square bracket. The square brackets denote arrays, and the curly braces denote dictionaries. The keys in the dictionary correspond to the properties in a `JournalEntry` instance, and the values correspond to the values assigned to those properties.

To learn more about using JSON with Swift types, see `https://developer.apple.com/documentation/foundation/archives_and_serialization/using_json_with_custom_types`.

To learn more about parsing JSON files, watch the video available here: `https://devstreaming-cdn.apple.com/videos/wwdc/2017/212vz78e2gzl2/212/212_hd_whats_new_in_foundation.mp4`.

A custom Swift type needs to conform to the `Codable` protocol before it can be converted to and from JSON.

To learn more about `Codable`, see `https://developer.apple.com/documentation/swift/codable`.

JSON supports dates, strings, numbers, Boolean values, and `null` values, but it does not support images. To conform to the `Codable` protocol, you will modify the `JournalEntry` class to use types that are supported by JSON and modify the rest of your app to work with the updated `JournalEntry` instance. Follow these steps:

1. In the Project navigator, click the **JournalEntry** file. Modify the `JournalEntry` class declaration as shown to adopt the `Codable` protocol:

```
class JournalEntry: NSObject, MKAnnotation, Codable {
```

2. An error will appear because the `UIImage` type does not conform to `Codable`. Modify the `photo` property as shown to make `JournalEntry` conform to `Codable`:

```
let date: Date
let rating: Int
let entryTitle: String
let body: String
let photoData: Data?
let latitude: Double?
let longitude: Double?
```

The error will disappear, but another error will appear in the initializer.

3. Modify the initializer as shown:

```
self.date = Date()
```

```
self.rating = rating
self.entryTitle = title
self.body = body
self.photoData = photo?.jpegData(compressionQuality: 1.0)
```

This converts the value in the `photo` argument into a `Data` instance and assigns it to `photoData`.

To learn more about the `Data` type, see `https://developer.apple.com/documentation/foundation/data`.

All the errors in the initializer are gone, but if you build your app now, you'll see other errors appear. Let's fix them now. Follow these steps:

1. In the Project navigator, click the **JournalListViewController** file. Modify the `tableView(_:cellForRowAt:)` method in the `JournalListViewController` class as shown:

   ```
   let journalEntry = SharedData.shared.getJournalEntry(index: indexPath.
   row)
   if let photoData = journalEntry.photoData {
     journalCell.photoImageView.image = UIImage(data: photoData)
   }
   journalCell.dateLabel.text = journalEntry.date.formatted(
     .dateTime.month().day().year()
   )
   journalCell.titleLabel.text = journalEntry.entryTitle
   return journalCell
   ```

   The updated code converts the data stored in `photoData` back into a `UIImage` and assigns it to the image view in `journalCell`.

2. In the Project navigator, click the **JournalEntryDetailViewController** file. Modify the `viewDidLoad()` method as shown:

   ```
   override func viewDidLoad() {
     super.viewDidLoad()
     dateLabel.text = selectedJournalEntry?.date.formatted(
       .dateTime.day().month(.wide).year()
     )
     titleLabel.text = selectedJournalEntry?.entryTitle
   ```

```
    bodyTextView.text = selectedJournalEntry?.entryBody
    if let photoData = selectedJournalEntry?.photoData {
      photoImageView.image = UIImage(data: photoData)
    }
    getMapSnapshot()
  }
```

The updated code converts the data stored in photoData to a UIImage instance and assigns it to the image property of photoImageView. There should be no more errors in your app at this point.

3. Build and run your app. Verify that the simulated location has been set and add a new journal entry. Your app should work the way it did before.

Note that the image is now black instead of blue. This is due to the image conversion process and will not be noticeable when you use images from your camera or photo library. You will learn how to do so in *Chapter 20, Getting Started with the Camera and Photo Library*.

You have successfully modified the JournalEntry class to conform to the Codable protocol, and you have addressed all the errors in your app. In the next section, you'll implement saving and loading app data, so it will not be lost when you quit your app.

Loading and saving JSON data

Now that you have modified the JournalEntry class to conform to the Codable protocol, you are ready to implement loading data from and saving data to JSON files.

To make it easier for you to work with JSON files, Apple provides JSONDecoder and JSONEncoder classes.

A JSONDecoder instance decodes instances of a data type from JSON objects, and you will use it when loading files from your device storage.

To learn more about JSONDecoder, see https://developer.apple.com/documentation/foundation/jsondecoder.

A JSONEncoder instance encodes instances of a data type to JSON objects, and you will use it when saving files to your device storage.

To learn more about JSONEncoder, see https://developer.apple.com/documentation/foundation/jsonencoder.

You'll now implement the methods to load data from a file and save data to a file in the `SharedData` class. Follow these steps:

1. In the Project navigator, click the **SharedData** file. In the `SharedData` class, implement a method to get the location where you can load or save a file on your device storage before the closing curly brace:

   ```
   // MARK: - Persistence
   func documentDirectory() -> URL {
     FileManager.default.urls(for: .documentDirectory,
     in: .userDomainMask).first!
   }
   ```

   This is analogous to getting the path to the Documents directory in your home directory on your Mac.

   To learn more about accessing the iOS file system, see `https://developer.apple.com/documentation/foundation/filemanager`.

2. Implement a method to load journal entries from a file on your device storage after the `documentDirectory()` method:

   ```
   func loadJournalEntriesData() {
     let pathDirectory = documentDirectory()
     let fileURL = pathDirectory.
     appendingPathComponent("journalEntriesData.json")
     do {
       let data = try Data(contentsOf: fileURL)
       let entries = try JSONDecoder().decode(
       [JournalEntry].self, from: data)
       journalEntries = entries
     } catch {
       print("Failed to read JSON data:
       \(error.localizedDescription)")
     }
   }
   ```

   This method uses the `documentDirectory()` method to get the location where files can be loaded from. Then it specifies a file name, `journalEntriesData.json`, where the data is saved and appends that to the path. It then attempts to load the file. If successful, it attempts to decode the data into an array of `JournalEntry` instances and assign it to the `journalEntries` array.

3. Implement a method to save journal entries to a file on your device storage after the `loadJournalEntriesData()` method:

```
func saveJournalEntriesData() {
  let pathDirectory = documentDirectory()
  do {
    try? FileManager().createDirectory(at: pathDirectory,
    withIntermediateDirectories: true)
    let filePath =   pathDirectory.appendingPathComponent(
    "journalEntriesData.json")
    let json = try JSONEncoder().encode(journalEntries)
    try json.write(to: filePath)
  } catch {
    print("Failed to write JSON data:
    \(error.localizedDescription)")
  }
}
```

   This method uses the `documentDirectory()` method to get the location where files can be saved. Then it specifies a file name where the data is to be saved and appends that to the path. It then attempts to use a `JSONEncoder` instance to encode the `journalEntries` array to JSON format and subsequently write it to the file specified earlier.

You have implemented the methods to load and save journal entries in your app. Now you'll modify the `JournalListViewController` class to call these methods at appropriate times. Follow these steps:

1. In the Project navigator, click the **JournalListViewController** file. In the `JournalListViewController` class, modify `viewDidLoad()` to call the `loadJournalEntriesData()` method:

```
override func viewDidLoad() {
  super.viewDidLoad()
  SharedData.shared.loadJournalEntriesData()
}
```

   This will load any saved journal entries as the app is starting up.

2. Modify the `tableView(_:commit:forRowAt:)` method to call the `saveJournalEntriesData()` method after a row has been removed from the table view:

```
if editingStyle == .delete {
  SharedData.shared.removeJournalEntry(at: indexPath.row)
  SharedData.shared.saveJournalEntriesData()
  tableView.reloadData()
}
```

3. Modify the `unwindNewEntrySave(segue:)` method to call `saveJournalEntriesData()` after a new journal entry has been added:

```
if let sourceViewController = segue.source as?
AddJournalEntryViewController, let newJournalEntry =
sourceViewController.newJournalEntry {
  SharedData.shared.addJournalEntry(newJournalEntry)
  SharedData.shared.saveJournalEntriesData()
  tableView.reloadData()
}
```

4. Build and run your app. Verify a simulated location has been set and add a new journal entry. Your app should work the way it did before.
5. Stop your app and run it again. The journal entry you added earlier should still be present:

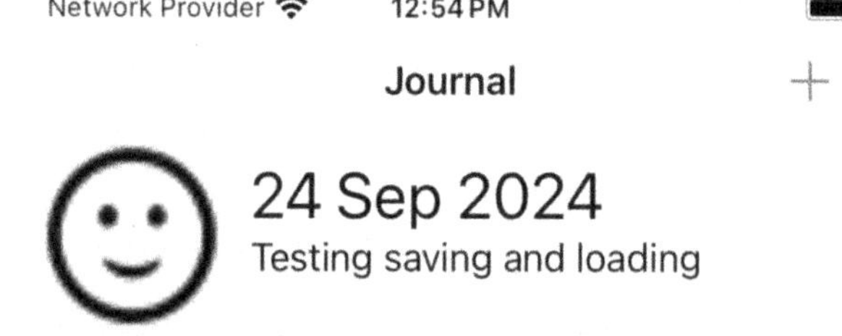

Figure 18.6: Simulator showing persistent app data in your app

If you are running your app in Simulator, you can use a `print(filePath)` statement in the `saveJournalEntriesData()` method to print the file path to the Debug area. This will tell you where the `journalEntriesData.json` file is saved on your Mac.

You have successfully implemented saving and loading using JSON files for your app! Fantastic job!

Summary

In this chapter, you created a singleton, `SharedData`, and configured your app to use it. Next, you modified the `JournalEntry` class to be compatible with the JSON format, so you can save journal entries to a JSON file and load journal entries from a JSON file. After that, you added methods to save journal entry data when you add or delete journal entries, and to load journal entry data when your app is starting up.

You now know how to create a class to store, load, and save data from JSON files for use in your own apps.

In the next chapter, you'll implement a custom user interface element that allows you to set star ratings for journal entries.

Join us on Discord!

Read this book alongside other users, experts, and the author himself. Ask questions, provide solutions to other readers, chat with the author via Ask Me Anything sessions, and much more. Scan the QR code or visit the link to join the community.

`https://packt.link/ios-Swift`

19

Getting Started with Custom Views

At this point, your *JRNL* app is functional. All the screens are working, but as the rating user interface element is missing, you can't set a star rating for a journal entry as shown in the app tour. You also can't set a custom picture, but that will be addressed in *Chapter 20*, *Getting Started with the Camera and Photo Library*.

You have been using Apple's standard UI elements so far. In this chapter, you'll create a **custom view** subclass of the `UIStackView` class that displays a journal entry rating in the form of stars, and you'll modify this subclass so users can set a rating for a journal entry by tapping it. After that, you'll implement it on the Add New Journal Entry screen. Finally, you'll implement it on the Journal Entry Detail screen.

By the end of this chapter, you'll have learned how to create custom views for your own apps.

The following topics will be covered in this chapter:

- Creating a custom `UIStackView` subclass
- Adding your custom view to the Add New Journal Entry screen
- Adding your custom view to the Journal Entry Detail screen

Technical requirements

You will continue working on the `JRNL` project that you modified in the previous chapter.

The completed Xcode project for this chapter is in the `Chapter19` folder of the code bundle for this book, which can be downloaded here:

`https://github.com/PacktPublishing/iOS-18-Programming-for-Beginners-Ninth-Edition`

Check out the following video to see the code in action:

`https://youtu.be/Y21q6voqYuk`

Let's start by learning how to create a custom `UIStackView` subclass that will display a star rating on the screen.

Creating a custom UIStackView subclass

You've only used Apple's predefined UI elements so far, such as labels and buttons. All you had to do was click the Library button, search for the object you want, and drag it into the storyboard. However, there will be cases where the objects provided by Apple are either unsuitable or don't exist. In such cases, you will need to build your own. Let's review the Add New Journal Entry screen that you saw in the app tour:

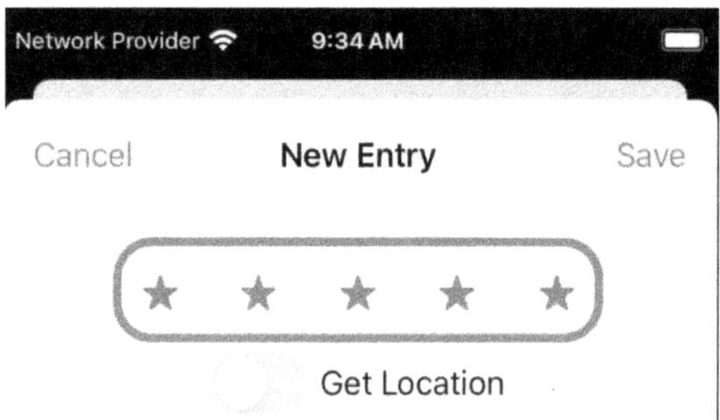

Figure 19.1: Add New Journal Entry screen showing the star rating

You can see a group of five stars just above the **Get Location** switch. Currently, the **New Entry Scene** and the **Entry Detail Scene** in the `Main` storyboard file have placeholder view objects where the stars should be. You will create the `RatingView` class, a custom subclass of the `UIStackView` class, which you will use in both scenes. An instance of this class will display ratings as stars.

For the rest of this chapter, an instance of the `RatingView` class will be referred to as a rating view (the same way an instance of the `UIButton` class is referred to as a button).

Let's begin by creating a subclass of the `UIStackView` class. Follow these steps:

1. In the Project navigator, right-click on the **JRNL** folder and choose **New Group** from the pop-up menu. Name this group **Views**. Move this group to a new location below the **Model** group:

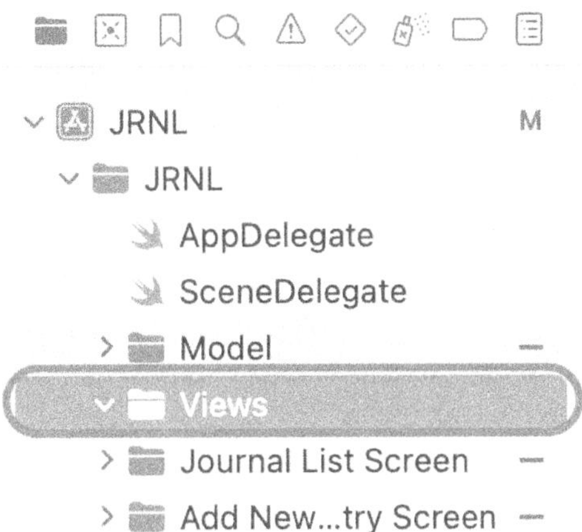

Figure 19.2: Project navigator showing Views group below the Model group

2. Right-click the **Views** folder and select **New File from Template...** from the pop-up menu.
3. **iOS** should already be selected. Choose **Cocoa Touch Class** and then click **Next**.
4. Configure the file as follows:

    - **Class:** `RatingView`
    - **Subclass:** `UIStackView`
    - **Language:** `Swift`

    Click **Next**.

5. Click **Create**. The `RatingView` file will appear in the Project navigator.
6. Remove all the commented code in this file, and type the following after the `RatingView` class declaration to declare the properties for the class:

```
// MARK: - Properties
private var ratingButtons = [UIButton()]
var rating = 0
private let buttonSize = CGSize(width: 44.0, height: 44.0)
private let buttonCount = 5
```

The `ratingButtons` property is an array that will hold all the buttons for this class.

The `rating` property is used to store a journal entry rating. It determines the number and types of stars that will be drawn. For instance, if `rating` contains 3, the rating view will display three filled stars and two empty stars.

The `buttonSize` property determines the height and the width of the buttons that will be drawn onscreen.

The `buttonCount` property determines the total number of buttons to be drawn onscreen.

7. Implement the initializer for this class after the property declarations:

```
// MARK: - Initialization
required init(coder: NSCoder) {
  super.init(coder: coder)
}
```

8. Implement a method to draw stars on the screen after the initializer:

```
// MARK: - Private methods
private func setupButtons() {
  for button in ratingButtons {
    removeArrangedSubview(button)
    button.removeFromSuperView()
  }
  ratingButtons.removeAll()
  let filledStar = UIImage(systemName:"star.fill" )
  let emptyStar = UIImage(systemName: "star")
  let highlightedStar =
  UIImage(systemName: "star.fill")?.withTintColor(.red,
  renderingMode: .alwaysOriginal)
  for _ in 0..<buttonCount {
    let button = UIButton()
    button.setImage(emptyStar, for: .normal)
    button.setImage(filledStar, for: .selected)
    button.setImage(highlightedStar, for: .highlighted)
    button.setImage(highlightedStar, for: [.highlighted,
    .selected])
    button.translatesAutoresizingMaskIntoConstraints =
    false
    button.heightAnchor.constraint(equalToConstant:
    buttonSize.height).isActive = true
    button.widthAnchor.constraint(equalToConstant:
    buttonSize.width).isActive = true
    addArrangedSubview(button)
```

```
      ratingButtons.append(button)
    }
  }
```

Let's break this down:

```
for button in ratingButtons {
  removeArrangedSubview(button)
  button.removeFromSuperView()
}
ratingButtons.removeAll()
```

These statements remove any existing buttons from the stack view and the `ratingButtons` array.

```
let filledStar = UIImage(systemName:"star.fill" )
let emptyStar = UIImage(systemName: "star")
let highlightedStar =
UIImage(systemName: "star.fill")?.withTintColor(.red,
renderingMode: .alwaysOriginal)
```

These statements create three `UIImage` instances from symbols in Apple's `SFSymbols` library. `filledStar` will store an image of a filled star, `emptyStar` will store an image of a star outline, and `highlightedStar` will store an image of a filled star that has been tinted red.

For more information on Apple's `SFSymbols` library, see `https://developer.apple.com/design/human-interface-guidelines/sf-symbols`.

```
for _ in 0..<buttonCount {
```

Since `buttonCount` is set to `5`, this `for` loop will repeat five times.

```
let button = UIButton()
```

This statement assigns an instance of `UIButton` to `button`.

For more information on `UIButton`, see `https://developer.apple.com/documentation/uikit/uibutton`.

```
button.setImage(emptyStar, for: .normal)
button.setImage(filledStar, for: .selected)
button.setImage(highlightedStar, for: .highlighted)
button.setImage(highlightedStar, for: [.highlighted, .selected])
```

These statements set the images for the different states of the `UIButton` instance. The `.normal` state displays a star outline. When in the `.selected` state, a filled star is displayed. If you tap the `UIButton` instance, it will be in either the `.highlighted` state or the `.highlighted` and `.selected` states, depending on whether it was in the `.normal` state or `.selected` state prior to being tapped. It then displays a red-tinted filled star.

```
button.translatesAutoresizingMaskIntoConstraints =
false
button.heightAnchor.constraint(equalToConstant:
buttonSize.height).isActive = true
button.widthAnchor.constraint(equalToConstant:
buttonSize.width).isActive = true
```

These statements set the size of the buttons. The first statement sets the `UIButton` instance's `translatesAutoresizingMaskIntoConstraints` property to `false`; otherwise, the system would create a set of constraints that duplicate the behavior specified by the view's auto-resizing mask, and you would not be able to set your own constraints. The next two statements set the instance's height and width by programmatically setting the height and width constraints using the value stored in `buttonSize`.

```
addArrangedSubview(button)
```

This statement adds the `UIButton` instance as a subview of the stack view programmatically.

```
ratingButtons.append(button)
```

This statement adds the `UIButton` instance to the `ratingButtons` array.

9. Call the `setupButtons()` method in the initializer:

```
required init(coder: NSCoder) {
  super.init(coder: coder)
  setupButtons()
}
```

This draws the rating view onscreen when the rating view is initialized.

You have created a custom `UIStackView` subclass named `RatingView`, and you have added code to make it draw five stars on the screen. Now let's add code to enable the user to change the rating when the stars in the rating view are tapped. Follow these steps:

1. Implement a method to change the rating view's `rating` property when a button in the `ratingButtons` array is tapped after the `setupButtons()` method:

```
@objc private func ratingButtonTapped(_ button: UIButton) {
  guard let index = ratingButtons.firstIndex(of: button)
  else {
```

```
      fatalError("The button, \(button), is not in the
      ratingButtons array: \(ratingButtons)")
    }
    let selectedRating = index + 1
    if selectedRating == rating {
      rating = 0
    } else {
      rating = selectedRating
    }
  }
```

When a button in the `ratingButtons` array is tapped, the `guard` statement assigns the index of the button to `index`. `selectedRating` is then set to the value stored in `index + 1`. If the `rating` property has the same value as `selectedRating`, it is set to `0`; otherwise, it is set to the same value as `selectedRating`.

For example, let's say you tap the third star in the rating view. Since the third star is the third element in the `ratingButtons` array, `index` would be set to `2` and `selectedRating` would be set to `2 + 1 = 3`. Assuming that the initial value of the `rating` property is `0`, `selectedRating == rating` would return `false`, and the `rating` property's value would be set to `3`.

2. Assign this method as the button action in the `for` loop of the `setupButtons` method after the statements setting the constraints:

```
button.widthAnchor.constraint(equalToConstant: starSize.width).isActive =
true
button.addTarget(self, action: #selector(RatingView.
ratingButtonTapped(_:)), for: .touchUpInside)
addArrangedSubview(button)
```

3. Add a method to change the button state according to the rating that was set before the closing curly brace:

```
private func updateButtonSelectionStates() {
  for (index, button) in ratingButtons.enumerated() {
    button.isSelected = index < rating
  }
}
```

To see how this works, let's say that the `rating` property is set to `3`. The default state for each button is `.normal`.

The first button is at index `0`, so `button.isSelected` is `0 < 3`, which returns `true`. Since the image for the `.selected` state is a filled star, this button's image is set to a filled star. The same is true for the next two buttons.

The fourth button is at index 3, so `button.isSelected` is 3 < 3, which returns `false`. This means that the button's state remains `.normal`. The image for the `.normal` state is a star outline, so this button's image is set to a star outline. The same is true for the fifth button.

In short, when the `rating` property is set to 3, the rating view displays the first three buttons with filled stars, and the remaining two buttons with star outlines.

4. You'll need to call the `updateButtonSelectionStates()` method every time the `rating` property's value changes. To do so, modify the `rating` property as shown:

```
var rating = 0 {
  didSet {
    updateButtonSelectionStates()
  }
}
```

This is known as a **property observer**, and every time the `rating` property's value changes, the `updateButtonSelectionStates()` method will be called.

You have completed the implementation of the rating view. In the next section, you'll add it to the Add New Journal Entry screen.

Adding your custom view to the Add New Journal Entry screen

So far, you have created a new `RatingView` class in your project and configured it to set its `rating` property when a star in it is tapped. In this section, you will set the identity of the stack view object above the **Get Location** switch in the **New Entry Scene** to the `RatingView` class, configure an outlet for it in the `AddJournalEntryViewController` class, and add code to use the `rating` property's value when creating a new journal entry. Follow these steps:

1. In the Project navigator, click the **AddJournalEntryViewController** file. Add a new outlet for a rating view in the `AddJournalEntryViewController` class after all other property declarations:

```
@IBOutlet var getLocationSwitch: UISwitch!
@IBOutlet var getLocationSwitchLabel: UILabel!
@IBOutlet var ratingView: RatingView!
```

2. Modify the `prepare(for:sender:)` method to get the rating view's `rating` property value when creating a new journal entry:

```
let photo = photoImageView.image
let rating = ratingView.rating
let lat = currentLocation?.coordinate.latitude
```

3. Click the **Main** storyboard file and select **New Entry Scene** in the document outline. Click the `UIStackView` object above the **Get Location** switch, as shown:

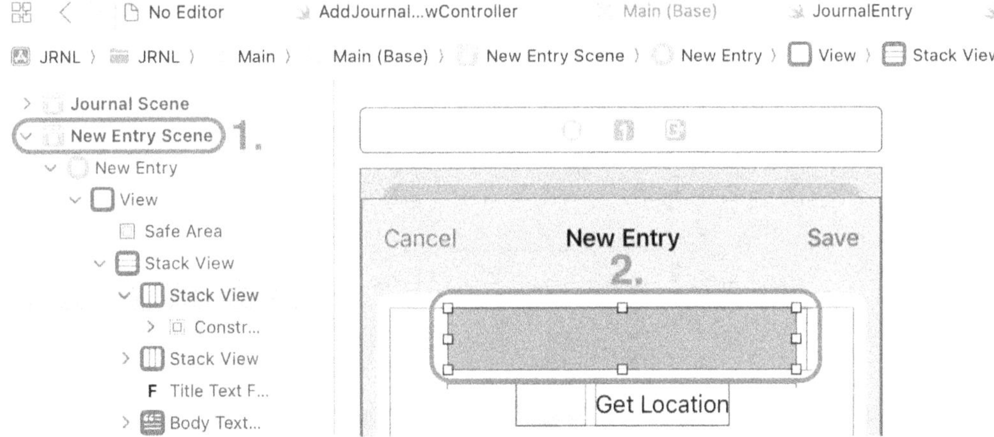

Figure 19.3: Editor area showing UIStackView object above the Get Location switch

4. Click the Identity inspector button. Under **Custom Class**, set **Class** to `RatingView`:

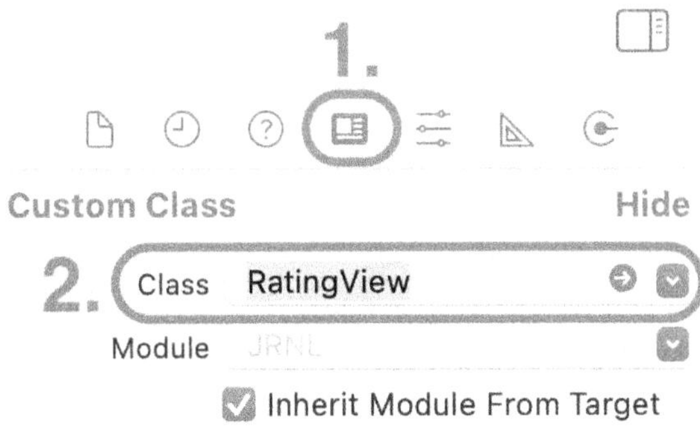

Figure 19.4: Identity inspector with Class set to RatingView

5. Click the Attributes inspector button. Verify the settings under **Stack View**, and under **View**, set **Background** to `Default`:

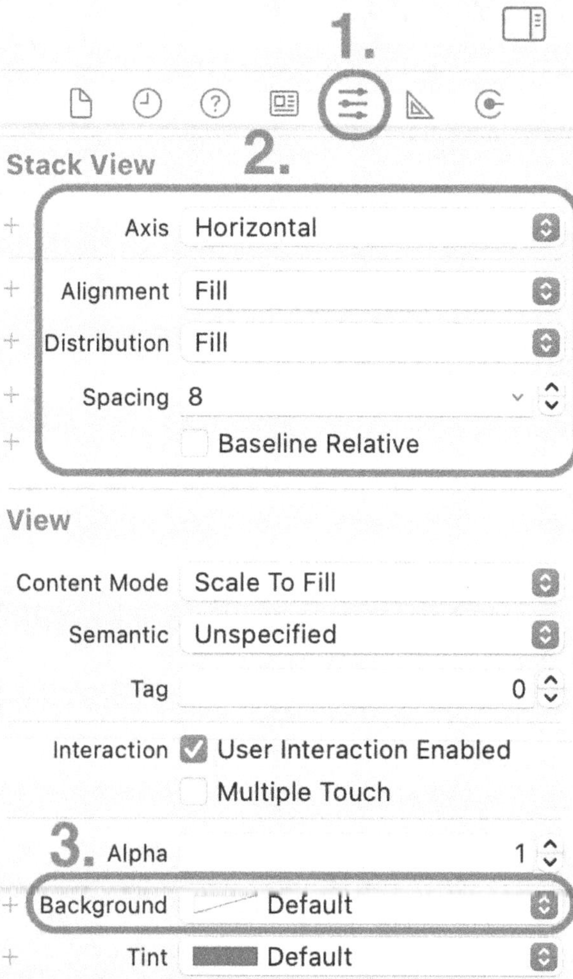

Figure 19.5: Attributes inspector with Background set to Default

6. Click **New Entry Scene** in the document outline, and click the Connections inspector button. Connect the `ratingView` outlet to the rating view in the **New Entry Scene**:

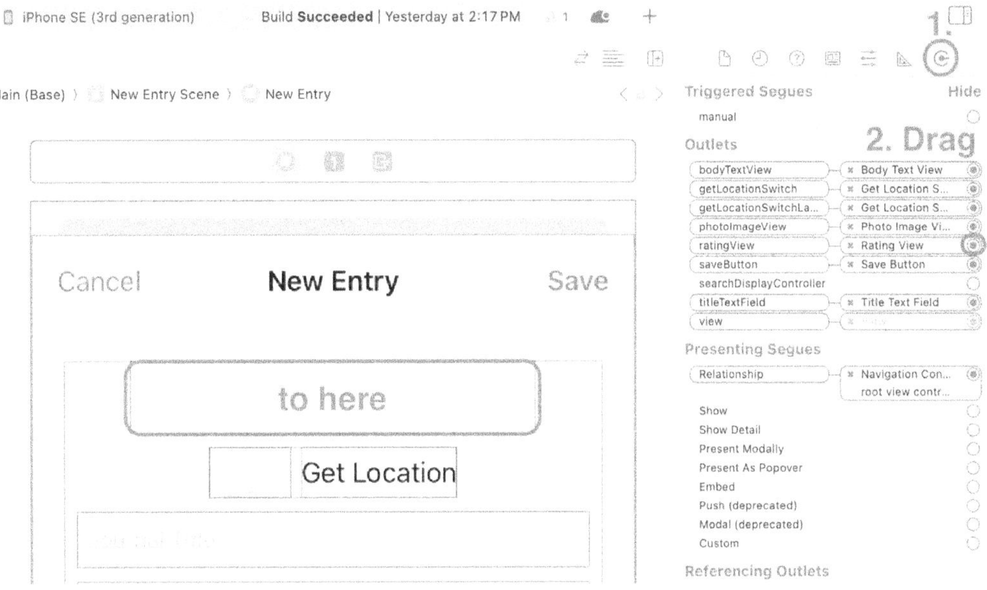

Figure 19.6: Connections inspector showing ratingView outlet

7. Build and run your app. Tap the + button to go to the Add New Journal Entry screen and you will see the rating view displayed above the **Get Location** switch. Add a journal entry title, body, and rating, and tap **Save**:

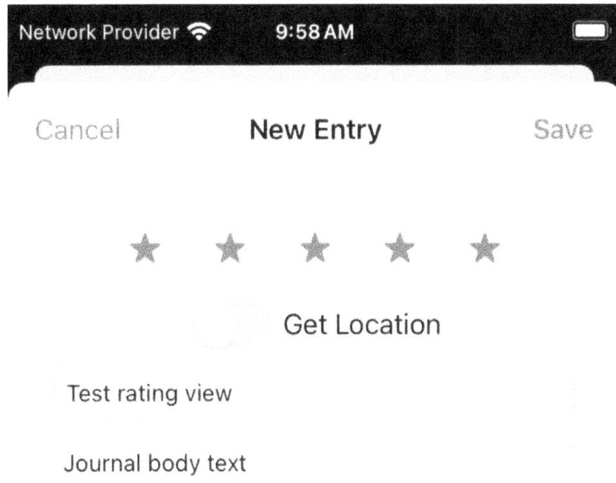

Figure 19.7: Simulator showing the rating view on the Add New Journal Entry screen

A new journal entry with a rating is now visible on the Journal List screen. In the next section, you'll modify the Journal Entry Detail screen to display the rating for this journal entry.

Adding your custom view to the Journal Entry Detail screen

At this point, you're able to set a rating when you create a new journal entry using the Add New Journal Entry screen, but the rating you set is not visible on the Journal Entry Detail screen. You'll add an outlet for a rating view and modify the code in the `JournalEntryDetailViewController` class, and you'll add a rating view to the **Entry Detail Scene**.

Follow these steps:

1. In the Project navigator, click the **JournalEntryDetailViewController** file. Add an outlet for a rating view after all other property declarations:

   ```
   @IBOutlet var photoImageView: UIImageView!
   @IBOutlet var mapImageView: UIImageView!
   @IBOutlet var ratingView: RatingView!
   ```

2. Modify the code in the `viewDidLoad()` method to set the value of the rating view's `rating` property:

   ```
   super.viewDidLoad()
   dateLabel.text = selectedJournalEntry?.date.formatted(
   .dateTime.day().month(.wide).year()
   )
   ratingView.rating = selectedJournalEntry?.rating ?? 0
   titleLabel.text = selectedJournalEntry?.entryTitle
   ```

3. Click the **Main** storyboard file and click **Entry Detail Scene** in the document outline. Select the stack view in the second table view cell. Click the Identity inspector button and, under **Custom Class**, set **Class** to `RatingView`:

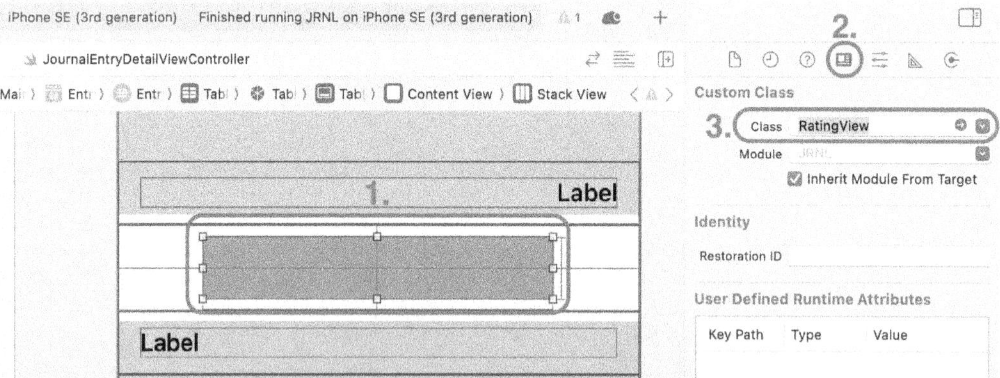

Figure 19.8: Identity inspector with Class set to RatingView

4. Click the Attributes inspector button and verify the settings under **Stack View**. Under **View**, untick the **User Interaction Enabled** checkbox (as the user should not be able to change the rating on the Journal Entry Detail screen), and set **Background** to `Default`:

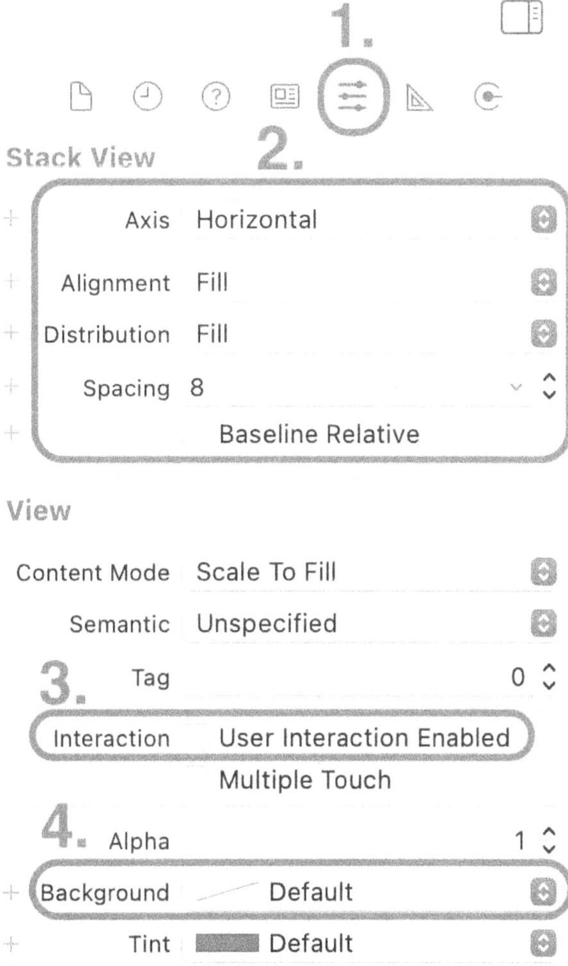

Figure 19.9: Attributes inspector with Background set to Default

5. Click **Entry Detail Scene** in the document outline, and click the Connections inspector button. Connect the **ratingView** outlet to the rating view in the **Entry Detail** scene:

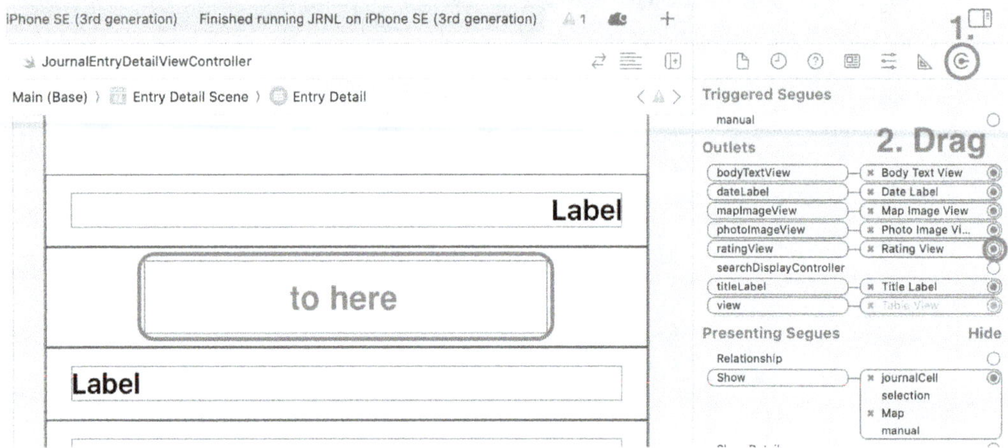

Figure 19.10: Connections inspector showing ratingView outlet

6. Build and run your app. Tap the journal entry you added in the previous section, and you'll see the rating displayed in a rating view on the Journal Entry Detail screen:

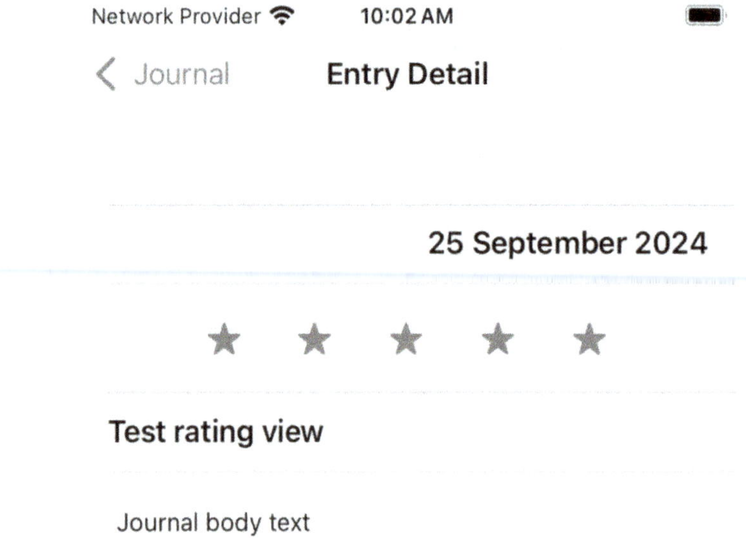

Figure 19.11: Simulator showing the rating view on the Journal Entry Detail screen

You've successfully added and configured a rating view on the Journal Entry Detail screen! Good job!

Summary

In this chapter, you created a custom subclass of the `UIStackView` class that displays a journal entry rating in the form of stars, and you modified this subclass so users can set a rating for a journal entry by tapping it. After that, you added it to the Add New Journal Entry screen. Finally, you implemented it on the Journal Entry Detail screen.

You now know how to create custom views for your own apps.

In the next chapter, you'll learn how to work with photos from the camera or photo library.

Join us on Discord!

Read this book alongside other users, experts, and the author himself. Ask questions, provide solutions to other readers, chat with the author via Ask Me Anything sessions, and much more. Scan the QR code or visit the link to join the community.

`https://packt.link/ios-Swift`

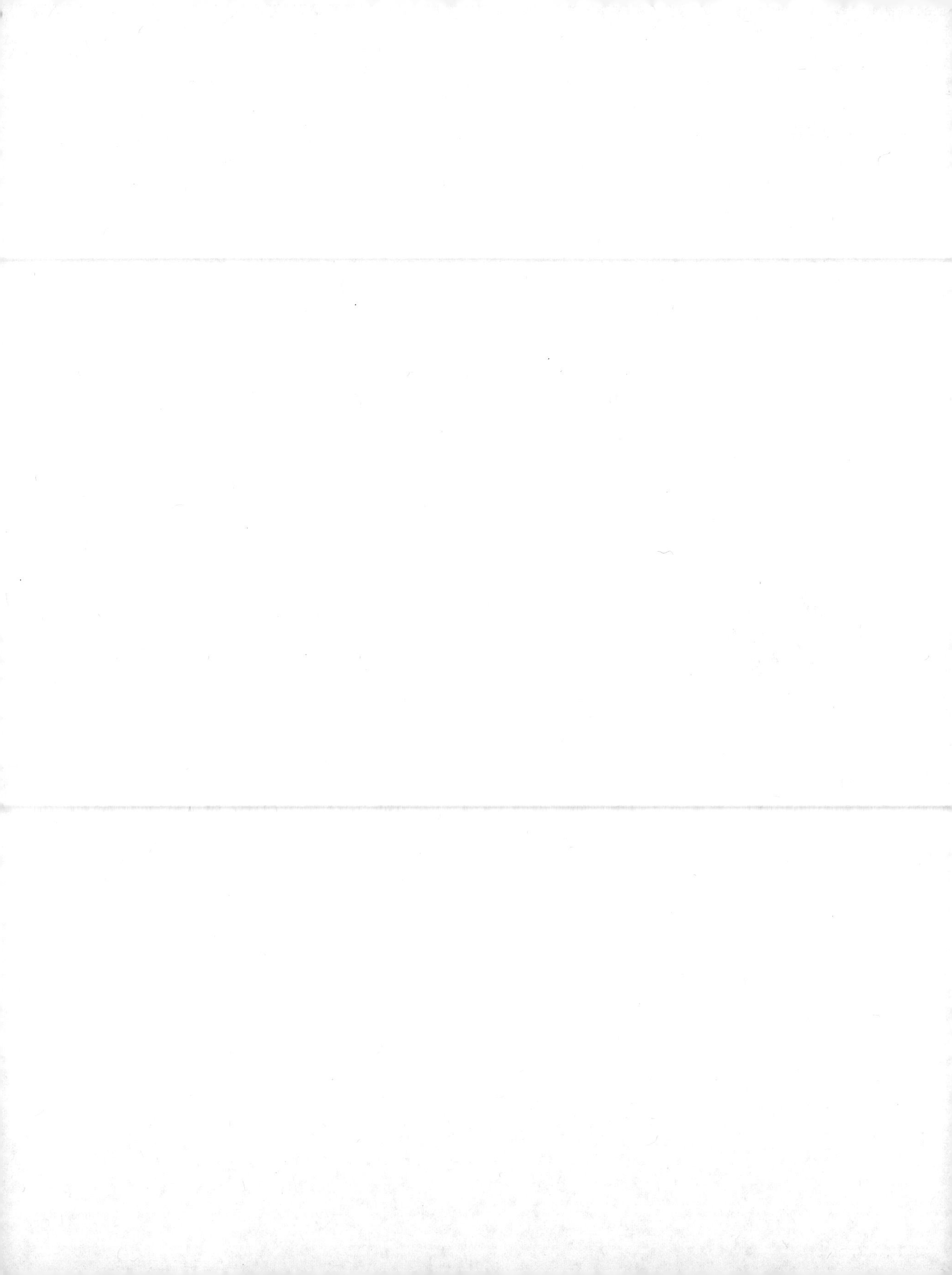

20

Getting Started with the Camera and Photo Library

In the previous chapter, you created the `RatingView` class and added it to the Add New Journal Entry and Journal Entry Detail screens.

In this chapter, you will complete the implementation of the Add New Journal Entry screen by adding a way for the user to get a photo from the **camera** or **photo library**, which they can then add to a new journal entry. You'll start by adding a **tap gesture recognizer** to the image view in the **New Entry Scene**, and configure it to display an **image picker controller** instance. Then, you will implement methods from the `UIImagePickerControllerDelegate` protocol, which allows you to get a photo from the camera or photo library and make the photo smaller before it is saved to the journal entry instance. You'll also modify the `Info.plist` file to allow you to access the camera or photo library.

By the end of this chapter, you'll have learned how to access the camera or photo library in your own apps.

The following topics will be covered in this chapter:

- Creating a new `UIImagePickerController` instance
- Implementing `UIImagePickerControllerDelegate` methods
- Getting permission to use the camera or photo library

Technical requirements

You will continue working on the `JRNL` project that you modified in the previous chapter.

The resource files and completed Xcode project for this chapter are in the `Chapter20` folder of the code bundle for this book, which can be downloaded here:

`https://github.com/PacktPublishing/iOS-18-Programming-for-Beginners-Ninth-Edition`

Check out the following video to see the code in action:

`https://youtu.be/MvOOKyBcVak`

Let's start by modifying the Add New Journal Entry screen to display an image picker controller, which allows you to use the device camera or select a photo from the user's photo library.

Creating a new UIImagePickerController instance

To make it easy for a user to use the camera or photo library, Apple implemented the `UIImagePickerController` class. This class manages the system interfaces for taking photos and choosing items from the user's photo library. An instance of this class is called an image picker controller, and it can display an image picker on the screen.

If you have ever added a photo to a social media post, you will have seen what the image picker looks like. It typically displays either the view from your camera or a grid of photos from your photo library, and you can then choose a photo to be added to your post:

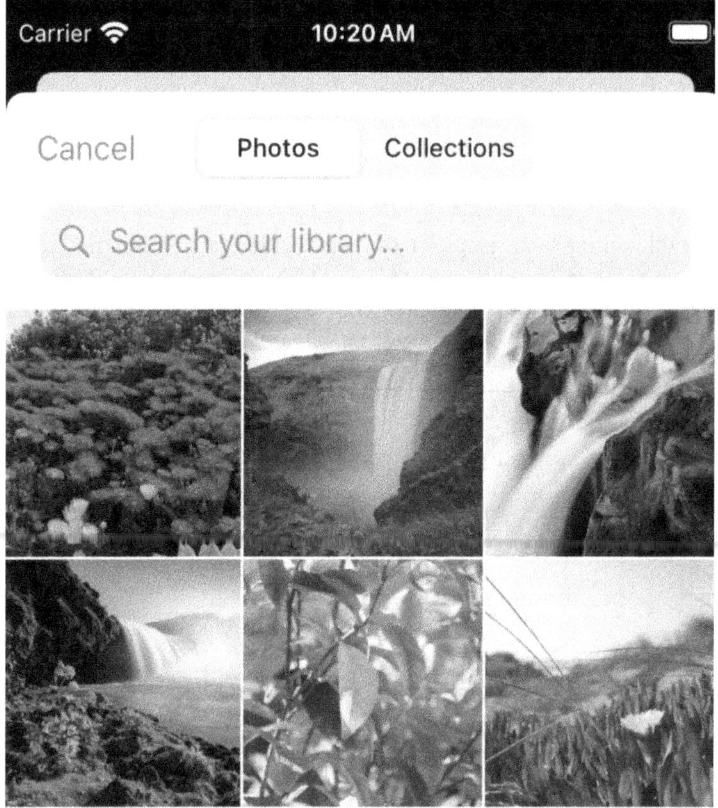

Figure 20.1: Simulator showing the image picker

To learn more about the `UIImagePickerController` class, see `https://developer.apple.com/documentation/uikit/uiimagepickercontroller`.

To display the image picker on the Add New Journal Entry screen, you'll add a tap gesture recognizer instance to the image view in the **New Entry Scene**, and you'll add a method to create and display an image picker controller when the image view is tapped. Follow these steps:

1. In the Project navigator, click the **Main** storyboard file. Click **New Entry Scene** in the document outline.
2. Click the image view in the **New Entry Scene**. Click the Attributes inspector button and, under **View**, tick the **User Interaction Enabled** checkbox:

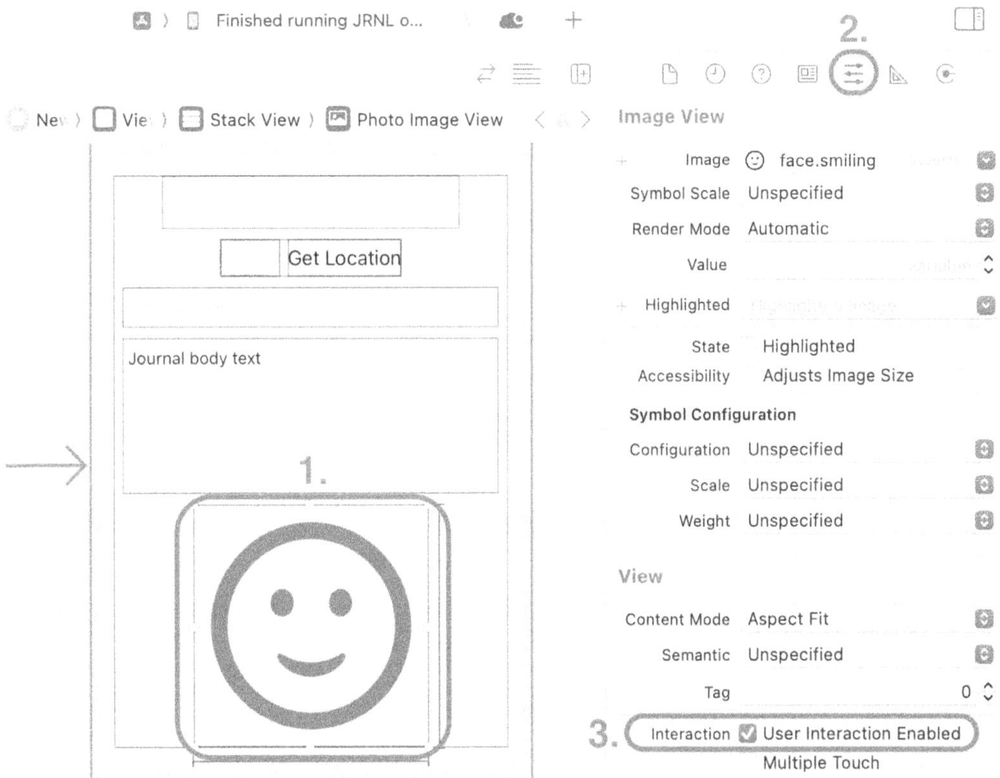

Figure 20.2: Attributes inspector showing the User Interaction Enabled checkbox

3. Click the Library button to display the library. Type `tap` in the filter field. A **Tap Gesture Recognizer** object will appear as one of the results. Drag it to the image view:

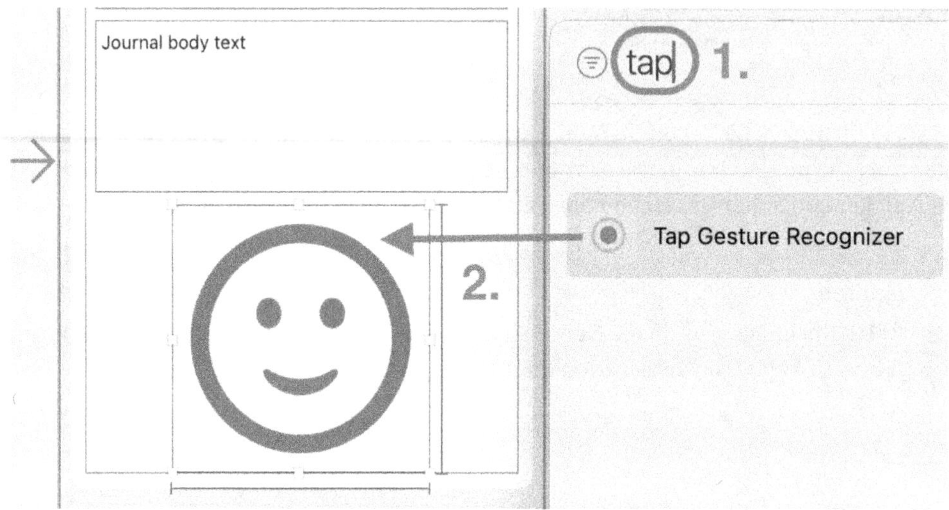

Figure 20.3: Library with Tap Gesture Recognizer object selected

4. Click the Navigator and Inspector buttons if you need more room to work. Click the Adjust Editor Options button and choose **Assistant** from the pop-up menu:

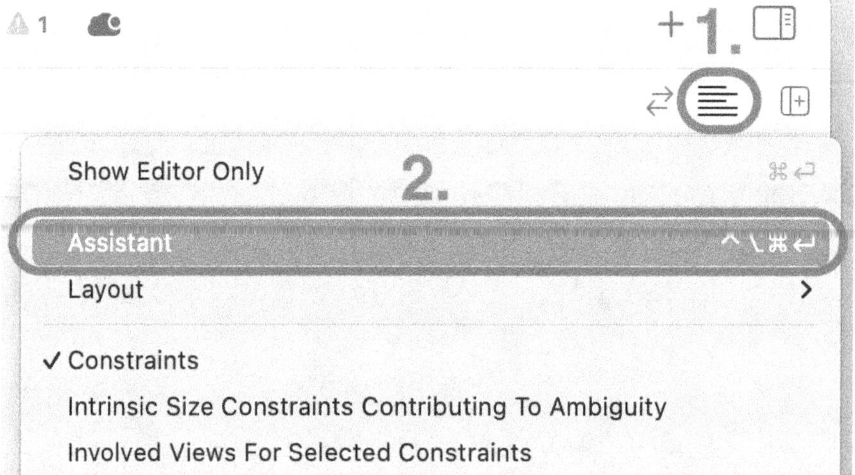

Figure 20.4: Adjust Editor Options menu with Assistant selected

5. You should see the contents of the `AddJournalEntryViewController` file in the assistant editor. *Ctrl + Drag* from the **Tap Gesture Recognizer** in the document outline to the space between the `locationSwitchValueChanged(_:)` method and the closing curly brace:

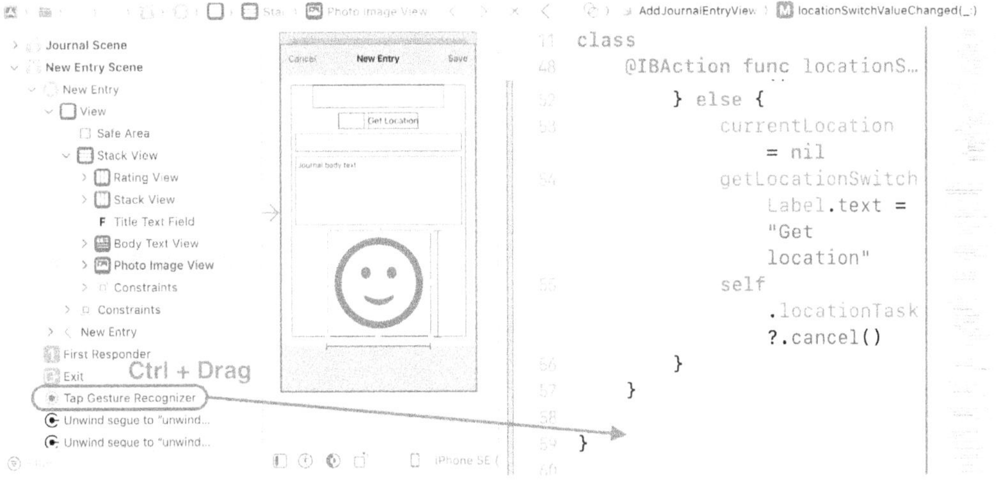

Figure 20.5: Editor area showing drag destination

6. In the pop-up dialog box, set **Name** to `getPhoto` and set **Type** to `UITapGestureRecognizer`. Click **Connect**:

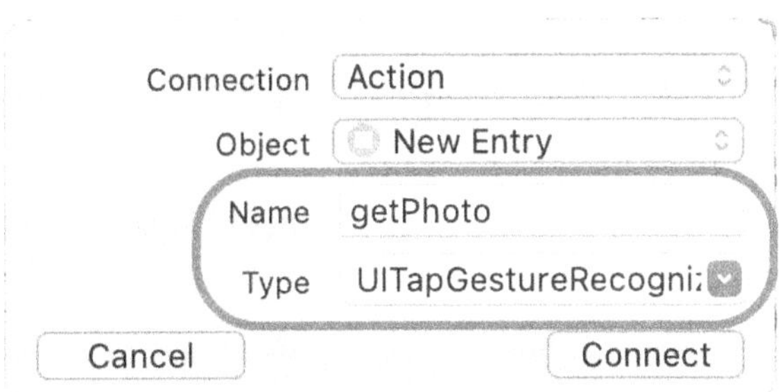

Figure 20.6: Pop-up dialog box for action creation

7. Verify that the `getPhoto(_:)` method has been created in the `AddJournalEntryViewController` class. Click the **x** to close the assistant editor window:

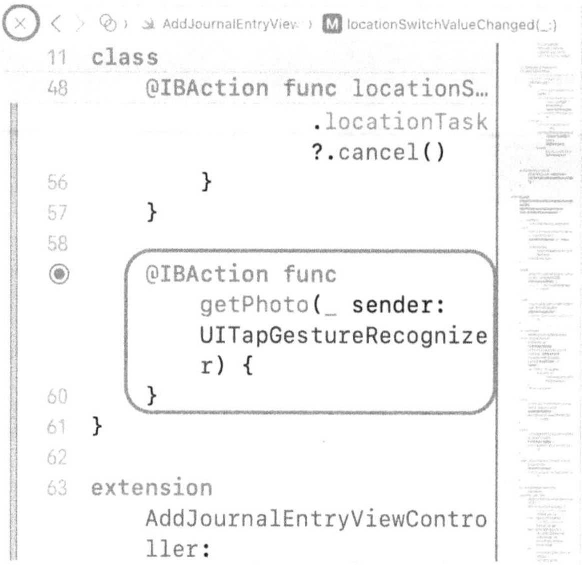

Figure 20.7: Assistant editor showing getPhoto(_:) method

You have successfully added a tap gesture recognizer to the image view in the **New Entry Scene** and linked it to a `getPhoto()` method in the `AddJournalEntryViewController` class. Now you'll modify the `getPhoto()` method to create and display a `UIImagePickerController` instance.

Follow these steps:

1. In the Project navigator, click the **AddJournalEntryViewController** file. Add a new extension after all other code in the file to make the `AddJournalEntryViewController` class declaration conform to the `UIImagePickerControllerDelegate` and `UINavigationControllerDelegate` protocols:

```
extension AddJournalEntryViewController: UIImagePickerControllerDelegate,
UINavigationControllerDelegate {

}
```

2. Modify the `getPhoto()` method as shown:

```
@IBAction func getPhoto(_ sender: UITapGestureRecognizer) {
  let imagePickerController = UIImagePickerController()
  imagePickerController.delegate = self
  #if targetEnvironment(simulator)
  imagePickerController.sourceType = .photoLibrary
  #else
```

```
        imagePickerController.sourceType = .camera
        imagePickerController.showsCameraControls = true
        #endif
        present(imagePickerController, animated: true)
    }
```

Let's break this down:

```
let imagePickerController = UIImagePickerController()
```

This statement creates an instance of the `UIImagePickerController` class and assigns it to `imagePickerController`.

```
imagePickerController.delegate = self
```

This statement sets the image picker controller's `delegate` property to the `AddJournalEntryViewController` instance.

```
#if targetEnvironment(simulator)
imagePickerController.sourceType = .photoLibrary
#else
imagePickerController.sourceType = .camera
imagePickerController.showsCameraControls = true
#endif
```

This block of code is known as a conditional compilation block. It starts with an `#if` compilation directive and ends with an `#endif` compilation directive. If you're running in Simulator, only the statement setting the image picker controller's `sourceType` property to the photo library is compiled.

If you're running on an actual device, the statements setting the image picker controller's `sourceType` property to the camera and displaying the camera controls are compiled. This means that the image picker controller will use the photo library when running in Simulator and will use the camera when running on an actual device.

You can learn more about conditional compilation blocks at this link: `https://docs.swift.org/swift-book/documentation/the-swift-programming-language/statements/#Conditional-Compilation-Block`.

```
present(imagePickerController, animated: true)
```

This statement presents the image picker controller on the screen.

You've implemented all the code required to present the image picker controller when the image view is tapped. In the next section, you'll implement the `UIImagePickerControllerDelegate` methods that will be called when the user chooses an image or cancels.

Implementing UIImagePickerControllerDelegate methods

The `UIImagePickerControllerDelegate` protocol has a set of methods that you must implement in your delegate object to interact with the image picker controller interface.

To learn more about the `UIImagePickerControllerDelegate` protocol, see `https://developer.apple.com/documentation/uikit/uiimagepickercontrollerdelegate`.

When the image picker controller appears on screen, the user has the option of selecting a photo or canceling. If the user cancels, the `imagePickerControllerDidCancel(_:)` method is triggered, and if the user selects a photo, the `imagePickerController(_:didFinishPickingMediaWithInfo:)` method is triggered.

You'll implement these methods in your `AddJournalEntryViewController` class now. In the Project navigator, click the **AddJournalEntryViewController** file. Type the following code in the `UIImagePickerControllerDelegate` extension:

```
// MARK: - UIImagePickerControllerDelegate
func imagePickerControllerDidCancel(_ picker: UIImagePickerController) {
  dismiss(animated: true)
}
func imagePickerController(_ picker: UIImagePickerController,
didFinishPickingMediaWithInfo info: [UIImagePickerController.InfoKey : Any]) {
  guard let selectedImage =
  info[UIImagePickerController.InfoKey.originalImage]
  as? UIImage else {
    fatalError("Expected a dictionary containing an image,
    but was provided the following: \(info)")
  }
  let smallerImage = selectedImage.preparingThumbnail(
  of: CGSize(width: 300, height: 300))
  photoImageView.image = smallerImage
  dismiss(animated: true)
}
```

The `imagePickerControllerDidCancel(_:)` method is triggered when the user cancels. The image picker controller is dismissed and the user is returned to the Add New Journal Entry screen.

The `imagePickerController(_:didFinishPickingMediaWithInfo:)` method is triggered when the user selects a photo. This photo is then assigned to `selectedImage`. Next, the `selectedImage` instance's `preparingThumbnail(of:)` method will be used to create a smaller image with a width and height of 300 points, the same as the size of the image view on the Journal Entry Detail screen. This image will then be assigned to the `photoImageView` property and the image picker controller will be dismissed

You can learn more about the `preparingThumbnail(of:)` method at this link: `https://developer.apple.com/documentation/uikit/uiimage/3750835-preparingthumbnail`.

All the required `UIImagePickerController` delegate methods have been implemented. In the next section, you'll modify the `Info.plist` file so that your app will ask for permission to use the camera or photo library.

Getting permission to use the camera or photo library

Apple stipulates that your app must inform the user if it wishes to access the camera or photo library. If you don't do this, your app will be rejected and will not be allowed on the App Store.

You'll modify the `Info.plist` file in your project to make your app display messages when it tries to access the camera or photo library. Follow these steps:

1. Click the **Info** file in the Project navigator. Move your pointer over the **Information Property List** row and click the + button to create a new row.
2. In the new row, set the **Key** to **Privacy – Photo Library Usage Description** and set the **Value** to `This app uses photos in your photo library when creating journal entries.`
3. Add a second row using the + button. This time, set the **Key** to **Privacy – Camera Usage Description** and set the **Value** to `This app uses your camera when creating journal entries.` Your `Info.plist` file should look like the following when done:

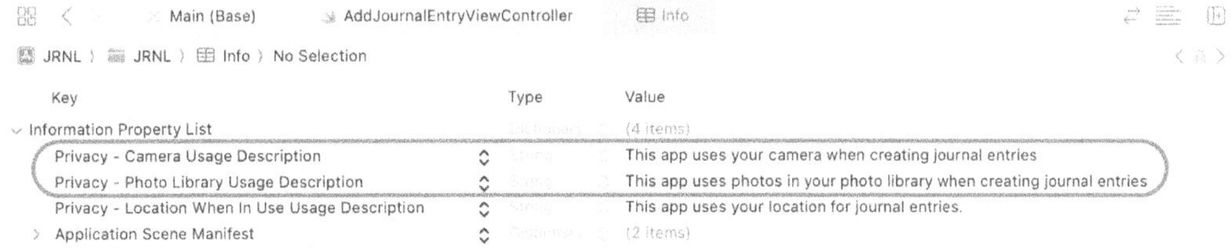

Figure 20.8: Info.plist with additional keys added

4. Build and run your app. Go to the Add New Journal Entry screen and tap the image view. The image picker will appear:

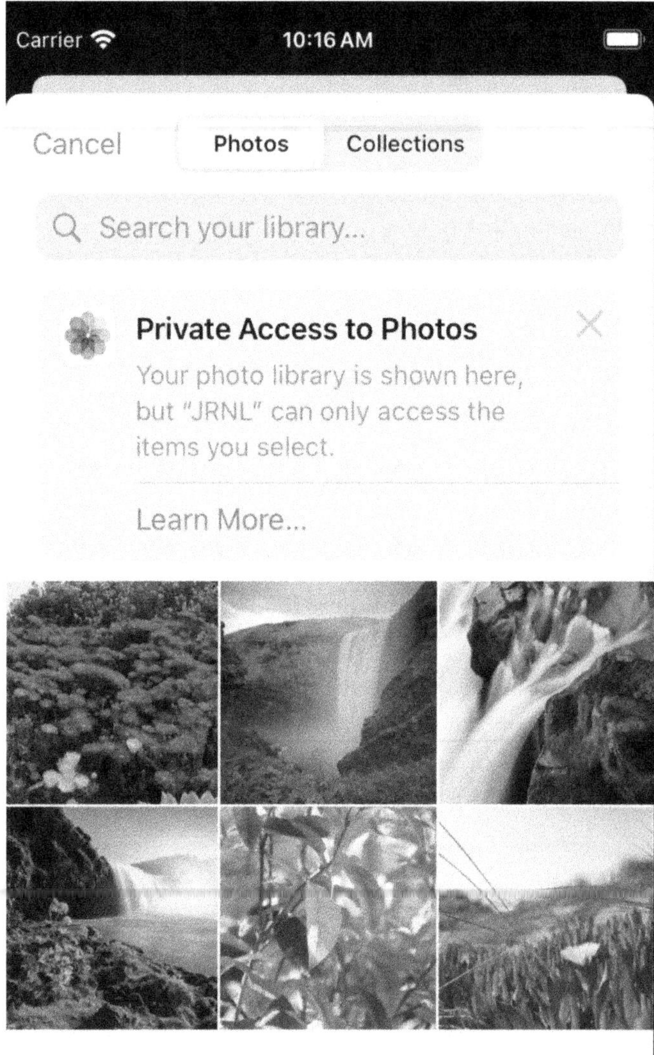

Figure 20.9: Simulator showing image picker

If you run the app on an actual iOS device, a dialog box will appear asking for permission to use the camera. Click **OK** to continue.

5. Select a photo and it will appear in the image view on the Add New Journal Entry screen. Enter sample details for the journal entry and click **Save**:

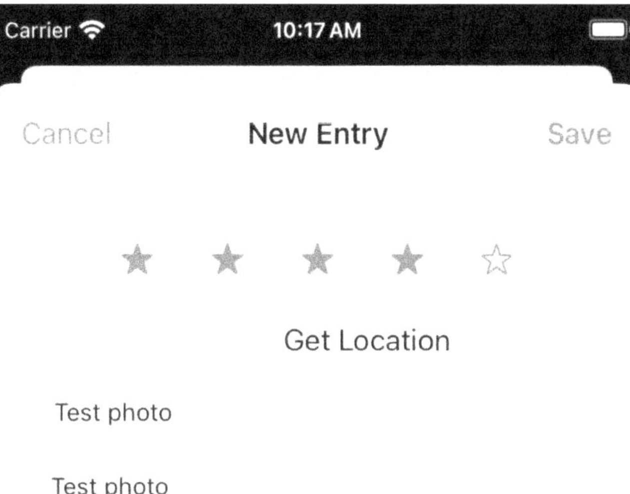

Figure 20.10: Simulator showing photo on Add New Journal Entry screen

6. You will be returned to the Journal List screen. Tap the newly added journal entry. You will see the photo is displayed on the Journal Entry Detail screen:

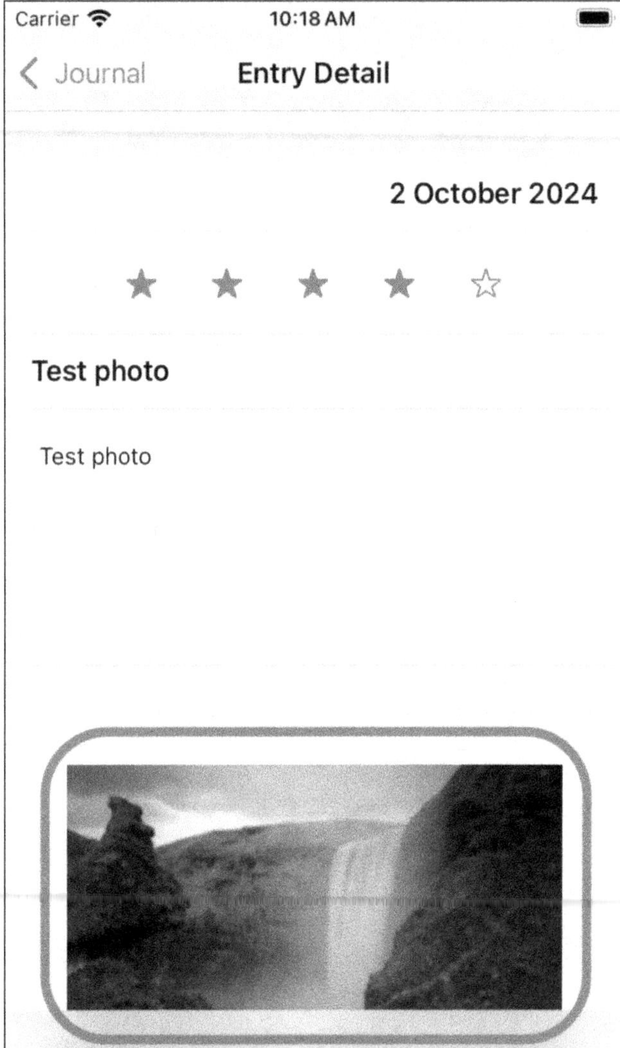

Figure 20.11: Simulator showing photo on the Journal Entry Detail screen

You can now add photos to a new journal entry on the Add New Journal Entry screen and display them on the Journal Entry Detail screen. Fantastic!

Summary

In this chapter, you completed the implementation of the Add New Journal Entry screen by adding a way for the user to get a photo from the camera or photo library, which then can be added to a new journal entry. First, you added a tap gesture recognizer to the image view in the New Entry scene and configured it to display an image picker controller. Then, you implemented the `UIImagePickerDelegate` protocol, which allows you to get a photo from the camera or the photo library, and made the photo smaller before it is saved to the journal entry instance. You also modified the `Info.plist` file to allow you to access the camera and photo library.

You are now able to write your own apps that import photos from your camera or photo library.

In the next chapter, you'll implement a way for the user to search through the journal entries on the Journal List screen.

Join us on Discord!

Read this book alongside other users, experts, and the author himself. Ask questions, provide solutions to other readers, chat with the author via Ask Me Anything sessions, and much more. Scan the QR code or visit the link to join the community.

`https://packt.link/ios-Swift`

21

Getting Started with Search

In the previous chapter, you added a way for the user to get a photo from the camera or photo library, which can be added to a new journal entry.

In this chapter, you will implement a **search bar** for the Journal List screen. You'll start by modifying the `JournalListViewController` class to conform to the `UISearchResultsUpdating` protocol and display a search bar on the Journal List screen. Next, you'll modify the data source methods to display the correct journal entries when the user types in a search term. After that, you'll modify the `prepare(for:sender:)` method to ensure that the correct journal entry details are displayed on the Journal Entry Detail screen. Finally, you'll modify the method used to delete a journal entry.

By the end of this chapter, you'll have learned how to implement a search bar for your own apps. To name one example, if you were creating a contacts app, you could use a search bar to search for a particular contact.

The following topics will be covered in this chapter:

- Implementing a search bar on the Journal List screen
- Modifying table view data source methods
- Modifying the `prepare(for:sender:)` method
- Modifying the method to remove journal entries

Technical requirements

You will continue working on the `JRNL` project that you modified in the previous chapter.

The resource files and completed Xcode project for this chapter are in the `Chapter21` folder of the code bundle for this book, which can be downloaded here:

`https://github.com/PacktPublishing/iOS-18-Programming-for-Beginners-Ninth-Edition`

Check out the following video to see the code in action:

`https://youtu.be/hPwG2gfv448`

Let's start by modifying the `JournalListViewController` class to conform to the `UISearchResultsUpdating` protocol and to display a search bar on the Journal List screen.

Implementing a search bar for the Journal List screen

At present, you only have a few entries on the Journal List screen. But the longer you use the app, the more entries you'll have, and it's going to be hard to find a specific entry. To make it easier to look for a journal entry, you'll implement a search bar in the navigation bar of the Journal List screen. You will use Apple's `UISearchController` class to do this. This class incorporates a `UISearchBar` class that you can install in your user interface. To perform the search, you will adopt the `UISearchResultsUpdating` protocol and implement the `updateSearchResults(for:)` method required for this protocol.

To learn more about the `UISearchController` class, see `https://developer.apple.com/documentation/uikit/uisearchcontroller`.

You will now add an instance of the `UISearchController` class to the `JournalListViewController` class, adopt the `UISearchResultsUpdating` protocol, and implement the `updateSearchResults(for:)` method. Follow these steps:

1. In the Project navigator, click the **JournalListViewController** file. Add a new extension after all other code in this file to make the `JournalListViewController` class conform to the `UISearchResultsUpdating` protocol:

```
extension JournalListViewController: UISearchResultsUpdating {
}
```

2. You'll see an error because the method required to conform to the `UISearchResultsUpdating` protocol has not been implemented. Add the following code to the newly added extension to implement it:

```
// MARK: - Search
func updateSearchResults(for searchController: UISearchController) {
  guard let searchBarText = searchController.searchBar.text
  else { return }
  print(searchBarText)
}
```

Any text you type into the search bar will be printed to the Debug area.

3. Declare the following properties in the JournalListViewController class after the tableView property:

```
@IBOutlet var tableView: UITableView!
private let search = UISearchController(searchResultsController: nil)
private var filteredTableData: [JournalEntry] = []
```

The search property will store an instance of the UISearchController class.

The filteredTableData property will store an array of JournalEntry instances that match the search text entered by the user.

4. Modify the viewDidLoad() method in the JournalListViewController class as shown:

```
override func viewDidLoad() {
  super.viewDidLoad()
  SharedData.shared.loadJournalEntriesData()
  search.searchResultsUpdater = self
  search.obscuresBackgroundDuringPresentation = false
  search.searchBar.placeholder = "Search titles"
  navigationItem.searchController = search
}
```

Let's break this down:

```
search.searchResultsUpdater = self
```

This statement sets the JournalListViewController instance as the object responsible for updating the search results.

```
search.obscuresBackgroundDuringPresentation = false
```

This statement obscures the view controller containing the search content when the user interacts with the search bar. Since you're using the table view on the Journal List screen to display the search results, this value is set to false, otherwise, you will obscure the search results.

```
search.searchBar.placeholder = "Search titles"
```

This statement sets the placeholder text for the search bar.

```
navigationItem.searchController = search
```

This statement adds the search bar to the navigation bar on the screen.

5. Build and run your app, and you'll see a search bar on the Journal List screen. Type some text into the search bar:

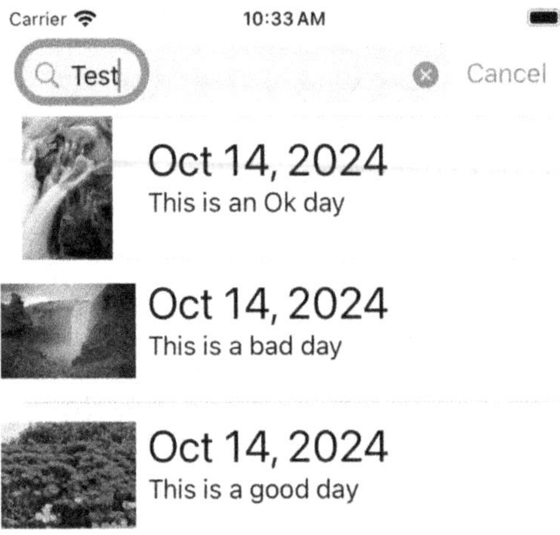

Figure 21.1: Simulator showing a search bar on the Journal List screen

6. Note that the text you type into the search bar appears in the Debug area:

Figure 21.2: Debug area showing the search text

You've added a search bar to the Journal List screen. Great! In the next section, you'll modify the `JournalListViewController` file to display journal entries whose titles match the search text typed into the search bar.

Modifying table view data source methods

As you learned in *Chapter 14, Getting Started with MVC and Table Views*, you can use `UITableViewDataSource` methods to determine how many table view rows to display, and what to put in each row.

In the previous section, you added a new property, `filteredTableData`, to hold an array of `JournalEntry` instances that match the search text. You'll modify the `updateSearchResults(for:)` method to populate `filteredTableData` with `JournalEntry` instances that match the search text, and you'll modify the `UITableViewDataSource` methods to display the contents of `filteredTableData` on the Journal List screen while the search bar is active. Follow these steps:

1. In the Project navigator, click the **JournalListViewController** file. Modify the `updateSearchResults(for:)` method in the `UISearchResultsUpdating` extension as shown:

```
//MARK: - Search
func updateSearchResults(for searchController: UISearchController) {
  guard let searchBarText = searchController.searchBar.text
  else {
    return
  }
  filteredTableData = SharedData.shared.allJournalEntries()
  .filter { entry in
    entry.entryTitle.lowercased().contains(searchBarText
    .lowercased())
  }
  tableView.reloadData()
}
```

   This method gets a copy of the `journalEntries` array and then adds only those `JournalEntry` instances matching the search text to the `filteredTableData` array. When done, the table view is reloaded.

2. Modify the `tableView(_:numberOfRowsInSection:)` method to get the number of `JournalEntry` instances from the `filteredTableData` array when the search bar is in use:

```
//MARK: - UITableViewDataSource
func tableView(_ tableView: UITableView, numberOfRowsInSection section:
Int) -> Int {
  if search.isActive {
    return filteredTableData.count
  } else {
    return SharedData.shared.numberOfJournalEntries()
  }
}
```

3. Modify the `tableView(_:cellForRowAt:)` method to get the `JournalEntry` instance for the specified row from the `filteredTableData` array when the search bar is in use:

```
func tableView(_ tableView: UITableView, cellForRowAt indexPath:
IndexPath) -> UITableViewCell {
  let journalCell =
  tableView.dequeueReusableCell(withIdentifier:
  "journalCell", for: indexPath) as!
  JournalListTableViewCell
  let journalEntry: JournalEntry
  if search.isActive {
```

```
      journalEntry = filteredTableData[indexPath.row]
    } else {
      journalEntry = SharedData.shared.journalEntry(
      at: indexPath.row)
    }
    if let photoData = journalEntry.photoData {
      journalCell.photoImageView.image =
      UIImage(data: photoData)
    }
    journalCell.dateLabel.text = journalEntry.date.formatted(
    .datetime.month().day().year()
  )
    journalCell.titleLabel.text = journalEntry.entryTitle
    return journalCell
  }
```

4. Build and run your app, and type some text that matches the title of one of your journal entries into the search bar. The journal entries with titles that match the search text will be displayed:

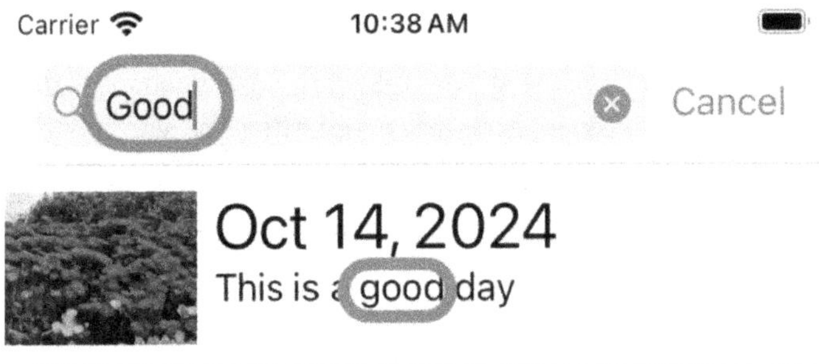

Figure 21.3: Simulator showing journal entries with titles that match the search text

You can now display journal entries with titles that match the search text, but when you tap on them, the Journal Entry Detail screen may or may not display the details for the tapped journal entry. You will fix this issue in the next section.

Modifying the prepare(for:sender:) method

When you enter text in the search bar, the journal entries with titles matching the search text will appear in the table view on the Journal List screen. But if you were to tap one of them, the Journal Entry Detail screen may or may not display the details for the tapped journal entry. This is because the `prepare(for:sender:)` method will reference the `journalEntries` array in the `SharedData.shared` instance instead of the `filteredTableData` array. To fix this, follow these steps:

1. Modify the `prepare(for:sender:)` method in the `JournalListViewController` class as shown to assign the appropriate `JournalEntry` instance from the `filteredTableData` array to the `journalEntry` property for the destination view controller when the search bar is active:

```
//MARK: - Navigation
override func prepare(for segue: UIStoryboardSegue, sender: Any?) {
  super.prepare(for: segue, sender: sender)
  guard segue.identifier == "entryDetail" else {
    return
  }
  guard let journalEntryDetailViewController =
  segue.destination as?
  JournalEntryDetailViewController,
  let selectedJournalEntryCell = sender as?
  JournalListTableViewCell,
  let indexPath = tableView.indexPath(for:
  selectedJournalEntryCell)
  else {
    fatalError("Could not get indexpath")
  }
  let selectedJournalEntry: JournalEntry
  if search.isActive {
    selectedJournalEntry = filteredTableData[indexPath.row]
  } else {
    selectedJournalEntry =
    SharedData.shared.journalEntry(at: indexPath.row)
  }
  journalEntryDetailViewController.selectedJournalEntry =
  selectedJournalEntry
}
```

2. Build and run your app, and type some text that matches the title of one of your journal entries into the search bar. The journal entries with titles that match the search text will be displayed:

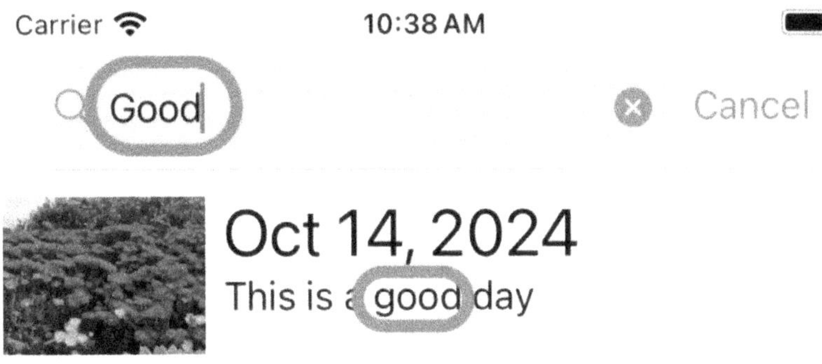

Figure 21.4: Simulator showing journal entries with titles that match the search text

3. Tap one of the journal entries and the details displayed on the Journal Entry Detail screen now match the journal entry that was tapped:

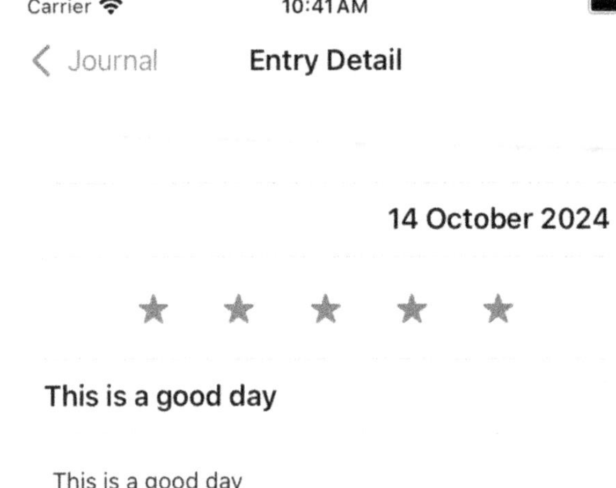

Figure 21.5: Simulator showing details of the tapped journal entry on the Journal Entry Detail screen

Your app now correctly displays the details of a tapped journal entry in the Journal Entry Detail screen. Cool! In the next section, you'll modify the methods used to remove a journal entry in the `JournalListViewController` class.

Modifying the method to remove journal entries

Up to this point, the method used to remove JournalEntry instances from the journalEntries array in the SharedData instance uses the table view row to identify the index of the JournalEntry instance to be removed. However, when the search bar is active, the table view row may not match the index of the JournalEntry instance to be removed. You will add a property to the JournalEntry class to store a value that will identify a JournalEntry instance, and modify methods in the SharedData and JournalListViewController classes to use this property to determine the JournalEntry instance to be removed. Follow these steps:

1. In the Project navigator, click the **JournalEntry** file. Add a new property to the JournalEntry class to store what's known as a **UUID** string:

```
class JournalEntry: NSObject, MKAnnotation, Codable {
  // MARK: - Properties
  var key = UUID().uuidString
  let date: Date
  let rating: Int
```

   When a new JournalEntry instance is created, the key property is assigned a string, which is generated by the UUID class, and is guaranteed to be unique.

To learn more about the UUID class, see https://developer.apple.com/documentation/foundation/uuid.

2. Click the **SharedData** file in the Project navigator. Add a method to the SharedData class after the removeJournalEntry(at:) method to remove a JournalEntry instance with the UUID string that matches the UUID string of the JournalEntry instance passed into it:

```
func removeSelectedJournalEntry(_ selectedJournalEntry: JournalEntry) {
  journalEntries.removeAll {
    $0.key == selectedJournalEntry.key
  }
}
```

3. Click the **JournalListViewController** file in the Project navigator. Modify the tableView(_:commit:forRowAt:) method in the JournalListViewController class as shown:

```
//MARK: - TableViewDelegate
func tableView(_ tableView: UITableView, commit editingStyle:
UITableViewCell.EditingStyle, forRowAt indexPath: IndexPath) {
  if editingStyle == .delete {
    if search.isActive {
```

```
            let selectedJournalEntry = filteredTableData[
            indexPath.row]
            filteredTableData.remove(at: indexPath.row)
            SharedData.shared.removeSelectedJournalEntry(
            selectedJournalEntry)
          } else {
            SharedData.shared.removeJournalEntry(at:
            indexPath.row)
          }
          SharedData.shared.saveJournalEntriesData()
          tableView.reloadData()
        }
      }
```

This method now checks to see if the search bar is active. If it is, the `JournalEntry` instance in the `filteredTableData` array corresponding to the row that was tapped is assigned to `selectedJournalEntry`. This instance is then removed from the `filteredTableData` array and passed as an argument to the `removeSelectedJournalEntry(_:)` method. The `journalEntry` instance with the same `UUID` string as the one passed into the `removeSelectedJournalEntry(_:)` method is removed from the `journalEntries` array in the `SharedData` instance.

4. Build and run your app. Since you made changes to the `JournalEntry` class, the previously saved journal entries stored in the JSON file will not load, so you will need to create new sample journal entries:

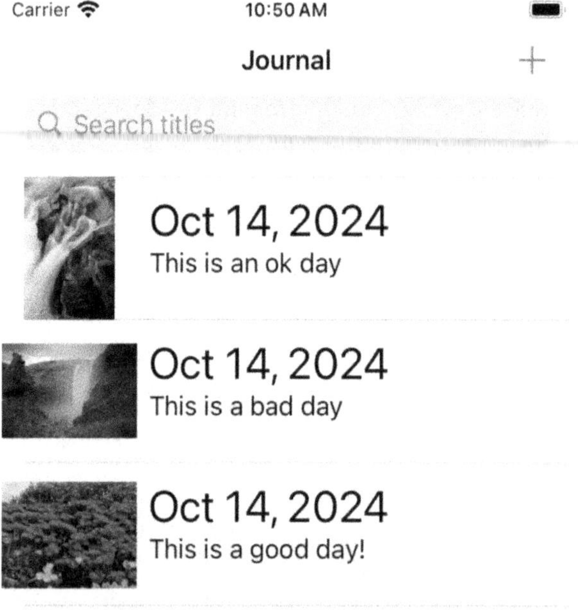

Figure 21.6: Simulator showing journal entries on the Journal List screen

5. Type some text that matches the title of one of the journal entries into the search bar. Journal entries with titles that match the search text will be displayed. Swipe left on a row and click **Cancel** to exit the search:

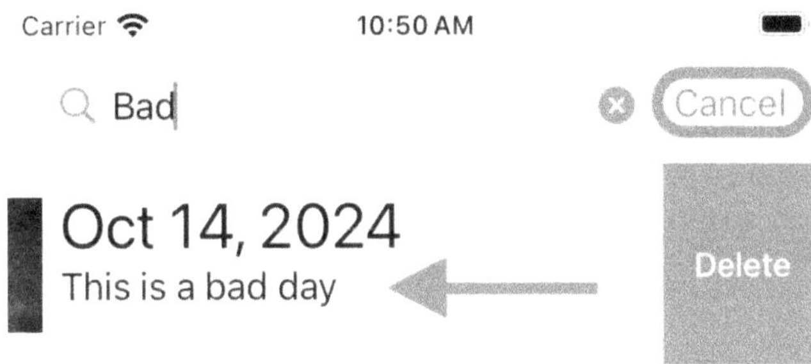

Figure 21.7: Simulator showing a row about to be deleted

Swiping left on a table view row when the search bar is active will delete it from the table view and delete the corresponding journal entry from the `journalEntries` array in the `SharedData` instance.

6. Verify that the deleted journal entry no longer appears on the Journal List screen:

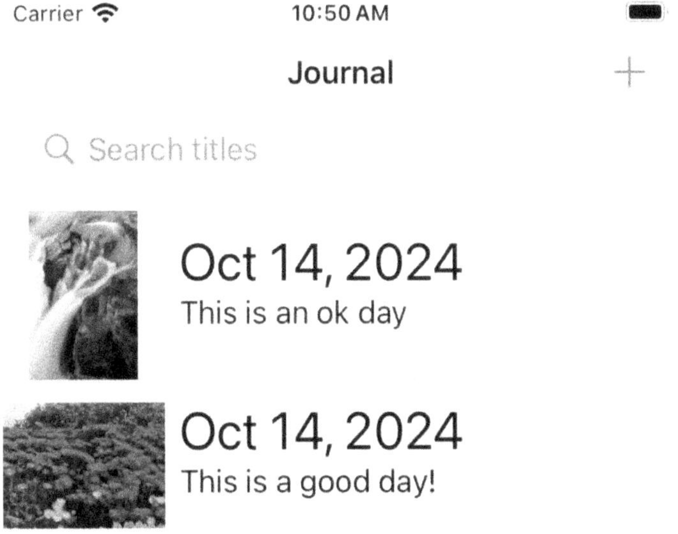

Figure 21.8: Simulator showing remaining journal entries on the Journal List screen

You have successfully modified the method to delete journal entries, and the implementation of a search bar on the Journal List screen is now complete. Excellent!

Summary

In this chapter, you implemented a search bar for the Journal List screen. First, you modified the `JournalListViewController` class to conform to the `UISearchResultsUpdating` protocol and display a search bar on the Journal List screen. Next, you modified the data source methods to display the correct journal entries when the user types in a search term. After that, you modified the `prepare(for:sender:)` method to ensure that the correct journal entry details are displayed on the Journal Entry Detail screen. Finally, you modified the method used to remove journal entries.

You have now learned how to implement a search bar for your own apps, and you have also completed the *JRNL* app. Fantastic job!

In the next chapter, you'll learn how to get your app ready for iPads and Macs.

Join us on Discord!

Read this book alongside other users, experts, and the author himself. Ask questions, provide solutions to other readers, chat with the author via Ask Me Anything sessions, and much more. Scan the QR code or visit the link to join the community.

`https://packt.link/ios-Swift`

22

Getting Started with Collection Views

In the previous chapter, you implemented a search bar for the Journal List screen, and your app is now complete. However, your app is designed to suit the iPhone's screen, and if you were to run it on an iPad or Mac, you'd see that it does not make the best use of the larger screen size.

In this chapter, you will replace the table view on the Journal List screen with a **collection view**, which will make better use of the extra screen space available when you run your app on an iPad or Mac. You'll also dynamically modify the number of columns and the collection view cell sizes when your device is rotated using size classes.

First, in the `Main` storyboard file, you'll replace the table view on the Journal List screen with a collection view and configure the collection view cell to display the same information the table view cell used to display. Next, you'll refactor the `JournalListViewController` and `JournalListTableViewCell` classes to work with the collection view and collection view cells you added. After that, you'll add code to dynamically change the collection view cell size to suit the display your app is running on. Finally, you'll test your app on different devices.

By the end of this chapter, you'll have learned about collection views, how to use collection view delegate and data source protocols, and how to dynamically modify your app's interface based on size classes.

The following topics will be covered in this chapter:

- Understanding collection views
- Modifying the Journal List screen to use a collection view
- Dynamically modifying collection view cell size using size classes
- Testing your app on different devices

Technical requirements

You will continue working on the `JRNL` project that you modified in the previous chapter.

The resource files and completed Xcode project for this chapter are in the `Chapter22` folder of the code bundle for this book, which can be downloaded here:

`https://github.com/PacktPublishing/iOS-18-Programming-for-Beginners-Ninth-Edition`

Check out the following video to see the code in action:

`https://youtu.be/yIJpBHzAHCU`

Let's start by learning about collection views in the next section.

Understanding collection views

A collection view is an instance of the `UICollectionView` class. It manages an ordered collection of elements and presents them using customizable layouts.

To learn more about collection views, visit `https://developer.apple.com/documentation/uikit/uicollectionview`.

The data displayed by a collection view is usually provided by a view controller. A view controller providing data for a collection view must adopt the `UICollectionViewDataSource` protocol. This protocol declares a list of methods that tells the collection view how many cells to display and what to display in each cell.

To learn more about the `UICollectionViewDataSource` protocol, visit `https://developer.apple.com/documentation/uikit/uicollectionviewdatasource`.

To provide user interaction, a view controller for a collection view must also adopt the `UICollectionViewDelegate` protocol, which declares a list of methods triggered when a user interacts with the collection view.

To learn more about the `UICollectionViewDelegate` protocol, visit `https://developer.apple.com/documentation/uikit/uicollectionviewdelegate`.

The way the collection view is laid out is specified by a `UICollectionViewLayout` object. This determines the cell placement, supplementary views, and decoration views inside the collection view's bounds.

You'll use the `UICollectionViewFlowLayout` class, a subclass of the `UICollectionViewLayout` class, for your app. Collection view cells in the collection view flow from one row or column to the next, with each row containing as many cells as will fit.

To learn more about the `UICollectionViewFlowLayout` class, visit `https://developer.apple.com/documentation/uikit/uicollectionviewflowlayout`.

A flow layout works with the collection view's delegate object to determine the size of items, headers, and footers in each section and grid. That delegate object must conform to the `UICollectionViewDelegateFlowLayout` protocol. This allows you to adjust layout information dynamically.

To learn more about the `UICollectionViewFlowLayoutDelegate` protocol, visit `https://developer.apple.com/documentation/uikit/uicollectionviewdelegateflowlayout`.

Now that you have a basic understanding of collection views, you'll modify the Journal List screen by replacing the table view with a collection view in the next section.

Modifying the Journal List screen to use a collection view

At present, the Journal List screen in the *JRNL* app uses a table view. A table view presents table view cells using rows arranged in a single column. This works great on an iPhone, but if you were to run the app on an iPad, you'd see there is a lot of wasted screen space on the Journal List screen, as shown in the following figure:

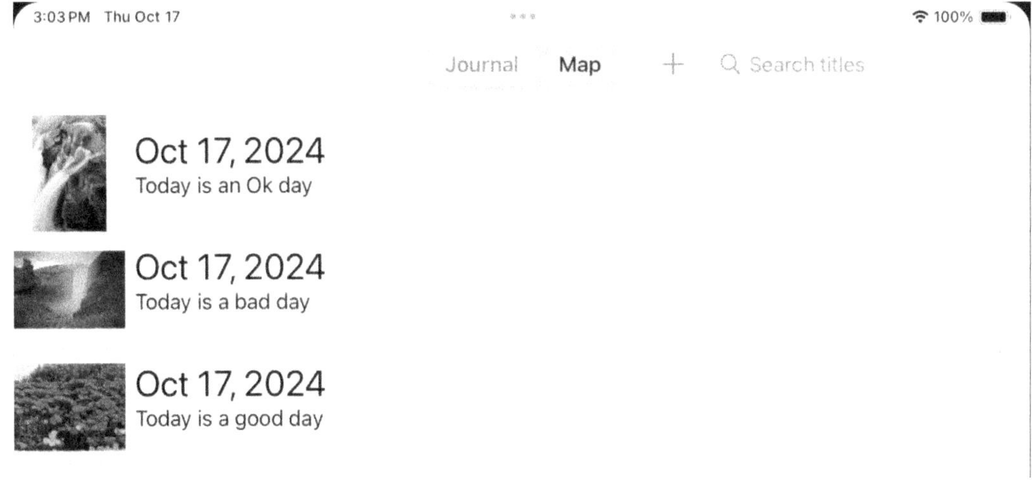

Figure 22.1: Simulator showing Journal List screen containing a table view on an iPad

To address this, you'll replace the table view with a collection view, which will allow you to more effectively use the available screen space, as shown in the following figure:

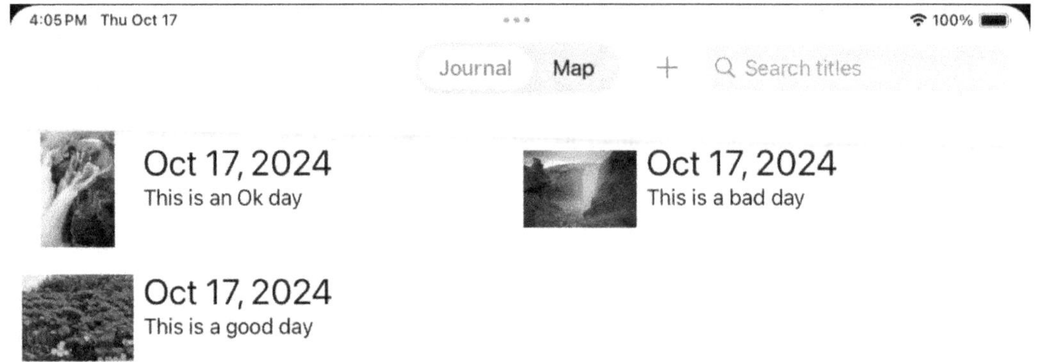

Figure 22.2: Simulator showing Journal List screen containing a collection view on an iPad

To implement a collection view on the Journal List screen, you'll need to do the following:

1. In the `Main` storyboard file, replace the table view in the **Journal Scene** with a collection view.
2. Add UI elements to the collection view cell.
3. Modify the `JournalListTableViewCell` class to manage the collection view cell's contents.
4. Modify the `JournalListViewController` class to manage what the collection view displays.
5. Add methods to dynamically change the collection view cell's size based on device screen size and orientation.

You'll start by modifying the **Journal Scene** in the `Main` storyboard file to use a collection view instead of a table view in the next section.

Replacing the table view with a collection view

At present, the **Journal Scene** in the `Main` storyboard file contains a table view. You'll replace this with a collection view. Follow these steps:

1. Open the `JRNL` project you modified in the previous chapter and choose **iPad (10th generation)** as the destination from the Destination menu:

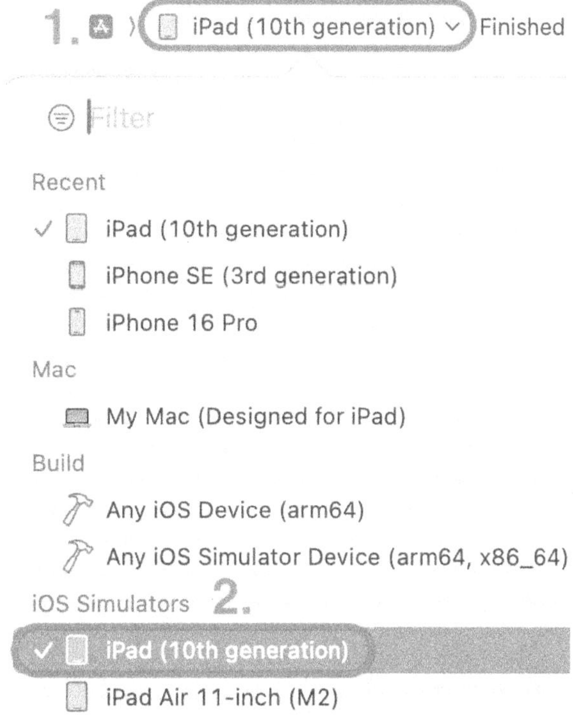

Figure 22.3: Destination menu showing iPad (10th generation) selected

2. Build and run your app, and note how it appears on the iPad's screen:

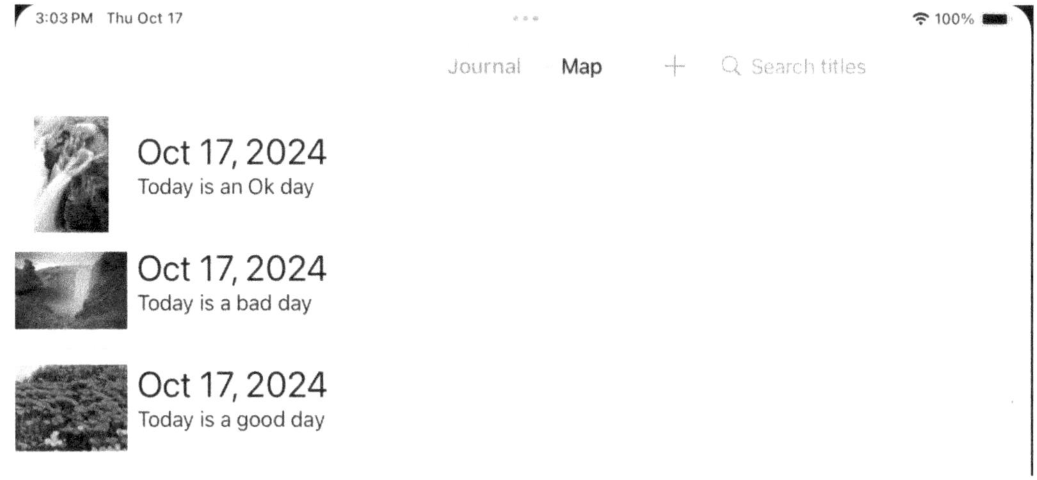

Figure 22.4: Simulator showing iPad screen

Although the app works as it should, note that a lot of space is wasted on the right side of the Journal List screen.

3. Click the Stop button. Click the **Main** storyboard file in the Project navigator. In the document outline, click the table view under **Journal Scene**. Press the *Delete* key to remove it:

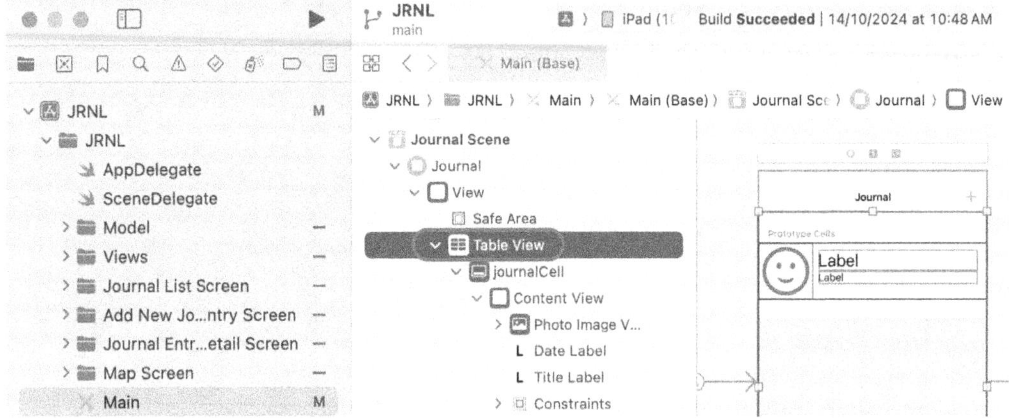

Figure 22.5: Editor area showing table view selected in the document outline

4. Click the Library button to display the library. Type `collec` in the filter field. A **Collection View** object will appear as one of the results. Drag it to the middle of the view of the **Journal Scene**:

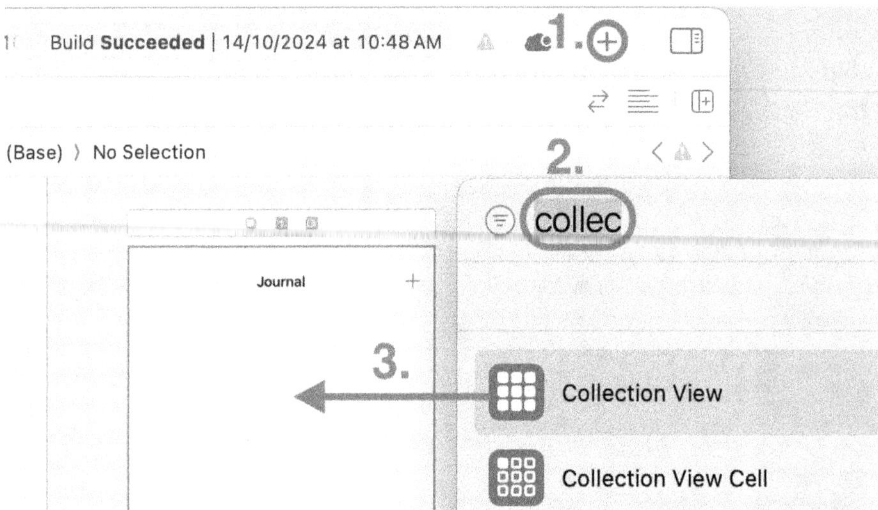

Figure 22.6: Library with Collection View object selected

5. Make sure the collection view is selected and click the Auto Layout Add New Constraints button:

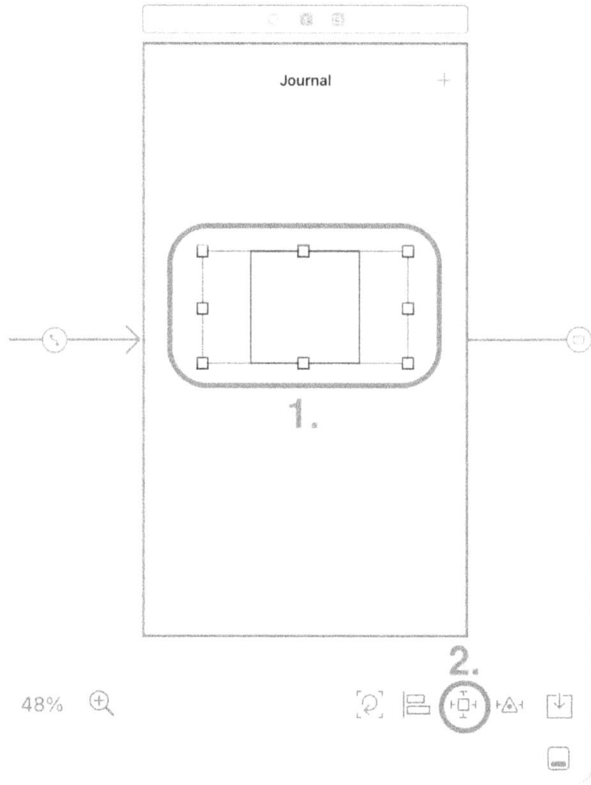

Figure 22.7: Journal scene with collection view selected

6. Type `0` in the top, left, right, and bottom edge constraint fields and click all the pale red struts. Make sure all the struts have turned bright red and **Constrain to margins** is not checked. Then, click the **Add 4 Constraints** button:

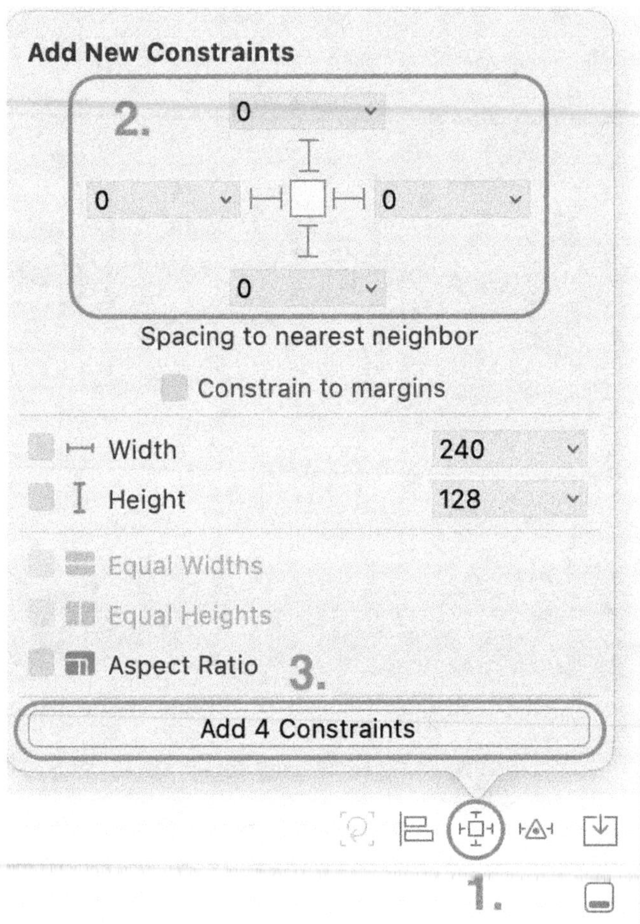

Figure 22.8: Auto Layout – Add New Constraints pop-up dialog box

This sets the space between the edges of the collection view and the edges of the enclosing view to 0, binding the collection view's edges to those of the enclosing view.

7. Verify that all four sides of the collection view are now bound to the edges of the screen as shown:

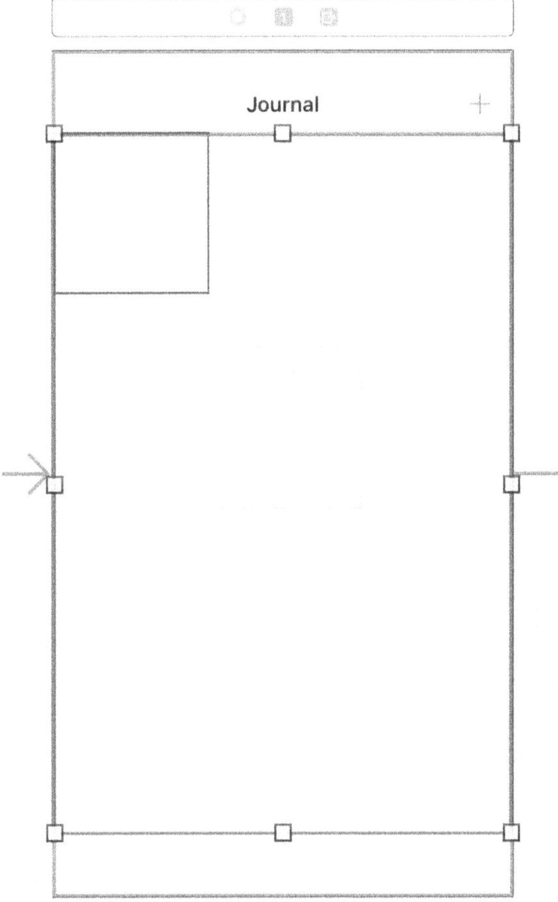

Figure 22.9: Journal Scene with collection view filling the screen

8. With the collection view still selected, click the Size inspector button. Under **Collection View**, set **Estimate Size** to **None**.

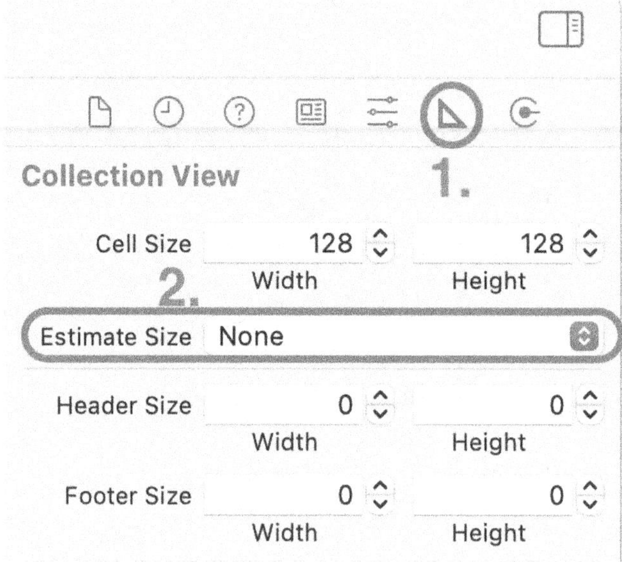

Figure 22.10: Size inspector with Estimate Size highlighted

You'll be adding code to determine the collection view size dynamically later.

9. You'll need to re-establish the segue between the Journal List screen and the Journal Entry Detail screen. *Ctrl + Drag* from **Collection View Cell** in the document outline to the **Entry Detail Scene** and choose **Show** from the pop-up menu.

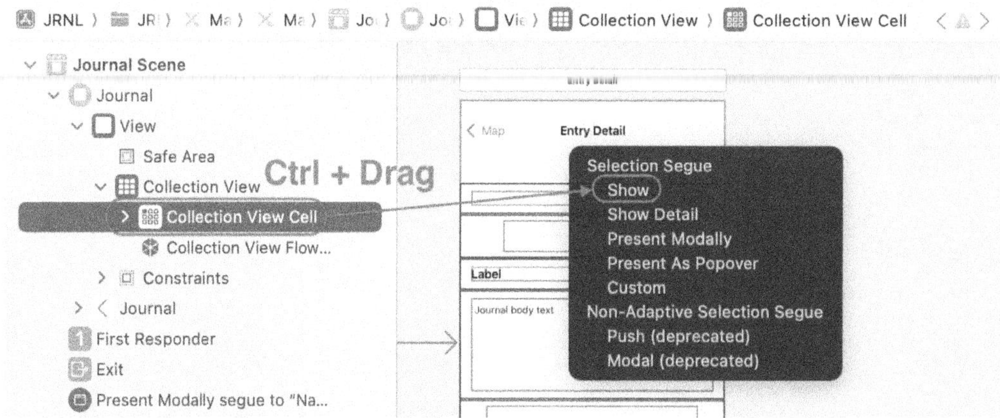

Figure 22.11: Editor area showing drag destination

10. Click the newly added storyboard segue and click the Attributes inspector button. Under **Storyboard Segue**, set **Identifier** to `entryDetail`.

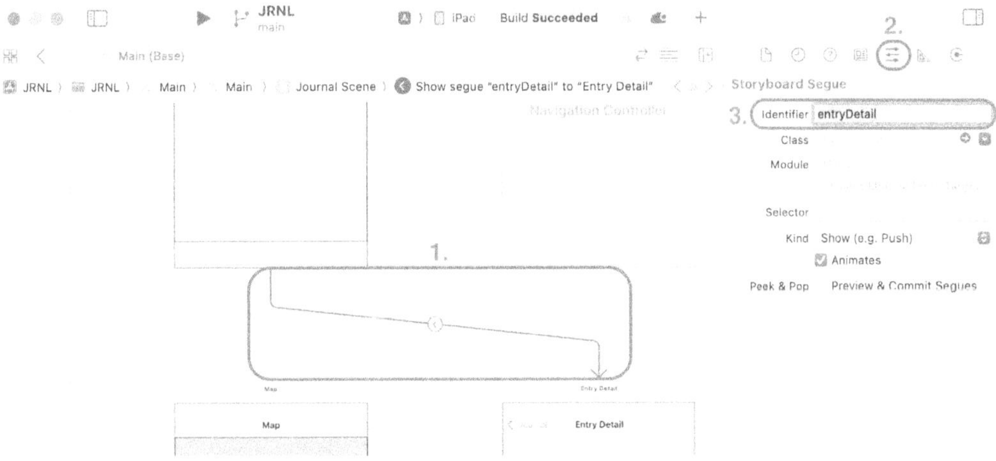

Figure 22.12: Attributes inspector with Identifier set to entryDetail

You have added a collection view to the Journal Scene and used Auto Layout constraints to make it fill the screen, but the prototype collection view cell is currently empty. You'll add UI elements to the collection view cell in the next section.

Adding UI elements to the collection view cell

You have replaced the table view inside the **Journal Scene** with a collection view, but the prototype collection view cell inside the collection view is empty. You'll need to add an image view and two labels to the prototype collection view cell and set up the constraints for them. This will make it match the table view cell used previously. Follow these steps:

1. Select **Collection View Cell** for the **Journal Scene** in the document outline. Drag the right edge of the collection view cell to the right until it reaches the right side of the screen:

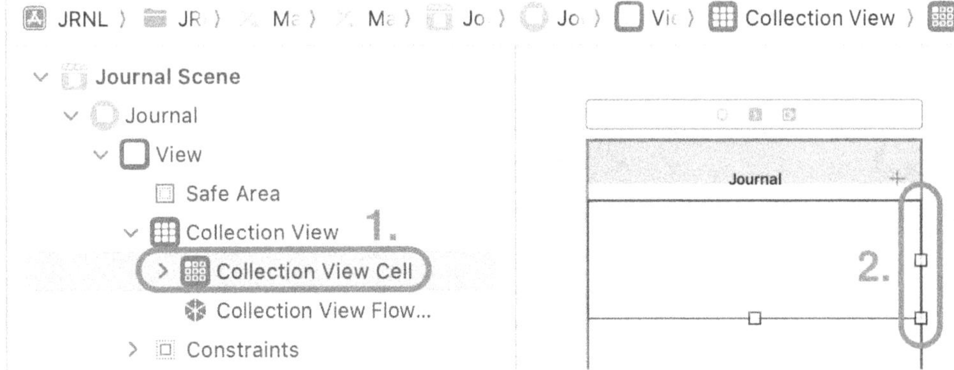

Figure 22.13: Editor area showing Collection View Cell

2. Click the Size inspector button, and under **Collection View Cell**, set **Height** to `90`.

3. To add an image view to the table view cell, click the Library button. Type `imag` into the filter field. An **Image View** object will appear in the results. Drag it into the prototype cell:

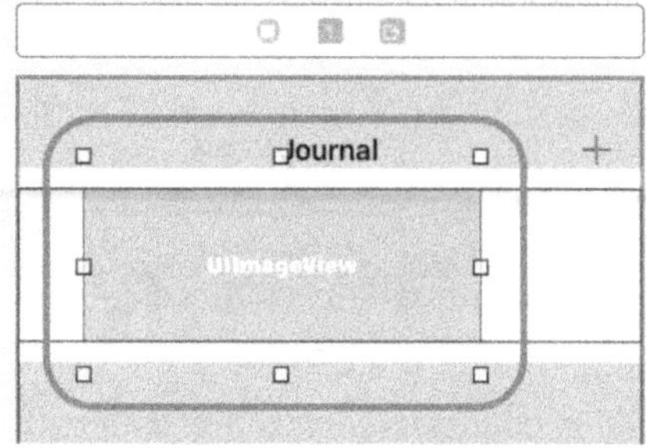

Figure 22.14: Prototype cell with image view added

4. With the image view selected, click the Add New Constraints button and enter the following values to set the constraints for the newly added image view:

   - Top: `0`
   - Left: `0`
   - Bottom: `0`
   - Width: `90`

   Constrain to margins should not be checked. When done, click the **Add 4 Constraints** button:

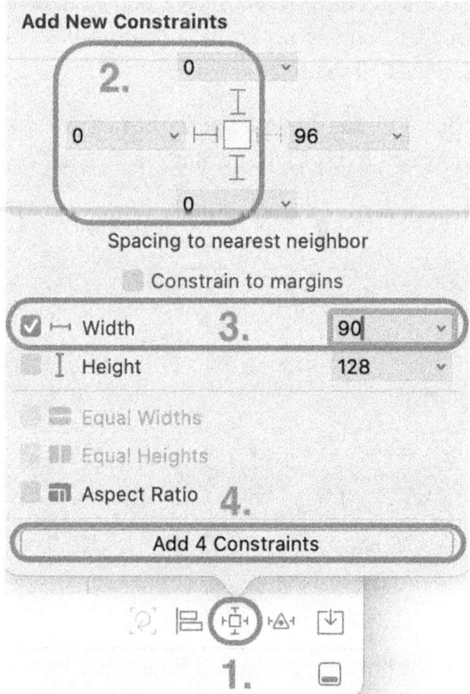

Figure 22.15: Constraints for image view

5. Click the Attributes inspector button. Under **Image View**, set **Image** to `face.smiling`:

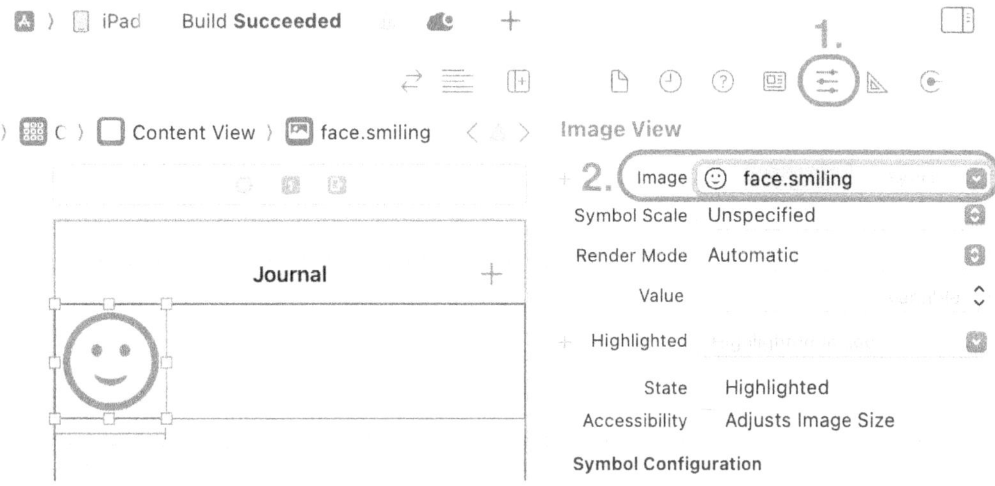

Figure 22.16: Image View with Image set to face.smiling

6. Next, you'll add a label to display the journal entry date. Click the Library button. Type `label` into the filter field. A **Label** object will appear in the results. Drag it to the space between the image view you just added and the right side of the cell:

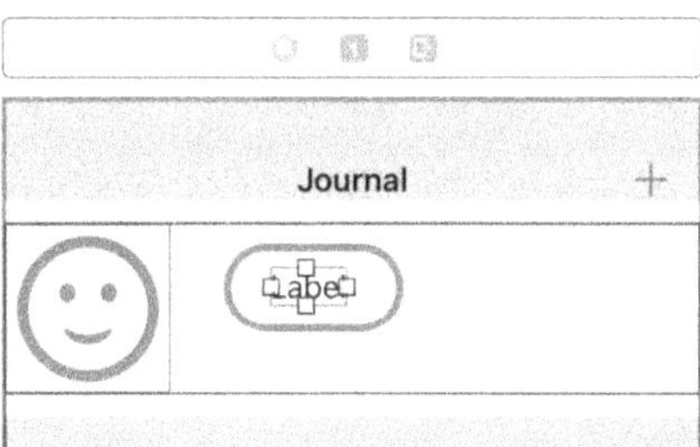

Figure 22.17: Prototype cell with label added

7. In the Attributes inspector, under **Label**, set **Font** to **Title 1** using the **Font** menu:

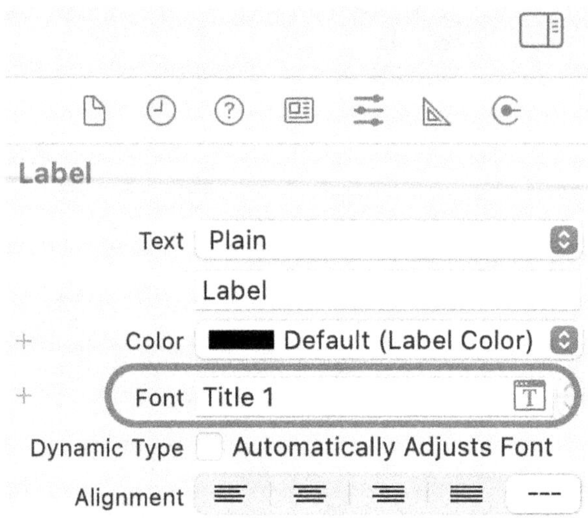

Figure 22.18: Attributes inspector for Label

8. Click the Add New Constraints button and enter the following values to set the constraints for the label:

   - Top: `0`
   - Left: `8`
   - Right: `0`

   Constrain to margins should be checked, which sets a standard margin of 8 points. When done, click the **Add 3 Constraints** button.

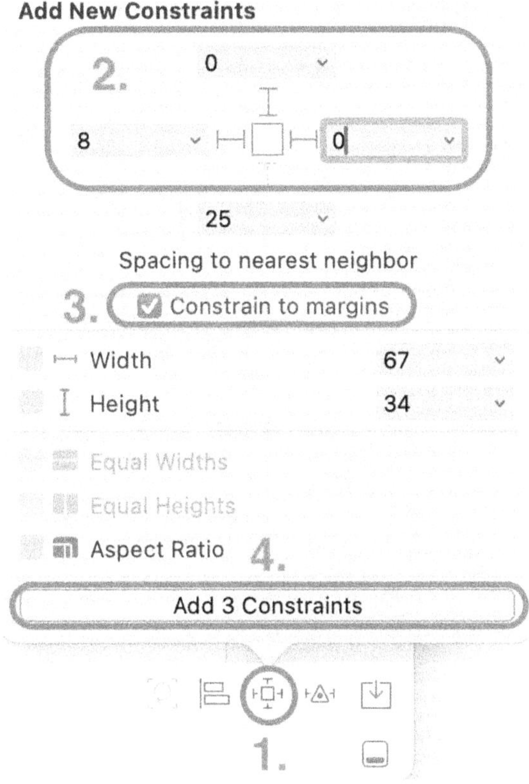

Figure 22.19: Constraints for label

9. Finally, you'll add a label to display the journal entry title. Click the Library button. Type `label` into the filter field. A **Label** object will appear in the results. Drag it to the space between the label you just added and the bottom of the cell:

Figure 22.20: Prototype cell with second label added

10. In the Attributes inspector, under **Label**, set **Font** to **Body** using the **Font** menu, and set **Lines** to 2:

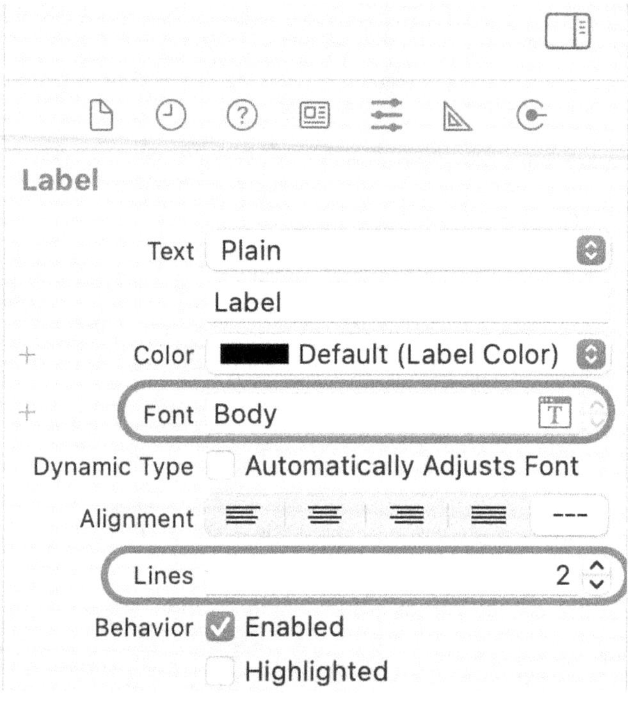

Figure 22.21: Attributes inspector for Label

11. Click the Add New Constraints button and enter the following values to set the constraints for the label:

    - Top: `0`
    - Left: `8`
    - Right: `0`

    Constrain to margins should be checked, which sets a standard margin of 8 points. When done, click the **Add 3 Constraints** button.

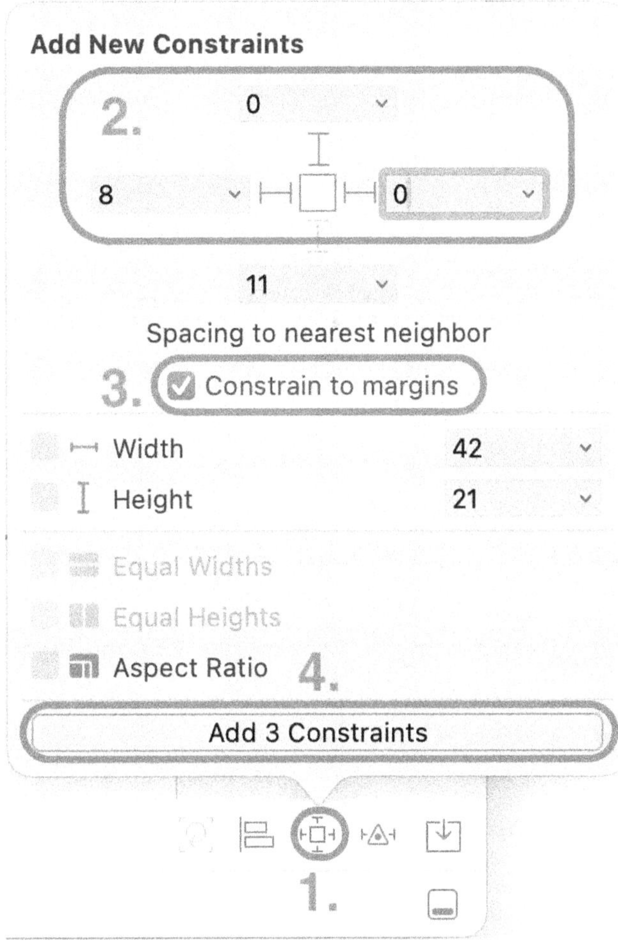

Figure 22.22: Constraints for second label

The prototype collection view cell now has an image view and two labels, and all the necessary constraints have been added. Fantastic! In the next section, you'll modify the `JournalListTableViewCell` class to manage the collection view cell's contents.

Modifying the JournalListTableViewCell class

Originally, the `JournalListTableViewCell` class was used to manage the table view instance's table view cells in the Journal List screen. Since you have replaced the table view with a collection view, all the connections between the `JournalListTableViewCell` class will need to be re-established to work with the UI elements that you just added to the collection view cell. Follow these steps:

1. First, you'll change the `JournalListTableViewCell` class name to more accurately describe its new role. Click the `JournalListTableViewCell` file in the Project navigator. Right-click on the class name in the file and choose **Refactor** | **Rename...** from the pop-up menu:

   Figure 22.23: Pop-up menu with Refactor | Rename... selected

2. Change the name to `JournalListCollectionViewCell` and click **Rename**:

   Figure 22.24: Editor area showing the new name

3. You'll modify the class declaration as this class is now used to manage a collection view cell. Change the superclass to `UICollectionViewCell`:

   ```
   class JournalListCollectionViewCell: UICollectionViewCell {
   ```

4. You'll now assign this class as the identity of the collection view cell. Click the **Main** storyboard file in the Project navigator and click **Collection View Cell** under **Journal Scene** in the document outline:

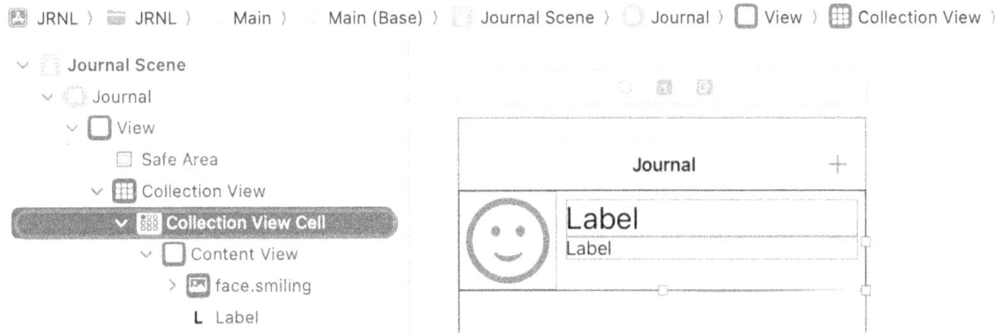

Figure 22.25: Editor area showing Collection View Cell inside Journal Scene

5. Click the Identity inspector button. Under the **Custom Class** section, set **Class** to `JournalListCollectionViewCell`. This sets a `JournalListCollectionViewCell` instance as the custom collection view subclass for the collection view cell. Press *Return* when done:

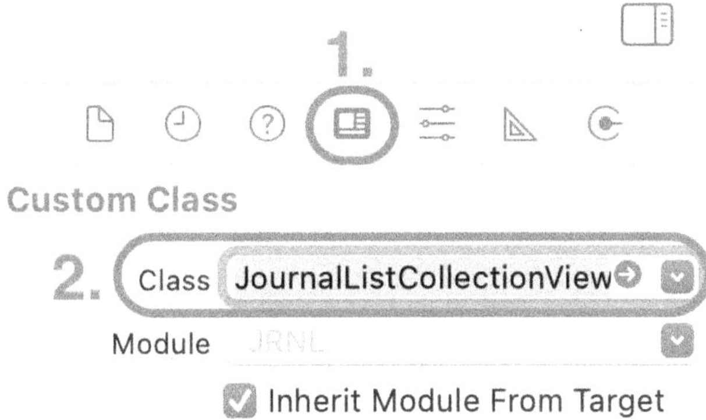

Figure 22.26: Identity inspector showing Class set to JournalListCollectionViewCell

6. Click the Attributes inspector button. Under **Collection Reusable View**, set **Identifier** to `journalCell`:

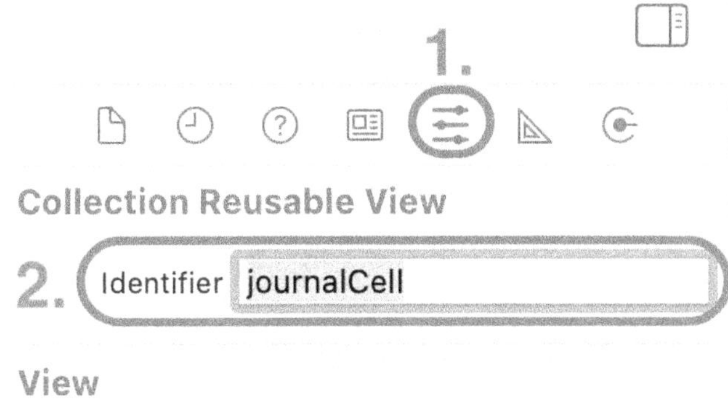

Figure 22.27: Attributes inspector showing Identifier set to journalCell

Note that the name of the collection view cell in the document outline has changed to **journalCell.**

7. With **journalCell** selected in the document outline, click the Connections inspector button to display the outlets for **journalCell.**
8. Drag from the **photoImageView** outlet to the image view in the table view cell.
9. Drag from the **dateLabel** outlet to the top label in the table view cell.
10. Drag from the **titleLabel** outlet to the bottom label in the table view cell.
11. Verify that the connections look like the following screenshot when done:

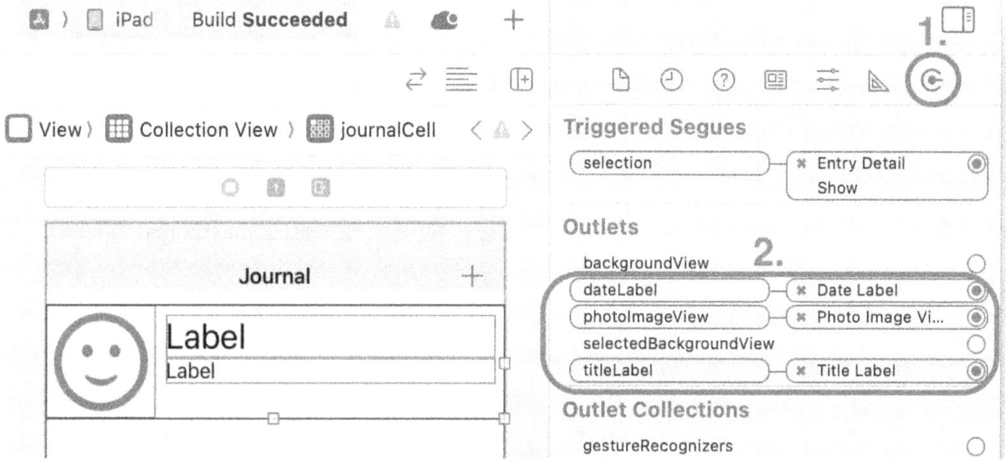

Figure 22.28: Connections inspector showing the connections for journalCell

Remember that if you make a mistake, you can click the **x** to break the connection and drag from the outlet to the UI element once more.

The `journalCell` collection view cell in the `Main` storyboard file has now been set up with the `JournalCollectionTableViewCell` class. The outlets for the collection view cell's image view and labels have also been assigned. In the next section, you'll update the `JournalListViewController` class to work with a collection view instead of a table view.

Modifying the JournalListViewController class

At present, the `JournalListViewController` class has an outlet for a `UITableView` object and implemented data source and delegate methods to manage a table view. You'll modify this class to work with a collection view instead. Follow these steps:

1. First, you'll modify the class declaration to work with a `UICollectionView` instance. Click the **JournalListViewController** file in the Project navigator. Modify the class declaration as shown:

```
class JournalListViewController: UIViewController,
```

```
UICollectionViewDataSource, UICollectionViewDelegate,
UICollectionViewDelegateFlowLayout {
```

Here, you've changed the data source and delegate protocols to the collection view equivalents and added conformance to a new protocol, `UICollectionViewDelegateFlowLayout`. This protocol is used to determine how the collection view cells in a collection view are laid out. You will see an error because the collection view data source methods have not been implemented. Don't worry about the error as you'll fix it in a later step in this section.

2. To change the name for the `tableview` outlet, right-click it and choose **Refactor | Rename...** from the pop-up menu:

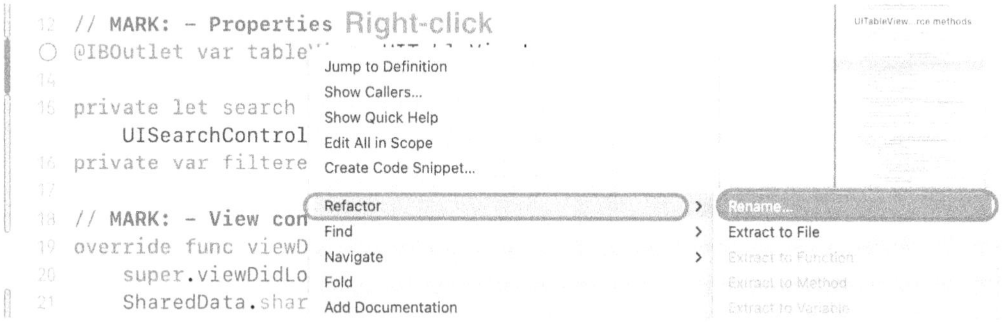

Figure 22.29: Pop-up menu with Refactor | Rename... selected

3. Change the name to `collectionView` and click **Rename**:

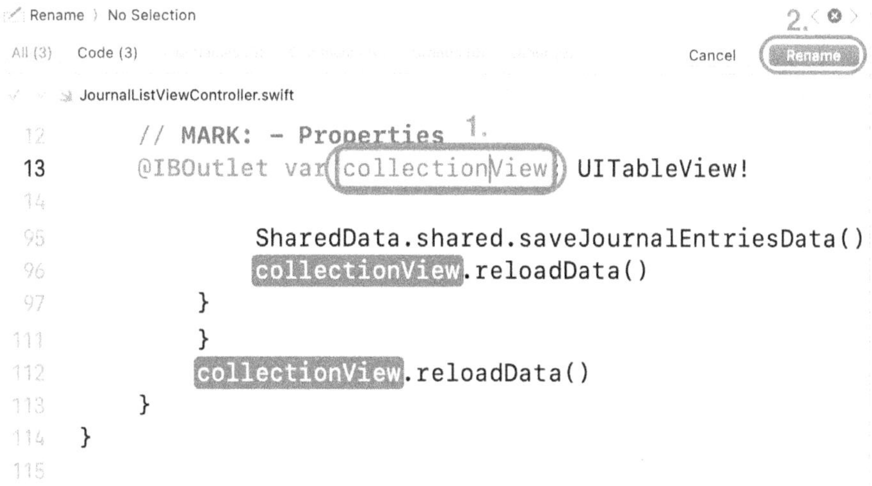

Figure 22.30: Editor area showing the new name

4. Since the view controller will be managing a collection view, change the outlet type from `UITableView` to `UICollectionView`:

```
@IBOutlet var collectionView: UICollectionView!
```

5. To establish the connections between the UI element and your code, click the **Main** storyboard file in the Project navigator and click the first **Journal Scene** in the document outline.
6. Click the Connections inspector button and drag from the **collectionView** outlet to the **Collection View** in the document outline:

Figure 22.31: Connections inspector showing the connections for JournalListViewController

7. Click **Collection View** in the document outline. Drag from the **dataSource** and **delegate** outlets to the view controller (shown as **Journal**) in the document outline.

Figure 22.32: Connections inspector showing the connections for collectionView

8. Now you'll fix the errors in the `JournalListViewController` class. Click the **JournalListViewController** file in the Project navigator and replace the table view data source methods in your code with these collection view data source methods:

```
// MARK: - UICollectionViewDataSource
func collectionView(_ collectionView: UICollectionView,
numberOfItemsInSection section: Int) -> Int {
  if search.isActive {
    return filteredTableData.count
  } else {
    return SharedData.shared.numberOfJournalEntries()
  }
}
```

```
func collectionView(_ collectionView: UICollectionView, cellForItemAt
indexPath: IndexPath) -> UICollectionViewCell {
  let journalCell = collectionView.dequeueReusableCell(
  withReuseIdentifier: "journalCell", for: indexPath) as!
  JournalListCollectionViewCell
  let journalEntry: JournalEntry
  if search.isActive {
    journalEntry = filteredTableData[indexPath.row]
  } else {
    journalEntry = SharedData.shared.journalEntry(at:
    indexPath.row)
  }
  if let photoData = journalEntry.photoData {
    journalCell.photoImageView.image = UIImage(data:
    photoData)
  }
  journalCell.dateLabel.text = journalEntry.date.formatted(
   .dateTime.month().day().year()
  )
  journalCell.titleLabel.text = journalEntry.entryTitle
  return journalCell
}
```

As you can see, they are very similar to the table view data source methods that you used previously.

9. As you are now using a collection view, `tableView(_:commit:forRowAt:)` can no longer be used to remove cells. Replace the `tableView(_:commit:forRowAt:)` method with the following method:

```
// MARK: - UICollectionView delete method
func collectionView(_ collectionView: UICollectionView,
contextMenuConfigurationForItemsAt indexPaths: [IndexPath], point:
CGPoint) -> UIContextMenuConfiguration? {
  guard let indexPath = indexPaths.first else {
    return nil
  }
  let config = UIContextMenuConfiguration(
  previewProvider: nil)
  { (elements) -> UIMenu? in
    let delete = UIAction(title: "Delete") { (action) in
      if self.search.isActive {
        let selectedJournalEntry = self.filteredTableData[
        indexPath.item]
```

```
          self.filteredTableData.remove(at: indexPath.item)
          SharedData.shared.removeSelectedJournalEntry(
          selectedJournalEntry)
        } else {
          SharedData.shared.removeJournalEntry(at:
          indexPath.item)
        }
        SharedData.shared.saveJournalEntriesData()
        collectionView.reloadData()
      }
      return UIMenu(children: [delete])
    }
    return config
  }
```

Instead of swiping left to delete, this method implements a contextual menu with a single option, **Delete**, that appears when you tap and hold on a collection view cell.

10. You'll also see an error in the `prepare(for:sender:)` method. Modify the `guard` statement in the `prepare(for:sender:)` method as shown:

```
guard let journalEntryDetailViewController = segue.destination as?
JournalEntryDetailViewController,
let selectedJournalEntryCell = sender as? JournalListCollectionViewCell,
let indexPath = collectionView.indexPath(for: selectedJournalEntryCell)
else {
  fatalError("Could not get indexpath")
}
```

All the errors in `JournalListViewController` have been resolved. Cool! In the next section, you'll add code to change the size of the collection view cells based on device screen size and orientation.

Dynamically modifying collection view cell size using size classes

As you saw earlier, the table view on the Journal List screen presents table view cells using rows arranged in a single column. This works great on an iPhone, but as you have seen, this results in a lot of wasted space if you run the app on an iPad. Even though you can use the same UI for both iPhone and iPad, it would be better if you could customize it to suit each device.

To do this, you'll add some code so your app can identify the size of the screen it's running on, and you'll dynamically modify the size of the collection view cells in the collection view to suit. You can identify the current screen size using size classes; you'll learn about them in the next section.

Understanding size classes

To determine the size of the screen your app is running on, you must consider the effects of device orientation on your UI. It can be challenging to do this as there is a wide variety of screen sizes, in both portrait and landscape orientations. To make this easier, instead of using the physical resolution of the device, you will use size classes.

For more information on size classes, see this link: `https://developer.apple.com/design/human-interface-guidelines/layout`.

Size classes are traits that are automatically assigned to a view by the operating system. Two classes are defined, which describe the height and width of a view: regular (expansive space) and compact (constrained space). Let's look at size classes for a full-screen view on different devices:

Device	Portrait	Landscape
iPad	Regular width Regular height	Regular width Regular height
iPhone 16 Pro Max	Compact width Regular height	Regular width Compact height
iPhone SE (3rd Gen)	Compact width Regular height	Compact width Compact height

Figure 22.33: Size classes for different iOS devices

For the *JRNL* app, you will configure the collection view in the Journal List screen to use a single column of collection view cells if the size class is compact, and two columns if the size class is regular.

You'll add code to your app to determine the current size class. Once you know the size class, you'll be able to set the number of columns to use and the size of the collection view cells in the collection view. You'll learn how to do this in the next section.

Modifying the JournalListViewController class

You have already made the `JournalListViewController` class adopt the `UICollectionViewDelegateFlowLayout` protocol. Now you will create and set the collection view's layout using a `UICollectionViewFlowLayout` instance and implement methods to dynamically set the collection view cell size.

Follow these steps:

1. Click the **JournalListViewController** file in the Project navigator. In the `JournalListViewController` class, add the following method to the class definition before the closing curly brace:

```
func setupCollectionView() {
  let flowLayout = UICollectionViewFlowLayout()
  flowLayout.sectionInset = UIEdgeInsets(top: 10, left: 10,
  bottom: 10, right: 10)
  flowLayout.minimumInteritemSpacing = 0
  flowLayout.minimumLineSpacing = 10
  collectionView.collectionViewLayout = flowLayout
}
```

   This method creates an instance of the `UICollectionViewFlowLayout` class, sets all the edge insets for the collection view to 10 points, sets the minimum inter-item spacing to 0 points, sets the minimum line spacing to 10 points, and assigns it to the collection view. Section insets reflect the spacing at the outer edges of the section. Minimum inter-item spacing is the minimum spacing to use between items in the same row. Minimum line spacing is the minimum spacing to use between lines of items in the grid.

2. Add the following `UICollectionViewDelegateFlowLayout` method after the `setupCollectionView()` method:

```
// MARK: - UICollectionViewDelegateFlowLayout
func collectionView(_ collectionView: UICollectionView, layout
collectionViewLayout: UICollectionViewLayout, sizeForItemAt indexPath:
IndexPath) -> CGSize {
  let numberOfColumns: CGFloat
  if (traitCollection.horizontalSizeClass == .compact) {
    numberOfColumns = 1
  } else {
    numberOfColumns = 2
  }
  let viewWidth = collectionView.frame.width
  let inset = 10.0
  let contentWidth = viewWidth - inset * (
  numberOfColumns + 1)
  let cellWidth = contentWidth / numberOfColumns
  let cellHeight = 90.0
  return CGSize(width: cellWidth, height: cellHeight)
}
```

This method determines the number of columns to be displayed and sets the height and width of the collection view cells.

Let's break it down:

```
func collectionView(_ collectionView: UICollectionView, layout
collectionViewLayout: UICollectionViewLayout, sizeForItemAt indexPath:
IndexPath) -> CGSize {
```

This method returns a `CGSize` instance that the collection view cell size should be set to.

```
let numberOfColumns: CGFloat
if (traitCollection.horizontalSizeClass == .compact) {
  numberOfColumns = 1
} else {
  numberOfColumns = 2
}
```

This code sets the number of columns to display.

```
let viewWidth = collectionView.frame.width
```

This statement gets the width of the screen and assigns it to `viewWidth`.

```
let inset = 10.0
let contentWidth = viewWidth - inset * (
numberOfColumns + 1)
```

This code subtracts the space used for the edge insets so the cell size can be determined.

```
let cellWidth = contentWidth / numberOfColumns
```

This statement calculates the width of the cell by dividing `contentWidth` by the number of columns and assigns it to `cellWidth`.

```
let cellHeight = 90.0
```

This statement assigns `90` to `cellHeight`, which will be used to set the cell height.

```
return CGSize(width: cellWidth, height: cellHeight)
}
```

This returns the `CGSize` instance containing the cell size.

Assume you're running on iPhone 16 Pro Max in portrait mode. The horizontal size class would be `.compact`, so `numberOfColumns` is set to `1`. `viewWidth` would be assigned the width of the iPhone screen, which is `414` points. `contentWidth` is set to `414 - (10 x 2) = 394`. `cellWidth` is set to `contentWidth / numberOfColumns = 394`, and `cellHeight` is set to `90`, so the `CGSize` instance returned would be `(394, 90)`, enabling one cell to fit in a row.

When you rotate the same iPhone to landscape mode, the horizontal size class would be `.regular`, so `numberOfColumns` is set to 2. `viewWidth` would be assigned the height of the iPhone screen, which is 896 points. `contentWidth` is set to `896 - (10 x 3) = 866`. `cellWidth` is set to `contentWidth / numberOfColumns = 433`, and `cellHeight` is set to `90`, so the `CGSize` instance returned would be `(433, 90)`, enabling two cells to fit in a row.

3. Modify the `viewDidLoad()` method to call the `setupCollectionView()` method:

```
override func viewDidLoad() {
  super.viewDidLoad()
  SharedData.shared.loadJournalEntriesData()
  setupCollectionView()
  search.searchResultsUpdater = self
  search.obscuresBackgroundDuringPresentation = false
  search.searchBar.placeholder = "Search titles"
  navigationItem.searchController = search
}
```

4. Add the following method after the `viewDidLoad()` method to recalculate the number of columns and size of the collection view cells when the device is rotated:

```
override func viewWillLayoutSubviews() {
  super.viewWillLayoutSubviews()
  collectionView.collectionViewLayout.invalidateLayout()
}
```

You have implemented all the code required to change the collection view cell size based on size classes. Excellent! In the next section, you'll test your app on different simulated devices and on your Mac.

Testing your app on different devices

Now that you have implemented all the code required to dynamically set collection view cell size, you'll test your app on different simulated devices and on your Mac. Follow these steps:

1. Simulator should still be set to iPad. Build and run your app. It will display two columns, as shown:

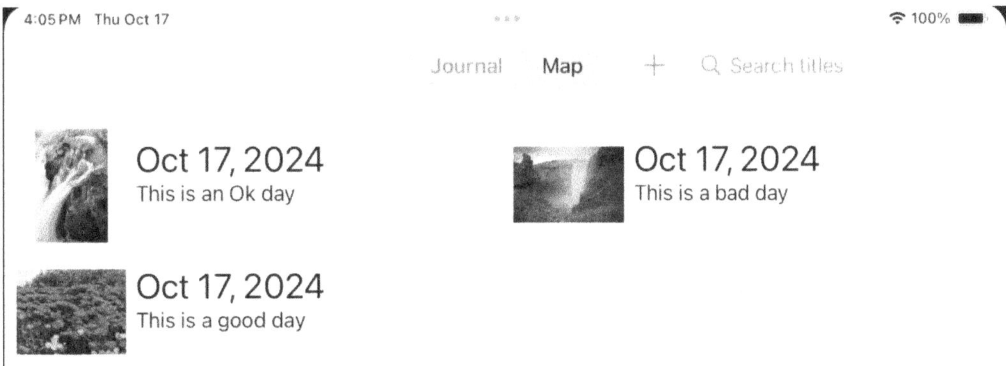

Figure 22.34: Simulator showing iPad screen with two columns

2. Choose **Rotate Left** from the **Device** menu, and you'll still see two columns, but the cells have expanded in size to fill the screen:

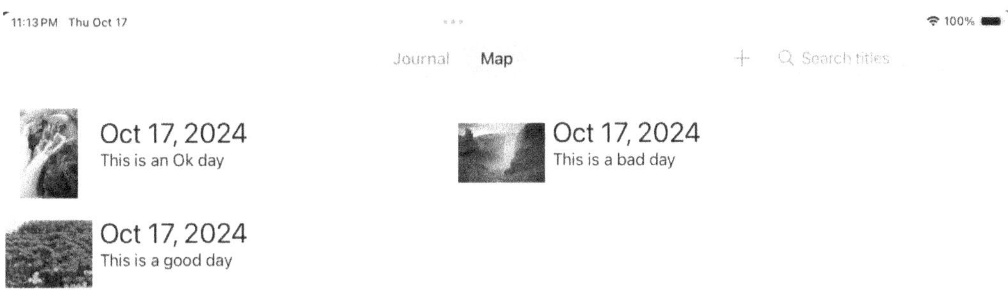

Figure 22.35: Simulator showing iPad screen rotated to the left with two columns

3. Stop your app and choose **iPhone SE (3rd Generation)** from the Destination menu. Run your app on Simulator again, and it will display a single column, as shown:

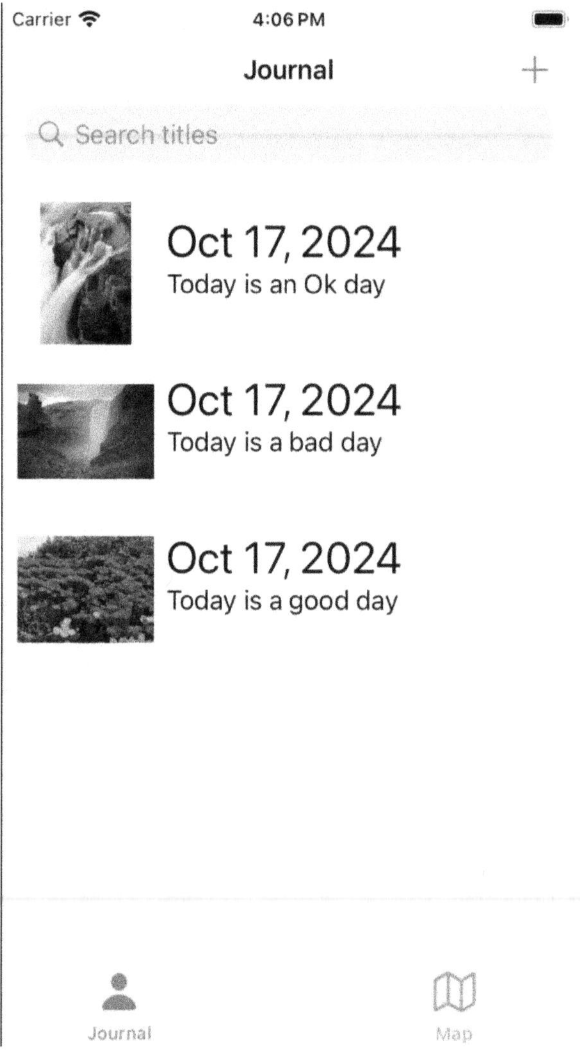

Figure 22.36: Simulator showing iPhone screen with a single column

Simulator will not automatically close the iPad instance when it launches the iPhone instance. Close the iPad instance manually for better performance.

4. Choose **Rotate Left** from the **Device** menu, and you'll still see a single column, but the cell size has expanded to fill the screen:

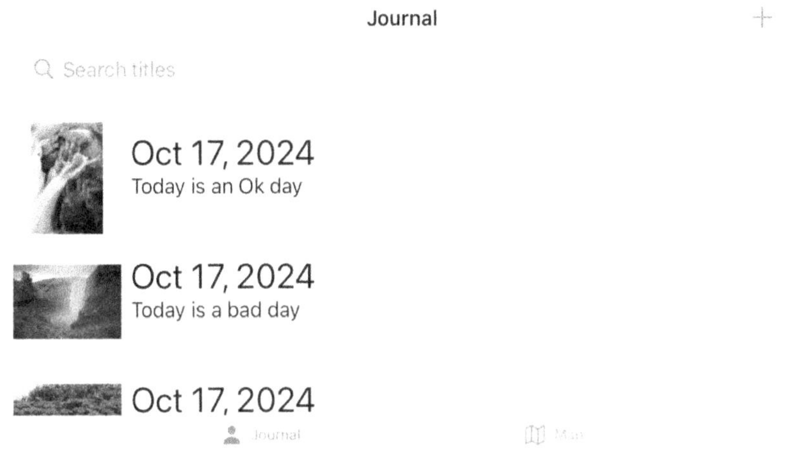

Figure 22.37: Simulator showing iPhone screen rotated with a single column

5. Stop your app and choose **iPhone 16 Pro Max** from the Destination menu. Run your app on Simulator again, and it will display a single column, as shown:

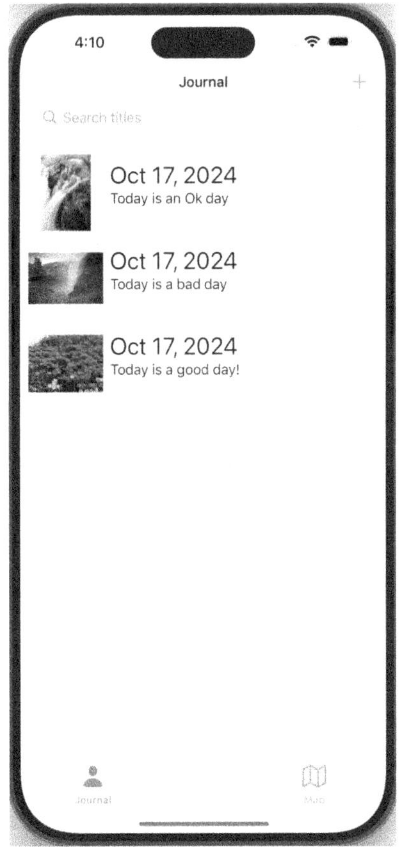

Figure 22.38: Simulator showing iPhone screen with a single column

6. Choose **Rotate Left** from the **Device** menu, and you'll see two columns:

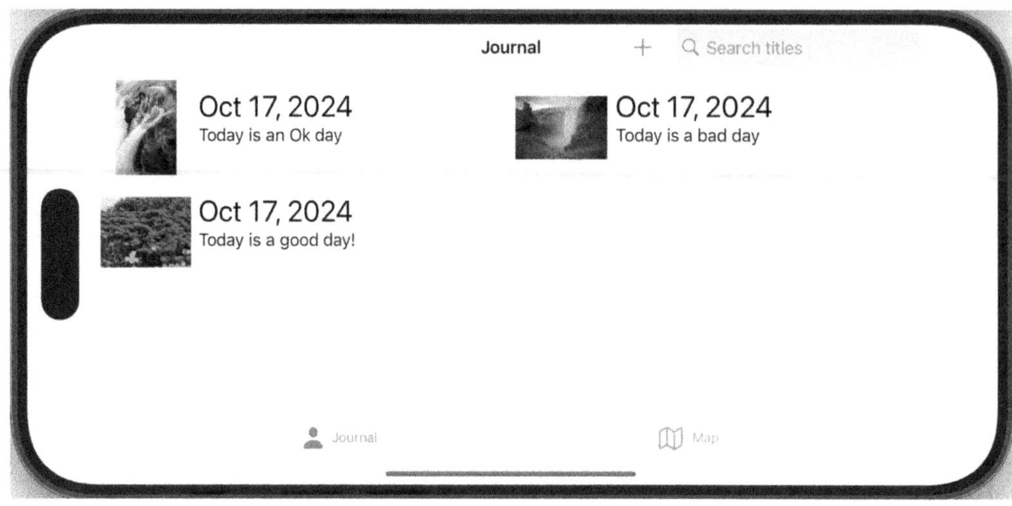

Figure 22.39: Simulator showing iPhone screen rotated with two columns

7. Stop your app and choose **MyMac (Designed for iPad)** in the Destination menu. Run your app and it should display two columns:

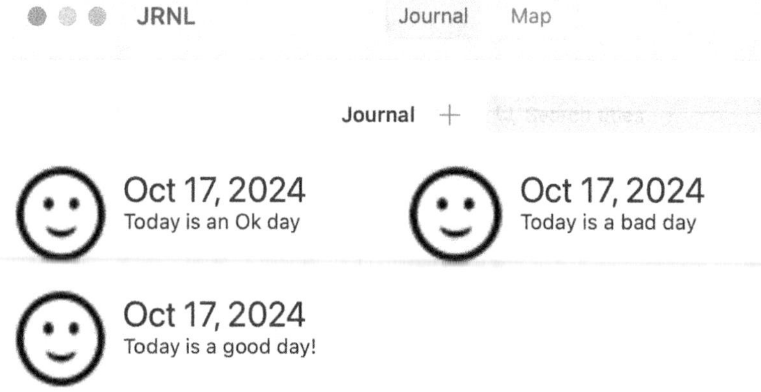

Figure 22.40: Mac app with two columns

You'll need a free or paid Apple Developer account to run your app on your Mac.

At the time of writing, clicking the camera button while the app is running on the Mac will cause the app to crash. To work around this, choose **Product | Scheme | Edit Scheme...** from the menu bar, select **Run** from the sidebar, click the **Diagnostics** tab, and uncheck the **Metal API Validation** check box.

You have modified your app to use a collection view in place of a table view, and you have enabled it to dynamically modify the collection view cell size when it runs on different devices. Great job!

Summary

In this chapter, you replaced the table view on the Journal List screen with a collection view, which makes better use of the extra screen space available when you run your app on an iPad or Mac. You also made your app dynamically modify the number of columns and the collection view cell sizes when your device is rotated using size classes.

First, in the `Main` storyboard file, you replaced the table view on the Journal List screen with a collection view and configured the collection view cell to display the same information the table view cell used to display. Next, you modified the `JournalListTableViewController` and `JournalListTableViewCell` classes to work with the collection view and collection view cells. After that, you added code to dynamically change the collection view cell size to suit the display your app is running on. Finally, you created and tested your app on Simulator and your Mac.

You should now be able to use collection views in your app and know how to dynamically modify your app's interface based on size classes.

This is the end of *Part 3* of this book. You'll learn more about the cool new features Apple introduced during WWDC 2024 in the next part, starting with SwiftUI.

Leave a review!

Thank you for purchasing this book from Packt Publishing—we hope you enjoy it! Your feedback is invaluable and helps us improve and grow. Once you've completed reading it, please take a moment to leave an Amazon review; it will only take a minute, but it makes a big difference for readers like you. Scan the QR code below or visit the link to receive a free ebook of your choice.

`https://packt.link/NzOWQ`

Part 4

Features

Welcome to *Part 4* of this book. In this part, you will implement the latest iOS 18 features. First, you will learn how to persist your app data using Apple's new SwiftData framework. Next, you will learn how to develop SwiftUI apps, a great new way of developing apps for all Apple platforms. After that, you'll learn how to test your code using Swift Testing, and how to implement Apple Intelligence features in your app. Finally, you'll see how to test your app with internal and external testers and upload it to the App Store.

This part comprises the following chapters:

- *Chapter 23, Getting Started with SwiftData*
- *Chapter 24, Getting Started with SwiftUI*
- *Chapter 25, Getting Started with Swift Testing*
- *Chapter 26, Getting Started with Apple Intelligence*
- *Chapter 27, Testing and Submitting Your App to the App Store*

By the end of this part, you'll be able to implement cool iOS 18 features in your own apps. You'll also be able to test and publish your own apps to the App Store. Let's get started!

23

Getting Started with SwiftData

During Apple's **World Wide Developer Conference** (WWDC) in 2023, they introduced **SwiftData**, an all-new framework for saving app data. Previously, developers had to use an editor to create data models, but SwiftData allows developers to describe data models and manipulate model instances using regular Swift code. Features such as relationship management, undo/redo support, iCloud synchronization, and more are provided automatically. In 2024, Apple added new APIs that enable developers to build custom data stores, work with transaction history, model indices and compound uniqueness constraints, and more.

In this chapter, you're going to modify the *JRNL* app that you completed in *Chapter 16, Passing Data between View Controllers,* to save journal entries using SwiftData. This means that when you add new journal entries to the app, they will reappear the next time the app is launched.

First, you'll learn about SwiftData and its components. Next, you'll modify the `JournalEntry` class to make it compatible with SwiftData and modify the `JournalListViewController` class to work with the modified `JournalEntry` class. After that, you'll implement SwiftData by adding code that will allow you to read, write, and delete journal entries; and finally, you'll modify the `JournalViewController` class to read, save, and delete stored journal entries.

By the end of this chapter, you'll have learned how to save app data using SwiftData and will be able to implement it in your own apps.

The following topics will be covered in this chapter:

- Introducing SwiftData
- Modifying the `JournalEntry` class
- Implementing SwiftData components
- Modifying the `JournalListViewController` class

Technical requirements

You will continue working on the `JRNL` project that you modified in *Chapter 16, Passing Data between View Controllers*.

The resource files and completed Xcode project for this chapter are in the `Chapter23` folder of the code bundle for this book, which can be downloaded here:

`https://github.com/PacktPublishing/iOS-18-Programming-for-Beginners-Ninth-Edition`

Check out the following video to see the code in action:

`https://youtu.be/VFYb8Yohh6g`

Let's start by learning more about SwiftData.

Introducing SwiftData

SwiftData is Apple's all-new framework for saving app data to your device. It automatically provides relationship management, undo/redo support, iCloud synchronization, and more. You can model your data using regular Swift types, and SwiftData will then build a custom schema using your specified model and map its fields to device storage. You can query and filter your data using expressions that are type-checked by the compiler, resulting in fewer typos or mistakes.

You can learn more about SwiftData at `https://developer.apple.com/documentation/swiftdata`.

During WWDC 2024, Apple added new APIs to SwiftData. These enable you to model indices and compound unique constraints using macros, build custom data stores using your own document format, keep track of transaction history, and more.

You can learn more about what's new in SwiftData at `https://developer.apple.com/videos/play/wwdc2024/10137/`.

Several steps are required to implement SwiftData for an app. First, existing classes are turned into models with the `@Model` macro. Primitive types such as `Bool`, `Int`, and `String` are supported, as well as complex value types such as structures and enumerations. Next, model attributes are customized as required using annotations such as `@Attribute(.unique)` to ensure the property's value is unique and `@Attribute(.externalStorage)` to store the property's value adjacent to model storage as binary data. Then, models to be persisted are specified and a `ModelContainer` instance is created, which manages an app's schema and model storage configuration.

After that, a `ModelContext` instance is used to fetch, insert, and delete model instances and save any changes to device storage while the app is running. Finally, to fetch specific instances from device storage, a `FetchDescriptor` instance containing a search predicate and a sort order is used.

You can learn more about preserving your app's model data across launches at `https://developer.apple.com/documentation/swiftdata/preserving-your-apps-model-data-across-launches`.

Before you implement SwiftData for the *JRNL* app, here's an example to help you visualize what you need to do to save journal entries.

Imagine you're saving a journal entry using Microsoft Word. You first create a new Word document template with the relevant fields for a journal entry. You then create new Word documents based on the templates and fill in the data. You make whatever changes are necessary, perhaps changing the text of the journal entry, or changing the photo. When you are happy with your document, you save it to the hard disk of your computer. The next time you want to view your journal entry, you search your hard disk for the relevant document and double-click it to open it in Word so you can see it once more.

Now that you have an idea of what you need to do, let's review the steps required to implement SwiftData for your app.

First, you turn the existing `JournalEntry` class into a model, which is like a Microsoft Word template. You do this by annotating the `JournalEntry` class with the `@Model` macro. The properties of the `JournalEntry` class are like fields in the Microsoft Word template and you will customize properties with the `@Attribute` macro if required.

Next, you will create a `ModelContainer` instance, which will be used to store `JournalEntry` model instances on your device's storage, and create a `ModelContext` instance, which will be used to store `JournalEntry` model instances in memory. This is like the way Microsoft Word files created from Microsoft Word templates can be stored on your computer's hard disk or kept in memory as they are being edited.

After that, you will add code so that when you create a new journal entry, a `JournalEntry` model instance is created and added to the `ModelContext` instance, which then coordinates with the `ModelContainer` instance to save it to device storage. This is like saving Word documents to your hard disk when you're done with them.

You can learn more about how to build an app with SwiftData by watching the following video: `https://developer.apple.com/videos/play/wwdc2023/10154/`.

Now that you have a basic understanding of how SwiftData works, you'll use the `JournalEntry` class to create the `JournalEntry` model in the next section.

Modifying the JournalEntry class

Currently, when you create a new journal entry using the Add New Journal Entry screen and click the **Save** button, the entry will appear on the Journal List screen:

Figure 23.1: Simulator showing new entries added to the Journal List screen

If you quit and restart your app, the newly added entries will disappear:

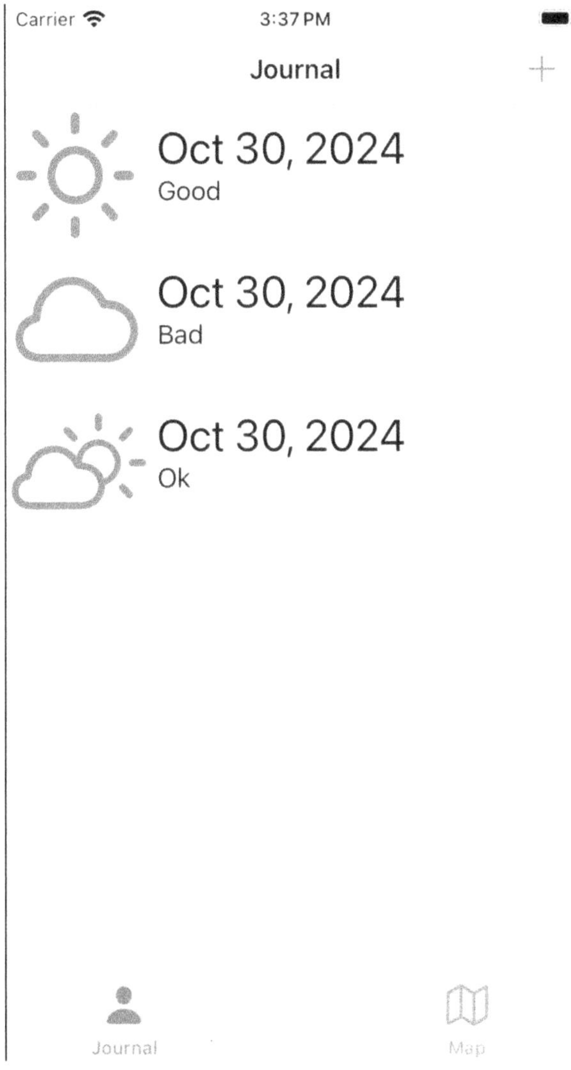

Figure 23.2: Simulator showing new entries disappearing after the app is relaunched

This is because the contents of the `journalEntries` array are only kept in memory and not saved to device storage when the app is closed. To resolve this, you will implement SwiftData for your app. The first step in implementing SwiftData is to create model objects from your existing `JournalEntry` class, modifying this and other classes in your app as needed.

Follow these steps:

1. In the Project navigator, click the **JournalEntry** file (located in the **Journal List Scene** | **Model** group). Import the `SwiftData` framework and annotate the `JournalEntry` class with the `@Model` macro, as shown:

```
import UIKit
import SwiftData
@Model
class JournalEntry {
    //MARK: - Properties
    let date: Date
```

2. Choose **Build** from Xcode's **Product** menu. You will see error messages appear in the Navigator area. Click any one of the first three error messages to expand the macro and display the error:

Figure 23.3: Editor area showing errors

These errors appear because SwiftData at present does not support the `UIImage` class. To fix this, you will need to modify the `JournalEntry` class to use `Data` instances in place of `UIImage` instances.

3. Replace the `photo` property in the `JournalEntry` class with a `photoData` property of type `Data?`, as shown:

```
let body: String
@Attribute(.externalStorage) let photoData: Data?
let latitude: Double?
```

The `@Attribute(.externalStorage)` annotation will store the data for the photo in a binary file adjacent to the model data, which makes it more efficient.

4. Next, you will address the other error messages, which say **Cannot expand accessors on variable declared with 'let'**:

Figure 23.4: Navigator area showing errors

5. To resolve this, replace all the `let` keywords for the `JournalEntry` class properties with `var`, as shown:

```
// MARK: - Properties
var date: Date
var rating: Int
var title: String
var body: String
@Attribute(.externalStorage) var photoData: Data?
var latitude: Double?
var longitude: Double?
```

6. There will be an error in the initializer for the `JournalEntry` class. To fix it, modify the code in the initializer, as shown:

```
self.body = body
self.photoData = photo?.jpegData(compressionQuality: 1.0)
self.latitude = latitude
```

This statement will convert a `UIImage` instance into a `Data` instance using JPEG encoding, which can be stored in the `photoData` property.

7. If you build the project now, more errors will appear. Click the **JournalEntryDetailViewController** file in the Project navigator. Modify the `viewDidLoad()` method as shown:

```
titleLabel.text = selectedJournalEntry?.title
```

```
bodyTextView.text = selectedJournalEntry?.body
if let photoData = selectedJournalEntry?.photoData {
  photoImageView.image = UIImage(data: photoData)
}
```

This code checks to see if the `JournalEntry` instance's `photoData` property has a value. If this is the case, it is converted into a `UIImage` instance and assigned to the `photoImageView` instance's `image` property.

8. Click the **JournalListViewController** file in the Project navigator. Modify the `tableView(_:cellForRowAt:)` method as follows:

```
let journalEntry = journalEntries[indexPath.row]
if let photoData = journalEntry.photoData {
  Task {
    journalCell.photoImageView.image = UIImage(data:
    photoData)
  }
}
journalCell.dateLabel.text = journalEntry.date.formatted(
  .dateTime.month().day().year()
)
```

This code checks to see if the `JournalEntry` instance's `photoData` property has a value. If this is the case, it is converted into a `UIImage` instance and assigned to `journalCell.photoImageView.image`. Since this process is repeated for every row in the table view and decoding JPEG data into a `UIImage` instance can be slow, a `Task` block is used to make this process asynchronous.

All errors should now be resolved. You may need to quit and reopen your project before all the errors disappear.

If you were to build and run your project now, you would get an error because the SwiftData `ModelContainer` instance has not been created. You'll learn how to do that in the next section.

Implementing SwiftData components

Now that you have used the `JournalEntry` class to create a `JournalEntry` model, you will create a **singleton** class that contains a `ModelContainer` object and a `ModelContext` instance. You will then add methods to manipulate instances of the `JournalEntry` model stored in `ModelContext`.

The term **singleton** means that there is only one instance of this class in your app.

Follow these steps:

1. Right-click the `JRNL` folder in the Project navigator and choose **New File from Template...** from the pop-up menu.
2. **iOS** should already be selected. Choose **Swift File** and then click **Next**.
3. Name this file `SharedData`. Click **Create**. The `SharedData` file appears in the Project navigator. Move the file so that it is under the `SceneDelegate` file.
4. Add the following code after the `import` statement:

   ```
   import SwiftData
   ```

   This lets you use the SwiftData framework.

5. Add the following code after the `import` statements to declare and define the `SharedData` class:

   ```
   class SharedData {
     // MARK: - Properties
     @MainActor static let shared = SharedData()
     let container: ModelContainer
     let context: ModelContext

     // MARK: - Initialization
     private init() {
       do {
         container = try ModelContainer(for:
         JournalEntry.self)
         context = ModelContext(container)
       } catch {
         fatalError("Could not create SwiftData model
         container or context")
       }
     }
   }
   ```

   This class creates a singleton instance that will be available throughout your app and assigns it to the `shared` static variable. It also creates and initializes `ModelContainer` and `ModelContext` instances and assigns them to the `container` and `context` properties, respectively.

Next, you'll add methods for loading, adding, and deleting `JournalEntry` model instances. Follow these steps:

1. Add the following code after the initializer to implement the `loadJournalEntries()` method:

   ```
   func loadJournalEntries() -> [JournalEntry] {
     let descriptor = FetchDescriptor<JournalEntry>(sortBy:
   ```

```
    [SortDescriptor<JournalEntry>(\.date, order: .reverse)])
    do {
      let journalEntries = try context.fetch(descriptor)
      return journalEntries
    } catch {
      return []
    }
}
```

Let's break this down:

```
func loadJournalEntries() -> [JournalEntry] {
```

This method returns an array of `JournalEntry` instances.

```
let descriptor = FetchDescriptor<JournalEntry>(sortBy:
[SortDescriptor<JournalEntry>(\.date, order: .reverse)])
```

This statement creates a `FetchDescriptor` instance that specifies that all `JournalEntry` model instances stored in the `ModelContext` instance are to be fetched and sorted by date, from newest to oldest.

```
do {
  let journalEntries = try context.fetch(descriptor)
  return journalEntries
} catch {
  return []
  }
}
```

This block of code gets all the `JournalEntry` model instances specified by the `FetchDescriptor` instance from the `ModelContext` instance and assigns them to `journalEntries`, a constant of type `[JournalEntry]`, which is then returned. If the operation fails, an empty array will be returned.

2. Implement the `saveJournalEntry(_:)` method by adding the following code before the final curly brace:

   ```
   func saveJournalEntry(_ journalEntry: JournalEntry) {
     context.insert(journalEntry)
     try? context.save()
   }
   ```

   This method inserts the `journalEntry` instance passed to it into the `ModelContext` instance as a `JournalEntry` model instance.

3. Implement the `deleteJournalEntry(_:)` method by adding the following code before the final curly brace:

   ```
   func deleteJournalEntry(_ journalEntry: JournalEntry) {
     context.delete(journalEntry)
     try? context.save()
   }
   ```

   This method removes the corresponding `JournalEntry` model instance from the `ModelContext` instance.

You've created a `SharedData` class that creates a `ModelContainer` instance and implemented methods to fetch, add, and delete `JournalEntry` model instances from a `ModelContext` object. You can build your app to test for errors, but you can't run it yet.

At this point, you've implemented all the required SwiftData components in your app. In the next section, you'll configure the `JournalListViewController` class to fetch all journal entries stored in the `ModelContainer` instance when the app is run, add new `JournalEntry` model instances to the `ModelContext` instance when you add a new journal entry, and remove `JournalEntry` model instances from the `ModelContext` instance when you delete a journal entry.

Modifying the JournalListViewController class

Previously, when you added or removed journal entries from your app, the changes made would be gone when you stopped and ran the app again because the code in the `JournalListViewController` class had no way to save app data to your device. You will add code to the `JournalListViewController` class to save app data using SwiftData in this section.

You will update the `viewDidLoad()` method to fetch all journal entries from device storage when you run the app, update the `unwindNewEntrySave(segue:)` method to add a new journal entry to the `ModelContext` instance, and update the `tableView(_:commit:forRowAt:)` method to remove the specified journal entry from the `ModelContext` instance. Follow these steps:

1. Click the `JournalListViewController` file in the Project navigator and add a method to fetch all journal entries from device storage before the closing curly brace:

   ```
   func fetchJournalEntries() {
     journalEntries =
     SharedData.shared.loadJournalEntries()
     tableView.reloadData()
   }
   ```

   This method calls the `loadJournalEntries()` method in the `SharedData` singleton, which returns an array of `JournalEntry` instances. This array is then assigned to the `journalEntries` array, and the table view is reloaded.

2. Modify the `viewDidLoad()` method as follows to call the `fetchJournalEntries()` method:

```
override func viewDidLoad() {
  super.viewDidLoad()
  fetchJournalEntries()
}
```

Since the journal entries you add to the app will now be persistent, you no longer need to call the method used to create sample data for your app.

3. Modify the `unwindNewEntrySave(segue:)` method to add the new journal entry to device storage and update the table view:

```
@IBAction func unwindNewEntrySave(segue: UIStoryboardSegue) {
  if let sourceViewController = segue.source as?
  AddJournalEntryViewController, let newJournalEntry =
  sourceViewController.newJournalEntry {
    SharedData.shared.saveJournalEntry(newJournalEntry)
    fetchJournalEntries()
  }
}
```

The new `JournalEntry` instance is passed to the `saveJournalEntry` method in the `SharedData` singleton, where it is inserted into the `ModelContext` instance as a `JournalEntry` model instance. The `fetchJournalEntries()` method then updates the `journalEntries` array and reloads the table view.

4. Modify the `tableView(_:commit:forRowAt:)` method to delete the specified journal entry from device storage and update the table view:

```
func tableView(_ tableView: UITableView, commit editingStyle:
UITableViewCell.EditingStyle, forRowAt indexPath: IndexPath) {
  if editingStyle == .delete {
    SharedData.shared.deleteJournalEntry(
    journalEntries[indexPath.row])
    fetchJournalEntries()
  }
}
```

The `JournalEntry` instance to be removed is passed to the `deleteJournalEntry` method in the `SharedData` singleton, where the corresponding `JournalEntry` model instance is removed from the `ModelContext` instance. The `fetchJournalEntries()` method then updates the `journalEntries` array and reloads the table view.

5. You have made all the changes required for the `JournalListViewController` class. Build and run your app, and you should see a blank table view on the Journal List screen. Add a few journal entries using the Add New Journal Entry screen, and they will appear on the Journal List screen. Stop and run your app again. The journal entries you added will still be there:

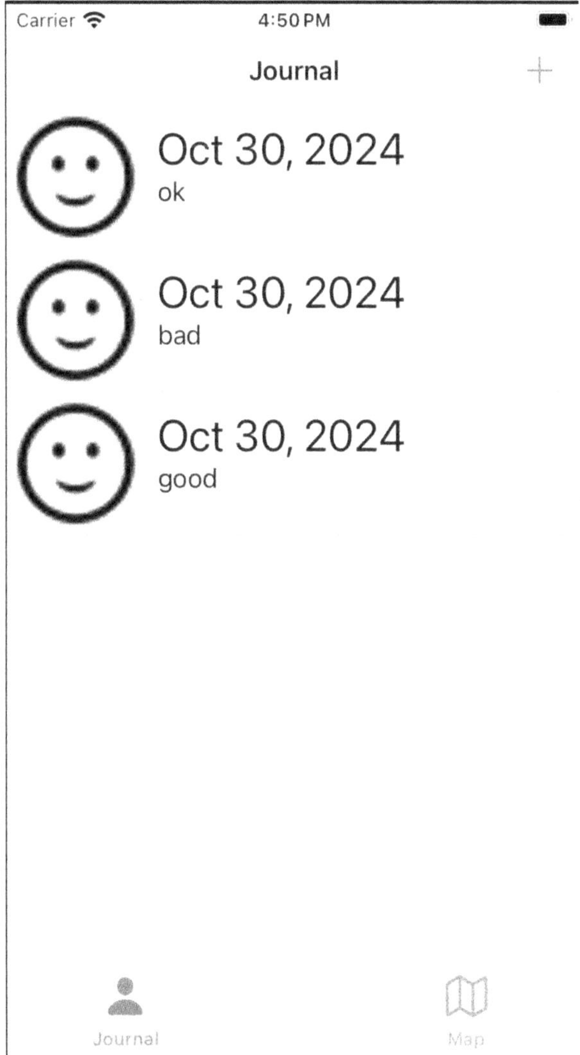

Figure 23.5: Simulator showing that new entries are still present after your app has been relaunched

6. Swipe a row to delete a journal entry. Stop and run your app again. The journal entry you removed will still be gone.

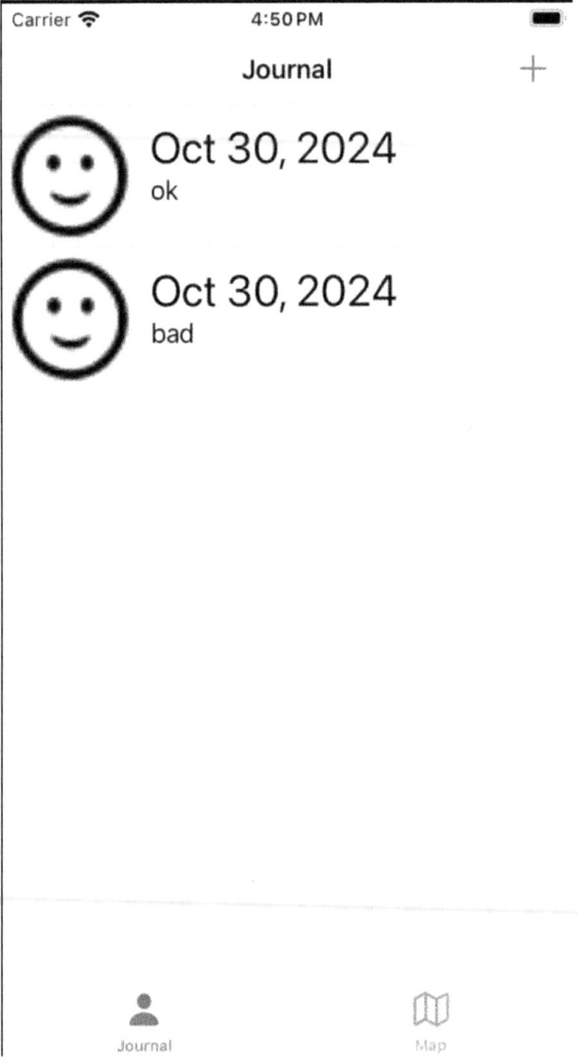

Figure 23.6: Simulator showing that deleted entries do not reappear after your app has been relaunched

You have successfully implemented SwiftData in your app, and now journal entries added to your app will still be there when your app is relaunched. Congratulations!

Summary

In this chapter, you modified the *JRNL* app that you completed in *Chapter 16*, *Passing Data between View Controllers*, to use SwiftData to save journal entries to your device storage, so that any changes you make will still be there when you next launch the app. First, you learned about SwiftData and its different components. Next, you modified the `JournalEntry` class to make it work with SwiftData and modified your `JournalListViewController` class to work with the modified `JournalEntry` class. After that, you added code that allowed you to fetch, add, and delete journal entries from a SwiftData model container, and finally, you modified the `JournalViewController` class so it can read, save, and delete stored journal entries.

You now have a basic understanding of how SwiftData works, and you will now be able to write your own apps that use SwiftData to save app data. Great job!

In the next chapter, you'll learn about the latest developments in **SwiftUI**.

Join us on Discord!

Read this book alongside other users, experts, and the author himself. Ask questions, provide solutions to other readers, chat with the author via Ask Me Anything sessions, and much more. Scan the QR code or visit the link to join the community.

`https://packt.link/ios-Swift`

24

Getting Started with SwiftUI

In previous chapters, you created the **user interface** (**UI**) for the *JRNL* app using storyboards. The process involved dragging objects representing views to a storyboard, creating outlets in view controller files, and connecting the two together.

This chapter will focus on **SwiftUI**, an easy and innovative way to create apps across all Apple platforms. Instead of specifying the user interface using storyboards, SwiftUI uses a declarative Swift syntax and works with new Xcode design tools to keep your code and design in sync. Features such as Dynamic Type, dark mode, localization, and accessibility are automatically supported.

Even though this book focuses on UIKit, a working knowledge of SwiftUI is beneficial as some iOS features, such as widgets, can only be implemented using SwiftUI. It also appears that SwiftUI is the way ahead for app development for all Apple platforms, but at present it does not have feature parity with UIKit.

In this chapter, you will build a simplified version of the *JRNL* app using SwiftUI. This app will contain just the Journal List and Journal Entry Detail screens. Since writing apps with SwiftUI is very different from what you have already done, you will not be modifying the *JRNL* project you have been working on thus far. You will create a new SwiftUI Xcode project instead.

You'll start by adding and configuring SwiftUI views to create the Journal List screen. Next, you'll add model objects to your app, and configure the navigation between the Journal List and Journal Entry Detail screens. After that, you'll learn how to use MapKit to build a map view for the Journal Entry Detail screen. Finally, you'll create the Journal Entry Detail screen.

By the end of this chapter, you'll have learned how to build a SwiftUI app that reads model objects, presents them in a list, and allows navigation to a second screen containing a map view. You will then be able to implement this functionality in your own projects.

The following topics will be covered:

- Creating a SwiftUI Xcode project
- Creating the Journal List screen
- Adding model objects and configuring navigation

- Using MapKit for SwiftUI
- Creating the Journal Entry Detail screen

Technical requirements

You will create a new SwiftUI Xcode project for this chapter.

The resource files and completed Xcode project for this chapter are in the `Chapter24` folder of the code bundle for this book, which can be downloaded here:

`https://github.com/PacktPublishing/iOS-18-Programming-for-Beginners-Ninth-Edition`

Check out the following video to see the code in action:

`https://youtu.be/VIbBcmHmf8k`

Let's start by creating a new SwiftUI Xcode project for your SwiftUI app in the next section.

Creating a SwiftUI Xcode project

A SwiftUI Xcode project is created in the same way as a regular Xcode project, but you configure it to use SwiftUI instead of storyboards to create the user interface. As you will see, the user interface is generated entirely in code, and you'll be able to see changes in the user interface immediately as you modify your code.

You can watch a video of Apple's SwiftUI presentation during WWDC20 at `https://developer.apple.com/videos/play/wwdc2020/10119`.

You can watch a video showing what's new in SwiftUI during WWDC24 at `https://developer.apple.com/videos/play/wwdc2024/10144/`.

Apple's official SwiftUI documentation can be found online at `https://developer.apple.com/xcode/swiftui/`.

Let's begin by creating a new SwiftUI Xcode project. Follow these steps:

1. Launch Xcode and create a new Xcode project.
2. Click **iOS**. Select the **App** template, and then click **Next**.

3. The **Choose options for your new project:** screen appears:

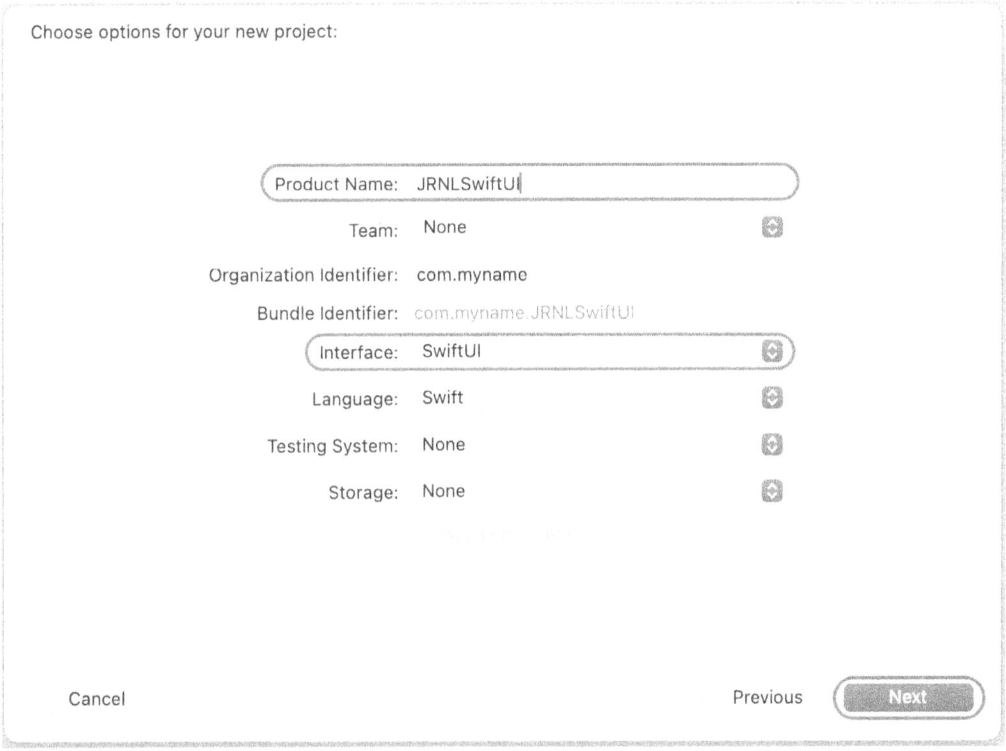

Figure 24.1: Project options screen

Configure this screen as follows:

- **Product Name:** `JRNLSwiftUI`
- **Interface:** `SwiftUI`

The other settings should already be set. Click **Next** when done.

4. Choose a location to save the `JRNLSwiftUI` project and click **Create**.

5. Your project appears on the screen, with the `ContentView` file selected in the Project navigator. You'll see the content of this file on the left side of the Editor area, and a canvas containing a preview on the right side:

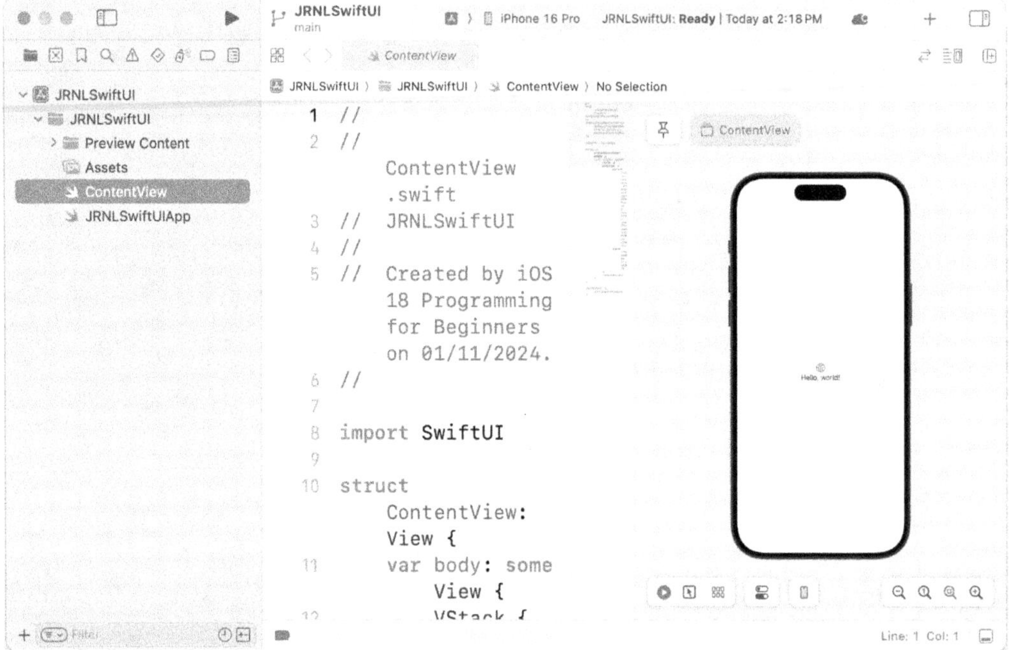

Figure 24.2: Xcode showing JRNLSwiftUI project

6. If you see a **Preview paused** box in the canvas, click the circular arrow to display the preview:

Figure 24.3: Preview paused box with circular arrow

7. The `ContentView` file contains code that declares and defines the initial view for your app. If you need more room to work, click the Navigator button to hide the Navigator, and drag the border in the Editor area to resize the canvas:

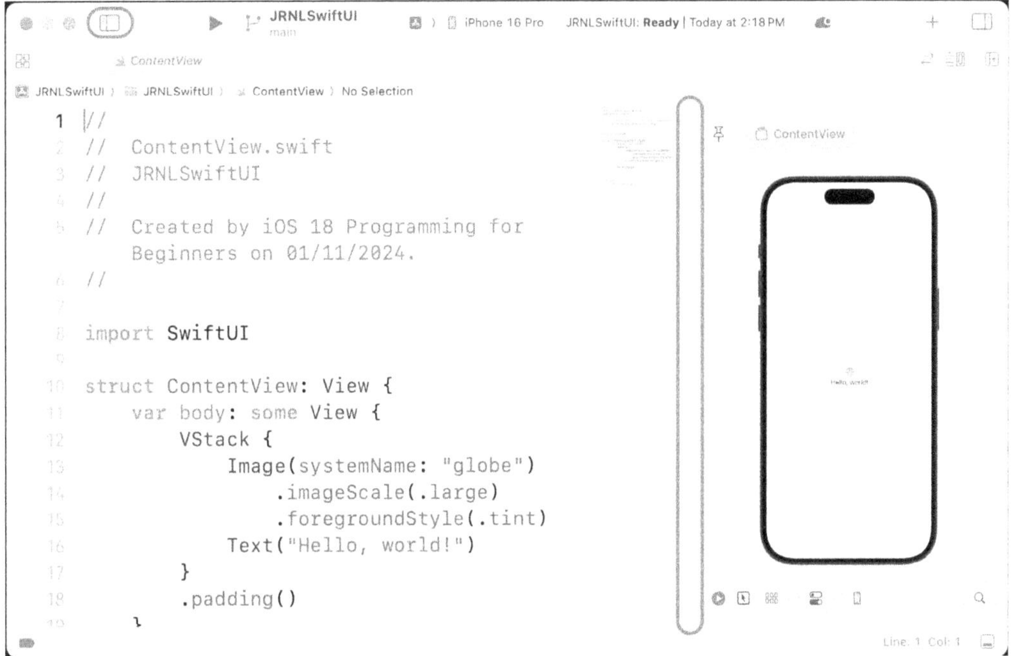

Figure 24.4: Xcode interface showing Navigator button and border in the Editor area

You have successfully created your first SwiftUI Xcode project! Great! Now you'll see how changing the code in the Editor area will update the preview on the canvas.

Let's look at the `ContentView` file. This file contains a `ContentView` structure and a `#Preview` macro. The `ContentView` structure describes the view's content and layout and conforms to the `View` protocol. The `#Preview` macro generates source code that declares a preview for the `ContentView` structure, which is displayed on the canvas.

To view the code generated by a macro, right-click the macro and choose **Expand Macro** from the pop-up menu.

To see this in action, change the `Hello, World!` text to `JRNL` as shown:

```
struct ContentView: View {
  var body: some View {
    VStack {
      Image(systemName: "globe")
        .imageScale(.large)
        .foregroundStyle(.tint)
      Text("JRNL")
    }
    .padding()
  }
}
```

The preview in the canvas updates to reflect your changes:

Figure 24.5: Canvas showing app preview with updated text view

In the next section, you'll create the Journal List screen, starting with a view that will display the data of a particular journal entry.

Creating the Journal List screen

When using storyboards, you modify attributes of a view using the Attributes inspector. In SwiftUI, you can modify either your code or the preview in the canvas. As you have seen, changing the code in the `ContentView` file will immediately update the preview, and modifying the preview will update the code.

Let's customize the `ContentView` structure to display the data of a particular restaurant. Follow these steps:

1. Click the Library button. Type `tex` in the filter field, then drag a **Text** view to the Editor area and drop it under the text view containing the "JRNL" string:

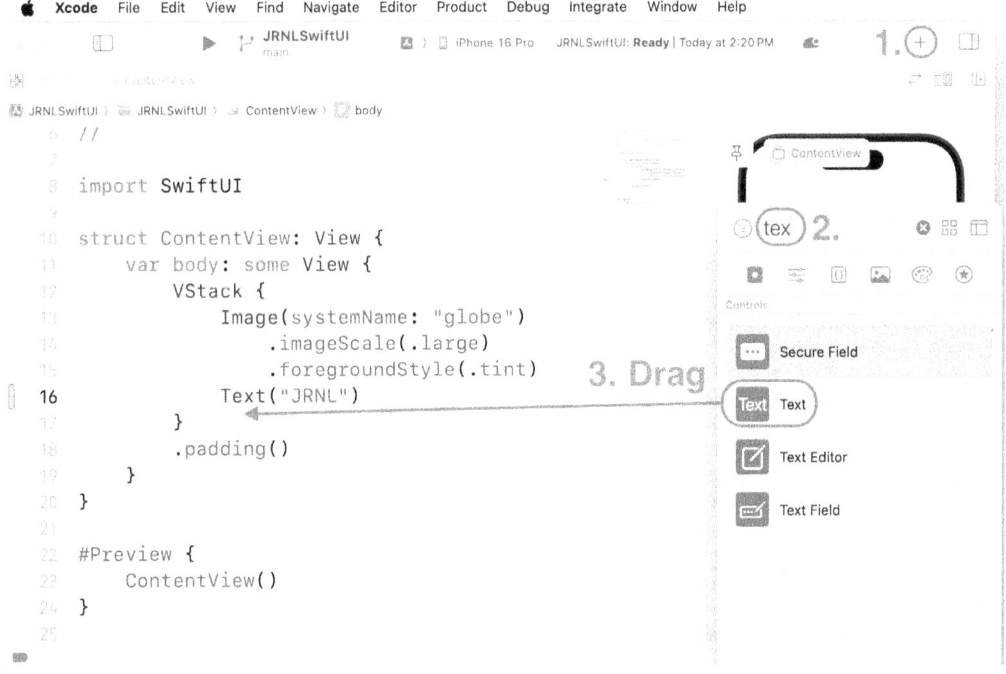

Figure 24.6: Library with Text object to be dragged

2. Xcode has automatically added code to the `ContentView` file for this text view. Verify that your code looks like this:

```
struct ContentView: View {
  var body: some View {
    VStack {
      Image(systemName: "globe")
        .imageScale(.large)
        .foregroundStyle(.tint)
      Text("JRNL")
      Text("Placeholder")
    }
    .padding()
  }
}
```

As you can see, a second text view has been added after the text view containing the "JRNL" string, and both text views and an image view are enclosed in a VStack view. A VStack view contains subviews that are arranged vertically, and it is like a vertically oriented stack view in a storyboard. Note that the image view has a systemName property. This property can be set to one of the images in Apple's **SF Symbols** library.

You can learn more about the SF Symbols library here: https://developer.apple.com/sf-symbols/.

3. Right-click on the VStack view and choose **Embed in HStack** from the pop-up menu.

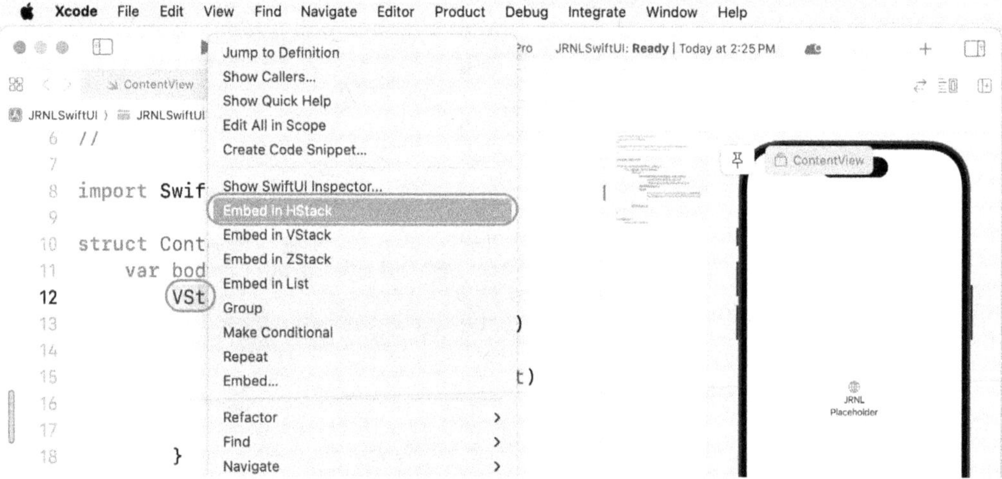

Figure 24.7: Pop-up menu showing Embed in HStack

4. Verify that your code looks like this:

```
struct ContentView: View {
  var body: some View {
    HStack {
      VStack {
        Image(systemName: "globe")
          .imageScale(.large)
          .foregroundStyle(.tint)
        Text("JRNL")
        Text("Placeholder")
      }
      .padding()
```

```
        }
      }
    }
```

As you can see, the `VStack` view is now enclosed in an `HStack` view. An `HStack` view contains subviews that are arranged horizontally, and it is like a horizontally oriented stack view in a storyboard.

5. Modify the code as shown here to display a sample journal entry and to reposition the image view to the left of the two text views:

```
struct ContentView: View {
  var body: some View {
    HStack {
      Image(systemName: "face.smiling")
        .imageScale(.large)
        .foregroundStyle(.tint)
      VStack {
        Text("18 Aug 2024")
        Text("Today is a good day")
      }
      .padding()
    }
  }
}
```

6. Verify that the changes are reflected in the preview:

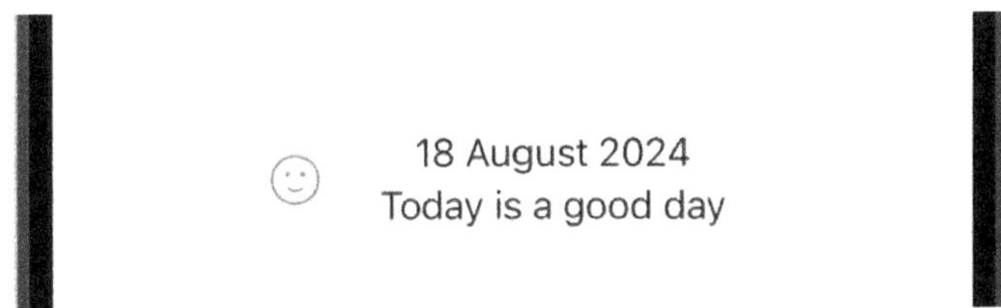

Figure 24.8: App preview showing sample journal entry

7. To change the appearance of user interface elements, you use **modifiers** instead of the Attributes inspector. These are methods that change how your objects look or behave. Note that the image view already has modifiers. Update your code as shown here to set the style and color of your text views and set the size for the image view:

```
struct ContentView: View {
  var body: some View {
    HStack {
      Image(systemName: "face.smiling")
```

```
        .resizable()
        .frame(width: 90, height: 90)
      VStack {
        Text("18 Aug 2023")
          .font(.title)
          .fontWeight(.bold)
          .frame(maxWidth: .infinity, alignment: .leading)
        Text("Today is a good day")
          .font(.title2)
          .foregroundStyle(.secondary)
          .frame(maxWidth: .infinity, alignment: .leading)
      }.padding()
    }.padding()
  }
}
```

8. Verify that the changes are reflected in the preview:

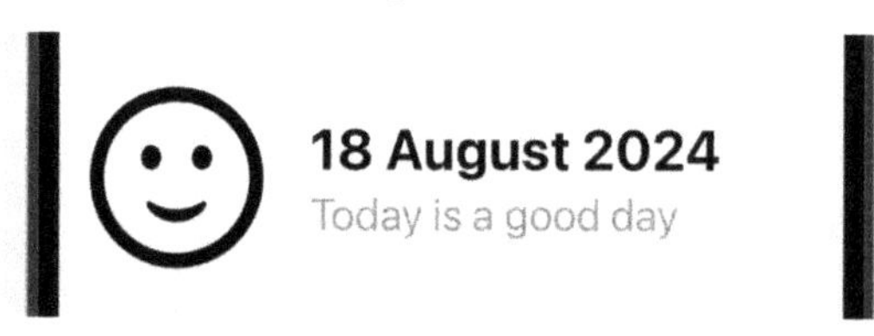

Figure 24.9: App preview showing sample journal entry

Your view is now complete. You will use this view as a cell on the Journal List screen in the next section.

Adding model objects and configuring navigation

You now have a view that can be used to represent a journal entry on the Journal List screen. You'll use this view as a cell in a SwiftUI list, which is a container that presents data in a single column. You'll also configure model objects to populate this list. Follow these steps:

1. Right-click on the `HStack` view and choose **Embed in VStack** from the pop-up menu. This keeps all the views together when you embed them in a list.
2. Right-click on the outer `VStack` view and choose **Embed in List** to display a list containing five cells in the canvas. Remove the padding modifiers as well.
3. Verify that your code now looks like this:

```
struct ContentView: View {
  var body: some View {
    List(0 ..< 5) { item in
      VStack {
        HStack {
          Image(systemName: "face.smiling")
```

```
                .resizable()
                .frame(width: 90, height: 90)
              VStack {
                Text("18 Aug 2024")
                  .font(.title)
                  .fontWeight(.bold)
                  .frame(maxWidth: .infinity,
                  alignment: .leading)
                Text("Today is a good day")
                  .font(.title2)
                  .foregroundStyle(.secondary)
                  .frame(maxWidth: .infinity,
                  alignment: .leading)
              }
            }
          }
        }
      }
    }
```

As you can see in the canvas, the view you created in the previous section is now enclosed in a list configured to display five items. Note that no delegates and data sources are required to display data in the list.

4. Open the `resources` folder contained in the `Chapter24` folder of the code bundle you downloaded from `https://github.com/PacktPublishing/iOS-18-Programming-for-Beginners-Ninth-Edition`. Drag the `JournalEntry` file to the Project navigator and click **Finish** when prompted to add it to your project.
5. Click the `JournalEntry` file in the Project navigator and you should see the following code inside it:

```
import UIKit
struct JournalEntry: Identifiable, Hashable {
  // MARK: - Properties
  let id = UUID()
  let date = Date()
  let rating: Int
  let entryTitle: String
  let entryBody: String
  let photo: UIImage?
  let latitude: Double?
  let longitude: Double?
}
```

```
//MARK: - Sample data
let testData = [
  JournalEntry(rating: 5, entryTitle: "Today is a good
  day", entryBody: "I got top marks in my exam today!
  Great!", photo: UIImage(systemName: "sun.max"),
  latitude: 37.3346, longitude: -122.0090),
  JournalEntry(rating: 0, entryTitle: "Today is a bad
  day", entryBody: "I wasn't feeling very well today.",
  photo: UIImage(systemName: "cloud"), latitude: nil,
  longitude: nil),
  JournalEntry(rating: 3, entryTitle: "Today is an OK
  day", entryBody: "Just having a nice lazy day at home",
  photo: UIImage(systemName: "cloud.sun"), latitude: nil,
  longitude: nil)
]
```

The `JournalItem` file contains a structure, `JournalItem`, and an array, `testData`.

The `JournalItem` structure is like the `JournalItem` class that you used in your `JRNL` project. To use this structure in a list, you must make it conform to the `Identifiable` protocol. This protocol specifies that a list item must have an `id` property that can identify a particular item. A `UUID` instance is assigned to each `JournalEntry` instance upon creation to ensure the value stored in each `id` property is unique.

Note that this structure also conforms to the `Hashable` protocol. This will be used later to determine the data to be displayed when you tap on a cell.

`testData` is an array containing three `JournalItem` instances, which you will use to populate the Journal List screen.

You can learn more about the Identifiable protocol at this link: `https://developer.apple.com/documentation/swift/identifiable`.

You can learn more about the Hashable protocol at this link: `https://developer.apple.com/documentation/swift/hashable`.

6. Click the `ContentView` file in the Project navigator. Add a `journalEntries` property to your view and assign the `testData` array to it, after the opening curly brace of the `ContentView` structure:

```
struct ContentView: View {
  var journalEntries = testData
  var body: some View {
```

7. Modify your code as shown here to display a journal entry's photo, date, and title in each view:

```
struct ContentView: View {
  var journalEntries = testData
```

```
    var body: some View {
      List(journalEntries) { journalEntry in
        VStack {
          HStack {
            Image(uiImage: journalEntry.photo ?? UIImage(
              systemName: "face.smiling")!)
              .resizable()
              .frame(width: 90, height: 90)
            VStack {
              Text(journalEntry.date.formatted(
              .dateTime.day().month().year()))
                .font(.title)
                .fontWeight(.bold)
                .frame(maxWidth: .infinity,
                alignment: .leading)
              Text(journalEntry.entryTitle)
                .font(.title2)
                .foregroundStyle(.secondary)
                .frame(maxWidth: .infinity,
                alignment: .leading)
            }
          }
        }
      }
    }
  }
```

Let's see how this code works.

The `ContentView` structure stores an array of `JournalEntry` instances in the `journalEntries` property. This array is passed to the list. For every item in the `journalEntries` array, a view is created and assigned with data from the item's properties.

The image for each journal entry is converted from a `UIImage` instance stored in `journalEntry.photo`, and a default value is provided if the `photo` property is `nil`. The date is converted into a text string using the `formatted()` method.

Since there are three items in the array, three `VStack` views appear in the canvas.

When you make major changes to your code, the automatic updating of the canvas is paused. Click the circular arrow in the **Preview paused** box to resume if required.

Next, you'll implement navigation so that when a cell is tapped, a second screen is presented that will show details of a particular journal entry. Follow these steps:

1. Right-click the `List` view, choose **Embed...** from the pop-up menu, and replace the placeholder text with `NavigationStack`.
2. Verify that your code now looks like this:

```
struct ContentView: View {
  var journalEntries = testData
  var body: some View {
    NavigationStack {
      List(journalEntries) { journalEntry in
```

   In the preceding code, the navigation stack works like an instance of the `UINavigationController` class, which you've used before in your app.

3. Add a `navigationTitle()` modifier at the location shown here to set the list view's `title` property to display `Journal List` at the top of the screen:

```
              alignment: .leading
              }
            }
          }
        }.navigationTitle("Journal List")
      }
    }
}
```

4. Embed the cell in a navigation link view as shown here and add a `.navigationDestination(for:destination:)` modifier after the `.navigationTitle()` modifier to display the title of the journal entry in a new screen when the `VStack` view is tapped:

```
List(journalEntries) { journalEntry in
  NavigationLink(value: journalEntry) {
    VStack {
      HStack {
        Image(uiImage: journalEntry.photo ?? UIImage(
          systemName: "face.smiling")!)
          .resizable()
          .frame(width: 90, height: 90)
        VStack {
          Text(journalEntry.date.formatted(
          .dateTime.day().month().year()))
```

```
                .font(.title)
                .fontWeight(.bold)
                .frame(maxWidth: .infinity,
                alignment: .leading)
              Text(journalEntry.entryTitle)
                .font(.title2)
                .foregroundStyle(.secondary)
                .frame(maxWidth: .infinity,
                alignment: .leading)
            }
          }
        }
      }.navigationTitle("Journal List")
       .navigationDestination(for:
         JournalEntry.self) {
           journalEntry in
           Text(journalEntry.entryTitle)
         }
    }
```

5. Note that the list in the canvas has automatically displayed disclosure arrows:

Figure 24.10: App preview showing disclosure arrows

To see this working as it should in an app, make sure the Live Preview button in the canvas is selected:

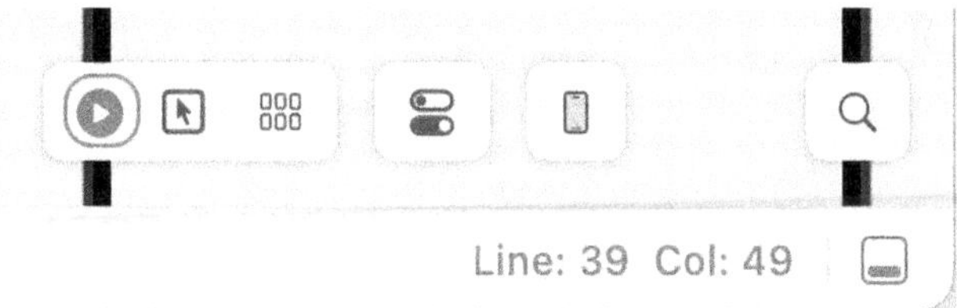

Figure 24.11: Canvas showing Live Preview button

Click on any cell in the preview to display text containing the title of the tapped journal entry:

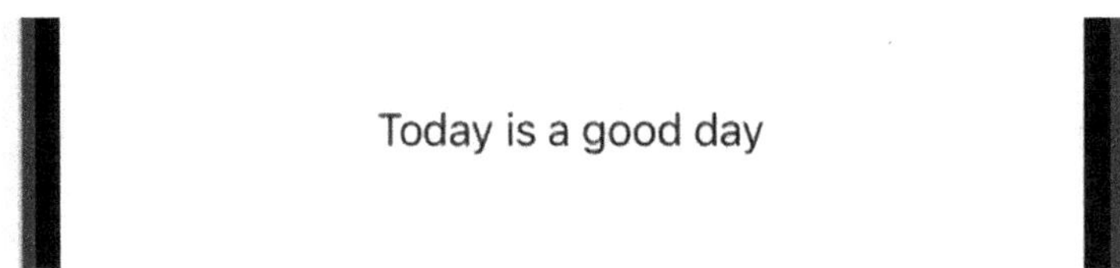

Figure 24.12: App preview showing the title of a tapped journal entry

This is a great way of ensuring your list works as expected.

6. The view code is starting to look cluttered, so you'll extract the `VStack` view into its own separate view. Right-click the `NavigationLink` view and choose **Extract Subview** from the pop-up menu.
7. Verify that all the view code for the `VStack` view has been moved into a separate view named `ExtractedView`. Your code will now look like this:

```
struct ContentView: View {
  var journalEntries = testData
  var body: some View {
    NavigationStack {
      List(journalEntries) { journalEntry in
        ExtractedView()
      }.navigationTitle("Journal List")
        .navigationDestination(for:
          JournalEntry.self) {
            journalEntry in
              Text(journalEntry.entryTitle)
          }
    }
  }
}

#Preview {
```

```
        ContentView()
    }

    struct ExtractedView: View {
      var body: some View {
        NavigationLink(value: journalEntry) {
```

8. Right-click the `ExtractedView` view and choose **Refactor | Rename** from the pop-up menu.
9. Change the name of the extracted view to `JournalCell` and click **Rename** when done.

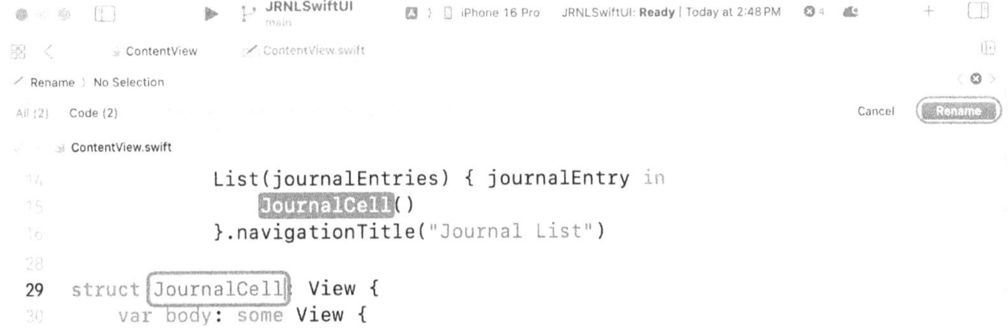

Figure 24.13: ExtractedView renamed to JournalCell

10. Verify that your code now looks like this:

```
    struct ContentView: View {
      var journalEntries = testData
      var body: some View {
        NavigationStack {
          List(journalEntries) { journalEntry in
            JournalCell()
          }.navigationTitle("Journal List")
             .navigationDestination(for:
               JournalEntry.self) {
                   journalEntry in
                       Text(journalEntry.entryTitle)
               }
        }
      }
    }

    #Preview {
        ContentView()
    }
```

```
struct JournalCell: View {
  var body: some View {
    NavigationLink(value: journalEntry) {
```

Don't worry about the error; you'll fix it in the next two steps.

11. Add a property to the `JournalCell` view to hold a `JournalEntry` instance:

```
struct JournalCell: View {
  var journalEntry: JournalEntry
```

12. Add code to the `ContentView` structure to pass the `JournalEntry` instance to the `JournalCell` view as shown:

```
struct ContentView: View {
  var journalEntries: = testData
  var body: some View {
    NavigationStack {
      List(journalEntries) { journalEntry in
        JournalCell(journalEntry: journalEntry)
      }.navigationTitle("Journal List")
          .navigationDestination(for: JournalEntry.self) {
              journalEntry in
              Text(journalEntry.entryTitle)
          }
    }
  }
}
```

13. Verify that the preview still works the way it did before.

With that, you've completed the implementation of the Journal List screen. Cool! In the next section, you'll see how you can use **MapKit for SwiftUI** to create a map view that you'll use in the Journal Entry Detail screen.

Using MapKit for SwiftUI

During WWDC23, Apple introduced expanded MapKit support for SwiftUI, which makes it easier than ever to integrate Maps into your app. Using SwiftUI, you can easily add annotations and overlays to a map, control the camera, and more.

To watch Apple's Meet MapKit for SwiftUI video from WWDC23, refer to this link: `https://developer.apple.com/videos/play/wwdc2023/10043/`.

At this point, you have created the Journal List screen, and tapping each cell on this screen displays the journal entry's title on a second screen. You'll modify your app to display a Journal Entry Detail screen when a cell on the Journal List screen is tapped, but before that, you'll create a SwiftUI view that displays a map.

Follow these steps:

1. Choose **File** | **New** | **File from Template...** to open the template selector.
2. iOS should already be selected. In the **User Interface** section, click **SwiftUI View** and click **Next**.
3. Name the new file `MapView` and click **Create**. The `MapView` file will appear in the Project navigator.
4. In the `MapView` file, import `MapKit` and replace the `Text` view with a `Map` view:

```
import SwiftUI
import MapKit

struct MapView: View {
  var body: some View {
    Map()
  }
}
```

5. Verify that a map is displayed in the canvas:

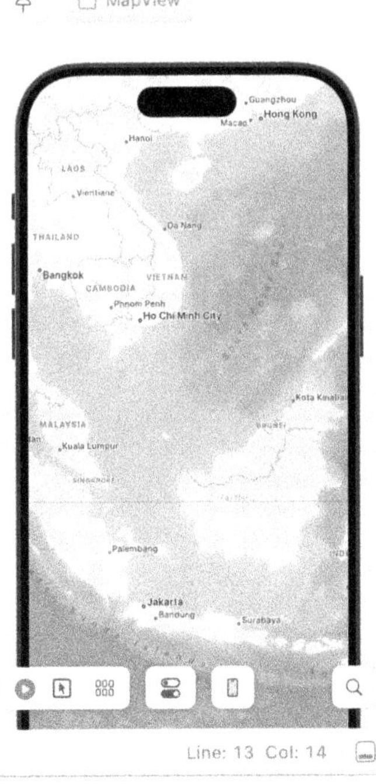

Figure 24.14: Canvas displaying a map

6. Add a `journalEntry` property of type `JournalEntry` to the `MapView` structure:

```
struct MapView: View {
    var journalEntry: JournalEntry
    var body: some View {
        Map()
    }
}
```

7. Modify the `#Preview` macro as shown to assign a journal entry from the `testData` array to the `journalEntry` property:

```
#Preview {
    MapView(journalEntry: testData[0])
}
```

8. Add a `Marker` instance to the `Map` view as shown:

```
Map() {
  Marker(journalEntry.entryTitle, coordinate:
  CLLocationCoordinate2D(latitude:
  journalEntry.latitude ?? 0.0, longitude:
  journalEntry.longitude ?? 0.0))
}
```

The values for the `title` and `coordinate` properties of the `Marker` instance are obtained from the `journalEntry` instance's `entryTitle`, `latitude`, and `longitude` properties, and the Marker instance's `coordinate` property will determine the center point of the map region to be displayed.

9. The map is currently zoomed all the way in. To set the zoom level, add the following code to the **Map** view:

```
Map(bounds: MapCameraBounds(minimumDistance: 4500)) {
  Marker(journalEntry.entryTitle, coordinate:
  CLLocationCoordinate2D(latitude:
  journalEntry.latitude ?? 0.0, longitude:
  journalEntry.longitude ?? 0.0))
}
```

10. Verify that the map is currently displaying a map of Apple Park:

Figure 24.15: Canvas displaying a map of Apple Park

You've created a SwiftUI map view that shows a journal entry location. Now, let's see how to make the complete Journal Entry Detail screen in the next section.

Completing the Journal Entry Detail screen

You now have a SwiftUI map view displaying a map. Now, you'll create a new SwiftUI view to represent the Journal Entry Detail screen and add the map view to it. Follow these steps:

1. Choose **File** | **New** | **File from Template...** to open the template selector.
2. **iOS** should already be selected. In the **User Interface** section, click **SwiftUI View** and click **Next**.
3. Name the new file `JournalEntryDetail` and click **Create**. The `JournalEntryDetail` file appears in the Project navigator.
4. Declare and define the `JournalEntryDetail` structure in this file as shown here:

```
import SwiftUI
struct JournalEntryDetail: View {
  var selectedJournalEntry: JournalEntry
  var body: some View {
    ScrollView {
      VStack(spacing: 30) {
        Text(selectedJournalEntry.date.formatted(
        .dateTime.day().month().year()))
          .font(.title)
          .fontWeight(.bold)
          .frame(maxWidth: .infinity, alignment: .trailing)
        Text(selectedJournalEntry.entryTitle)
          .font(.title)
          .fontWeight(.bold)
          .frame(maxWidth: .infinity, alignment: .leading)
        Text(selectedJournalEntry.entryBody)
          .font(.title2)
          .frame(maxWidth: .infinity, alignment: .leading)
        Image(uiImage: selectedJournalEntry.photo ??
        UIImage(systemName: "face.smiling")!)
          .resizable()
          .frame(width: 300, height: 300)
        if (selectedJournalEntry.longitude != nil &&
          selectedJournalEntry.latitude != nil) {
            MapView(journalEntry: selectedJournalEntry)
              .frame(width: 300, height: 300)
        }
      }.padding()
        .navigationTitle("Entry Detail")
    }
  }
```

```
}

#Preview {
  NavigationView {
    JournalEntryDetail(selectedJournalEntry: testData[0])
  }
}
```

The `JournalEntryDetail` structure contains a `selectedJournalEntry` property of the type `JournalEntry` and a `ScrollView` view enclosing a `Vstack` view. The `VStack` view contains `Text` views that display the selected journal entry's date, title, and body, an `Image` view that displays the selected journal entry's photo, and a `MapView` view that displays a map showing the selected journal entry's location, provided that the selected journal entry's `longitude` and `latitude` properties are not `nil`.

To create the preview in the canvas, the first `JournalEntry` instance in the `testData` array is assigned to the `selectedJournalEntry` property. Note that the `JournalEntryDetail` instance is enclosed in a `NavigationView` instance to make the navigation bar appear in the preview.

5. Verify that the canvas now displays a scrollable Journal Entry Detail screen with a rendered map:

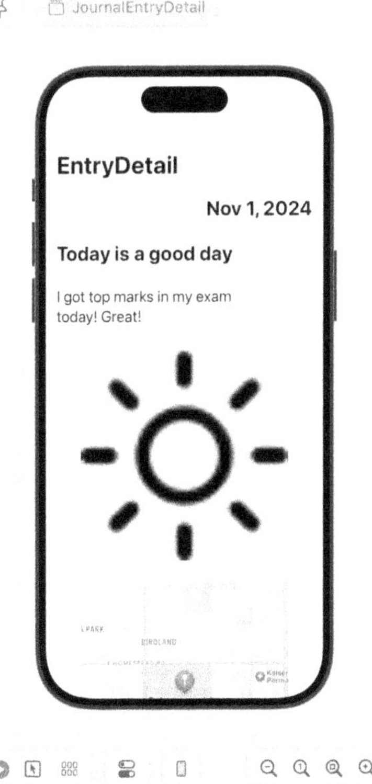

Figure 24.16: App preview showing Journal Entry Detail screen

Now that you've completed the implementation of the Journal Entry Detail screen using SwiftUI, you'll modify the list on the Journal List screen so that the Journal Entry Detail screen will be displayed when a cell is tapped.

Follow these steps:

1. Click the `ContentView` file in the project navigator and modify the `navigationDestination(for:destination:)` modifier to use the `JournalEntryDetail` structure as the destination when a cell is tapped:

```
.navigationDestination(for: JournalEntry.self) {
  journalEntry in
  JournalEntryDetail(selectedJournalEntry: journalEntry)
}
```

2. The Live Preview button in the canvas should already be selected. Tap a cell on the Journal List screen. You'll see the Journal Entry Detail screen for that restaurant appear:

Figure 24.17: App preview showing Journal Entry Detail screen

3. Build and run your app to test it in Simulator:

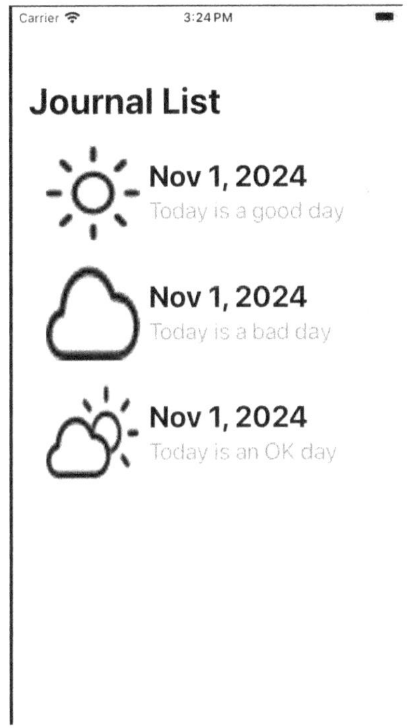

Figure 24.18: Simulator showing Journal List screen

You have completed building a simple version of the *JRNL* app using SwiftUI! Awesome!

Summary

In this brief introduction to SwiftUI, you've seen how to build a simplified version of the *JRNL* app using SwiftUI.

You started by adding and configuring SwiftUI views to create the Journal List screen. You then added the model objects to your app and configured the navigation between the Journal List and Journal Entry Detail screens. After that, you used MapKit to build a map view for the Journal Entry Detail screen. Finally, you created the Journal Entry Detail screen and added to it the map view you created earlier.

You now know how to use SwiftUI to create an app that reads model objects, presents them in a list, and enables navigation to a second screen containing a map view. You are now able to implement this in your own projects.

If you wish to learn more about SwiftUI, you can refer to Apple's Develop in Swift tutorials here:

`https://developer.apple.com/tutorials/develop-in-swift`

Packt Publishing also has a book on SwiftUI. You can find out more here:

`https://www.amazon.com/SwiftUI-Cookbook-building-beautiful-interactive/dp/1805121731`

In the next chapter, you will learn about **Swift Testing**, which lets you test your Swift code with ease.

Join us on Discord!

Read this book alongside other users, experts, and the author himself. Ask questions, provide solutions to other readers, chat with the author via Ask Me Anything sessions, and much more. Scan the QR code or visit the link to join the community.

`https://packt.link/ios-Swift`

25

Getting Started with Swift Testing

Apple introduced **Swift Testing** during WWDC24. It is a new framework that makes it easy for you to test your Swift code using expressive and intuitive APIs.

In this chapter, you'll create and run tests for the `JournalEntry` class to make sure it works as intended.

You'll start by adding a new **Unit Testing** target to your app. Next, you'll write some tests for the `JournalEntry` class, and finally, you'll run the tests on your `JournalEntry` class to make sure it works as it should.

By the end of this chapter, you'll have learned how to write tests for the classes in your app to make sure that they work as intended. This will be useful for larger projects involving many people, where you are not able to view the source code for all the classes in your project.

The following topics will be covered:

- Introducing Swift Testing
- Adding a Unit Testing target to your app
- Writing tests for the `JournalEntry` class
- Testing the `JournalEntry` class

Technical requirements

You will continue working on the `JRNL` project that you modified in *Chapter 23, Getting Started with SwiftData*.

The completed Xcode project for this chapter is in the `Chapter25` folder of the code bundle for this book, which can be downloaded here:

`https://github.com/PacktPublishing/iOS-18-Programming-for-Beginners-Ninth-Edition`

Check out the following video to see the code in action:

`https://youtu.be/se9ae9wrYC8`

Let's start by learning about Swift Testing and how it works.

Introducing Swift Testing

If you are the lead developer of a large project involving many developers, it would not be practical for you to review everyone's source code in detail, and in some cases, you would not be able to view the source code at all. Instead, you would issue specifications on what a class is supposed to do, and it would be the developer's job to write the class for you.

As an example, let's look at the initializer code for the `JournalEntry` class in the *JRNL* app:

```
init?(rating: Int, title: String, body: String,
photo: UIImage? = nil, latitude: Double? = nil,
longitude: Double? = nil) {
  if title.isEmpty || body.isEmpty || rating < 0 ||    rating > 5 {
    return nil
  }
  self.date = Date()
  self.rating = rating
  self.title = title
  self.body = body
  self.photoData = photo?.jpegData(compressionQuality: 1.0)
  self.latitude = latitude
  self.longitude = longitude
}
```

As you can see, you can only create valid `JournalEntry` instances if `entryTitle` and `entryBody` are not empty and `rating` is between 0 and 5 inclusive. If these requirements are not met, the initializer will return `nil`.

Assume that you're not able to view the source code for the `JournalEntry` class. How would you know that this class works as intended? This is where Swift Testing comes in.

Swift Testing has a clear and expressive API built using macros, which makes it easy to write tests to determine that your code is working properly. It works on all major platforms supported by Swift, and it is open source.

For more information on Swift Testing, you can watch Apple's WWDC24 video on it here: `https://developer.apple.com/videos/play/wwdc2024/10179/`.

In the next section, you'll see how to add a Unit Testing target to your project.

Adding a Unit Testing target to your app

In order to be able to test the `JournalEntry` class, you'll add a Unit Testing target to your app. You will be able to write all the tests for classes in your app here. Follow these steps:

1. In Xcode, choose **File** | **New** | **Target** to open the template selector.
2. **iOS** should already be selected. In the **Test** section, click **Unit Testing Bundle** and click **Next**:

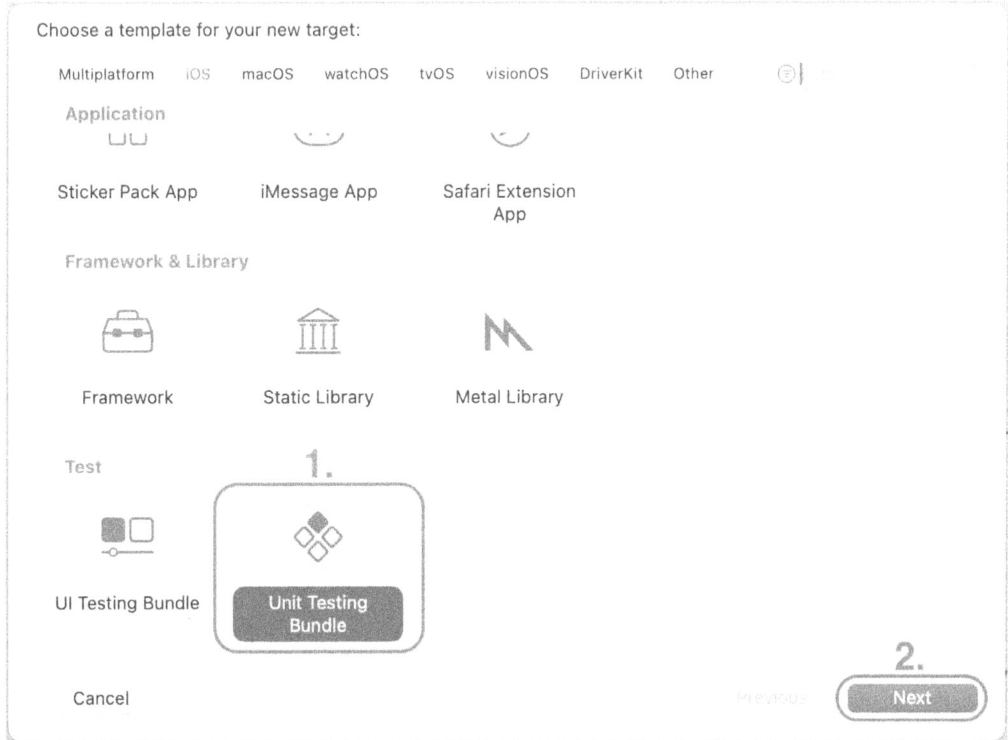

Figure 25.1: Template chooser window

3. In the **Choose options for your new target:** window, leave all the options at their default values and click **Finish:**

Choose options for your new target:

Product Name: JRNLTests
Team: None
Organization Identifier: com.myname
Bundle Identifier: com.myname.JRNLTests
Language: Swift
Testing System: Swift Testing
Project: JRNL
Target to be Tested: JRNL

Cancel Previous Finish

Figure 25.2: Options window

4. Verify that the **JRNLTests** folder is visible in the Project navigator:

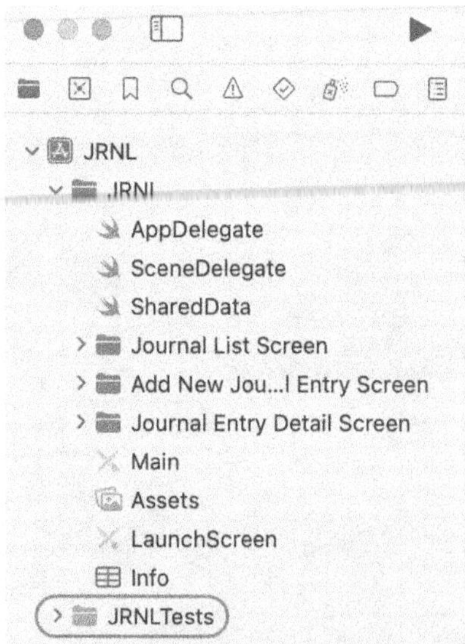

Figure 25.3: Project navigator showing JRNLTests folder

You have successfully added the Unit Testing target to your app! In the next section, you'll write some tests to ensure that the `JournalEntry` class works as it should.

Writing tests for the JournalEntry class

As you have seen, the initializer for the `JournalEntry` class will only return a `JournalEntry` instance if `title` and `body` are not empty, and `rating` is between 0 and 5 inclusive. If these conditions are not met, the initializer will return `nil`. To test this, you will write tests to confirm that a valid `JournalEntry` instance is created when the above conditions are met, and `nil` is returned when they are not. Follow these steps:

1. Click the **JRNLTests** file inside the **JRNLTests** folder in the Project navigator.
2. Modify the contents of this file as follows:

```
import Testing
@testable import JRNL

struct JRNLTests {

  @Test func JournalEntryInitializationSucceeds() {
    let zeroRatingJournalEntry = JournalEntry(rating: 0,
    title: "Zero", body: "Zero rating entry")
    #expect(zeroRatingJournalEntry != nil)

    let positiveRatingJournalEntry = JournalEntry(
    rating: 5, title: "Highest", body:
    "Highest rating entry")
    #expect(positiveRatingJournalEntry != nil)
  }

  @Test func JournalEntryInitializationFails() {
    let entryTitleEmptyJournalEntry = JournalEntry(
    rating: 3, title: "", body: "No title")
    #expect(entryTitleEmptyJournalEntry == nil)

    let entryBodyEmptyJournalEntry = JournalEntry(
    rating: 3, title: "No body", body: "")
    #expect(entryBodyEmptyJournalEntry == nil)

    let negativeRatingJournalEntry = JournalEntry(
    rating: -1, title: "Negative", body:
    "Negative rating entry")
    #expect(negativeRatingJournalEntry == nil)

    let invalidRatingJournalEntry = JournalEntry(
    rating: 6, title: "Invalid", body:
```

```
      "Invalid rating entry")
      #expect(invalidRatingJournalEntry == nil)
    }
}
```

Let's break this down:

```
import Testing
```

This imports the Swift Testing framework.

```
@testable import JRNL
```

This makes all the code inside the `JRNL` project available for testing.

```
  @Test func JournalEntryInitializationSucceeds() {
    let zeroRatingJournalEntry = JournalEntry(rating: 0,
    title: "Zero", body: "Zero rating entry")
    #expect(zeroRatingJournalEntry != nil)

    let positiveRatingJournalEntry = JournalEntry(
    rating: 5, title: "Highest", body:
    "Highest rating entry")
    #expect(positiveRatingJournalEntry != nil)
  }
```

This function is used to confirm that valid instances of the `JournalEntry` class are created when `entryTitle` has a value, `entryBody` has a value, and `rating` is between 0 and 5 inclusive. The `#expect` macro checks that both `zeroRatingJournalEntry` and `positiveRatingJournalEntry` are not `nil`, therefore confirming that a valid `JournalEntry` instance has been created.

```
  @Test func JournalEntryInitializationFails() {
    let entryTitleEmptyJournalEntry = JournalEntry(
    rating: 3, title: "", body: "No title")
    #expect(entryTitleEmptyJournalEntry == nil)

    let entryBodyEmptyJournalEntry = JournalEntry(
    rating: 3, title: "No body", body: "")
    #expect(entryBodyEmptyJournalEntry == nil)

    let negativeRatingJournalEntry = JournalEntry(
    rating: -1, title: "Negative", body:
    "Negative rating entry")
    #expect(negativeRatingJournalEntry == nil)

    let invalidRatingJournalEntry = JournalEntry(
```

```
        rating: 6, title: "Invalid", body:
        "Invalid rating entry")
        #expect(invalidRatingJournalEntry == nil)
    }
```

This function is used to confirm that `nil` will be returned when `entryTitle` is empty, `entryBody` is empty, and `rating` is not between 0 and 5 inclusive. The `#expect` macro will determine if `entryTitleEmptyJournalEntry`, `entryBodyEmptyJournalEntry`, `negativeRatingJournalEntry`, and `invalidRatingJournalEntry` are all `nil`, therefore confirming that a `JournalEntry` instance has not been created.

Since this is a brief introduction to Swift Testing, multiple conditions are checked in a single function, but for clarity, professional programmers typically use one function to perform one test.

You've completed writing all the tests for the `JournalEntry` class. In the next section, you'll run the tests to confirm that the `JournalEntry` class works as it should.

Testing the JournalEntry class

Since you have completed writing all the tests for the `JournalEntry` class in the previous section, you will now run the test to see if the `JournalEntry` class works as expected. Follow these steps:

1. In Xcode, choose **Test** from the **Product** menu.
2. Xcode will automatically run all your tests. Upon completion, you will see the following results:

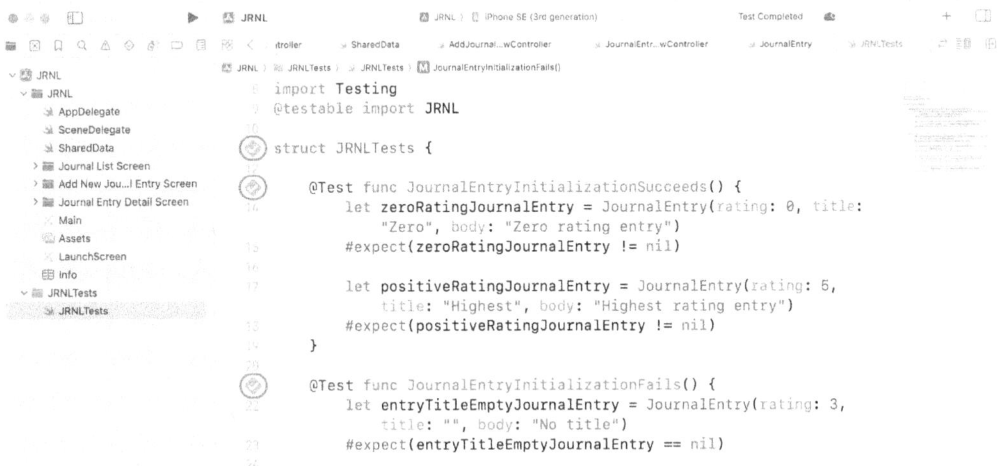

Figure 25.4: Canvas showing widget preview

The green squares with tick marks show that all the tests have been completed successfully and the `JournalEntry` class works as expected. Great job!

Summary

In this chapter, you tested the JournalEntry class to determine if it works as expected.

First, you added a new **Unit Testing** target to your app. Next, you wrote some tests for the `JournalEntry` class, and finally, you ran tests on your `JournalEntry` class to make sure it works as it should.

You have now learned how to write tests for the classes in your app to make sure that they work as intended. This will be useful for larger projects involving many people, where you are not able to view the source code for all the classes in your project, and also ensure that changes made anywhere in the app do not break existing functionality.

In the next chapter, you'll learn about **Apple Intelligence**, Apple's implementation of AI technology introduced during WWDC24.

Join us on Discord!

Read this book alongside other users, experts, and the author himself. Ask questions, provide solutions to other readers, chat with the author via Ask Me Anything sessions, and much more. Scan the QR code or visit the link to join the community.

`https://packt.link/ios-Swift`

26

Getting Started with Apple Intelligence

During Apple's Worldwide Developers Conference in 2024, Apple introduced **Apple Intelligence**, a personal intelligence system that puts powerful generative models in Apple devices, enabling new AI-driven features in your apps. These features include **Writing Tools**, **Image Playground**, **Genmoji**, and **Siri with App Intents**. However, only Writing Tools is available at the time of writing.

In this chapter, you're going to explore Apple Intelligence features using the *JRNL* app that you completed in *Chapter 22, Getting Started with Collection Views.*

First, you'll learn about Apple Intelligence and what it can do. Next, you'll see how predictive code completion can help you write your apps. Finally, you'll learn about Writing Tools and see how it works in your app.

By the end of this chapter, you'll have learned how to use Apple Intelligence features in Xcode and your apps.

The following topics will be covered in this chapter:

- Introducing Apple Intelligence
- Using predictive code completion in Xcode
- Implementing Writing Tools in your app

Technical requirements

You will continue working on the `JRNL` project that you modified in *Chapter 22, Getting Started with Collection Views.*

The resource files and completed Xcode project for this chapter are in the `Chapter26` folder of the code bundle for this book, which can be downloaded here:

`https://github.com/PacktPublishing/iOS-18-Programming-for-Beginners-Ninth-Edition`

Check out the following video to see the code in action:

`https://youtu.be/vraSf4dPHfc`

Let's start by learning about Apple Intelligence in the next section.

Introducing Apple Intelligence

Apple Intelligence is an artificial intelligence platform that consists of on-device and server processing that will enable incredible new features to help users communicate, work, and express themselves. These features include:

- Predictive code completion, which helps you write code using Xcode.
- Writing Tools, which helps users proofread, rewrite, and summarize text.
- Image Playground, which allows users to create fun and playful images.
- Genmoji, which lets users create emojis to suit any occasion.
- Siri with App Intents, which allows developers to give Siri the ability to take actions within your app.

Apple Intelligence will be free to all Apple users and will be available in the fall of 2024. At the time of writing, only predictive code completion and Writing Tools are available.

To learn about what Apple Intelligence can do, watch this video: `https://www.youtube.com/watch?v=Q_EYoV1kZWk`.

You can view Apple Developer documentation on Apple Intelligence at this link: `https://developer.apple.com/apple-intelligence/`.

In the next section, you will learn how Apple Intelligence will help you write code for your app.

Using predictive code completion in Xcode

As you know, when you type code, Xcode will try to help you by displaying suggestions in a pop-up menu. Code completion takes this to another level, providing more thorough code suggestions via an on-device AI coding model. This model is specifically trained on Swift and Apple SDKs and will be able to infer what you are trying to do based on the surrounding code context, like function names and comments.

To see this in action, you'll add a new file to your project and create some example structures and functions. Follow these steps:

1. Open the completed Xcode project in the `Chapter22` folder of the code bundle for this book, which can be downloaded from `https://github.com/PacktPublishing/iOS-18-Programming-for-Beginners-Ninth-Edition`.
2. Click **Settings** in the **Xcode** menu and click the **Text Editing** tab. Click the **Editing** tab and tick the **Predictive code completion** check box:

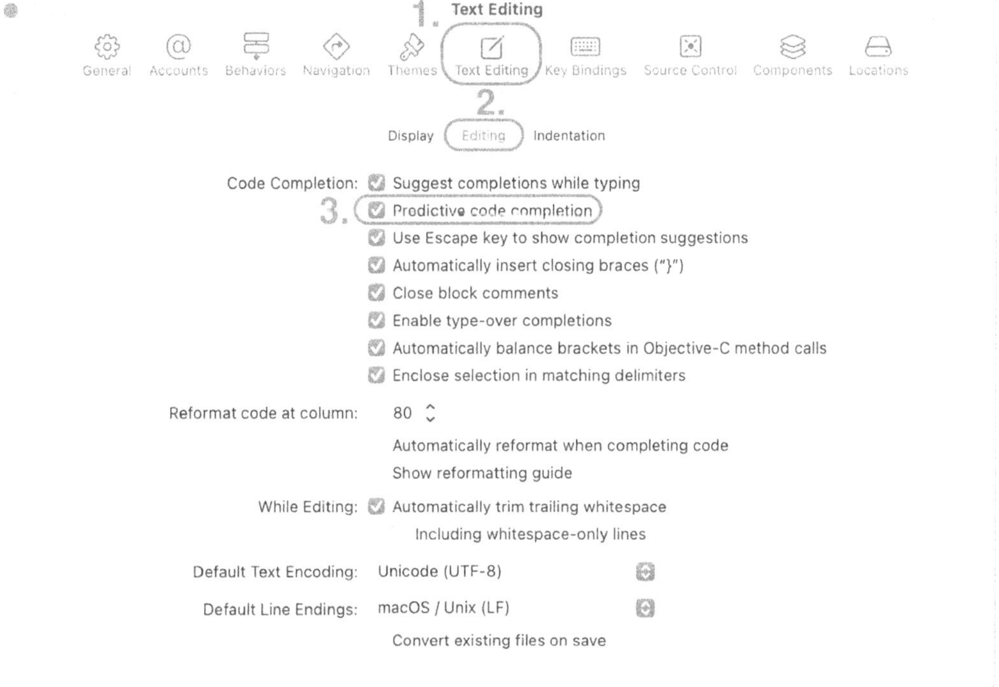

Figure 26.1: Settings window showing the Predictive code completion check box

3. A **Download predictive code completion model?** alert will appear. Click the **Download** button to download and install the language model, and wait for it to complete:

Download predictive code completion model?

Predictive code completion requires an on-device language model, which is a 2 GB download.

Figure 26.2: The Download predictive code completion model? alert

4. Click the **Components** tab in the **Settings** window and verify the **Predictive Code Completion Model** is present:

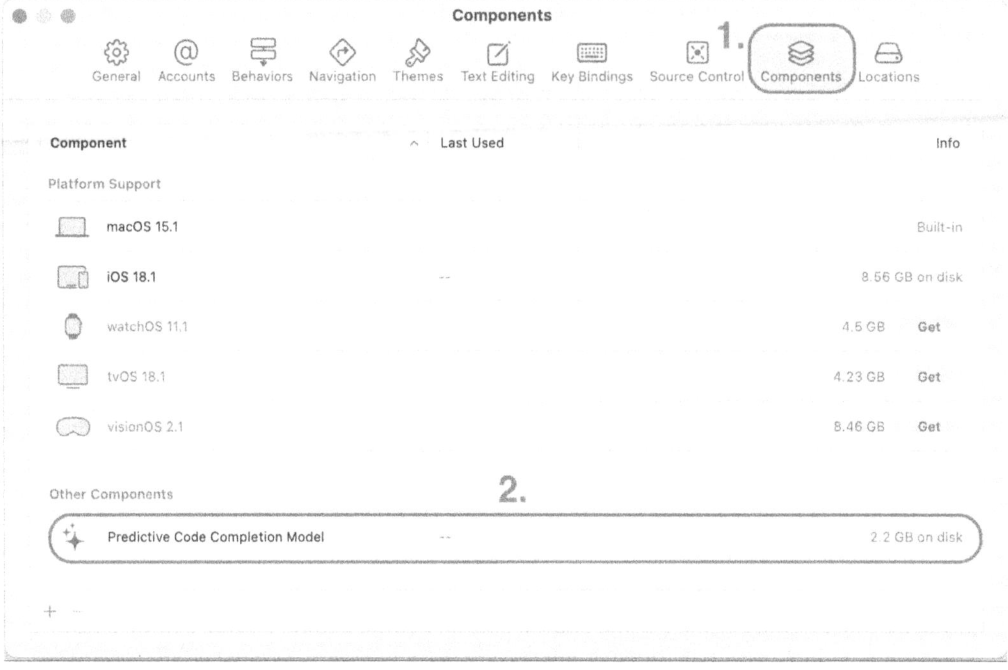

Figure 26.3: Settings window showing the Components tab

5. Right-click the **JRNL** folder in the Project navigator and choose **New File from Template...**. Choose **Swift File** from the template chooser and name the file `Employee`. It will appear in the Project navigator.
6. Type in the following comment and code after the `import` statement:

```
// This file contains the definition of the Employee structure, a method
that will generate sample data, an EmployeeDatabase structure containing
an array of Employee instances and methods to add, delete and find
employees in an array.

struct Employee
```

7. Xcode will display a predictive code suggestion. Press *Tab* to accept it:

```
import Foundation

// This file contains the definition of the Employee structure, a method that will generate
    sample data, an EmployeeDatabase structure containing an array of Employee instances and
    methods to add, delete and find employees in an array.

struct Employee: Codable, Identifiable { ••• }
```

Expected identifier in struct declaration

Figure 26.4: Editor area showing a predictive code suggestion

8. Xcode will display a list of possible properties for the `Employee` structure. Press *Tab* to accept it:

```
struct Employee: Codable, Identifiable {        Type 'Employee' does not conform to protocol 'Identifiable'
    let id: Int
    let name: String
    let age: Int
    let salary: Double
}
```

Figure 26.5: Editor area showing a predictive code suggestion

Note that Xcode has automatically created the `Employee` structure for you. Cool!

9. Type ex after the `Employee` structure definition and press *Tab* to accept the predictive code suggestion.
10. Verify that Xcode has automatically created a method to generate sample data:

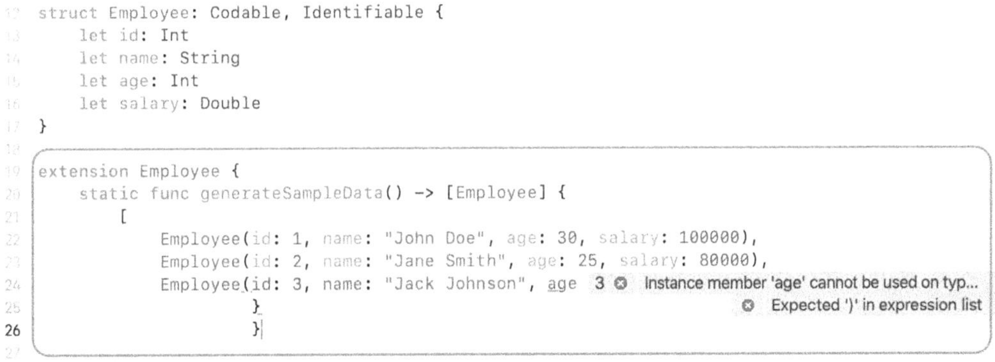

```
struct Employee: Codable, Identifiable {
    let id: Int
    let name: String
    let age: Int
    let salary: Double
}

extension Employee {
    static func generateSampleData() -> [Employee] {
        [
            Employee(id: 1, name: "John Doe", age: 30, salary: 100000),
            Employee(id: 2, name: "Jane Smith", age: 25, salary: 80000),
            Employee(id: 3, name: "Jack Johnson", age  3  Instance member 'age' cannot be used on typ...
                 }                                       Expected ')' in expression list
                 }
```

Figure 26.6: Editor area showing syntax errors

Note that there may be a few syntax errors that you need to fix. This will probably improve as Apple updates the language model over time.

After fixing the syntax errors, your code should look similar to this:

```
struct Employee: Codable, Identifiable {
  let id: Int
  let name: String
  let age: Int
  let salary: Double
}

extension Employee {
  static func generateSampleData() -> [Employee] {
    [
      Employee(id: 1, name: "John Doe", age: 25,
```

```
        salary: 100000),
        Employee(id: 2, name: "Jane Smith", age: 28,
        salary: 120000),
        Employee(id: 3, name: "Jimmy Johnson", age: 30,
        salary: 150000),
      ]
    }
}
```

11. Type `struct Emp` after the extension and keep pressing *Tab* to accept the suggestions until no more suggestions appear. The generated code will be similar to this:

```
struct EmployeeDatabase: Codable {
  var employees: [Employee]
}
```

12. Type `func` after the `employees` property declaration and keep pressing *Tab* to accept the suggestions until no more suggestions appear. The generated code will be similar to this:

```
mutating func add(_ employee: Employee) {
  employees.append(employee)
}
```

13. Type *Return* twice after the definition of the `add(_:)` method and wait until the code suggestion appears. Keep pressing *Tab* to accept the suggestions until no more suggestions appear. The generated code will be similar to this:

```
func find(by id: Int) -> Employee? {
  employees.first(where: { $0.id == id })
}
```

14. Repeat *Step 14* and keep pressing *Tab* to accept the suggestions until no more suggestions appear. The generated code will be similar to this:

```
mutating func delete(by id: Int) {
  employees.removeAll(where: { $0.id == id })
}
```

15. Verify that the generated code is similar to the code shown here:

```
// This file contains the definition of the Employee structure, a method
that will generate sample data, an EmployeeDatabase structure containing
an array of Employee instances and methods to add, delete and find
employees in an array.

struct Employee: Codable, Identifiable {
  let id: Int
```

```
    let name: String
    let age: Int
    let salary: Double
  }

  extension Employee {
    static func generateSampleData() -> [Employee] {
       [
         Employee(id: 1, name: "John Doe", age: 25,
         salary: 100000),
         Employee(id: 2, name: "Jane Smith", age: 28,
         salary: 120000),
         Employee(id: 3, name: "Jimmy Johnson", age: 30,
         salary: 150000),
      ]
    }
  }

  struct EmployeeDatabase: Codable {
    var employees: [Employee]

    mutating func add(_ employee: Employee) {
      employees.append(employee)
    }

    func find(by id: Int) -> Employee? {
      employees.first(where: { $0.id == id })
    }

    mutating func delete(by id: Int) {
      employees.removeAll(where: { $0.id == id })
    }
  }
```

With the help of predictive code completion, you have successfully created the classes and methods described in the comment, with very little typing required! Awesome!

However, do note that the code generated is not perfect, and you will need to fix errors and other issues as needed.

In the next section, you'll learn how to implement Writing Tools to proofread, rewrite, and summarize text in your apps.

Implementing Writing Tools in your app

Writing Tools is an Apple Intelligence feature that is available system-wide and can help you proofread, rewrite, and summarize text. Writing Tools appears automatically when your app is running in a supported environment, as long as you are using a `UITextView`, `NSTextView`, or `WKWebView` in your app. Apple has also introduced text view delegate methods and properties so that your app may take appropriate action when Writing Tools is in use.

To view Apple's WWDC24 video on Writing Tools, see: `https://developer.apple.com/videos/play/wwdc2024/10168/`.

To see Writing Tools in action, you'll run the *JRNL* app on your Mac and use Writing Tools to modify the text in the Add New Journal Entry screen. You'll also explore the new text view delegate methods and properties introduced by Apple. Follow these steps:

1. Open **System Settings** on your Mac, choose **Apple Intelligence & Siri** in the sidebar, and switch on **Apple Intelligence**:

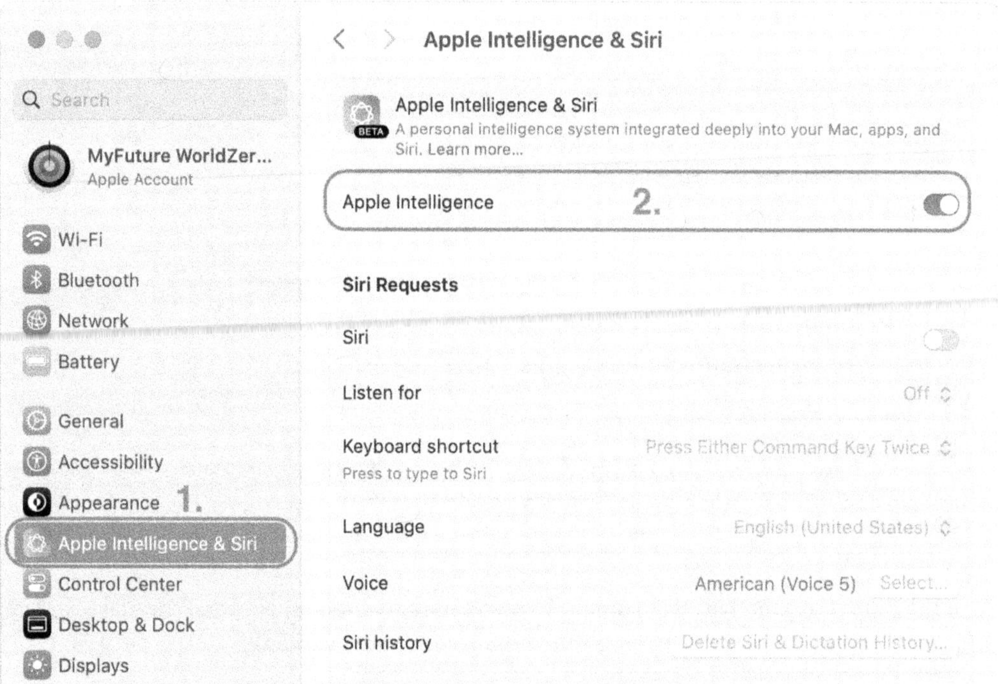

Figure 26.7: System Settings window with Apple Intelligence turned on

2. In Xcode, open the `JRNL` project and choose **My Mac (Designed for iPad)** from the destination menu in the toolbar:

Figure 26.8: Destination menu with My Mac (Designed for iPad) selected

3. Click the **JRNL** icon at the top of the Project navigator, click the **JRNL** target, and click the **Signing & Capabilities** tab. Set **Team** to a free or paid Apple Developer account, and modify the **Bundle Identifier** as needed, until there are no more errors in the provisioning profile:

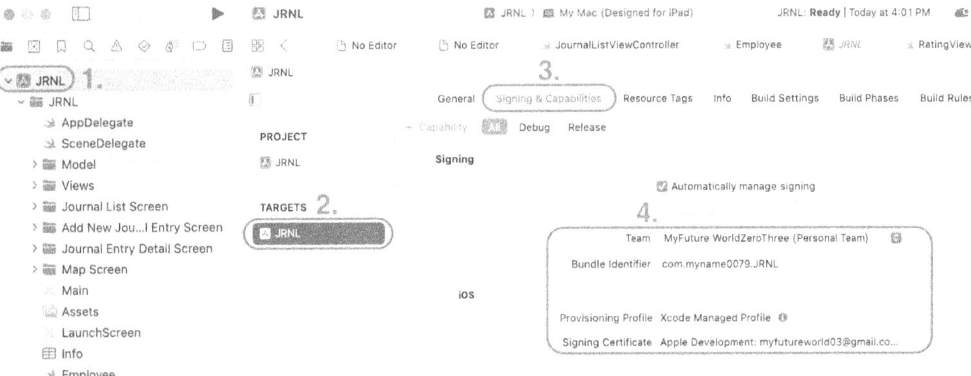

Figure 26.9: Signing and capabilities screen

Running your app on a device is covered in more detail in *Chapter 1, Exploring Xcode*.

4. Build and run the app on your Mac and click the + button to display the Add New Journal Entry screen:

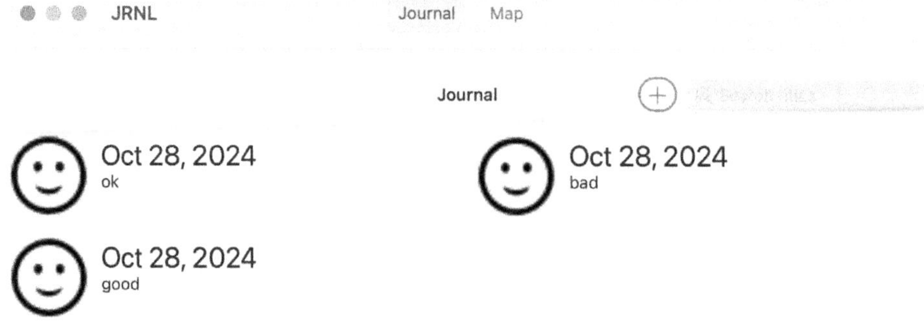

Figure 26.10: The Journal screen with the + button highlighted

5. Type in a few paragraphs of text into the text view. Select all the text, right-click, and choose **Writing Tools | Proofread**:

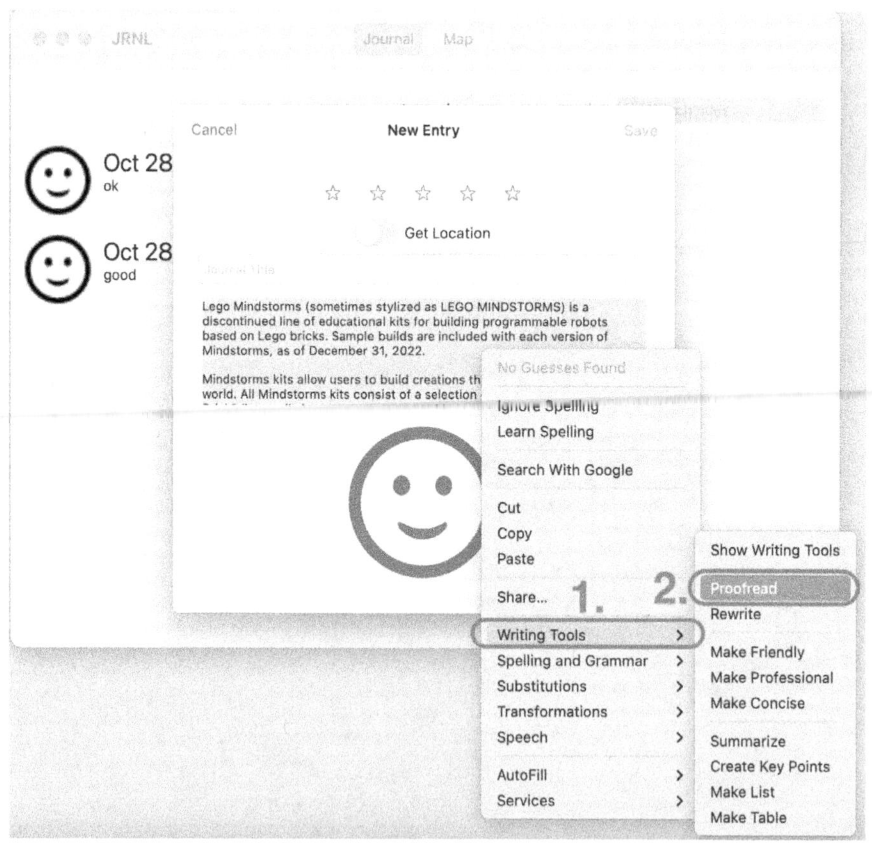

Figure 26.11: Text view with a pop-up menu showing Proofread selected

6. Click the left and right arrow buttons to step through the changes, and click **Done** when you have finished reviewing them:

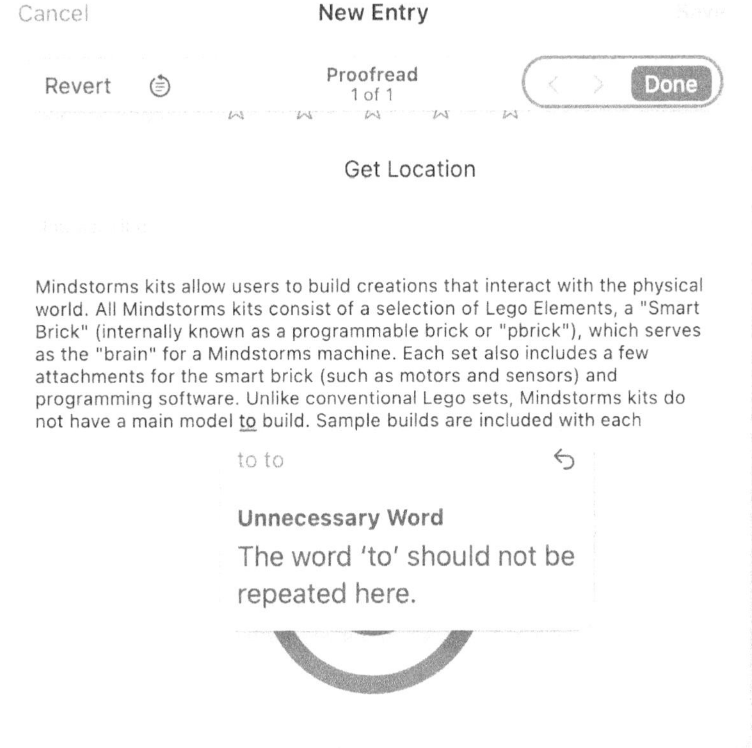

Figure 26.12: Reviewing Writing Tools changes in the text view

7. Try the other Writing Tools features and observe what they do:

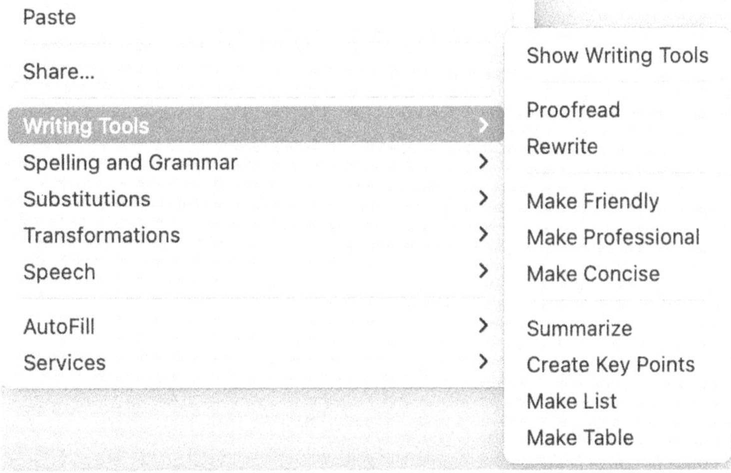

Figure 26.13: Writing Tools submenu showing available features

8. You may wish to disable editing in the text view while Writing Tools is active. Click the **AddJournalViewController** file in the Project navigator. Add the following extension after all other code in the file:

```
extension AddJournalEntryViewController {
  // MARK: - Writing Tools Delegate methods
  func textViewWritingToolsWillBegin(_ textView:
  UITextView) {
    textView.isEditable = false
  }
  func textviewWritingToolsDidEnd(_ textView:
  UITextView) {
    textView.isEditable = true
  }
}
```

   This code disables editing in the text view while Writing Tools is active.

9. In some cases, you may wish to disable Writing Tools altogether. In the `viewDidLoad()` method of the `AddJournalEntryViewController` class, add this line after all the other code:

```
bodyTextView.writingToolsBehavior = .none
```

   This code disables Writing Tools, and the Writing Tools menu item will no longer appear.

You have successfully explored how Writing Tools works in your app. Excellent!

Summary

In this chapter, you modified the *JRNL* app that you completed in *Chapter 22, Getting Started with Collection Views,* to work with Apple Intelligence.

First, you learned about Apple Intelligence and what it can do. Next, you created a new structure and related functions with the help of predictive code completion. Finally, you learned about Writing Tools and how it works in your app.

You now have learned how to use Apple Intelligence features in Xcode and your apps. Great!

In the next chapter, you'll learn how to test and submit your app to the App Store.

Join us on Discord!

Read this book alongside other users, experts, and the author himself. Ask questions, provide solutions to other readers, chat with the author via Ask Me Anything sessions, and much more. Scan the QR code or visit the link to join the community.

`https://packt.link/ios-Swift`

27

Testing and Submitting Your App to the App Store

Congratulations, you have reached the final chapter of this book!

Over the course of this book, you have learned about the Swift programming language and how to build an entire app using Xcode. However, so far, you've only been running your app in Simulator or on your own device using a free Apple Developer account.

In this chapter, you will start by learning how to obtain a **paid Apple Developer account**. Next, you'll learn about **certificates**, **identifiers**, **test device registration**, and **provisioning profiles**. After that, you'll learn how to create an App Store listing and submit your app to the App Store. Finally, you'll learn how to conduct testing for your app using internal and external testers.

By the end of this chapter, you'll know how to test and how to submit your own apps to the App Store.

The following topics will be covered:

- Getting an Apple Developer account
- Exploring your Apple Developer account
- Submitting your app to the App Store
- Conducting internal and external testing

Technical requirements

You will need an Apple Account and a paid Apple Developer account to complete this chapter.

There are no project files for this chapter as it should be used as a reference on how to submit apps and is not specific to any particular app.

To see the latest updates to the App Store, visit `https://developer.apple.com/app-store/whats-new/`.

To see what's new in App Store Connect, watch this video: `https://developer.apple.com/videos/play/wwdc2024/10063/`.

Let's start by learning how to get a paid Apple Developer account, which is required for App Store submission, in the next section.

Getting an Apple Developer account

As you saw in earlier chapters, all you need to test your app on a device is a free Apple Account. However, the apps will only work for a few days, and you will not be able to add advanced features, such as signing in with Apple or uploading your app to the App Store. For that, you need a paid Apple Developer account. Follow these steps to purchase an individual/sole proprietorship Apple Developer account:

1. Go to `https://developer.apple.com/programs/` and click on the **Enroll** button.
2. Click **Start your enrollment**.
3. Enter your Apple Account and password when prompted.
4. On the **Trust this browser?** screen, click **Trust** only if you are the only person using this browser; otherwise, click **Not Now**. This is to safeguard your account information.
5. Click **Continue enrollment on the web >**.
6. On the **Confirm your personal information** screen, enter your personal information and click **Continue**.
7. On the **Select your entity type** screen, choose **Individual/Sole Proprietor**. Click **Continue**.
8. On the **Review and Accept** screen, check the checkbox at the bottom of the page and click **Continue**.
9. On the **Complete your purchase** screen, click **Purchase**.
10. Follow the onscreen directions to complete your purchase. Once you have purchased your account, go to `https://developer.apple.com/account/` and sign in with the same Apple Account that you used to purchase your Developer account. You should see something like the following:

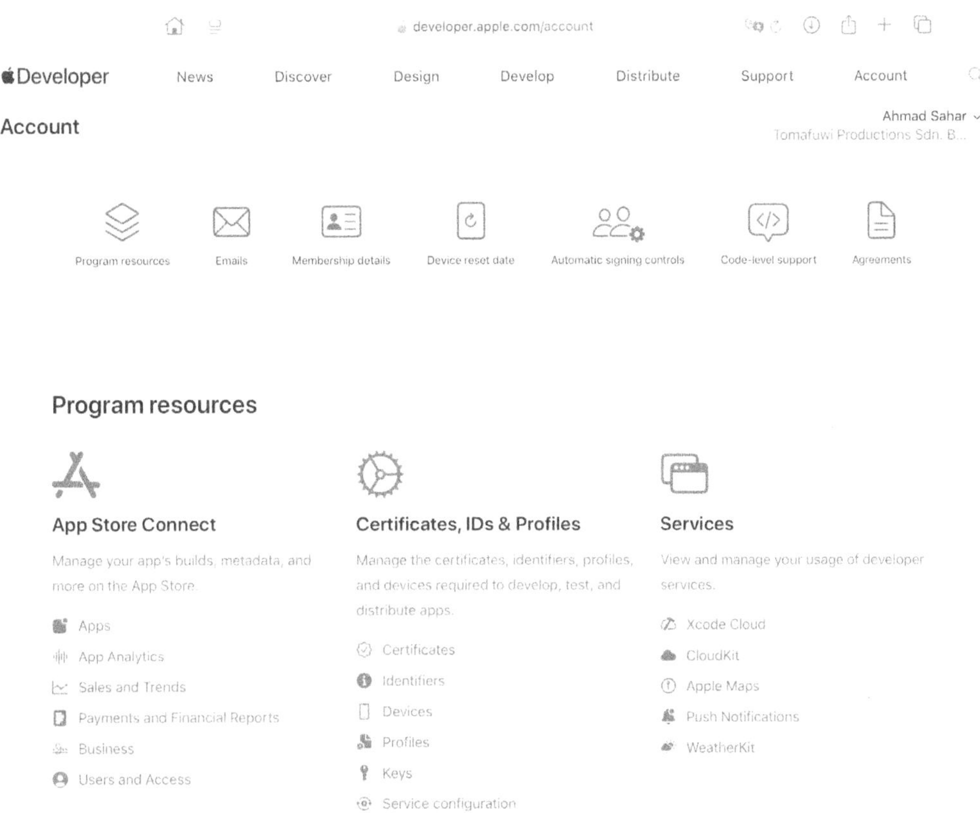

Figure 27.1: Apple Developer website with a paid Apple Developer account logged in

Now that you have a paid Apple Developer account, let's learn how to configure the various settings required for your app in the next section.

Exploring your Apple Developer account

Your Apple Developer account has everything you need to develop and submit apps. You can view your membership status, add and organize members of your development team, access developer documentation, download beta software, and more. All these features are beyond the scope of this book, though, and this section will only cover what you need to do to get your app on the App Store.

First, you'll get Apple Developer certificates that you'll install on your Mac. These certificates will be used to digitally sign your app. Next, you'll need to register your app's Apple Account and the devices that you'll be testing your app on. After that, you'll be able to generate provisioning profiles that allow your apps to run on your test devices and allow you to submit apps to the App Store.

Xcode can automatically handle this process for you when you add the Apple Account and password of your Apple Developer account to **Xcode** | **Settings** | **Accounts.**

Let's start by learning about **certificate signing requests (CSRs)**, which are required to obtain the Apple Developer certificates that you will install on your Mac, in the next section.

Generating a certificate signing request

Before you write apps that will be submitted to the App Store, you need to install a **developer certificate** on the Mac that you're running Xcode on. Certificates identify the author of an app. To get this certificate, you'll need to create a **CSR**. Here's how to create a CSR:

1. Use Spotlight on your Mac to find **Keychain Access** and launch it.
2. Choose **Certificate Assistant** | **Request a Certificate From a Certificate Authority...** from the **Keychain Access** menu:

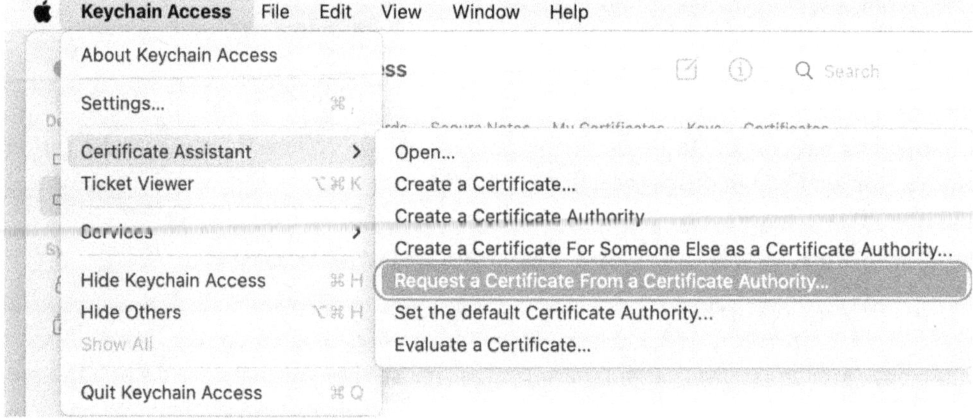

Figure 27.2: The Keychain Access application

3. For the **User Email Address** field, enter the email address of the Apple Account that you used to register your Apple Developer account. In the **Common Name** field, enter your name. Select **Saved to disk** under **Request is** and click **Continue**:

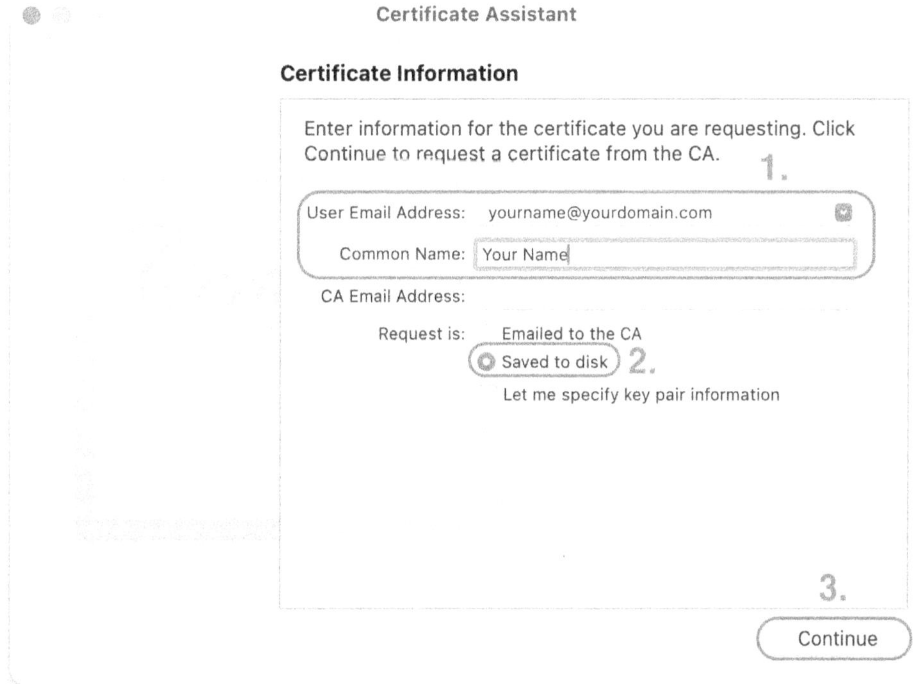

Figure 27.3: The Certificate Assistant screen

4. Save the CSR to your hard disk.
5. Click **Done**.

Now that you have a CSR, let's look at how you will use it to get development certificates (for testing on your own device) and distribution certificates (for App Store submission) in the next section.

Creating development and distribution certificates

Once you have a CSR, you can use it to create **development** and **distribution** certificates. Development certificates are used when you want to test your app on your test devices, and distribution certificates are used when you want to upload your app to the App Store.

Here's how to create development and distribution certificates:

1. Log in to your Apple Developer account and click **Certificates**:

Certificates, IDs & Profiles

Manage the certificates, identifiers, profiles, and devices required to develop, test, and distribute apps.

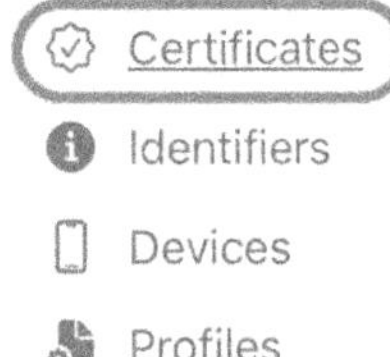

Figure 27.4: Apple Developer website with paid Apple Developer account logged in

2. You'll see the **Certificates** screen. Click the + button:

Certificates, Identifiers & Profiles

Certificates Certificates + All Types

Figure 27.5: Certificates screen showing the + button

3. Click the **Apple Development** radio button, and click **Continue**:

‹ All Certificates

Create a New Certificate 2. Continue

Software

1. Apple Development
Sign development versions of your iOS, iPadOS, macOS, tvOS, watchOS, and visionOS apps.

Apple Distribution
Sign your iOS, iPadOS, macOS, tvOS, watchOS, and visionOS apps for release testing using Ad Hoc distribution or for submission to the App Store.

Figure 27.6: Create a New Certificate screen showing the Apple Development radio button

4. Click **Choose File**:

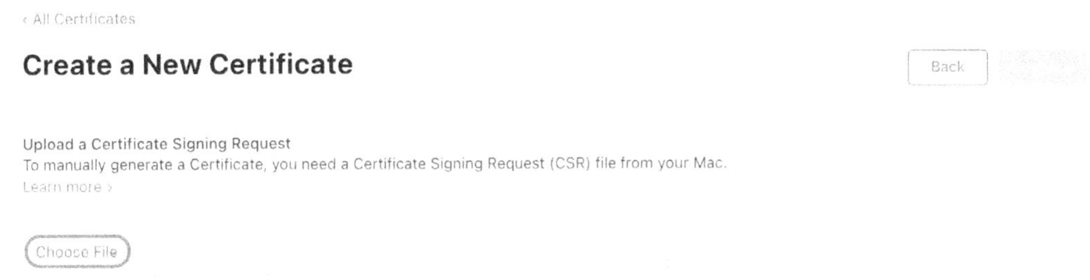

Figure 27.7: Upload a CSR screen

5. Upload your CSR by selecting the CSR file you saved earlier to your hard disk, and clicking **Choose**.
6. Click **Continue**:

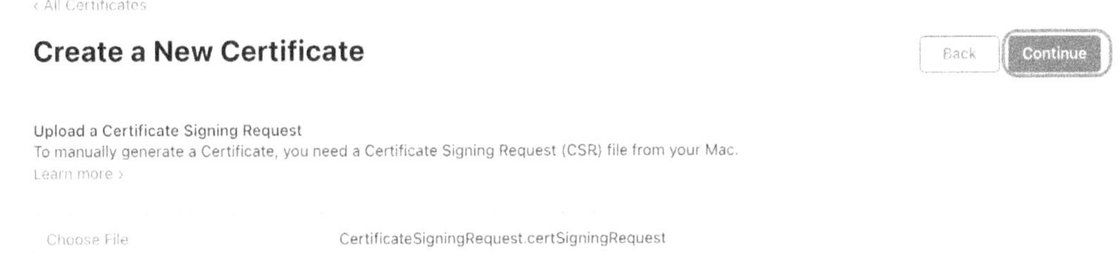

Figure 27.8: Upload a CSR screen with the certificate uploaded

7. Your certificate will be generated automatically. Click **Download** to download the generated certificate onto your Mac:

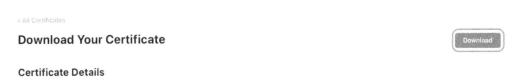

Figure 27.9: The Download Your Certificate screen

8. Double-click the downloaded certificate to install it on your Mac.
9. Repeat *Steps 3-8* again, but this time, choose the **Apple Distribution** radio button in *Step 3*:

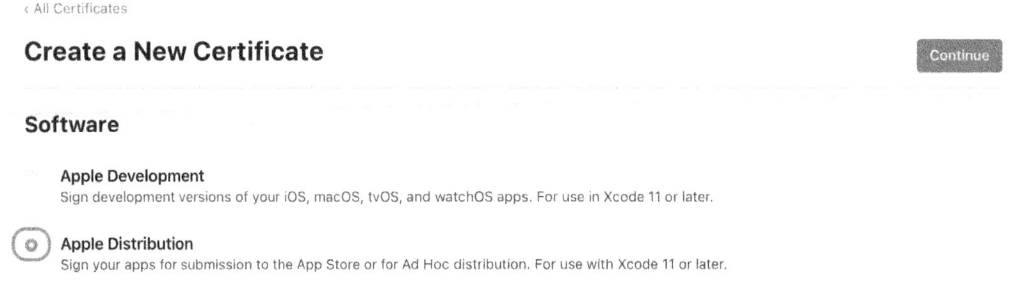

Figure 27.10: The Create a New Certificate screen showing the Apple Distribution radio button

Great! You now have development and distribution certificates. The next step is to register the **App ID** for your app to identify it on the App Store. You will learn how to do this in the next section.

Registering an App ID

When you created your project in *Chapter 1*, *Exploring Xcode*, you created a bundle identifier for it (also known as an App ID). An App ID is used to identify your app on the App Store. You'll need to register this App ID in your developer account prior to uploading your app to the App Store. Here's how to register your App ID:

1. Log in to your Apple Developer account and click **Identifiers**.
2. Click the + button:

Figure 27.11: The Identifiers screen

3. Click the **App IDs** radio button and click **Continue**:

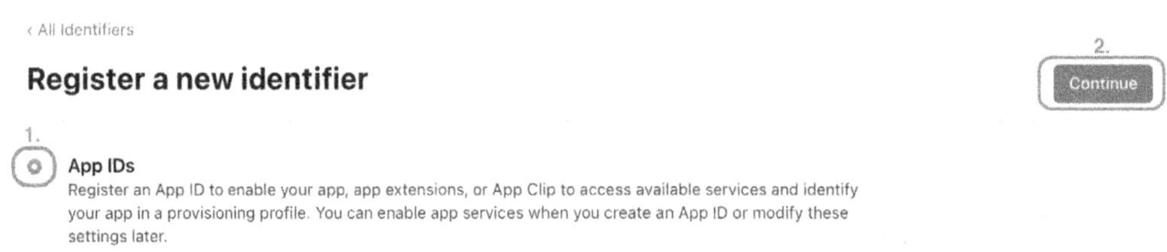

Figure 27.12: The Register a new identifier screen

4. Click **App** and click **Continue**:

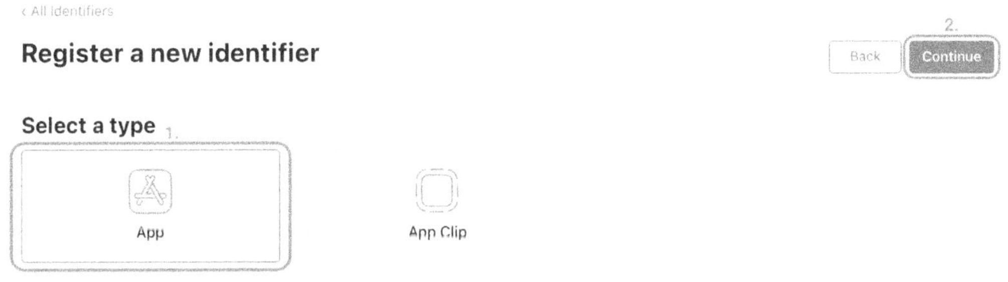

Figure 27.13: Identifier type screen

5. Enter a description for this App ID, such as `JRNL Packt Publishing App ID`. Check the **Explicit** button and enter your app's bundle ID in the field. Make sure that this value is the same as the bundle identifier you used when you created the project. Click the **Continue** button when you're done:

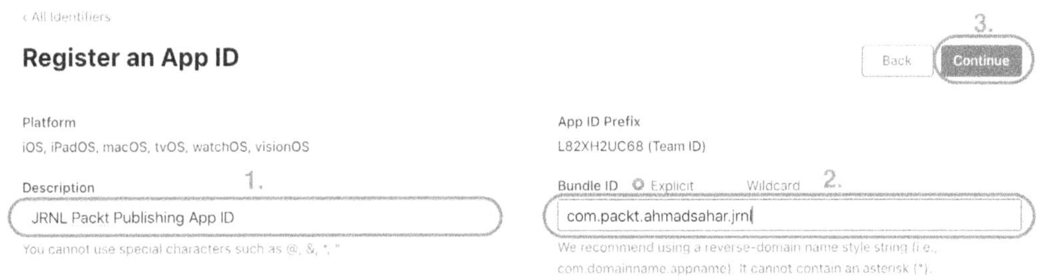

Figure 27.14: Description and bundle ID screen

Once your app has shipped, you will no longer be able to change the bundle ID for your app.

6. Click **Register**:

Figure 27.15: Register screen

Your App ID has now been registered. Cool! In the next section, you'll register the devices you'll be testing your app on.

Registering your devices

To run your apps on your personal devices for testing, you will need to register them in your developer account. Here's how to register your devices:

1. Log in to your Apple Developer account and click **Devices**.
2. Click the + button:

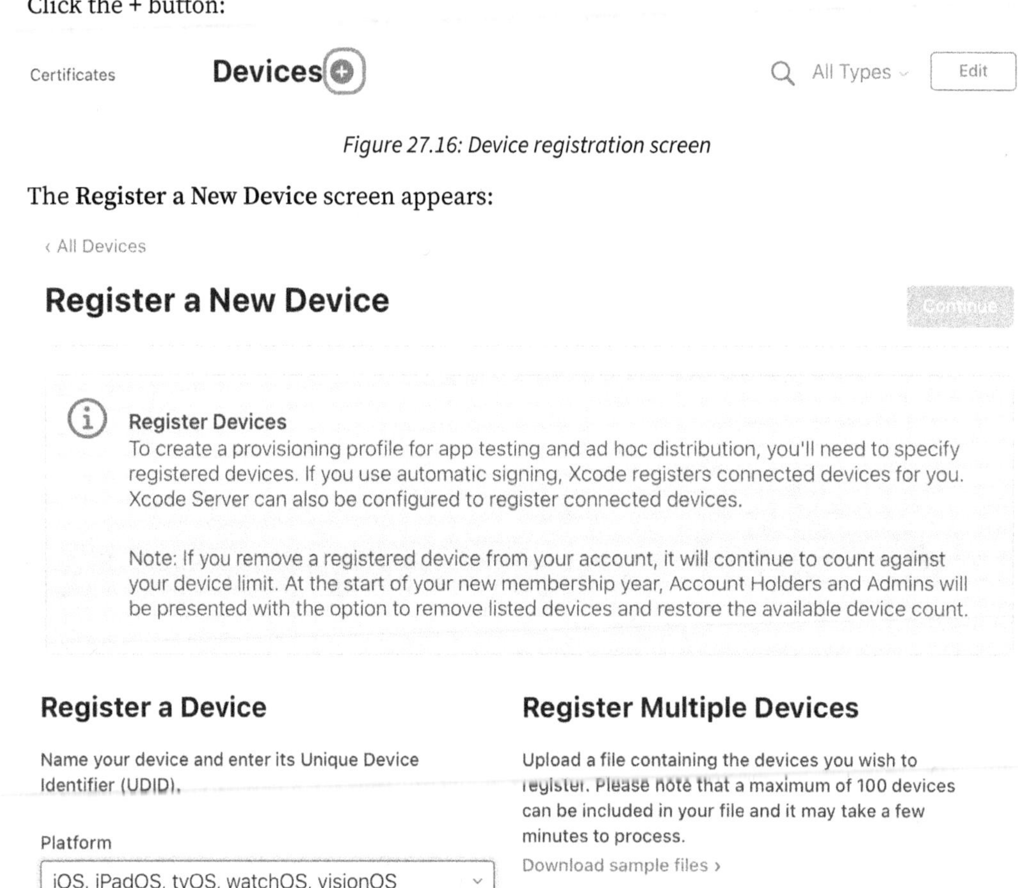
Certificates
Devices
All Types
Edit
‹ All Devices
Register a New Device
Continue
Register Devices
To create a provisioning profile for app testing and ad hoc distribution, you'll need to specify registered devices. If you use automatic signing, Xcode registers connected devices for you. Xcode Server can also be configured to register connected devices.
Note: If you remove a registered device from your account, it will continue to count against your device limit. At the start of your new membership year, Account Holders and Admins will be presented with the option to remove listed devices and restore the available device count.
Register a Device
Name your device and enter its Unique Device Identifier (UDID).
Platform
iOS, iPadOS, tvOS, watchOS, visionOS
Register Multiple Devices
Upload a file containing the devices you wish to register. Please note that a maximum of 100 devices can be included in your file and it may take a few minutes to process.
Download sample files ›
Device List

Figure 27.16: Device registration screen

3. The **Register a New Device** screen appears:

Figure 27.17: Register a New Device screen

You'll need to provide the **Device Name** and **Device ID** to register your device.

4. Connect your device to your Mac. Launch Xcode and choose **Devices and Simulators** from the **Window** menu. Choose the device in the left pane and copy the **Identifier** value:

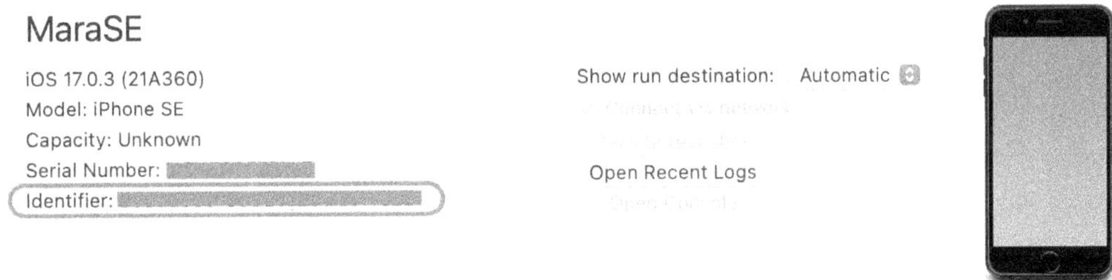

Figure 27.18: Devices and Simulators window

5. Type a name for the device in the **Device Name** field and paste the identifier value into the **Device ID (UDID)** field. Click **Continue**:

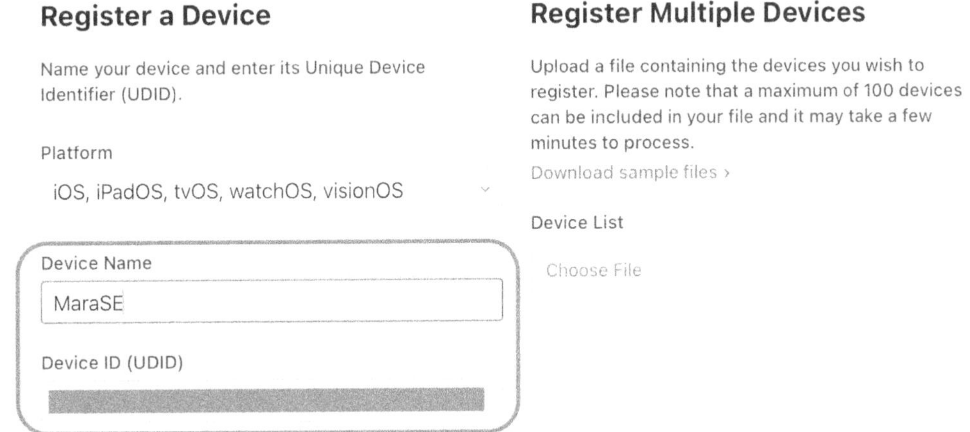

Figure 27.19: The Register a Device screen

You have successfully registered your test devices. Great! The next step is to create provisioning profiles. An **iOS App Development** profile is required so that your apps will be allowed to run on your test devices, and an **iOS App Store Distribution** profile is required for apps that will be uploaded to the App Store. You will create development and distribution profiles in the next section.

Creating provisioning profiles

You will need to create two provisioning profiles. An iOS app development profile is required for apps to run on test devices. An iOS App Store distribution profile is used to submit your app to the App Store. Here's how to create the development profiles:

1. Log in to your Apple Developer account and click **Profiles**.

2. Click the + button:

Certificates Profiles ⊕ All Types All Platforms

Identifiers

Figure 27.20: Profiles screen

3. Click the **iOS App Development** radio button and click **Continue**:

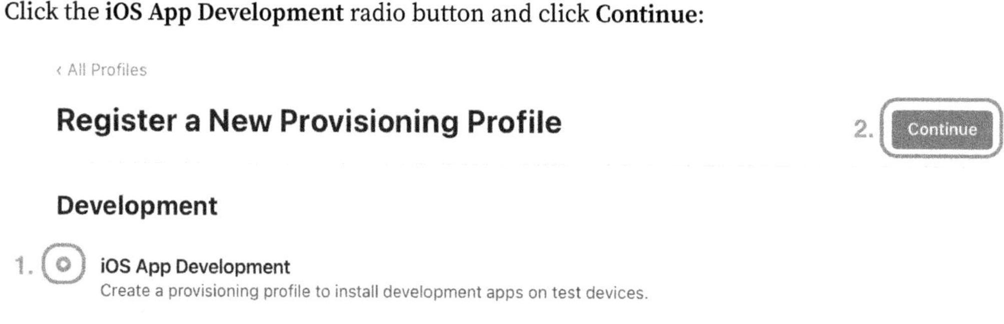
‹ All Profiles
Register a New Provisioning Profile
2.
Continue
Development
1.
iOS App Development
Create a provisioning profile to install development apps on test devices.

Figure 27.21: Register a New Provisioning Profile screen

4. Select the **App ID** for the app you want to test and click **Continue**:

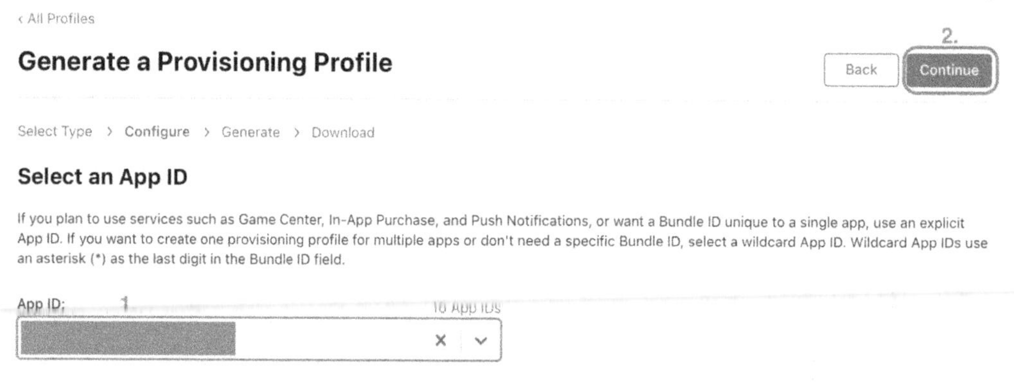
‹ All Profiles
Generate a Provisioning Profile
2.
Back
Continue
Select Type › Configure › Generate › Download
Select an App ID
If you plan to use services such as Game Center, In-App Purchase, and Push Notifications, or want a Bundle ID unique to a single app, use an explicit App ID. If you want to create one provisioning profile for multiple apps or don't need a specific Bundle ID, select a wildcard App ID. Wildcard App IDs use an asterisk (*) as the last digit in the Bundle ID field.
App ID:
1
10 App IDs

Figure 27.22: Selecting App ID screen

5. Select a development certificate checkbox and click **Continue**:

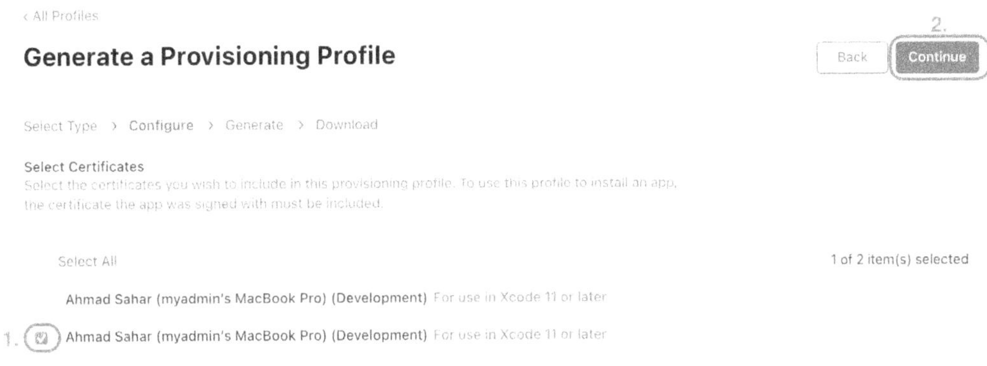

Figure 27.23: Selecting a development certificate

6. Check all the devices you will be testing this app on and click **Continue**:

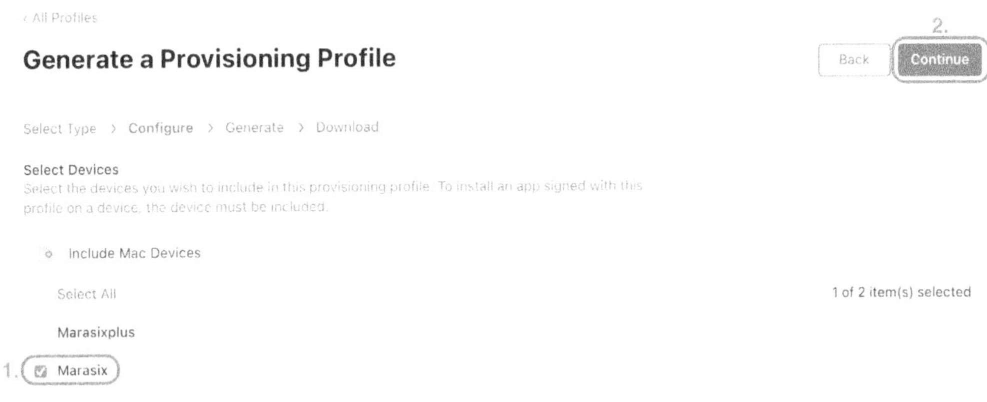

Figure 27.24: Selecting a device

7. Enter a name for the profile and click **Generate**:

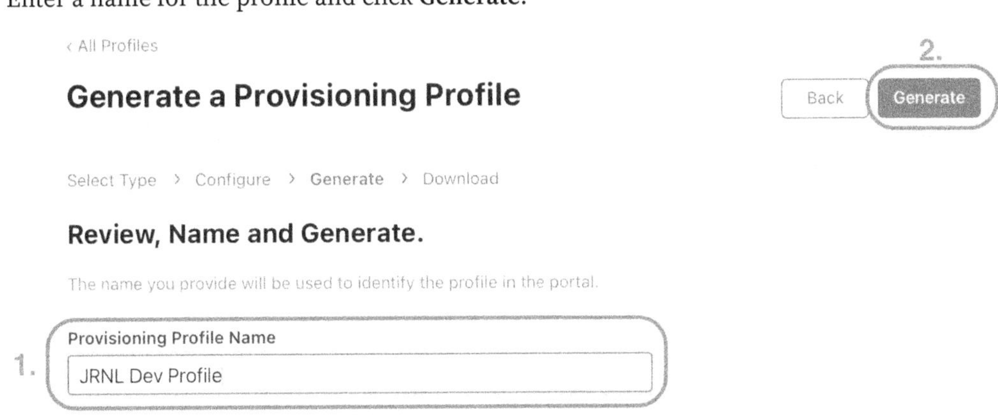

Figure 27.25: The Generate a Provisioning Profile screen

8. Click the **Download** button to download the profile.
9. Double-click the profile to install it.

Next, you'll create a distribution profile:

1. Click the **All Profiles** link to go back to the previous page:

Register a New Provisioning Profile

Figure 27.26: The All Profiles link

2. Click the + button:

Identifiers

All Types All Platforms

Figure 27.27: Profiles screen

3. Click the **App Store** radio button and click **Continue**:

Distribution

Ad Hoc
Create a distribution provisioning profile to install your app on a limited number of registered devices.

tvOS Ad Hoc
Create a distribution provisioning profile to install your app on a limited number of registered tvOS devices.

App Store Connect
Create a distribution provisioning profile to submit your app to App Store Connect.

tvOS App Store Connect
Create a distribution provisioning profile to submit your tvOS app to App Store Connect.

Figure 27.28: Registering a new provisioning profile

4. Select the **App ID** for the app you want to publish to the App Store and click **Continue**:

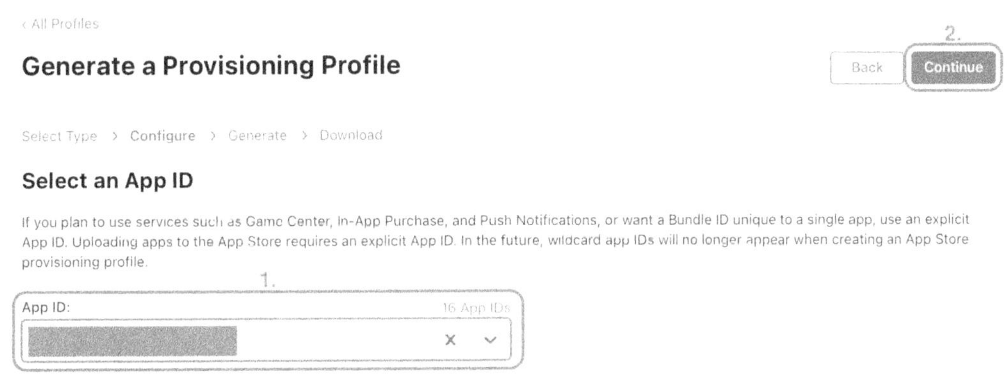

Figure 27.29: Selecting an App ID screen

5. Select a distribution certificate checkbox and click **Continue**:

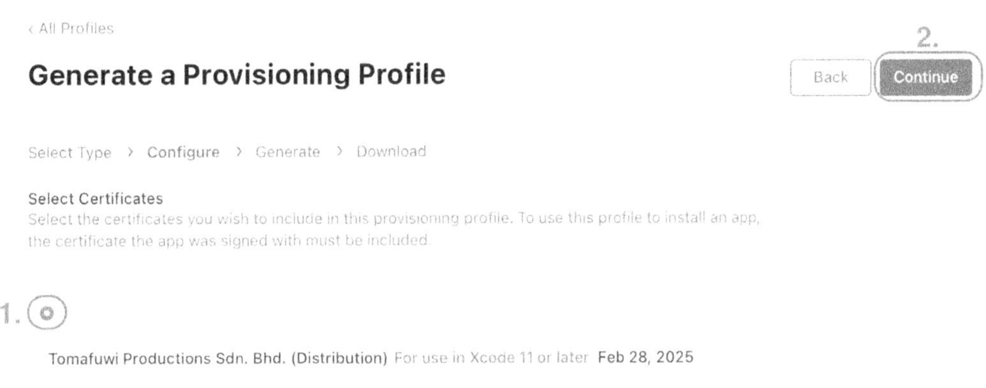

Figure 27.30: Selecting a distribution certificate

6. Enter a name for the profile and click **Generate**:

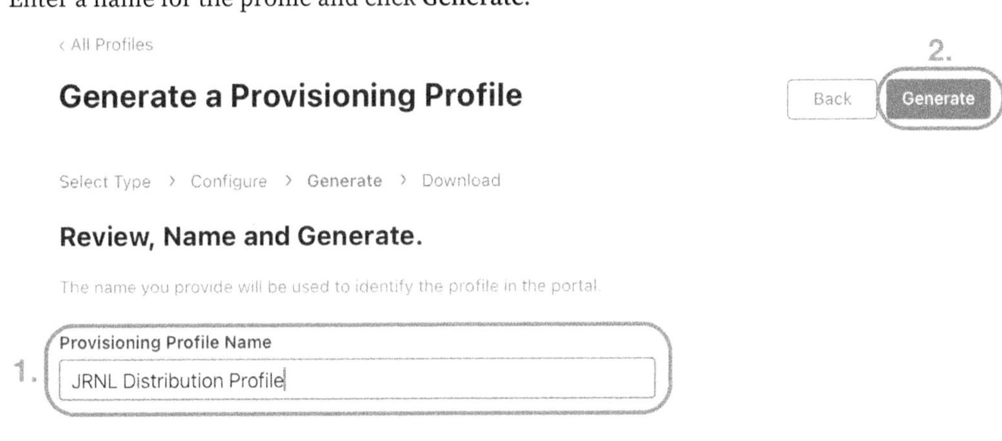

Figure 27.31: Generating a distribution profile

7. Click the **Download** button to download the profile.
8. Double-click the profile to install it.

You've completed all the steps necessary prior to submitting your app to the App Store. Let's learn more about the submission process in the next section, using the *ShareOrder* app as an example.

Submitting your app to the App Store

You are now ready to submit your app to the App Store! In this section, the *ShareOrder* app will be used as an example. Let's recap what you've done up to this point. You've created development and distribution certificates, registered your App ID and test devices, and generated development and distribution profiles.

To test your app on your test devices, you'll use the development certificate, App ID, registered test devices, and development profile. To submit your app to the App Store, you'll use the distribution certificate, App ID, and distribution profile. You'll configure Xcode to manage this automatically for you.

Before you submit your app, you must create your app's icons and get screenshots of your app. Then, you can create an App Store listing, generate an archive build to be uploaded, and complete the App Store Connect information. Apple will then review your app and, if all goes well, it will appear on the App Store.

To see more information on how to submit your apps, visit `https://developer.apple.com/app-store/submitting/`.

In the next section, let's see how to create icons for your app, which will appear on a device's screen when the app is installed.

Creating icons for your app

Before you upload your app to the App Store, you must create an icon set for it. Here's how to create an icon set for your app:

1. Create an icon for your app that is 1,024 x 1,024 pixels.
2. Click the **Assets** file in the Project navigator and drag the icon you created to the space shown in the following screenshot:

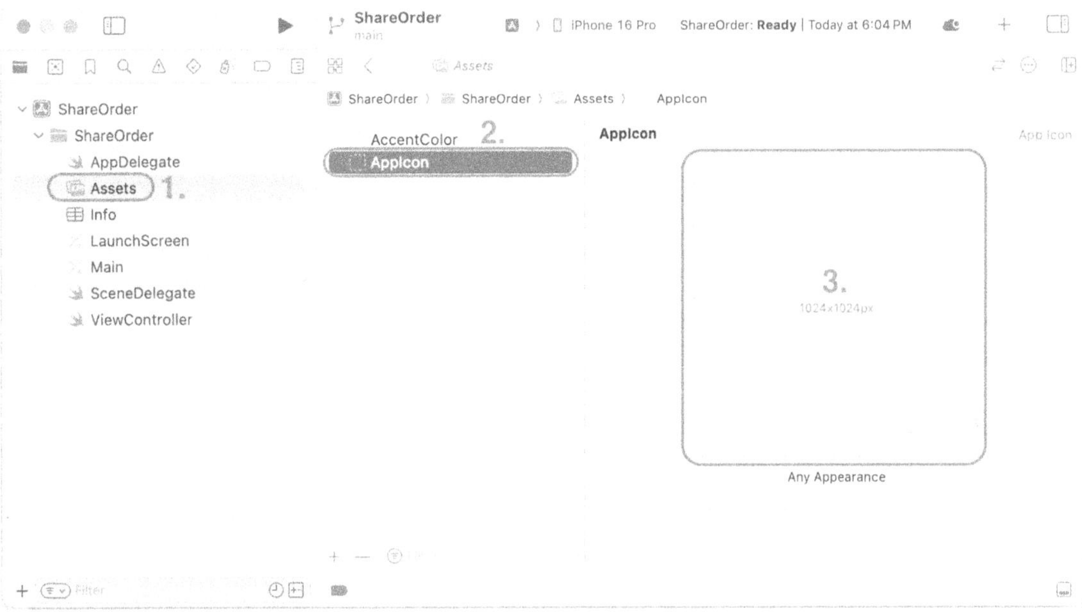

Figure 27.32: Assets file showing space for the app icon

When you run your app in Simulator or a device and quit your app, you should be able to see the app's icon on the home screen. Neat!

Let's look at how to create screenshots next. You'll need them for your App Store submission, so customers can see what your app looks like. You'll do this in the next section.

Creating screenshots for your app

You'll need screenshots of your app, which will be used in your App Store listing. To create them, run your app in Simulator and click the screenshot button. It will be saved to the desktop:

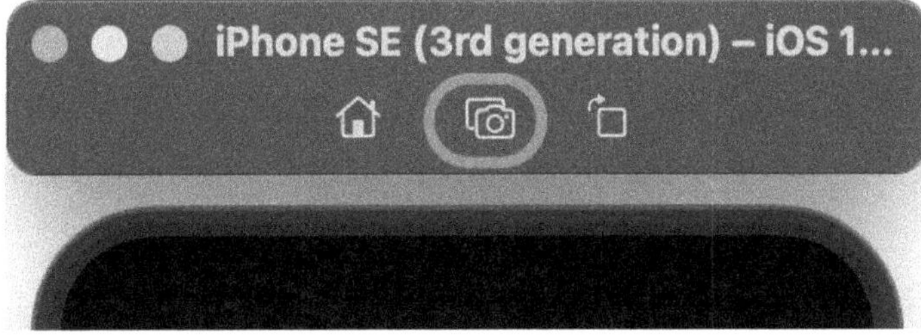

Figure 27.33: Simulator showing the screenshot button

Apple requires screenshots of your app running on an iPhone with a 6.5" display (such as the iPhone 16 Pro) and an iPhone with a 5.5" display (such as the iPhone 8 Plus). Optionally, you can also submit screenshots of your app running on an iPhone with a 6.7" display (such as the iPhone 16 Pro Max), iPad Pro (6th Gen) with a 12.9" display, and iPad Pro (2nd Gen) with a 12.9" display. You can use Simulator to simulate all these devices if you don't have the actual devices, and your screenshots should show your app's features and how your app looks in different screen sizes.

How to submit screenshots of your app will be discussed in more detail in the next section, where you will learn how to create an App Store listing. The App Store listing contains all the information about your app that will be displayed in the App Store, so customers can make an informed decision about downloading or purchasing your app.

Creating an App Store listing

Now that you have icons and screenshots of your app, you'll create the App Store listing. This allows customers to see information about your app before they download it. Follow these steps:

1. Go to `http://appstoreconnect.apple.com` and select **My Apps**:

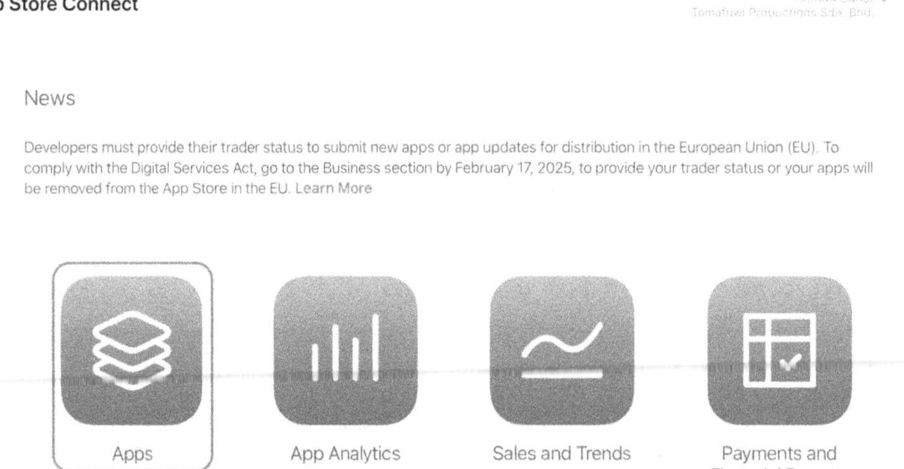

Figure 27.34: The App Store Connect website

2. Click the + button at the top left of the screen and select **New App**:

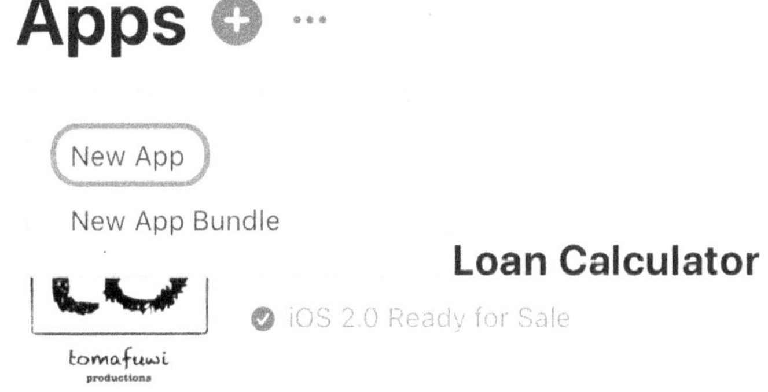

Figure 27.35: The New App button and menu

3. A **New App** screen displaying a list of fields will appear:

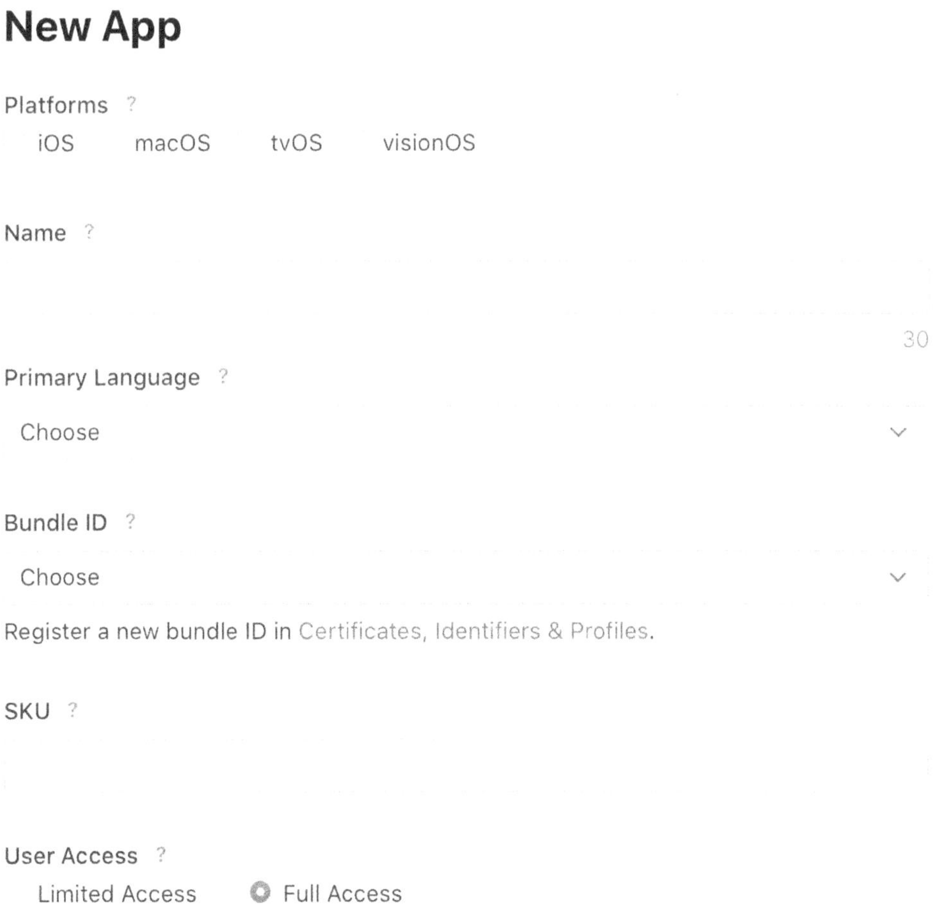

New App

Platforms ?
iOS macOS tvOS visionOS

Name ?
30

Primary Language ?
Choose

Bundle ID ?
Choose
Register a new bundle ID in Certificates, Identifiers & Profiles.

SKU ?

User Access ?
Limited Access Full Access

Figure 27.36: App details screen

4. Enter your app details:
    - **Platforms**: All the platforms your app supports (iOS, macOS, and/or tvOS).
    - **Name**: The name of your app.
    - **Primary Language**: The language your app uses.
    - **Bundle ID**: The bundle ID you created earlier.
    - **SKU**: Any reference number or string that you use to refer to your app. For instance, you could use something like `231020-1`, which would be a reference to the app version combined with the date you completed it. It can be any number that makes sense to you.
    - **User Access**: This manages who in your developer account team can see this app in App Store Connect. If you're the only one in your team, just set it to **Full Access**.
5. Click **Create** when you're done.

The app will now be listed in your account, but you still need to upload the app and all the information about it. To upload the app, you need to create an **archive build**, and you will learn how to do that in the next section.

Creating an archive build

You'll create an archive build, which will be submitted to Apple for placement on the App Store. This will also be used for your internal and external testing. Here are the steps to create an archive build:

1. Open Xcode, select the project name in the Project navigator, and select the **General** pane. In the **Identity** section, you can change the **Version** and **Build** number as you see fit. For instance, if this is the first version of your app and the first time you have built it, you can set **Version** to `1.0` and **Build** to `1`:

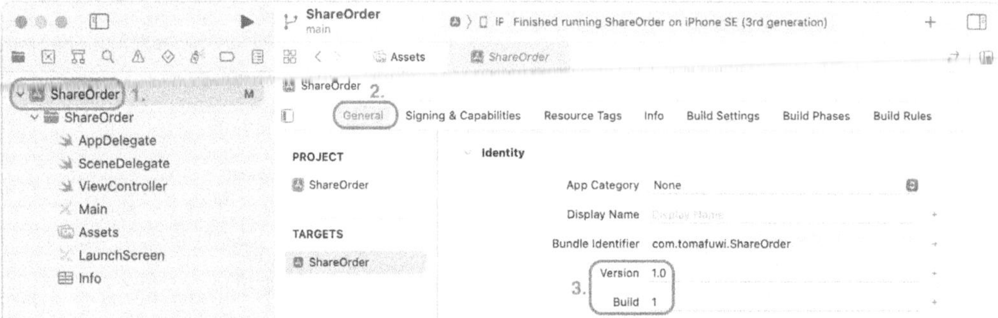

Figure 27.37: Editor area showing the General pane

2. Select the **Signing & Capabilities** pane. Make sure **Automatically manage signing** is checked. This will allow Xcode to create certificates, App IDs, and profiles, as well as register devices that are connected to your Mac. Select your paid developer account from the **Team** dropdown:

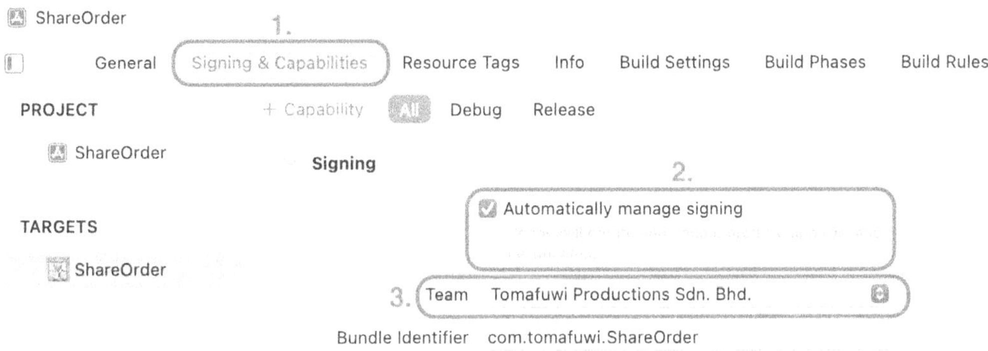

Figure 27.38: Editor area showing the Signing & Capabilities pane

3. Select **Any iOS Device** as the build destination:

Figure 27.39: Scheme menu with Any iOS Device selected

4. If your app does not use encryption, update your `Info.plist` file by adding `ITSAppUsesNonExemptEncryption`, making its type `Boolean`, and setting its value to `NO`. It should then appear in the Project navigator:

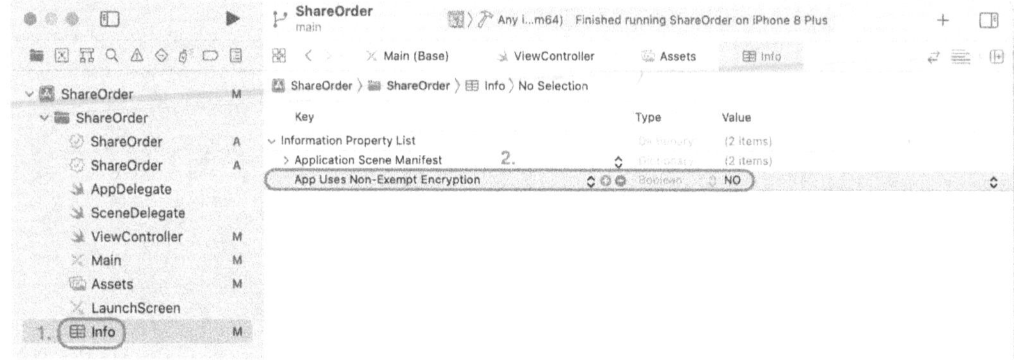

Figure 27.40: Project navigator with Info.plist selected

For more details, use this link: https://developer.apple.com/documentation/bundleresources/information_property_list/itsappusesnonexemptencryption.

5. Select **Archive** from the **Product** menu:

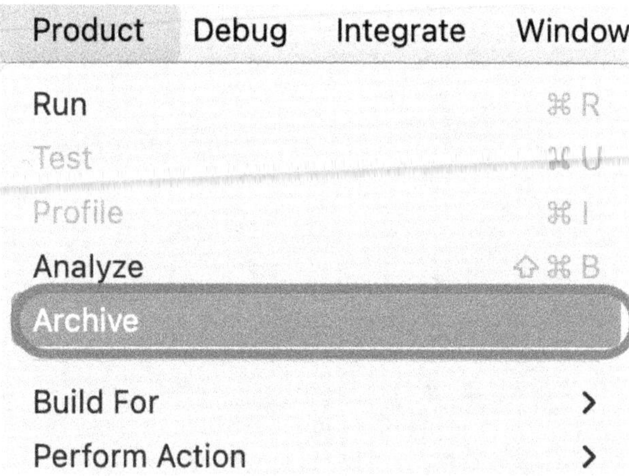

Figure 27.41: The Product menu with Archive selected

6. The **Organizer** window appears with the **Archives** tab selected. Your app will appear on this screen. Select it and click the **Distribute App** button:

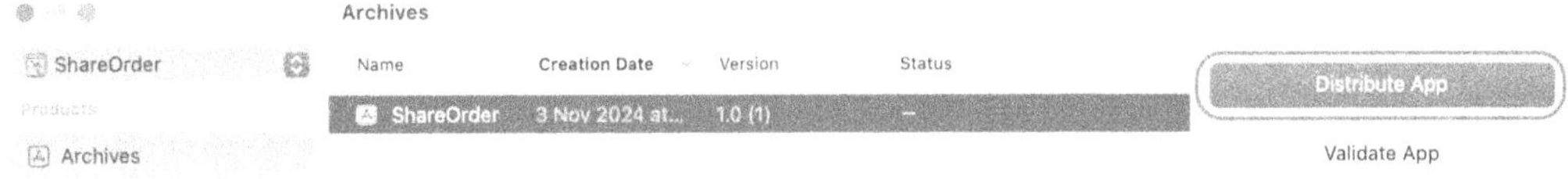

Figure 27.42: The Organizer window with the Distribute App button selected

7. Select **App Store Connect** and click **Distribute:**

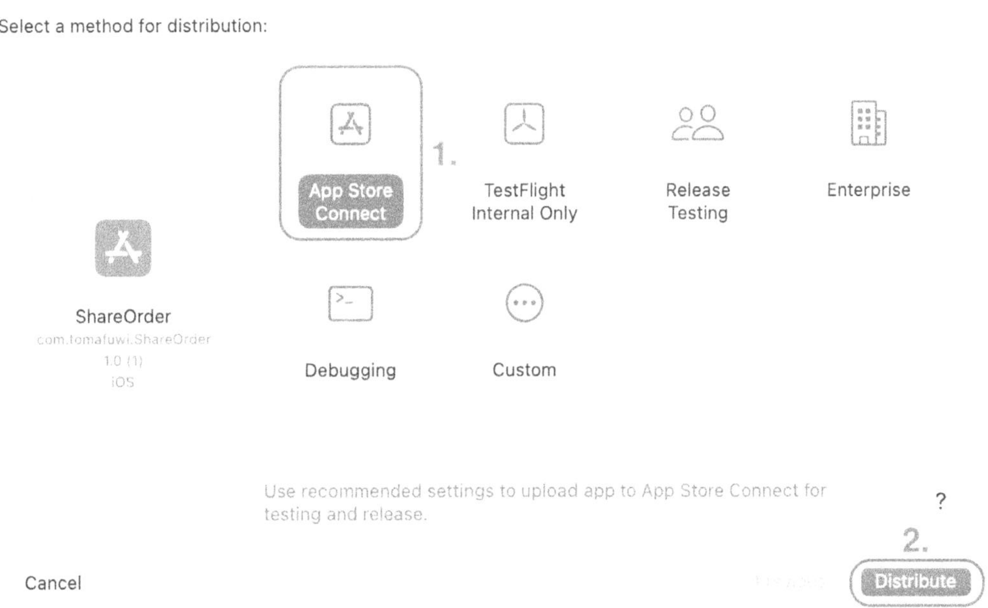

Figure 27.43: Selecting a method for the distribution

8. Wait for your upload to complete. If you're prompted for a password, enter your Mac account password and click **Always Allow**.

9. When your upload is complete, click **Done**:

App upload complete:

ShareOrder 1.0 (2) uploaded

Show in App Store Connect

Cancel Export... Previous Done

Figure 27.44: The App upload complete screen

At this point, the build of the app that will be distributed by the App Store has been uploaded. In the next section, you'll learn how to upload screenshots and complete the information about your app that will appear on the App Store along with the app.

Completing the information in App Store Connect

Your app has been uploaded, but you will still need to complete the information about your app in App Store Connect. Here are the steps:

1. Go to `http://appstoreconnect.apple.com` and select **My Apps**.
2. Select the app that you just created:

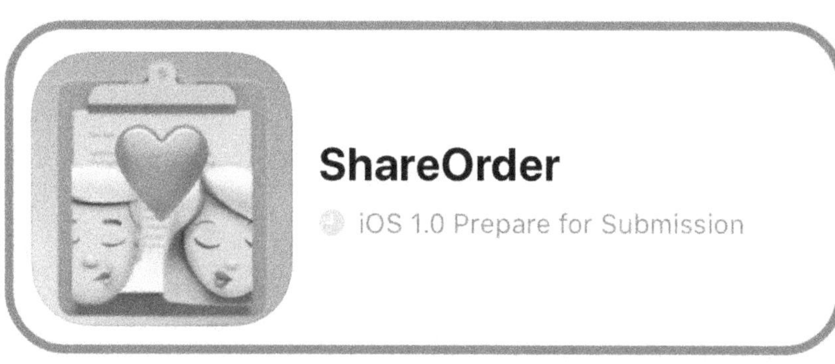

Figure 27.45: Apps screen with your app selected

3. Select **App Information** on the left side of the screen, and make sure all the information is correct:

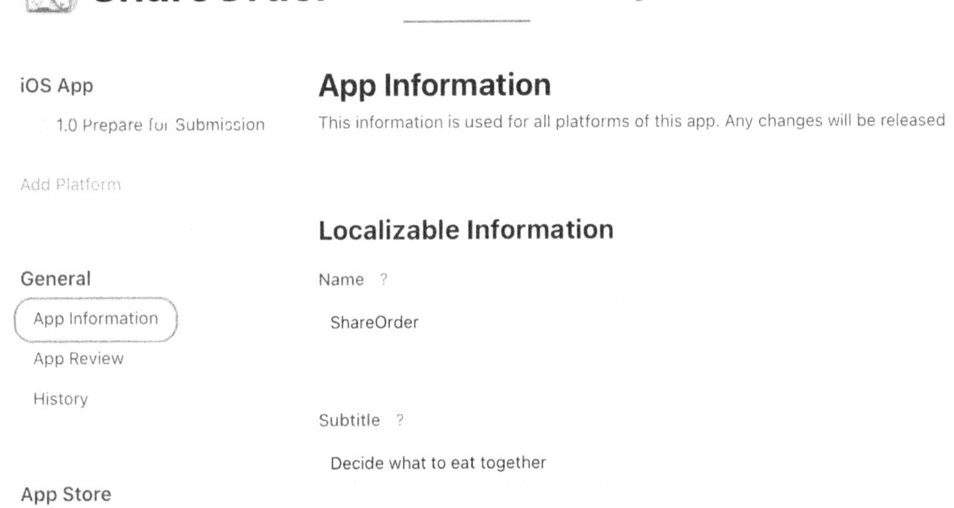

Figure 27.46: The App Information screen

4. Do the same for the **Pricing and Availability** and **App Privacy** sections:

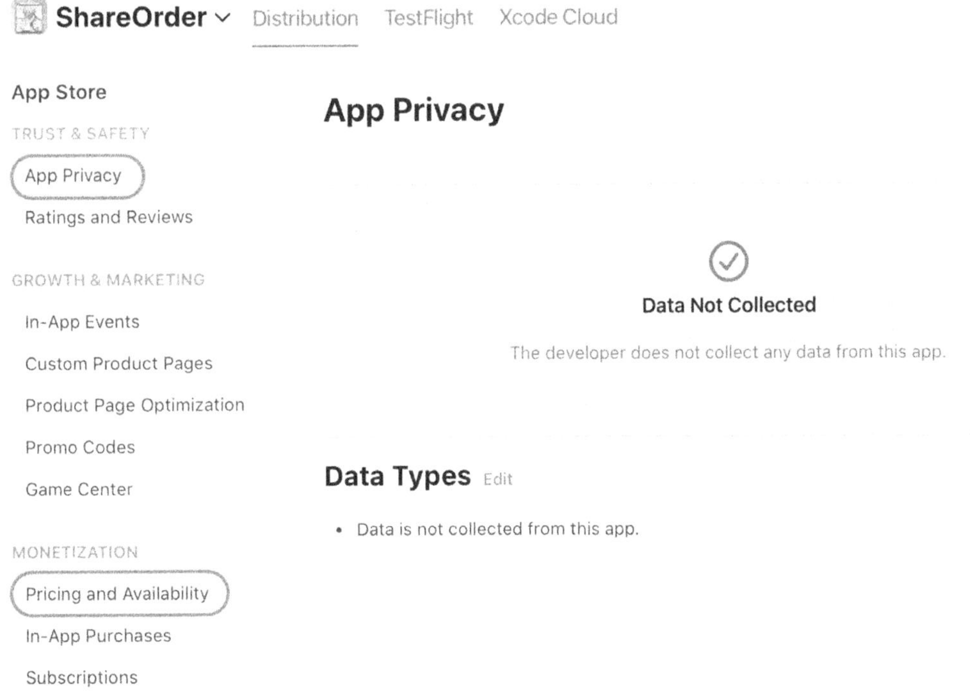

Figure 27.47: The App Privacy screen

5. Select **Prepare for Submission** on the left side of the screen. In the **App Preview** and **Screenshots** section, drag in the screenshots that you took earlier:

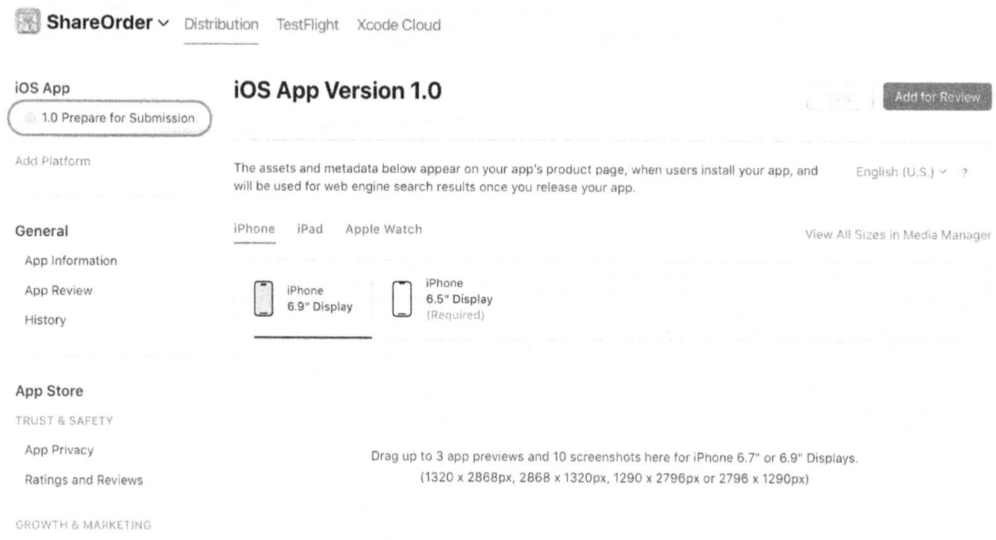

Figure 27.48: Prepare for Submission screen showing App Preview and Screenshots section

6. Scroll down and fill in the **Promotional Text, Description, Keywords, Support URL** (containing support information for your app), and **Marketing URL** (containing marketing information for your app) fields:

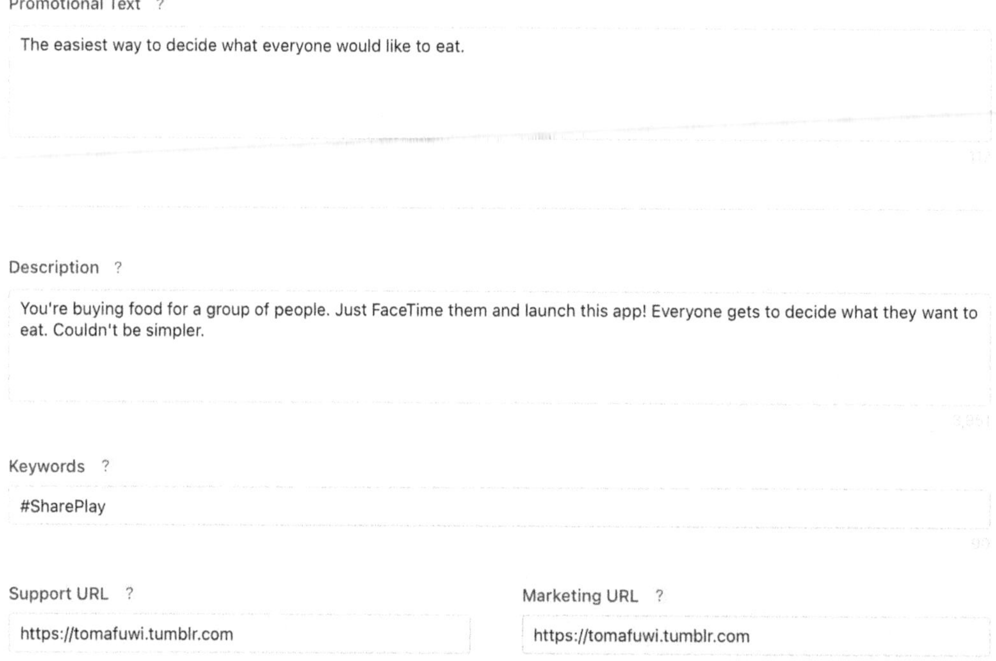

Figure 27.49: Version information section

7. Scroll to the general app information section and fill in all the required details:

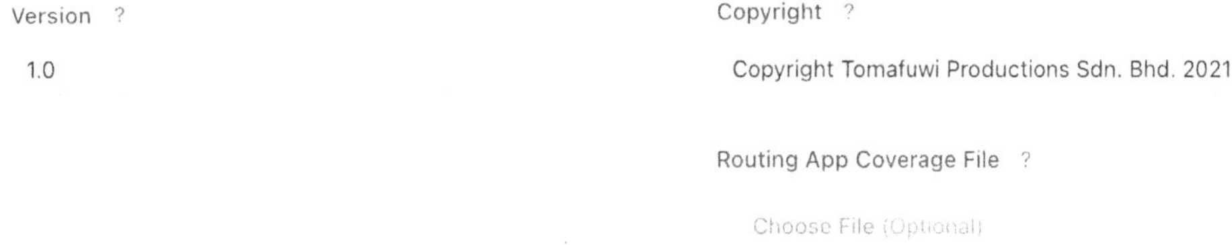

Figure 27.50: General app information section

8. Scroll down to the **Build** section and you'll see the archive build you uploaded earlier. If you don't see it, click the + or **Add Build** button, select a build, and click **Done**:

If your app uses encryption, you're required to upload export compliance documentation. You can submit this documentation before you submit your app for review in the Encryption section, or by uploading your app below.

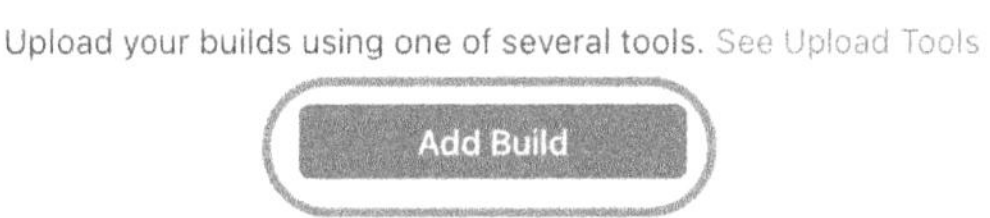

Figure 27.51: The Build selection screen

It may take as long as 30 minutes for Apple to process your submission.

9. Verify that your build is in the **Build** section:

Build
BUILD
VERSION
HAS APP CLIP
1
1.0
NO

Figure 27.52: The Build section

10. Scroll down to the **App Review Information** section. If you would like to provide any additional information to the app reviewer, put it here:

App Review Information

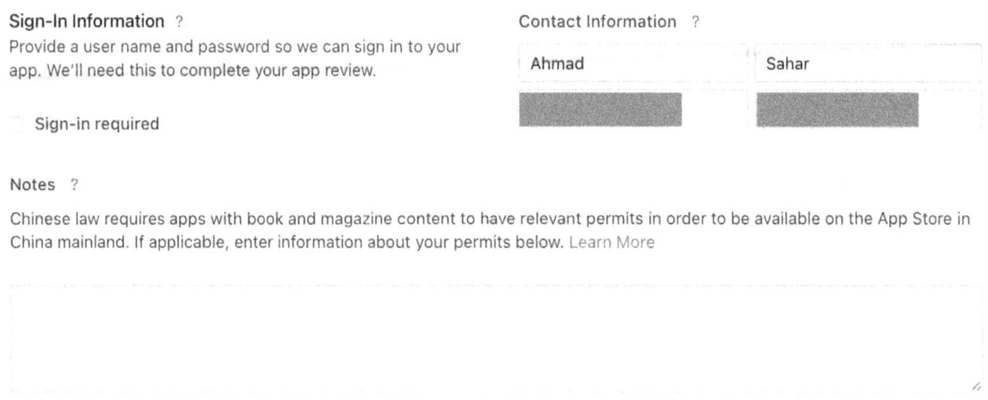

Figure 27.53: The App Review Information section

11. Scroll down to the **Version Release** section and maintain the default settings, so your app will be automatically released right after it has been approved by Apple:

App Store Version Release

To make your app available on the App Store, you can automatically release it after it's been approved by App Review. You can also manually release it on the App Store at a later date.

- Manually release this version
- Automatically release this version
- Automatically release this version after App Review, no earlier than

Your local date and time.

November 3, 2024 6:00 PM Nov 3, 2024 10:00 AM (GMT)

Figure 27.54: The Version Release section

12. Scroll back up to the top of the screen and click the **Add for Review** button:

Figure 27.55: The Add for Review button

13. Verify that the app status has changed to **Waiting for Review**:

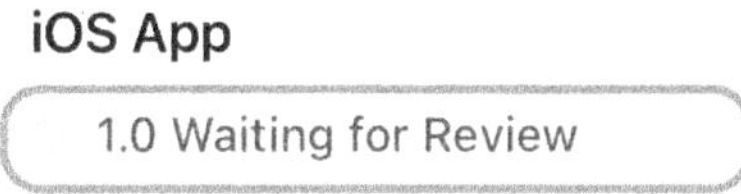

Figure 27.56: App status showing Waiting for Review

You will need to wait for Apple to review the app, and you will receive an email saying whether your app is approved or rejected. If your app is rejected, there will be a link that takes you to Apple's *Resolution Center* page, which describes why your app was rejected. After you have fixed the issues, you can then update the archive and resubmit.

You now know how to submit your app to the App Store! Awesome!

In the next section, you'll learn how to conduct internal and external testing for your app, which is important in ensuring that the app is high quality and bug-free.

Testing your app

Apple has a facility named **TestFlight** that allows you to distribute your apps to testers prior to releasing them to the App Store. You'll need to download the TestFlight app, available from `https://developer.apple.com/testflight/`, to test your app. Your testers can be both members of your internal team (internal testers) or the general public (external testers). First, let's see how to allow internal team members to test your app in the next section.

Testing your app internally

Internal testing should be performed when the app is in an early stage of development. It only involves members of your internal team; Apple does not review apps for internal testers. You can send builds to up to 100 testers for internal testing. To do so, follow these steps:

1. Go to `http://appstoreconnect.apple.com` and select **My Apps**.
2. Select the app that you want to test.
3. Click the **TestFlight** tab:

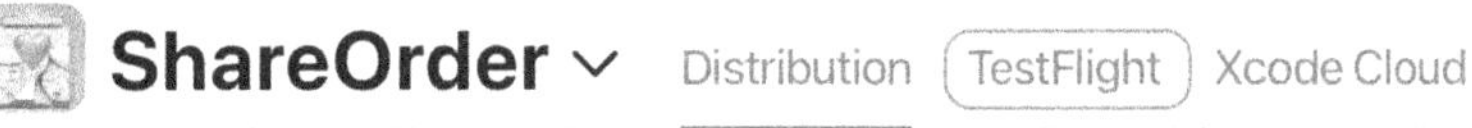

Figure 27.57: TestFlight tab

4. Click the + button next to **INTERNAL TESTING** to create a new internal test group:

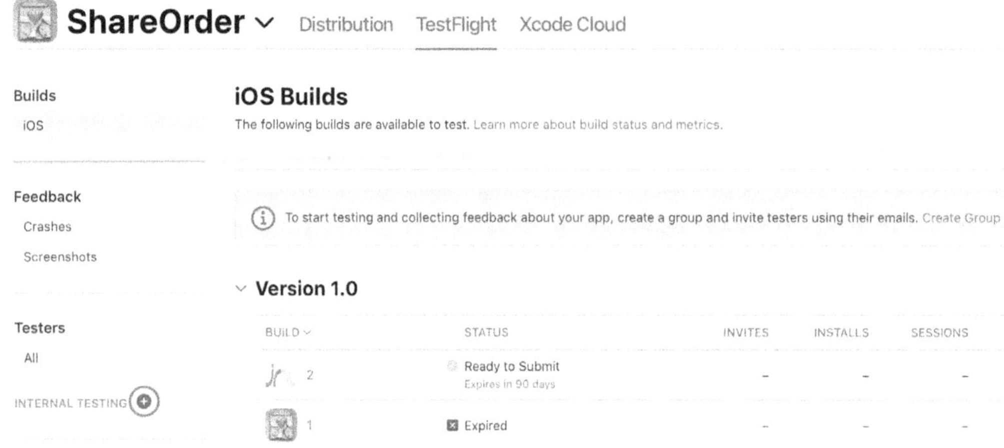

Figure 27.58: The TestFlight screen showing the + button

5. The **Create New Internal Group** dialog box will appear. Name your internal test group and click **Create**:

Create New Internal Group

You can add up to 100 testers, and they can test builds using the TestFlight app.

Select the "Enable automatic distribution" checkbox to automatically deliver all Xcode builds to everyone in the group. Xcode Cloud builds have to be added manually. This setting cannot be updated later.

☑ Enable automatic distribution

Figure 27.59: The Create New Internal Group dialog box

6. After your test group has been created, click the + button to add users to your group:

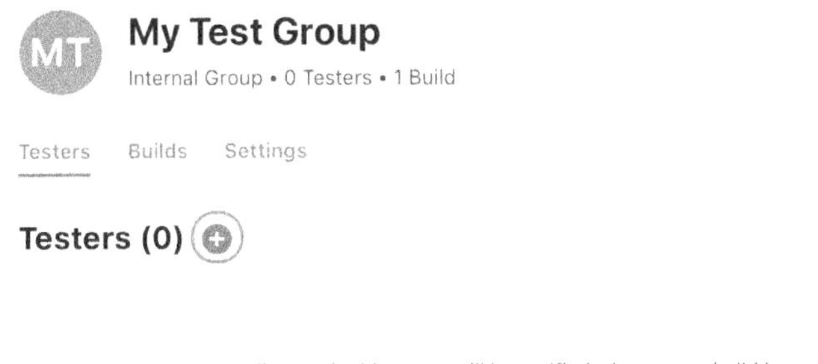
My Test Group
Internal Group • 0 Testers • 1 Build
Testers
Builds
Settings
Testers (0)
Testers in this group will be notified when a new build is available and will have access to all builds added to this group.

Figure 27.60: Test group screen showing the + button

7. Check all the users that you want to send test builds to and click **Add**. They'll be invited to test all available builds:

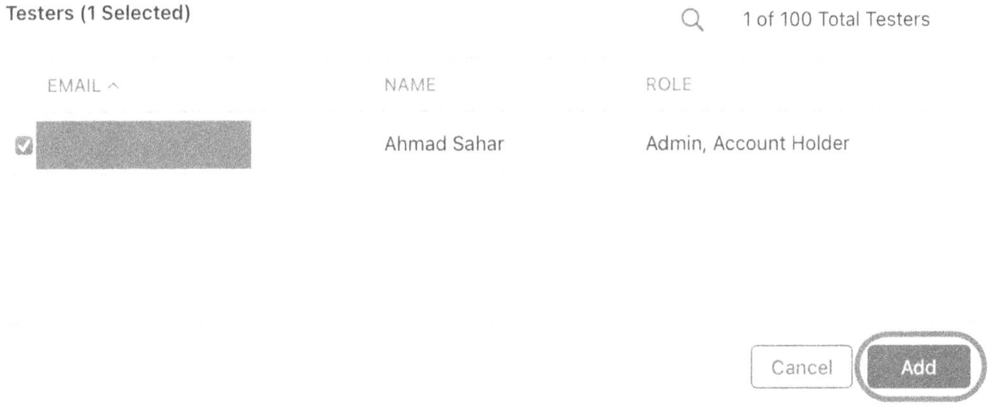
Add Testers to the Group "My Test Group"
Select up to 100 testers, and they'll be invited to test all available builds in the TestFlight app. They'll also be notified when new builds are added. If you'd like to add a tester you don't see, add them in Users and Access.
Testers (1 Selected)
1 of 100 Total Testers
EMAIL
NAME
ROLE
Ahmad Sahar
Admin, Account Holder
Cancel
Add

Figure 27.61: The Add Testers screen

8. Verify your testers have been added:

Tester (1)
Add Filter
TESTER
STATUS
SESSIONS
CRASHES
FEEDBACK
DEVICES
Ahmad Sahar
Invited

Figure 27.62: The TestFlight screen showing the Tester section

Remember that internal testing will only involve members of your team. If you want to conduct testing with more than 100 testers, you will need to do external testing, which is described in the next section.

Testing your app externally

External testing should be performed when the app is in the final stages of development. You can select anyone to be an external tester, and you can send builds to up to 10,000 testers. Apple may review apps for external testers. Here are the steps:

1. Go to `http://appstoreconnect.apple.com` and select **My Apps**.
2. Select the app that you want to test.
3. Click the **TestFlight** tab.
4. Click the + button next to **EXTERNAL TESTING**:

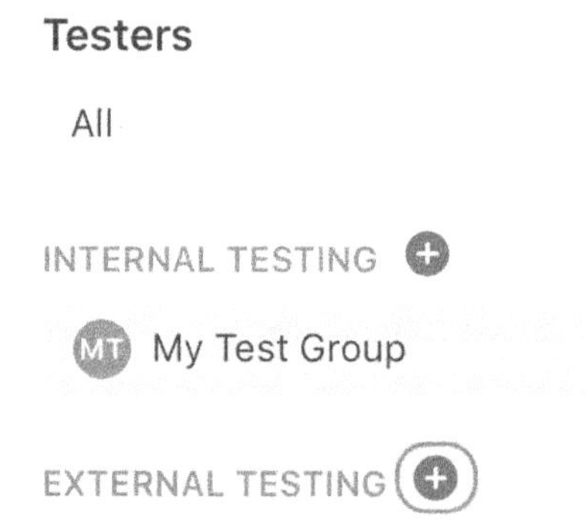

Figure 27.63: The TestFlight screen showing the + button

5. Type in a name for the test group and click **Create**:

Figure 27.64: The Create a New Group screen

6. Click the **Add Builds** link to choose a build that you want your testers to test:

ME
My External Test Group
External Group • 0 Testers • 0 Builds
To start testing with external testers, complete the following steps:
Add Builds
Choose which builds you want to add to this group.
Invite Testers
Choose how you want to invite testers to this group.

Figure 27.65: External test group showing Add Builds link

7. Choose one of your builds and click **Next**:

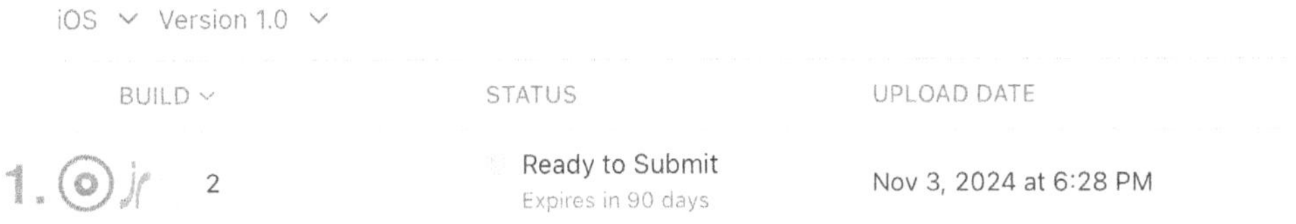
Select a Build to Test
Select a build, and we'll invite the group "My External Test Group" to start testing. Before your build can be tested, it may have to be approved by Beta App Review.
iOS Version 1.0
BUILD
STATUS
UPLOAD DATE
1.
2
Ready to Submit
Expires in 90 days
Nov 3, 2024 at 6:28 PM
2.
Cancel
Next

Figure 27.66: The Select a Build to Test screen

8. Your testers will need to know who to contact if they have issues. Type in your contact details and click **Next:**

Figure 27.67: The Test Information screen

9. Type in what you would like your testers to test in the box provided and click **Submit for Review:**

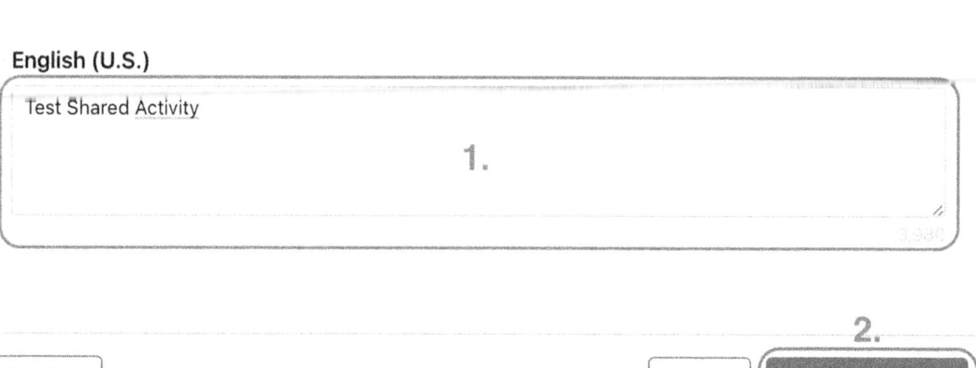

Figure 27.68: The What to Test screen

Apple may review the test build prior to making it available to your testers. If your app is rejected, you will need to fix the issues and resubmit.

10. Now, you will add external testers to your group. Click the **Testers** tab:

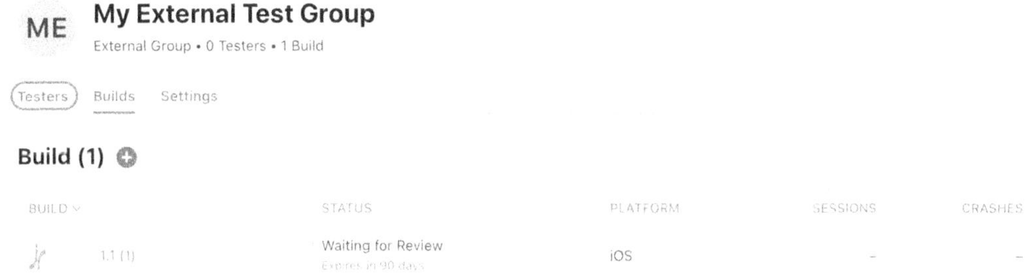

Figure 27.69: My External Test Group screen showing Testers link

11. Click the + button next to **Testers**:

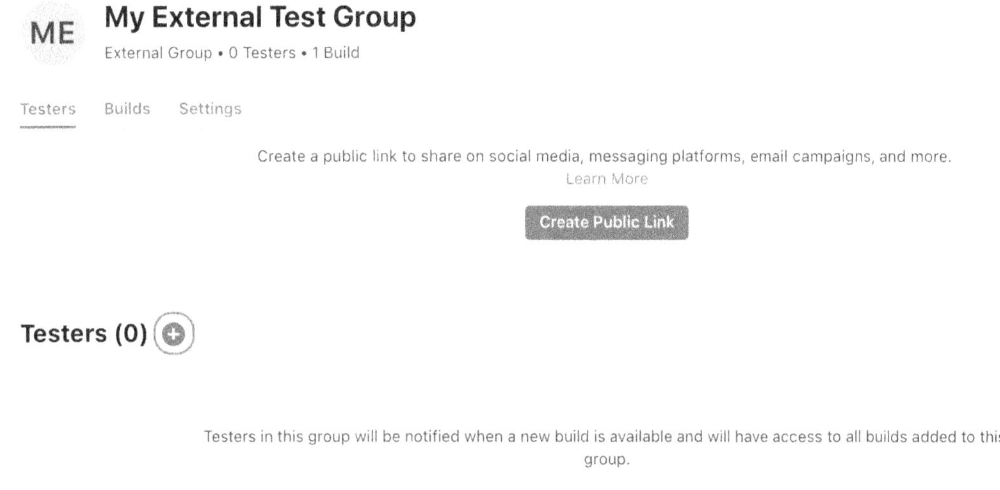

Figure 27.70: External test group showing + button to add new testers

12. On the **Invite Testers** screen, choose **Email** and click **Next:**

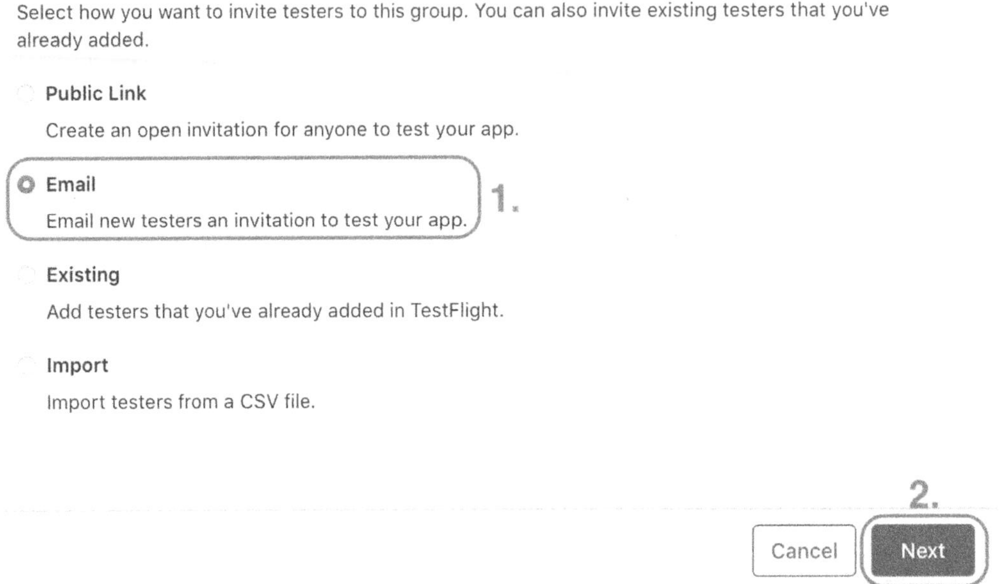

Figure 27.71: The Invite Testers screen

13. Enter the names and email addresses of your testers. Note that Apple will notify them automatically when a build is ready to be tested:

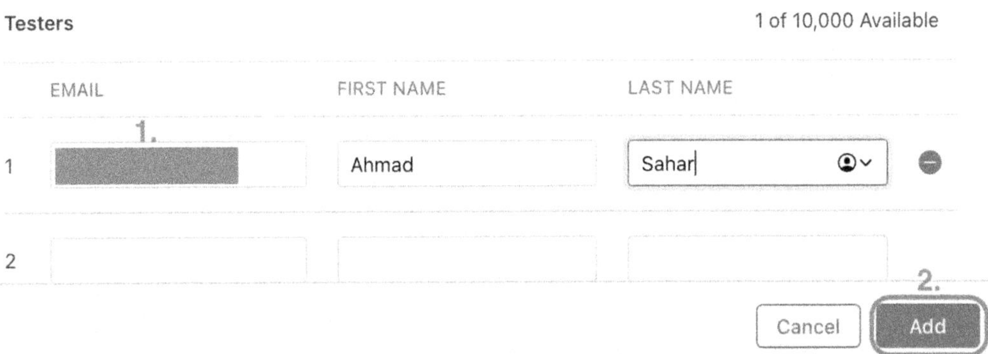

Figure 27.72: The Add New Testers to the Group screen

Great! You now know how to test your apps internally and externally, and you have reached the end of this book!

Summary

You have now completed the entire process of building an app and submitting it to the App Store. Congratulations!

You started by learning how to obtain an Apple Developer account. Next, you learned how to generate a CSR to create certificates that allow you to test apps on your own devices and publish them on the App Store. You learned how to create a bundle identifier to uniquely identify your app on the App Store and register your test devices. After that, you learned how to create development and production provisioning profiles to allow apps to run on your test devices and be uploaded to the App Store. Next, you learned how to create an App Store listing and submit your release build to the App Store. Finally, you learned how to conduct testing for your app using internal and external testers.

You now know how to build your own apps, conduct internal and external testing for them, and submit them to the App Store.

Once an app has been submitted for review, all you can do is wait for Apple to review your app. Don't worry if the app gets rejected—it happens to all developers. Work with Apple to resolve issues via the *Resolution Center* and do your research to know what is and what is not acceptable to Apple.

After your apps are on the App Store, feel free to reach out to me (`@shah_apple`) on Twitter to let me know—I would love to see what you have built.

Leave a review!

Thank you for purchasing this book from Packt Publishing—we hope you enjoyed it! Your feedback is invaluable and helps us improve and grow. Once you've completed reading it, please take a moment to leave an Amazon review; it will only take a minute, but it makes a big difference for readers like you.

`https://packt.link/r/1836204892`

Scan the QR code below or visit the link to receive a free ebook of your choice.

`https://packt.link/NzOWQ`

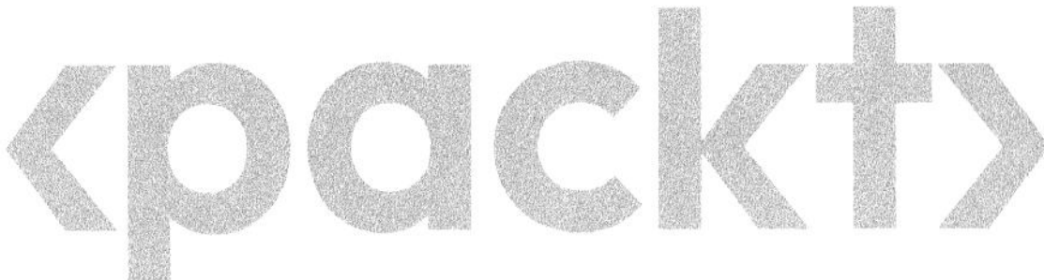

packt.com

Subscribe to our online digital library for full access to over 7,000 books and videos, as well as industry leading tools to help you plan your personal development and advance your career. For more information, please visit our website.

Why subscribe?

- Spend less time learning and more time coding with practical eBooks and Videos from over 4,000 industry professionals
- Improve your learning with Skill Plans built especially for you
- Get a free eBook or video every month
- Fully searchable for easy access to vital information
- Copy and paste, print, and bookmark content

At www.packt.com, you can also read a collection of free technical articles, sign up for a range of free newsletters, and receive exclusive discounts and offers on Packt books and eBooks.

Other Books You May Enjoy

If you enjoyed this book, you may be interested in these other books by Packt:

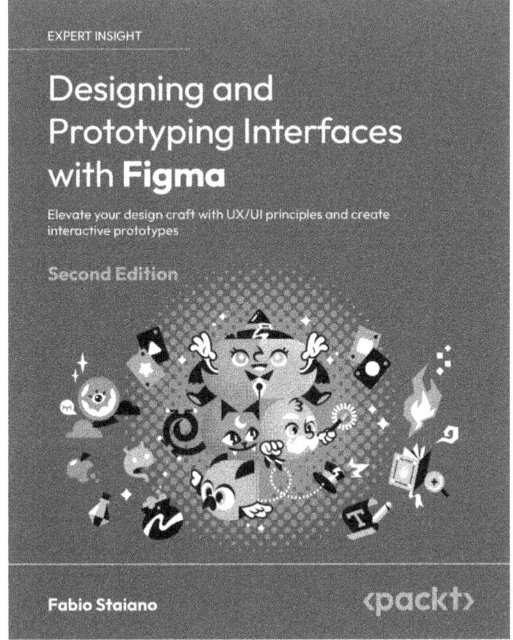

Designing and Prototyping Interfaces with Figma

Fabio Staiano

ISBN: 9781835464601

- Create high-quality designs that cater to your users' needs, providing an outstanding experience
- Mastering mobile-first design and responsive design concepts
- Integrate AI capabilities into your design workflow to boost productivity and explore design innovation
- Craft immersive prototypes with conditional prototyping and variables

- Communicate effectively to technical and non-technical audiences
- Develop creative solutions for complex design challenges
- Gather and apply user feedback through interactive prototypes

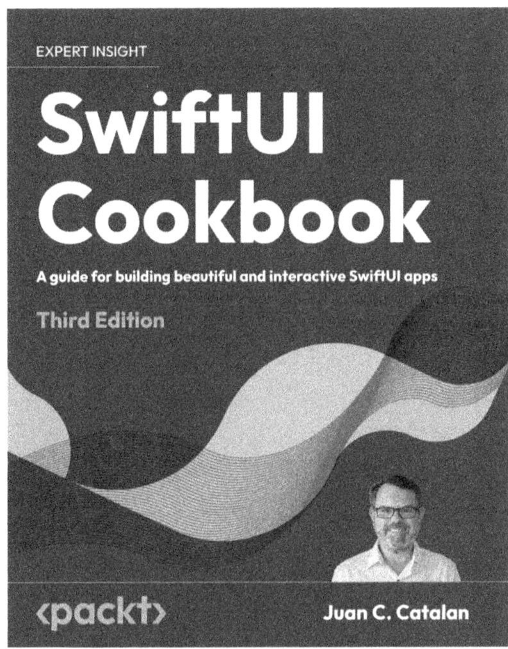

SwiftUI Cookbook

Juan C. Catalan

ISBN: 9781805121732

- Create stunning, user-friendly apps for iOS 17, macOS 14, and watchOS 10 with SwiftUI 5
- Use the advanced preview capabilities of Xcode 15
- Use async/await to write concurrent and responsive code
- Create powerful data visualizations with Swift Charts
- Enhance user engagement with modern animations and transitions
- Implement user authentication using Firebase and Sign in with Apple
- Learn about advanced topics like custom modifiers, animations, and state management
- Build multi-platform apps with SwiftUI

101 UX Principles – 2nd edition

Will Grant

ISBN: 9781803234885

- Work with user expectations, not against them
- Make interactive elements obvious and discoverable
- Optimize your interface for mobile
- Streamline creating and entering passwords
- Use animation with care in user interfaces
- How to handle destructive user actions

Packt is searching for authors like you

If you're interested in becoming an author for Packt, please visit `authors.packtpub.com` and apply today. We have worked with thousands of developers and tech professionals, just like you, to help them share their insight with the global tech community. You can make a general application, apply for a specific hot topic that we are recruiting an author for, or submit your own idea.

Index

A

L

M

T

Download a free PDF copy of this book

Thanks for purchasing this book!

Do you like to read on the go but are unable to carry your print books everywhere?

Is your eBook purchase not compatible with the device of your choice?

Don't worry, now with every Packt book you get a DRM-free PDF version of that book at no cost.

Read anywhere, any place, on any device. Search, copy, and paste code from your favorite technical books directly into your application.

The perks don't stop there, you can get exclusive access to discounts, newsletters, and great free content in your inbox daily.

Follow these simple steps to get the benefits:

1. Scan the QR code or visit the link below:

`https://packt.link/free-ebook/9781836204893`

2. Submit your proof of purchase.
3. That's it! We'll send your free PDF and other benefits to your email directly.

www.ingramcontent.com/pod-product-compliance
Lightning Source LLC
LaVergne TN
LVHW080110070326
832902LV00015B/2495
* 9 7 8 1 8 3 6 2 0 4 8 9 3 *